Ergebnisse der Mathematik und ihrer Grenzgebiete

Band 2

Herausgegeben von

P. R. Halmos · P. J. Hilton · R. Remmert · B. Szőkefalvi-Nagy

Unter Mitwirkung von

L. V. Ahlfors · R. Baer · F. L. Bauer · R. Courant
A. Dold · J. L. Doob · S. Eilenberg · M. Kneser · G. H. Müller
M. M. Postnikov · B. Segre · E. Sperner

Geschäftsführender Herausgeber: P. J. Hilton

Carlo Miranda

Partial Differential Equations
of Elliptic Type

Second Revised Edition

Translated by Zane C. Motteler

Springer-Verlag New York · Heidelberg · Berlin 1970

Carlo Miranda

Università di Napoli

Translation of
Equazioni alle derivate parziali di tipo ellittico, 1955
(Ergebnisse der Mathematik, Vol. 2)

ISBN-13: 978-3-642-87775-9   e-ISBN-13: 978-3-642-87773-5
DOI: 10.1007/ 978-3-642-87773-5

Subject-Classification No. QA-377

© by Springer-Verlag Berlin · Heidelberg 1970. Library of Congress Catalog Card Number 71-75930.
Softcover reprint of the hardcover 1st edition 1970
Title No. 4546

Preface to the First Edition

In the theory of partial differential equations, the study of elliptic equations occupies a preeminent position, both because of the importance which it assumes for various questions in mathematical physics, and because of the completeness of the results obtained up to the present time.

In spite of this, even in the more classical treatises on analysis the theory of elliptic equations has been considered and illustrated only from particular points of view, while the only expositions of the whole theory, the extremely valuable ones by LICHTENSTEIN and ASCOLI, have the character of encyclopedia articles and date back to many years ago.

Consequently it seemed to me that it would be of some interest to try to give an up-to-date picture of the present state of research in this area in a monograph which, without attaining the dimensions of a treatise, would nevertheless be sufficiently extensive to allow the exposition, in some cases in summary form, of the various techniques used in the study of these equations.

At present the interest of researchers seems to be directed mainly toward the study of mixed problems, of equations of higher order, and of systems of equations. This part of the theory is therefore in a phase of rapid and continual evolution, so that it does not seem that the moment has yet come to give it a systematic exposition. On the other hand, the methods used in this current research do not greatly differ, at least in their general conception, from those that have proved effective in the study of the DIRICHLET, NEUMANN, and oblique derivative problems for a single equation of second order. It therefore seemed to me opportune to dwell primarily on this initial part of the theory, which is the part which has by now reached a sufficiently stable state and whose study is without doubt indispensable to an understanding of the methods used in the most recent research.

Therefore the aspects of the theory to which I have decided to give the most complete treatment are: I. GIRAUD's method for transforming boundary value problems into integral equations of the second kind. II. The study of boundary value problems by means of various procedures from linear functional analysis which originated in some works by CACCIOPPOLI, PICONE, WEYL. III. The research of BERNSTEIN,

HOPF, SCHAUDER, LERAY, and CACCIOPPOLI based on the a priori majorization of the solutions of linear and nonlinear equations.

On the other hand, apart from references, I have not digressed either on the method of minima or on the so-called „kernel function" method, because the classical treatise of COURANT-HILBERT is authoritative on the first, and an exhaustive monograph by BERGMAN and SCHIFFER was recently devoted to the second.

The three aspects of the theory referred to above occupy Chapters III, IV, V, VI, of which the first three are devoted to linear equations, the last to nonlinear ones. On the other hand, Chapter I, in which we summarize various classical notions of a general character, and Chapter II, devoted to the study of generalized potentials, are of a preliminary character. Finally, in Chapter VII various other questions again concerning a single equation of second order or else equations of higher order, systems of equations, or problems depending on a parameter are taken into consideration. This last chapter, in contrast to the first six, and for reasons which were pointed out earlier has essentially the character of a bibliographic summary. The volume ends with an ample Bibliography comprising more than six hundred works, almost all published after 1924. For references to this Bibliography I adopted the convention of numbers in brackets; for the bibliography prior to 1924 I refer the reader to LICHTENSTEIN's article in the Encyclopedia or else I give some indication in footnotes at the bottom of the pages.

I should add informally that I have been almost exclusively concerned with the study of elliptic equations of general type and have hardly occupied myself at all with equations of special type. Consequently all the research which centres round the theory of harmonic functions, and within that theory the part, however interesting, relating to problems on free boundaries, does not occur in the Bibliography and remains outside the scope of our exposition. I have tried instead to compile the bibliography relating to polyharmonic functions and to the equations of elasticity, and to comment briefly on it. This part of the Bibliography is thus undoubtedly incomplete, because it contains none of the works of a predominantly mathematical-physical character. I cannot end without expressing my keenest gratitude to Springer-Verlag for accepting this book in their series, and for giving it a perfect typographical format.

Naples, 28 February 1954   CARLO MIRANDA

Preface to the Second Edition

When, upon the invitation of Springer Verlag, I began to prepare the second edition of this monograph, I was already conversant with the great progress made on the qualitative plane in the theory of elliptic equations in the fourteen years which had elapsed since the first edition. However, I would never have thought that on the quantitative plane the progress in this branch of analysis had reached that extent which now emerges from a careful examination of the bibliography. Suffice it to say that, while in the Bibliography to the first edition, which refers to the years between 1924 and 1953, somewhat more than 600 works were listed, yet from 1953 until now, namely in a period of time which is half the preceding one, more than 1600 works were published, for which a complete bibliography[1] would comprise more than 2200 items.

It was thus immediately clear that to illustrate this mass of work in a volume of the normal dimensions of an Ergebnisse volume would be entirely impossible and that it would be necessary to give up the treatment of certain topics which could be eliminated without modifying the general plan of the work. In agreement with the Editors of the Ergebnisse, it was therefore decided that in the new edition §§ 57 and 58 would be suppressed, the first because a nice monograph by S. BERGMAN [3] related to this topic has appeared in the meantime, the second because its theme, namely the spectral theory of elliptic equations, has acquired such dimensions that it can only be treated in a satisfactory way in a volume devoted exclusively to it.

By eliminating from the old and new bibliography mention of about 400 works relating to the topics of §§ 57 and 58, as well as many early notes and various expository works which have by now lost all interest, and also by referring to other texts for some particular matters for which sufficiently extensive bibliographic references have already been published, the list of works cited at the end of the volume has been reduced so as to contain little more than 1400 items.

Even with these reductions, a significant increase in the size of the volume has been inevitable, and the necessity to contain this increase

---

[1] Actually the bibliography which I have collected and from which I have deduced these numbers is almost complete up to 1965, but is somewhat less so for the last two years.

within acceptable bounds has made it impossible for me to enter into too much detail in the most recent part of the theory. I have therefore adhered to the principle, already followed in the first edition, of reporting in detail only the results obtained for equations of the second order, and limiting myself, for equations of higher order and systems of equations, to a general indication of the methods used in the various works. Even for equations of the second order I could only in any case place emphasis on proofs of the most recent theorems.

Here in summary are the modifications made to the first edition: Chapter I, apart from some additions, has remained almost the same. Chapter II has been modified here and there in order to take into account the theorems from potential theory due to SOBOLEV and to CALDERON and ZYGMUND. In Chapter III, § 19 has been completely revised in order to connect the problem of the existence of fundamental solutions with the validity of the uniqueness theorem for CAUCHY's problem and of the unique continuation property. In § 23 the results obtained in the study of the non-regular oblique derivative problems for equations in two variables have been discussed in rather more detail.

In Chapter IV, as well as various additions to §§ 26, 28, 29, 31, 32, the contents of §§ 27 and 30 are almost completely new, and, with changed titles, they are now devoted to local properties of the solutions of an elliptic equation and to the study of weak solutions of the boundary value problems, respectively.

In Chapter V, the title of which has been slightly changed, §§ 33, 34, 36 remain almost unaltered, § 35 has been supplemented with some new theorems, and §§ 37, 38, 39 have been completely revised. Of these three sections, the first two are devoted to the exposition of recent results on majorization, regularity, and existence in SOBOLEV spaces, while in § 39, which did not occur in the first edition, some results for the NEUMANN and oblique derivative problems are summarized. The contents of § 39 of the first edition have been divided among the preceding sections.

Finally, Chapters VI and VII are devoted, as in the first edition, to, first, nonlinear equations of second order, and second, to different problems from among the classical ones for equations of second order, equations of higher order, and systems of equations. Given the great quantity of results obtained in these areas in recent years, these chapters have had to be completely revised. In particular it has no longer been possible to put into Chapter VI a detailed exposition of the contributions due to CACCIOPPOLI, SCHAUDER, and LERAY to the theory of equations in two variables; however, these have been in large part absorbed by more recent results. Likewise in Chapter VII the section on degenerate elliptic equations has been placed at the end because its scope now includes equations of higher order.

Finally, I wish to thank Dr. Zane C. Motteler for the care and the dedication put into the work of translation and Springer Verlag for having taken this work into the Ergebnisse series in spite of its unusual dimensions.

Naples, 29 February 1968 CARLO MIRANDA

Translator's Preface

For many years I have been acquainted with the first edition of Professor Miranda's monograph, and I have found it a very useful reference book on several occasions. A couple of years ago it struck me that an English edition of this work would be quite useful, inasmuch as Italian is not well known among American mathematicians, primarily because French, German, and Russian are the usual languages required by graduate schools in this country. When I offered to translate this book Springer Verlag agreed that this should be done, and Professor Miranda agreed to produce an updated manuscript for this English second edition.

The translation was a difficult and demanding work for me, but at the same time it was a labor of love in my chosen field. Although many people have checked my manuscript for accuracy, I feel that I must be held solely responsible for any possible inaccuracies remaining in the text.

I wish to thank the Jesuit Research Council of Gonzaga University, which provided me with a grant which was very helpful in defraying the costs of typing and mailing the manuscript. Linda Balch did a fine job on the typing in spite of her many other duties. And most of all, Professor Miranda himself carefully inspected the manuscript, discovered many errors, and often supplied just the right turn of phrase to make a sentence read properly. I wish to thank Springer Verlag for the opportunity to work on this task with Professor Miranda, whose encyclopedic knowledge of the subject is sufficient to awe a younger mathematician.

Spokane, 4 July 1969 ZANE C. MOTTELER

Contents

CHAPTER I

Boundary value problems
for linear equations

In this chapter, after having introduced notation which will be constantly used in the course of our exposition, we shall indicate how to pose the various boundary value problems for linear elliptic equations of second order. Next we shall state uniqueness theorems for such problems which follow immediately from the maximum and minimum principles for the solutions. Introducing finally the concepts of *Levi functions, fundamental solutions*, and *Green's functions*, we shall establish some fundamental integral formulas.

1. Sets of points; functions. Let R_m be an m-dimensional Euclidean space. Then we shall use $x, y, \ldots$ to represent the generic points of R_m; $(x_1, x_2, \ldots, x_m)$, $(y_1, y_2, \ldots, y_m)$, $\ldots$, their coordinates; $r = \overline{xy} = |x - y|$, the distance between the two points x and y; $dx, dy, \ldots$, the volume element of R_m; $d_x \sigma$ (or, when confusion is not likely to arise, $d\sigma$), the element of hypersurface area at the point x of a generic hypersurface ($m - 1$ dimensional manifold) of R_m. Finally, we shall use ω_m to represent the hypersurface area

$$\omega_m = \frac{2 \pi^{m/2}}{\Gamma(m/2)}$$

of the unit sphere.

If A is a generic set of points in R_m, ∂A will represent its boundary and $\complement A$ its complement. The set $\overline{A} = A \cup \partial A$ will be called the *closure* of A. A set A is called a *region* if it is open and connected, and a *domain* if it is perfect, internally connected, and if each of its points is a point of accumulation of interior points. If A is a region, $\overline{A}$ is a domain.

By $\Gamma(x, \varrho)$ we shall mean the sphere of radius ϱ with center x, that is, the set of all points y for which $\overline{xy} \leq \varrho$.

If $0 < \lambda \leq 1$ we shall say that a function $u(x)$ defined (and continuous) on a set A is *HÖLDER continuous with exponent λ in A*, or, more briefly, *λ-Hölder continuous in A*, if the expression

$$\frac{|u(x) - u(y)|}{\overline{xy}^\lambda}$$

is bounded from above for all $x \neq y$ in A; the upper bound of this expression will be called the (*minimum*) HÖLDER coefficient of the function $u(x)$.

A function $u(x)$ defined in a bounded domain T and possessing there k [λ-Hölder] continuous derivatives is said to be of class $C^{(k)}[C^{(k,\lambda)}]$ in T; naturally $C^{(0)}[C^{(0,\lambda)}]$ will be the class of [λ-Hölder] continuous functions in T, $C^{(\infty)}$ the class of infinitely differentiable functions in T.

If $u(x)$ is instead defined in a set A which is a region, or more generally in a set whose closure is a domain, we shall say that it is of class $C^{(k)}$ or $C^{(k,\lambda)}$ in A if it is such in every bounded domain contained in A.

A function $f(x)$ defined in a measurable set T will be said to be of class L^p in T if $|f(x)|^p$ is LEBESGUE integrable in T, and of class L^∞ if it is measurable and bounded almost everywhere in T.

If $u(x)$ is a function defined and of class $C^{(n,\lambda)}$ in T, we denote by $U_k(T)$ and $U_{k,\lambda}(T)$, respectively, the sum of the maximum moduli of all the derivatives of u of order $k\ (\leq n)$, and the sum of the Hölder coefficients of exponent λ, of the same derivatives; likewise, we denote by $U_0(T)$ and $U_{0,\lambda}(T)$ the maximum modulus and the λ-Hölder coefficient of u in T. Whenever we consider several functions, we shall, for each function represented by a lower case letter, designate by the same upper case letter modified by the indices k or (k,λ), the quantity analogous to U_k and $U_{k,\lambda}$.

We shall agree to denote by $C^{(k)}(T)\ [C^{(k,\lambda)}(T)]$ the Banach space which has as elements the functions of class $C^{(k)}[C^{(k,\lambda)}(T)]$ in T, having defined the norm of an element u by means of the expression

$$\| u \|_{C^{(k)}(T)} = \sum_{n=0}^{k} U_n \left[\| u \|_{C^{(k,\lambda)}(T)} = U_{k,\lambda} + \sum_{n=0}^{k} U_n \right].$$

The subspace of $C^{(k)}(T)\ [C^{(k,\lambda)}(T)]$ consisting of all the functions of compact support in $T - \partial T$ will be denoted by $C_0^{(k)}(T)\ [C_0^{(k,\lambda)}(T)]$. Analogously, by $L^p(T)\ [L^\infty(T)]$ we shall mean the Banach space which has as elements the functions of class $L^p[L^\infty]$ in T, having defined the norm by means of the expression

$$\| u \|_{L^p(T)} = \left[\int_T |u|^p\, dx \right]^{1/p} \quad [\| u \|_{L^\infty(T)} = \text{ess. sup.}_T\, |u|].$$

Defining now for each m-tuple α of non-negative indices $(\alpha_1, \alpha_2, \ldots, \alpha_m)$:

$$|\alpha| = \sum_{k=1}^{m} \alpha_k,$$

$$D^\alpha u = \frac{\partial^{|\alpha|} u}{\partial x_1^{\alpha_1} \cdots \partial x_m^{\alpha_m}},$$

we agree to denote by $H^{k,p}(T)$ $[H_0^{k,p}(T)]$ the Banach space which is obtained as the completion of $C^{(k)}(T)$ $[C_0^{(k)}(T)]$ with respect to the norm

$$\| u \|_{H^{k,p}(T)} = \left[\sum_{|\alpha|=0}^{k} \| D^\alpha u \|_{L^p(T)}^p \right]^{1/p} .$$

For $k \geq 1$, every function u of $H^{k,p}$ or $H_0^{k,p}$ has, consequently, generalized derivatives up to order k all belonging to class L^p. It should be easy to see that for the functions of $H_0^{k,p}$ such generalized derivatives can be identified with the derivatives in the sense of the theory of distributions; analogously for the functions of $H^{k,p}$ if T is sufficiently regular. We note also that $H^{k,2}$ is a Hilbert space with inner product defined by

$$(u, v) = \sum_{|\alpha|=0}^{k} \int_T D^\alpha u \, D^\alpha \bar{v} \, dx .$$

We shall also agree, in what follows, to put $H^{0,p} = L^p$. Along with the space L^p we shall sometimes find it necessary to consider the space $L^{p,\lambda}(T)$ consisting of all the functions of class L^p for which the following norm is finite:

$$\| u \|_{L^{p,\lambda}(T)} = \sup_{\substack{x \in T \\ \varrho > 0}} \left[\varrho^{-\lambda} \int_{T \cap \Gamma(x,\varrho)} | u \, (y) |^p \, dy \right]^{1/p} .$$

More generally, we shall represent by $H^{k,p,\lambda}(T)$ $[H_0^{k,p,\lambda}(T)]$ the Morrey spaces that are obtained as completion of $C^{(k)}(T)$ $[C_0^{(k)}(T)]$ with respect to the norm

$$\| u \|_{H^{k,p,\lambda}(T)} = \left[\sum_{|\alpha|=0}^{k-1} \| D^\alpha u \|_{L^p(T)}^p + \sum_{|\alpha|=k} \| D^\alpha u \|_{L^{p,\lambda}(T)}^p \right]^{1/p} .$$

Now let T be a domain such that with each point x of ∂T it is possible to associate a hypersphere $\Gamma(x)$ with center x, in such a manner that the portion of ∂T contained in $\Gamma(x)$ can be represented by an equation of the form:

$$\xi_m = \zeta \, (\xi_1, \xi_2, \, ..., \, \xi_{m-1}) , \tag{1.1}$$

with respect to a system of axes $\xi_1, \xi_2, ..., \xi_m$ with origin at x, ζ being a function defined in a suitable region $\Omega(x)$, of class $C^{(1)}$ there at least, and vanishing together with its first derivatives at the point x.

Under these hypotheses there exists a hyperplane tangent to ∂T at x for every x on ∂T, namely the hyperplane $\xi_m = 0$; and there exists at x an *exterior normal* n whose direction cosines $X_i(x)$ are continuous functions on ∂T.

Such a domain will be said to be of class $A^{(k)}$ $[A^{(k,\lambda)}]$ if the functions ζ occurring in (1.1) are, for every x, of class $C^{(k)}$ $[C^{(k,\lambda)}]$.

In what follows, in order to represent a portion of ∂T we shall also use parametric equations of the type

$$x_i = x_i \, (t_1, t_2, \ldots, t_{m-1}) \tag{1.2}$$

with the x_i finite and continuous functions in a bounded domain D of the space $t_1, t_2 \ldots t_{m-1}$; admissible parametric representations will always be only those which satisfy the following conditions: 1°) (1.2) should constitute a one-to-one correspondence between D and the corresponding portion of ∂T. 2°) If T is of class $A^{(k)} [A^{(k,\lambda)}]$, the x_i should be of class $C^{(k)} [C^{(k,\lambda)}]$ in D. 3°) The following should hold:

$$J = [|| \, \partial x_i/\partial t_j \, ||^2]^{1/2} > 0 \, .$$

4°) The direction cosines of the exterior normal n should be given by

$$X_i = \frac{1}{J} \, \frac{\partial \, (x_{i+1}, \ldots, \, x_m, \, x_1, \ldots, \, x_{i-1})}{\partial \, (t_1, t_2, \ldots, t_{m-1})} \, .$$

Note that such conditions are always satisfied by representations of type (1.1) as long as in 4°) we consider the axis ξ_m to be oriented toward the exterior of T. It should also be borne in mind that the element of measure on ∂T is given by $d\sigma = J \, dt_1 \, dt_2 \ldots dt_{m-1}$.

In what follows, functions $u(x)$ defined only on ∂T will frequently occur. On every portion of ∂T admitting a parametric representation (1.2), such a $u(x)$ can be considered a function of the variables t_i: $u(t_1, t_2, \ldots, t_{m-1})$.

Now supposing that $T \in A^{(k)} [A^{(k,\lambda)}]$ with $k \geq 1$, if whatever portion of ∂T is considered, $u(t_1, t_2, \ldots, t_{m-1})$ turns out to be of class $C^{(k)} [C^{(k,\lambda)}]$ in D, we shall say that $u(x)$ is of class $C^{(k)} [C^{(k,\lambda)}]$ on ∂T. The set $C^{(k)} (\partial T) [C^{(k,\lambda)} (\partial T)]$ of such functions can also be considered a Banach space if we define the norm in some convenient way. For example, if $\mathcal{Z}_1, \mathcal{Z}_2, \ldots, \mathcal{Z}_s$ are portions of ∂T of the preceding type, such that every point of ∂T is interior to at least one of them, and if D_i is the basic domain of the parametric representation of $\mathcal{Z}_i$ we can put:

$$|| \, u \, ||_{C^{(k)} (\partial T)} = \sum_{i=1}^{s} || \, u \, ||_{C^{(k)} (D_i)} \, .$$

Analogously, we can consider the Banach space $H^{k,p} (\partial T)$ of functions of class $H^{k,p}$ on ∂T, defining the norm as follows:

$$|| \, u \, ||_{H^{k,p} (\partial T)} = \sum_{i=1}^{s} || \, u \, ||_{H^{k,p} (D_i)} \, .$$

Note that if $T \in A^{(1)}$ the norm in $L^p(\partial T)$ can be defined without recourse to the particular covering of ∂T by the sets Ξ_i by putting:

$$\| u \|_{L^p(\partial T)} = \left[\int_{\partial T} | u |^p \, d\sigma \right]^{1/p}.$$

Analogously, for any arbitrary domain T, we can define the space $C^{(0)}(\partial T)$ $[C^{(0,\lambda)}(\partial T)]$ defining the relative norm by

$$\| u \|_{C^{(0)}(\partial T)} = U_0(\partial T), \; [\| u \|_{C^{(0,\lambda)}(\partial T)} = U_0(\partial T) + U_{0,\lambda}(\partial T)].$$

Note finally that, if X is an open subset of ∂T, we can define analogously the spaces $C^{(k)}(\overline{X})$, $C^{(k,\lambda)}(\overline{X})$, $H^{k,p}(\overline{X})$.

2. Elliptic equations. Consider $m^2 + m + 1$ real functions $a_{ik}(x)$, $b_i(x)$, $c(x)$ $(i, k = 1, 2, \dots, m)$ defined in a region Ω. We shall denote by $\mathfrak{M}$ the linear differential operator of the second order:

$$\mathfrak{M} = \sum_{i,k=1}^{m} a_{ik} \frac{\partial^2}{\partial x_i \partial x_k} + \sum_{i=1}^{m} b_i \frac{\partial}{\partial x_i} + c.$$

Supposing, as we shall always do without notice to the contrary, that $a_{ik} = a_{ki}$, we shall say that $\mathfrak{M}$ is *of elliptic type* if the quadratic from $\sum a_{ik}(x) \, \xi_i \xi_k$ is always definite of the same sign, for all x in Ω.

To be precise, we shall suppose that this form is positive definite; consequently, if $A(x)$ is its determinant, $A(x) > 0$ for $x \in \Omega$.

We shall also say that $\mathfrak{M}$ is *uniformly elliptic* in Ω if the a_{ik} are measurable in Ω and if there exists a constant $a_0 > 0$ such that we have for $x \in \Omega$ and all real m-tuples $(\xi_1, \xi_2, \dots, \xi_m)$:

$$a_0 \sum_{i=1}^{m} \xi_i^2 \leq \sum_{i,k=1}^{m} a_{ik}(x) \, \xi_i \xi_k \leq a_0^{-1} \sum_{i=1}^{m} \xi_i^2. \tag{2.1}$$

Obviously, if Ω is bounded and the a_{ik} are continuous in $\overline{\Omega}$, the uniform ellipticity is a consequence of ellipticity. The constant a_0 is called the *constant of ellipticity* of the operator $\mathfrak{M}$.

If $f(x)$ is another function also defined in Ω, we propose to study the partial differential equation:

$$\mathfrak{M} u = f. \tag{2.2}$$

In the following we shall frequently denote the first and second derivatives of u by the symbols p_i, p_{ik}, so that (2.2) can also be written:

$$\sum_{i,k=1}^{m} a_{ik} p_{ik} + \sum_{i=1}^{m} b_i p_i + c u = f. \tag{2.3}$$

We shall say that a function $u(x)$ is *a regular solution* of (2.2) in Ω if it is of class $C^{(2)}$ in Ω and satisfies (2.2) at every point of Ω.

Besides regular solutions, in many contexts *generalized* solutions of (2.2) occur. We shall consider this starting in Chapter IV.

3. Maximum and minimum properties of the solutions of elliptic equations. Some properties concerning maxima and minima of regular solutions of (2.2) are of fundamental importance for the posing and the study of boundary value problems for (2.2).

3, I. *If in Ω: $c \leq 0$, $f < 0$ [$f > 0$] or else $c < 0$, $f \leq 0$ [$f \geq 0$], then every regular solution $u(x)$ of (2.2) is devoid in Ω of points of relative negative minimum [relative positive maximum].*

Indeed, at every point of relative minimum, for example, we have $p_i = 0$, $\Sigma p_{ik} \xi_i \xi_k \geq 0$, and hence $\mathfrak{M} u - c u = \Sigma a_{ik} p_{ik} \geq 0$. This is because the quadratic form $\Sigma a_{ik} \xi_i \xi_k$ can always be decomposed into the sum of squares of m linear forms:

$$\sum_{i,k=1}^{m} a_{ik} \xi_i \xi_k = \sum_{r=1}^{m} \left[\sum_{s=1}^{m} g_{rs} \xi_s \right]^2 ,$$

yielding

$$\mathfrak{M} u - c u = \sum_{r=1}^{m} \sum_{i,k=1}^{m} g_{ri} g_{rk} p_{ik} \geq 0 .$$

Furthermore, if $u < 0$ at this point as posed in the hypothesis, $\mathfrak{M} u - c u = f - c u < 0$, and from this contradiction the theorem follows.

More generally, E. HOPF [1] has proved that:

3, II. *Let the coefficients of (2.2) be bounded in Ω and let $A(x)$ admit in Ω a positive lower bound. If in Ω: $c \leq 0$, $f \leq 0$ [$f \geq 0$] then no regular solution $u(x)$ of (2.2) can have at a point x_0 of Ω a negative relative minimum [positive relative maximum] unless it is identically constant in every region Ω_0 containing x_0 in which $u(x) \geq u(x_0)$ [$\leq u(x_0)$].*

Let us suppose, to be precise, that x_0 is a point of negative relative minimum, and let I be the set of points of Ω_0 at which we have $u(x) = u(x_0)$. The set I is closed in Ω_0, and when we have shown that $\partial I \cup \Omega_0$ is empty, we shall have proved that $I \equiv \Omega_0$. Let us suppose, on the contrary, that this is not so; there then exists a point x of ∂I belonging to Ω_0 and having a positive distance d from $\partial \Omega_0$. If x' is a point of $\Omega_0 - I$ such that $\overline{x x'} < \frac{\delta}{2}$, then the entire hypersphere $\Gamma(x', \varrho)$ with center x' and radius $\varrho \leq \frac{\delta}{2}$ is contained in Ω_0; let r' be the upper bound of the values of ϱ for which $\Gamma(x', \varrho) \subset \Omega_0 - I$. On $\partial \Gamma(x', r')$ at least one point x_2 of I can be found; if now x_1 is any point of the radius $x' x_2$ distinct from x', and if $\overline{x_1 x_2} = \varrho$, we have:

$$u(x) > u(x_0) \quad for \quad x \in \Gamma(x_1, \varrho) - x_2, \quad u(x_2) = u(x_0) .$$

Choosing $\varrho_1 < \varrho$ such that $\Gamma(x_2, \varrho_1) \subset \Omega_0$ and putting:

$$r = \overline{x\,x_1}, \quad v(x) = e^{-k\varrho^2} - e^{-kr^2}, \tag{3.1}$$

choosing k sufficiently large, we have $\mathfrak{M}\,v < 0$ for $x \in \Gamma(x_2, \varrho_1)$. This is certainly possible since $e^{kr^2}\,\mathfrak{M}\,v$ is a trinomial of second degree in k, in which the coefficient of k^2 admits a negative upper bound in $\Gamma(x_2, \varrho_1)$ while the coefficient of k is bounded, and the known term is bounded from above. We now have for $\lambda > 0$: $\mathfrak{M}(u + \lambda v) < 0$ in $\Gamma(x_2, \varrho_1)$; and this contradicts theorem 3, I since λ can be chosen so small that $u + \lambda v > u(x_0)$ on $\partial\Gamma(x_2, \varrho_1)$, but $u(x_2) + \lambda v(x_2) = u(x_0)$, so that $u + \lambda v$ admits a negative relative minimum in $\Gamma(x_2, \varrho_1)$.

From this theorem, thus proved by *reductio ad absurdum*, the following corollary is immediate:

3, III. *Let the region Ω be bounded and let $u(x)$ be a regular solution in Ω of the homogeneous equation $\mathfrak{M}\,u = 0$ and suppose u is non-constant and continuous in $\overline{\Omega}$. If $c \leq 0$ then throughout Ω:*

$$|u| < \max_{\partial\Omega} |u|, \tag{3.2}$$

while if $c = 0$, then throughout Ω:

$$\min_{\partial\Omega} u < u < \max_{\partial\Omega} u. \tag{3.3}$$

Another important property of the solution of (2.2) is expressed in the following theorem of G. GIRAUD[1]:

3, IV. *Let T be a domain of class $A^{(1,\,\lambda)}$ and let $u(x)$ be a continuous, non-constant function in T and a regular solution in $T - \partial T$ of (2.2). If $c \leq 0, f \leq 0$ $[f \geq 0]$, $\min_{\partial T} u \leq 0$ $[\max_{\partial T} u \geq 0]$, then for every point y on ∂T at which u assumes its minimum [maximum] value, and for each axis l out of y such that $\cos(l, n) < 0$, there exists a positive constant L such that for $x \in l$ and $\overline{x\,y}$ sufficiently small we have:*

$$u(x) - u(y) < L\,\overline{x\,y}\,[< -L\,\overline{x\,y}]. \tag{3.4}$$

We shall not attempt to reproduce GIRAUD's rather deep proof; instead, we shall limit ourselves to proving the theorem under the more restrictive hypothesis that T is of class $A^{(2)}$. Under this hypothesis it is possible to select a point x_1 on the interior normal to ∂T at the point y, such that the hypersphere $\Gamma(x_1, \varrho)$ of center x_1 and radius $\varrho = \overline{x_1\,y}$ has in common with ∂T only the point y. Now choose $\varrho_1 < \varrho$ and let D be the domain $\Gamma(x_1, \varrho) \cap \Gamma(y, \varrho_1)$.

---

[1] See G. GIRAUD [9] p. 343 and [12] p. 50. See also O. A. OLEJNIK [2], E. HOPF [7], C. PUCCI [5], J. BOCHENEK [1].

By theorem 3, II, $u(x) - u(y) > 0$ for $x \in T - \partial T$; while the function
$v(x)$ defined by (3.1) is zero on $\partial D \cap \partial \Gamma (x_1, \varrho)$. Thus it is obvious that
for λ positive and sufficiently small:

$$- \lambda v(x) < u(x) - u(y) \quad \text{for} \quad x \in \partial D . \tag{3.5}$$

But for k sufficiently large we have, in D, $\mathfrak{M} v < 0$, and hence also:

$$\mathfrak{M} [u(x) - u(y) + \lambda v(x)] = f(x) - c(x) u(y) + \lambda \mathfrak{M} v < 0 ,$$

because by dint of theorem 3, I, (3.5) holds throughout D. From this the
theorem follows since:

$$- \lambda \left(\frac{dv}{dl} \right)_{x=y} = - 2 \lambda k \varrho \, e^{-k \varrho^2} \cos (l, n) > 0 .$$

The theorems with which we have until now been occupied here have
been variously extended by A. D. ALEKSANDROV[1], who has dedicated a
series of works [5, ..., 10] to the study of this question brought up in
what follows (§§ 37 and 56); here we restrict ourselves to mentioning one
of the theorems proved in [5]:

3, V. *Let the coefficients of (2.2) be of class $L^m(\Omega)$ and let $A(x)$
admit a positive lower bound in Ω. Then the conclusions of theorems 3, I,
II, III continue to hold when we suppose only that $u(x)$ is of class
$H^{2,m}(\Omega) \cap C^{(0)}(\Omega)$ and that (2.2) is satisfied almost everywhere in Ω.*

4. Various types of boundary value problems. The most important and lengthy question which is posed in the theory of elliptic
equations is the study of boundary value problems for such equations.

Given a bounded domain T of class $A^{(1)}$, let us consider at each point
of ∂T an axis l oriented toward the exterior of T, chosen such that we
always have $\cos (l, n) > 0$. Let α, β, and γ be three functions defined on
∂T, restricted for now solely by the condition $|\alpha| + |\beta| > 0$. If $\mathfrak{M}$ is
an elliptic operator defined in $T - \partial T$, the boundary value problems
for equation (2.2) with which we shall occupy ourselves, at least for now,
are mainly of the following type:

Seeking the existence of functions $u(x)$ satisfying the conditions

a) $u(x)$ is a regular solution of (2.2) in $T - \partial T$.

b) $u(x)$ is continuous in T.

*c) $u(x)$ is endowed with a derivative in the direction of the axis l at
every point of ∂T for which $\alpha \neq 0$, and satisfies at all points of ∂T the
boundary condition:*

$$\alpha \frac{du}{dl} + \beta u = \varphi . \tag{4.1}$$

---

[1] For some other extensions see M. KRZYŻAŃSKI [3], P. HARTMAN [1],
R. J. DUFFIN [2], C. PUCCI [5, 12].

The study of these problems under the most general conditions presents many difficulties and is still incomplete; we shall therefore point out those particular cases of special importance in mathematical physics toward which most analysts direct their attention.

A first particular case is that in which we suppose $\alpha = 0$, $\beta = 1$; in this case the boundary condition assumes the particularly simple form

$$u(x) = \varphi(x) \quad \text{for} \quad x \in \partial T, \tag{4.2}$$

and the problem considered is called the *Dirichlet problem* or *first boundary value problem*.

A second particular case occurs when $\alpha > 0$ and when the axis l coincides with the axis ν (the *conormal*) with direction cosines

$$Y_i = \frac{1}{a} \sum_{k=1}^{m} a_{ik} X_k,$$

$$a = \left[\sum_{i=1}^{m} \left(\sum_{k=1}^{m} a_{ik} X_k \right) \right]^{1/2}. \tag{4.3}$$

This problem is called the *Neumann problem* or *second boundary value problem*[1]. Note that in this case we can assume, without loss of generality, and shall from now on, that:

$$\alpha = a, \tag{4.4}$$

with which, if the first derivatives of u are also continuous on ∂T, (4.1) can be written:

$$\sum_{i,k=1}^{m} a_{ik} X_k \frac{\partial u}{\partial x_i} + \beta u = \varphi. \tag{4.5}$$

A third particular case is that in which, while we always suppose $\alpha > 0$, the axis l can be arbitrary provided that $\cos(l, n)$ has a positive lower bound on ∂T. This problem is called the *regular oblique derivative problem* or *third boundary value problem*. In this case we can always suppose without loss of generality that:

$$\alpha = a^{(l)}, \tag{4.6}$$

having posed by definition that:

$$a^{(l)} = a \frac{\cos(n, \nu)}{\cos(n, l)}. \tag{4.7}$$

---

[1] Translator's note: it is common in the United States to reserve the name *Neumann problem* for the case when $\beta = 0$. If $\beta \neq 0$, the problem is often called the *Robin problem*. See BERGMAN and SCHIFFER [2].

Using this, if the first derivatives of u are also continuous on ∂T, (4.1) assumes the form:

$$\sum_{i,k=1}^{m} a_{ik} X_k \frac{\partial u}{\partial x_i} + \sum_{i=1}^{m} \alpha_i \frac{\partial u}{\partial x_i} + \beta u = \varphi \qquad (4.8)$$

where, having put:

$$\alpha_i = a^{(l)} \cos(l, x_i) - a \cos(\nu, x_i) \qquad (4.9)$$

we have:

$$\sum_{i=1}^{m} \alpha_i X_i = 0 . \qquad (4.10)$$

Finally, we shall call any problem in which α, without being identically zero, vanishes on a portion of ∂T, a *mixed problem*.

In some cases, we shall also consider problems in which the axis l is tangent at some points of ∂T, or, more generally, is not bound by the condition $\cos(l, n) > 0$.

We shall also consider the same problems in an unbounded domain T. In this case it will be necessary to assign to u an *a priori* behavior at infinity. For example, we shall require that:

$$\lim_{x \to \infty} u(x) = 0 \qquad (4.11)$$

or that:

$$\overline{\lim_{x \to \infty}} \, |u(x)| < \infty . \qquad (4.12)$$

However, we shall occupy ourselves with these problems in unbounded domains only in § 49 of Chapter VII, and hence, in this chapter and the following ones, the domain T will always be assumed bounded unless notice is given to the contrary. Likewise, the study of mixed problems will be deferred to § 50 of Chapter VII, together with those other problems which will be considered there.

5. Uniqueness theorems. Later on we shall study various procedures for arriving at existence theorems for the above mentioned problems; meanwhile, the theorems of § 2 permit us to establish immediately some uniqueness theorems for the boundary value problems we are considering.

5, I. *If $c \le 0$ the Dirichlet problem admits at most one solution.*

In fact, if u_1 and u_2 are two solutions of the problem, for their difference $u = u_1 - u_2$ we have:

$$\mathfrak{M}u = 0 \quad \text{for} \quad x \in T - \partial T, \quad u = 0 \quad \text{for} \quad x \in \partial T,$$

from which, by theorem 3, III we have $u \equiv 0$ in T.

5, II. *If T is of class $A^{(1,\lambda)}$ and if $c \leq 0$, $\beta \geq 0$, provided c and β are not both identically zero, then the second and third boundary value problems admit at most one solution.*

Here we shall show that every solution of $\mathfrak{M}u = 0$ which satisfies the homogeneous equation (4.1) is identically zero. For, if u is not identically zero, it must (3, III) attain on ∂T either a positive maximum or a negative minimum, and this is impossible. This is because, since c and β are not both identically zero, u is not constant and hence can not have a positive maximum [negative minimum] at points of ∂T since at such points we would have to have (3, IV):

$$\frac{du}{dl} > 0 \left[\frac{du}{dl} < 0 \right],$$

and the boundary condition could not be satisfied. With the same reasoning we can prove:

5, III. *Under the same hypotheses as theorem 5, II, the mixed problem too admits at most one solution.*

5, IV. *If T is of class $A^{(1,\lambda)}$ and if $c \equiv 0$, $\beta \equiv 0$, then two distinct solutions of the boundary value problem differ by a constant.*

Apropos of theorem 5, IV we note that under the hypothesis $c \equiv \beta \equiv 0$, if u is a solution of the second or third boundary value problem, so also is $u + k$, with k an arbitrary constant.

Finally, regarding the case of unbounded domains, we see immediately that:

5, V. *The preceding theorems are also valid if T is not bounded, when condition (4.11) is imposed at infinity.*

This theorem will be frequently used in the following form:

5, VI. *Let the operator $\mathfrak{M}$ be defined in all of R_m, and, given a bounded domain T of class $A^{(1,\lambda)}$, suppose $c(x) \leq 0$ for $x \in \mathfrak{C}(T)$. Then every function $u(x)$ which is continuous in $\mathfrak{C}(T) \cup \partial T$ and is a regular solution of the equation $\mathfrak{M}u = 0$ in $\mathfrak{C}(T)$ which goes to zero at infinity and on ∂T is identically zero. If l is oriented toward the exterior of T and if $\beta \leq 0$ provided c and β are not both identically zero, then the same conclusion holds if for the condition of vanishing on ∂T we substitute*

$$a^{(l)} \frac{du}{dl} + \beta u = 0.$$

If $c \leq 0$ throughout R_m, then every function $u(x)$ which goes to zero at infinity and which is a regular solution in R_m of $\mathfrak{M}u = 0$ vanishes identically.

We observe now that if in a boundary value problem we effect a change of unknown function $u = v\,\omega$, with ω a fixed function, a problem for the unknown v of the same type arises. We can now try to determine ω in such a manner that this new problem satisfies hypotheses which allow the application of one of the preceding theorems. Thus, for example, if there exists a function $\omega > 0$ for which $\mathfrak{M} \omega < 0$, then the indicated change of unknown functions leads to an equation in which the coefficient of v is negative. The existence of the function ω thus ensures the uniqueness of the solution of the DIRICHLET problem. By this means it is possible to obtain various uniqueness theorems, such as, for example, theorems 19, IX, 35, IX, and 35, XI, which will be proved later, and others for which we refer to M. PICONE [1, 6, 7].

6. Green's formula. In this section we shall suppose that the functions a_{ik} and

$$e_i = b_i - \sum_{k=1}^{m} \frac{\partial a_{ik}}{\partial x_k} \tag{6.1}$$

are of class $C^{(1)}$ in the region Ω; under this hypothesis, $\mathfrak{M} u$ can be given the form:

$$\mathfrak{M} u = \sum_{i,k=1}^{m} \frac{\partial}{\partial x_k} \left(a_{ik} \frac{\partial u}{\partial x_i} \right) + \sum_{i=1}^{m} e_i \frac{\partial u}{\partial x_i} + c u . \tag{6.2}$$

The following operator is called the *adjoint operator* of $\mathfrak{M}$:

$$\mathfrak{N} v = \sum_{i,k=1}^{m} \frac{\partial}{\partial x_k} \left(a_{ik} \frac{\partial v}{\partial x_i} \right) - \sum_{i=1}^{m} \frac{\partial}{\partial x_i} (e_i v) + c v , \tag{6.3}$$

which, if we assume the additional hypotheses that the a_{ik} and b_i are respectively of classes $C^{(2)}$ and $C^{(1)}$ in Ω, can also be written:

$$\mathfrak{N} v = \sum_{i,k=1}^{m} \frac{\partial^2}{\partial x_i \partial x_k} (a_{ik} v) - \sum_{i=1}^{m} \frac{\partial}{\partial x_i} (b_i v) + c v . \tag{6.4}$$

From (6.3) it follows immediately that the adjoint of $\mathfrak{N} v$ is $\mathfrak{M} u$, so that $\mathfrak{M}$ and $\mathfrak{N}$ can be said to be adjoints of each other.

If $\mathfrak{M} \equiv \mathfrak{N}$, $\mathfrak{M}$ is called *self-adjoint*; and a necessary and sufficient condition for this to happen is that all the e_i be zero. Between two adjoint operators certain integral relations hold which are of fundamental importance for the theory of elliptic equations. Now, we have:

$$v \mathfrak{M} u - u \mathfrak{N} v = \sum_{i,k=1}^{m} \frac{\partial}{\partial x_k} \left[a_{ik} \left(v \frac{\partial u}{\partial x_i} - u \frac{\partial v}{\partial x_i} \right) \right] + \sum_{i=1}^{m} \frac{\partial}{\partial x_i} (e_i u v)$$

and hence, if u and v are of class $C^{(2)}$ in Ω, and if $T \subset \Omega$ is a domain of class $A^{(1)}$, it follows that:

$$\int_T [v \mathfrak{M} u - u \mathfrak{N} v] \, dx = \int_{\partial T} \left[a \left(v \frac{du}{dv} - u \frac{dv}{dv} \right) + b u v \right] d\sigma , \tag{6.5}$$

where v is the conormal, a is the function given by the second half of (4.8), and b is defined by:

$$b = \sum_{i=1}^{m} e_i X_i.$$

To (6.5) we give the name GREEN's *formula*. If T is of class $A^{(2)}$, this formula can be transformed[1] in such a way as to cause to appear, instead of the derivatives of u and v with respect to the conormal v, the derivatives of the same functions respectively in the directions of two axes l and λ, of which the first can be assigned arbitrarily except for the condition that $\cos (l, n) > 0$, and that its direction cosines are of class $C^{(1)}$ on ∂T.

In fact, for every axis l arbitrarily given under the above-mentioned conditions, we can construct a corresponding axis λ with direction cosines:

$$\cos (\lambda, x_k) = \frac{2\,a}{a^{(\lambda)}} \cos (v, x_k) - \frac{a^{(l)}}{a^{(\lambda)}} \cos (l, x_k).$$

Because, for all functions α_i defined by (4.9), one can give the expression:

$$\alpha_i = -a^{(\lambda)} \cos (\lambda, x_i) + a \cos (v, x_i),$$

we recognize immediately that:

$$a \left(v \frac{d u}{d v} - u \frac{d v}{d v} \right) = a^{(l)} v \frac{d u}{d l} - a^{(\lambda)} u \frac{d v}{d \lambda} - \sum_{i=1}^{m} \alpha_i \frac{d\,(uv)}{d x_i}. \qquad (6.6)$$

Now let S be a portion of ∂T which admits a parametric representation (1.2) in a domain D of class $A^{(1)}$; in virtue of (4.10) it is certainly possible to determine $m-1$ functions $\beta_i\,(t_1, t_2, \ldots, t_{m-1})$ of class $C^{(1)}$ in D, such that for any function w:

$$\sum_{i=1}^{m} \alpha_i \frac{\partial w}{\partial x_i} = \frac{1}{J} \sum_{i=1}^{m-1} \beta_i \frac{\partial w}{\partial t_i}.$$

If ∂S is the set of points of S corresponding to ∂D, we then have from (6.6) that:

$$\int_S a \left(v \frac{d u}{d v} - u \frac{d v}{d v} \right) d\sigma = \int_S \left(a^{(l)} v \frac{d u}{d l} - a^{(\lambda)} u \frac{d v}{d \lambda} \right) d\sigma + \int_S b'\, u v\, d\sigma$$

$$- \int_{\partial S} u v \sum_{i=1}^{m-1} \beta_i\, dt_{i+1} \ldots dt_{m-1}\, dt_1 \ldots dt_{i-1},$$

---

[1] Cf. G. GIRAUD [15] p. 371, M. PICONE and C. MIRANDA [1] and C. TOLOTTI [1].

where:

$$b' = \frac{1}{J} \sum_{i=1}^{m=1} \frac{\partial \beta_i}{\partial t_i}$$

is a function defined on ∂T, independent of the particular parametric representation adopted.

From this it is obvious that:

$$\int_{T} [v \, \mathfrak{M} \, u - u \, \mathfrak{N} \, v] \, dx = \int_{\partial T} \left[a^{(l)} \, v \, \frac{d u}{d l} - a^{(\lambda)} \, u \, \frac{d v}{d \lambda} + (b + b') \, u \, v \right] d\sigma. \quad (6.7)$$

We arrived at (6.6) and (6.7) with somewhat restrictive hypotheses on the functions u and v; however, it can be immediately seen by an obvious passage to the limit that:

6, I. *Retaining unchanged the hypotheses on the domain T and on the axis l, (6.5) and (6.7) also hold if u and v are defined only in the domain T, are of class $C^{(1)}$ there, and possess second derivatives continuous in $T - \partial T$ and* LEBESGUE *integrable in T.*

A more accurate analysis would show that:

6, II. *In theorem 6, I we can substitute for the hypotheses on the second derivatives of u and v that the first derivatives of u and v are absolutely* TONELLI *continuous.*

Similarly it is possible to reduce considerably the hypotheses on the domain T. If, for example, T is a domain whose boundary ∂T is decomposable into a finite union of a finite number of hypersurfaces, each one endowed at each point with a continuously varying tangent hyperplane, then (6.5) is still valid. An analogous extension also holds for (6.7).

Another observation which is sometimes useful is that (6.7) is valid even if T is only of class $A^{(1)}$, provided that the axis l is such that the functions a_i defined by (4.9) are of class $C^{(1)}$.

7. Compatibility conditions for the boundary value problems; other uniqueness theorems. From GREEN's formula it is possible to deduce immediately some consequences regarding the boundary value problems considered in § 4.

Let β be an arbitrary function defined and continuous on ∂T, and set:

$$\mathfrak{P} u = a^{(l)} \, \frac{d u}{d l} + \beta \, u, \quad (7.1)$$

$$\mathfrak{Q} v = a^{(\lambda)} \, \frac{d v}{d \lambda} + (\beta - b - b') \, v. \quad (7.2)$$

The operators $\mathfrak{P}$ and $\mathfrak{Q}$ are meaningful under the sole hypothesis that u and v are differentiable on ∂T in the directions of the axes l

and λ, respectively; if, further, u and v are of class $C^{(1)}$ in T, we can also write:

$$\mathfrak{P}\, u = \sum_{i,k=1}^{m} a_{ik}\, X_k\, \frac{\partial u}{\partial x_i} + \frac{1}{J} \sum_{i=1}^{m} \beta_i\, \frac{\partial u}{\partial t_i} + \beta\, u, \tag{7.3}$$

$$\mathfrak{Q}\, v = \sum_{i,k=1}^{m} X_k\, \frac{\partial}{\partial x_i}\, (a_{ik}\, v) - \frac{1}{J} \sum_{i=1}^{m-1} \frac{\partial}{\partial t_i}\, (\beta_i\, v) + \left(\beta - \sum_{k=1}^{m} b_k\, X_k\right) v. \tag{7.4}$$

Using this notation, (6.7) can be put in the form:

$$\int_{T} (v\, \mathfrak{M}\, u - u\, \mathfrak{N}\, v)\, dx = \int_{\partial T} (v\, \mathfrak{P}\, u - n\, \mathfrak{Q}\, v)\, d\sigma. \tag{7.5}$$

According to this, we observe that the boundary condition for the second or third boundary value problem for the equation $\mathfrak{M}\, u = f$ can always be written in the form:

$$\mathfrak{P}\, u = \varphi. \tag{7.6}$$

Note that in the case of the second boundary value problem, the β_i and b' are zero.

The *adjoint* of this boundary value problem is the one consisting of the quest for a regular solution v in $T - \partial T$ of the equation

$$\mathfrak{N}\, v = g \tag{7.7}$$

satisfying the boundary condition

$$\mathfrak{Q}\, v = \psi \quad \text{for} \quad x \in \partial T, \tag{7.8}$$

where g and ψ are two functions defined in T and ∂T, respectively. Also, the *adjoint* of the DIRICHLET problem for $\mathfrak{M}\, u = f$ is the DIRICHLET problem for (7.7), for which we write the boundary condition in the form

$$v = \psi \quad \text{for} \quad x \in \partial T. \tag{7.9}$$

We see immediately that every boundary value problem coincides with the adjoint of its adjoint. Furthermore, two adjoint problems are always of the same type; and the necessary and sufficient condition for the first and second boundary value problems to coincide with their adjoints, i.e. to be self-adjoint, is that the operator $\mathfrak{M}$ be self-adjoint. For the third boundary value problem it is necessary to add:

$$l \equiv \lambda \equiv v \tag{7.10}$$

and this alone allows the third boundary value problem to be self-adjoint.

This premise follows immediately from (7.5):

7, I. *Supposing T to be of class $A^{(1)}$ and f and φ to be continuous, a necessary condition that the* DIRICHLET *problem for* (2.2) *admit a solution of class $C^{(1)}$ in T is the following:*

$$\int_T f v \, dx + \int_{\partial T} \varphi \, a \, \frac{dv}{dv} \, d\sigma = 0 \tag{7.11}$$

for every solution v of the adjoint homogeneous problem which is also of class $C^{(1)}$ in T.

7, II. *Supposing T to be of class $A^{(1)}$ $[A^{(2)}]$ and f and φ to be continuous, a necessary condition that the second [third] boundary value problem* (2.2) *admit a solution of class $C^{(1)}$ in T is the following:*

$$\int_T f v \, dx - \int_{\partial T} \varphi \, v \, d\sigma = 0 \tag{7.12}$$

for every solution v of the adjoint homogeneous problem which is also of class $C^{(1)}$ in T.

We observe now that if we replace u by u^2 and v by 1 in (7.5) we obtain:

$$\int_T \left[u \, \mathfrak{M} \, u + \sum_{i,k=1}^m a_{ik} \frac{\partial u}{\partial x_i} \frac{\partial u}{\partial x_k} \right] dx = \frac{1}{2} \int_T (c + c^*) \, u^2 \, dx +$$

$$+ \int_{\partial T} u \left[\mathfrak{P} u - \left(\beta - \frac{b + b'}{2} \right) u \right] d\sigma, \tag{7.13}$$

where

$$c^* = c - \sum_{i=1}^m \frac{\partial e_i}{\partial x_i}$$

is the coefficient of v in the operator $\mathfrak{N} \, v$.

Applying this formula to the solution of a homogeneous problem, we easily extract the following uniqueness theorems:

7, III. *If the a_{ik} and e_i are of class $C^{(1)}$ in a domain T (of class $A^{(1)}$), and if $c + c^* \leq 0$ then the* DIRICHLET *problem for the equation $\mathfrak{M} u = f$ admits at most one solution of class $C^{(1)}$ in T.*

7, IV. *If the a_{ik} and e_i are of class $C^{(1)}$ in a domain T (of class $A^{(1)}$), and if $c + c^* \leq 0$ and $2\beta - b \geq 0$, without c and β both being identically zero, then the* NEUMANN *problem for the equation $\mathfrak{M} u = f$ admits at most one solution of class $C^{(1)}$ in T. This also holds for the third boundary value problem if T is if class $A^{(2)}$, the direction cosines of l are of class $C^{(1)}$ on ∂T, and $2\beta - b - b' \geq 0$.*

Referring to V. YA. SKOROBOGAT'KO [1, 2] for other uniqueness theorems of the same type for the DIRICHLET problem, we choose to finish this section by pointing out some other consequences of GREEN's formula. We have for example:

7, V. *If $\mathfrak{M}$ is self-adjoint and the equation $\mathfrak{M} u = 0$ admits a solution u_0 which is regular in T and zero on ∂T, then every other solution of the same equation must necessarily vanish at some point of T.*

The proof is immediate. Indeed, we have, by applying (7.3) to u_0:

$$\int_T \sum_{i,k=1}^m a_{ik} \frac{\partial u_0}{\partial x_i} \frac{\partial u_0}{\partial x_k} \, dx = \int_T c \, u_0^2 \, dx,$$

and this formula, in case $\mathfrak{M} u = 0$ admits a solution u positive in T, can be written:

$$\int_T \sum_{i,k=1}^m a_{ik} \frac{\partial u_0}{\partial x_i} \frac{\partial u_0}{\partial x_k} \, dx = -\int_T \frac{u_0^2}{u} \sum_{i,k=1}^m \left(a_{ik} \frac{\partial u}{\partial x_k} \right) dx,$$

from which, by integration by parts:

$$\int_T \sum_{i,k=1}^m a_{ik} \left(\frac{\partial u_0}{\partial x_i} - \frac{u_0}{u} \frac{\partial u}{\partial x_i} \right) \left(\frac{\partial u_0}{\partial x_k} - \frac{u_0}{u} \frac{\partial u}{\partial x_k} \right) dx = 0.$$

Reductio ad absurdum proves the theorem.

By analogous reasoning it is possible to establish comparison theorems even in certain cases in which u and u_0 satisfy two different equations. Along this line are the works of P. HARTMAN and A. WINTNER [6], R. M. REDHEFFER [2], M. H. PROTTER [1], C. CLARK and C. A. SWANSON [1]. For other theorems of comparison between solutions of different equations see also A. G. TETEREV [1] and P. HARTMAN and A. WINTNER [4].

8. Levi functions. Let us denote by A_{rs} the elements of the inverse matrix of a_{rs}, and by A the determinant of the matrix a_{rs}, and put:

$$H(x, y) = \begin{cases} \dfrac{1}{(m-2)\,\omega_m \sqrt{A(y)}} \left(\displaystyle\sum_{r,s=1}^m A_{rs}(y)\,(x_r - y_r)\,(x_s - y_s) \right)^{(2-m)/2} \\[2mm] \text{for} \quad m > 2, \\[4mm] \dfrac{1}{2\pi \sqrt{A(y)}} \log \left(\displaystyle\sum_{r,s=1}^m A_{rs}(y)\,(x_r - y_r)\,(x_s - y_s) \right)^{-1/2} \\[2mm] \text{for} \quad m = 2. \end{cases} \tag{8.1}$$

We easily verify the identity:

$$\sum_{i,k=1}^{m} a_{ik}(y) \frac{\partial^2 H}{\partial x_i \partial x_k} = 0, \qquad (8.2)$$

from which it follows in particular that $H(x, y)$, as a function of x, is a solution of the equation $\mathfrak{M} u = 0$ when the a_{ik} are constants and the b_i and c are zero. In the general case, the study of solutions of our equations which behave like $H(x, y)$ as $x \to y$ will be of great interest.

Before facing such research we need to introduce some notation.

We note that, in virtue of the hypotheses made on the quadratic form $\sum a_{ik} \xi_i \xi_k$, putting $\overline{x\,y} = r$, we have:

$$H = O\left(r^{2-m}\right), \qquad \frac{\partial H}{\partial x_i} = O\left(r^{1-m}\right), \qquad \frac{\partial^2 H}{\partial x_i\, \partial x_k} = O\left(r^{-m}\right), \qquad (8.3)$$

such bounds holding in every bounded domain T contained in Ω if, for example, the a_{ik} are assumed continuous[1].

Every function $L(x, y)$ continuous in the variables x and y for x and y in Ω and for $x \neq y$, together with its first and second derivatives with respect to the x_i, is called a *Levi function* if it satisfies bounds of the following type for some convenient $\lambda > 0$:

$$L - H = O\left(r^{\lambda+2-m}\right), \qquad \frac{\partial [L - H]}{\partial x_i} = O\left(r^{\lambda+1-m}\right), \qquad \frac{\partial^2 [L - H]}{\partial x_i\, \partial x_k} = O\left(r^{\lambda-m}\right), \qquad (8.4)$$

such bounds holding uniformly in every bounded domain contained in Ω.

Naturally $H(x, y)$ is itself a LEVI function; moreover, so is $H(y, x)$, with $\lambda = 1$, if the functions a_{ik} are of class $C^{(2)}$ in Ω.

If we take account of (8.2) and moreover suppose that the coefficients of the operator $\mathfrak{M}$ are bounded and that $\lambda \leq 1$, we have:

$$\mathfrak{M} L = \sum_{i,k=1}^{m} \left[a_{ik}(x) - a_{ik}(y)\right] \frac{\partial^2 H}{\partial x_i \partial x_k} + O\left(r^{\lambda-m}\right),$$

from which, with the final hypothesis that the a_{ik} are of class $C^{(0,\lambda)}$ in Ω, it follows that:

$$\mathfrak{M} L(x, y) = O\left(r^{\lambda-m}\right), \qquad (8.5)$$

uniformly in every bounded domain contained in Ω. In the case that (8.5) satisfies in particular the hypotheses of § 6, we have also:

$$\mathfrak{N} L(x, y) = O\left(r^{\lambda-m}\right). \qquad (8.6)$$

---

[1] By O and o we mean the symbols of LANDAU. In the first of (8.3) we should substitute $O\left(\log \frac{2R}{r}\right)$ for $O(r^{2-m})$, meaning by R the diameter of T, when $m = 2$.

Still with the hypotheses of § 6 and with $L = H$, (8.5) and (8.6) hold simply for $\lambda = 1$.

9. Stokes's formula. Let us posit the hypotheses of § 6 and let $I(y, \varrho)$ represent the neighborhood of y defined by the bound:

$$\sum_{r,s=1}^{m} A_{rs}(y)\, (x_r - y_r)\, (x_s - y_s) < \varrho^2. \tag{9.1}$$

Now if T is a domain contained in Ω for which (7.5) holds, let us choose y at will in $T - \partial T$ and ϱ so small that $I(y, \varrho) \subset T - \partial T$, and let us apply (7.5) assuming $T - I(y, \varrho)$ as domain of integration, the exterior normal to $T - I$ with direction cosines $X_k(x)$ as axis l, and any LEVI function $L(x, y)$ as the function $v(x)$. We have:

$$\int_{T-I} (L\,\mathfrak{M}\,u - u\,\mathfrak{N}\,L)\, dx = \int_{\partial T \cup \partial I} (L\,\mathfrak{P}\,u - u\,\mathfrak{Q}\,L)\, d\sigma. \tag{9.2}$$

Now an easy calculation shows that on ∂I:

$$L\,\mathfrak{P}\,u - u\,\mathfrak{Q}\,L = \frac{u(y)}{\varrho^m\,\omega_m\,\sqrt{A(y)}} \sum_{k=1}^{m} X_k(x)\, (x_k - y_k) + O\left(\varrho^{\lambda+1-m}\right),$$

from which:

$$\int_{\partial I} (L\,\mathfrak{P}\,u - u\,\mathfrak{Q}\,L)\, d\sigma = - \frac{m\,u(y)}{\varrho^m\,\omega_m\,\sqrt{A(y)}}\, \mathrm{mes}\,I + \mathrm{mes}\,(\partial I)\, O\left(\varrho^{\lambda+1-m}\right).$$

Since:

$$\mathrm{mes}\,I = \frac{\varrho^m\,\omega_m\,\sqrt{A(y)}}{m}, \qquad \mathrm{mes}\,(\partial I) = O\left(\varrho^{m-1}\right),$$

we have in passing to the limit as $\varrho \to 0$ in (9.2):

$$u(y) = \int_T (u\,\mathfrak{N}\,L - L\,\mathfrak{M}\,u)\, dx + \int_{\partial T} (L\,\mathfrak{P}\,u - u\,\mathfrak{Q}\,L)\, d\sigma, \tag{9.3}$$

a formula in which the first integral on the right-hand side is meaningful in virtue of (8.3), (8.4), and (8.6).

It is clear that in (9.2) the operators $\mathfrak{N}$ and $\mathfrak{Q}$ are applied to $L(x, y)$ considered as a function of x.

In analogous circumstances, when doubts may arise about the variable with respect to which an operator happens to be applied, we shall make this variable a subscript of the operator itself. Thus, for example, with this convention the symbols $\mathfrak{N}\,L$ and $\mathfrak{Q}\,L$ which occur in (9.2) could be replaced by $\mathfrak{N}_x\,L$ and $\mathfrak{Q}_x\,L$.

(9.3), called STOKES's *formula*, is valid under essentially the same hypotheses for the operator $\mathfrak{M}$, the function u, and the domain T, as those for which (7.5) holds; as for $L(x, y)$, it suffices to suppose that, (8.4) remaining true, it is defined in $T - y$ and continuous there with its first derivatives, and admitting second derivatives continuous in $(T - \partial T) - y$ and such that $\mathfrak{N}_x L$ is LEBESGUE integrable in T, for every $y \in T - \partial T$.

10. Fundamental solutions; Green's functions. Every LEVI function $L(x, y)$ which is a solution of the equation $\mathfrak{M}_x L = 0$ $[\mathfrak{N}_x L = 0]$ is called a *fundamental solution*[1] of the equation $\mathfrak{M} u = 0$ $[\mathfrak{N} u = 0]$. For example, if the a_{ik} are constants and the b_i and c are zero, the self-adjoint equation $\mathfrak{M} u = 0$ admits the function $H(x, y)$ as a fundamental solution.

The *Green's function* of a given boundary value problem is any $F(x, y)$ which, as a function of y, is a fundamental solution of the adjoint equation and which satisfies the boundary condition of the adjoint homogeneous problem[2].

Thus the GREEN's function $F(x, y)$ for the DIRICHLET problem for the equation $\mathfrak{M} u = f$ is a LEVI function of the point y and a solution of the equations:

$$\mathfrak{N}_y F(x, y) = 0 \quad \text{for} \quad y \in (T - \partial T) - x, \tag{10.1}$$

$$F(x, y) = 0 \quad \text{for} \quad y \in \partial T, \quad x \in T - \partial T, \tag{10.2}$$

while the GREEN's function of the second or third boundary value problem for the equation $\mathfrak{M} u = f$ is a LEVI function of the point y satisfying (10.1) and the boundary condition:

$$\mathfrak{D}_y F(x, y) = 0 \quad \text{for} \quad y \in \partial T, \quad x \in T - \partial T. \tag{10.3}$$

Analogously, the GREEN's function $\Phi(x, y)$ of a boundary value problem for the equation $\mathfrak{N} v = g$ is a LEVI function of the point y satisfying the equation $\mathfrak{M}_y \Phi = 0$ and satisfying the boundary condition $\Phi = 0$ in the case of the DIRICHLET problem or $\mathfrak{P}_y \Phi = 0$ in the case of the second or third boundary value problem.

The knowledge of the GREEN's function for a given boundary value problem permits us without further ado to write a formula for the solution, provided that this GREEN's function is continuous with its first derivatives with respect to the y_i in $T - x$.

---

[1] Or, with HADAMARD, *elementary solution*.

[2] Translator's note: Frequently the term "GREEN's function" is applied only in the case of the DIRICHLET problem; the terms "NEUMANN function" and "ROBIN's function" are used for the NEUMANN and ROBIN problem. Cf. BERGMAN & SCHIFFER [6].

In fact, from STOKES's formula, in which we exchange the roles of x and y, it follows that each solution $u(x)$ of the DIRICHLET problem for the equation $\mathfrak{M} u = f$, which is of class $C^{(1)}$ in T, is given by:

$$u(x) = - \int_T F(x, y) \, f(y) \, dy - \int_{\partial T} \mathfrak{Q}_y F(x, y) \, \varphi(y) \, d_y \sigma, \qquad (10.4)$$

while each solution of class $C^{(1)}$ in T of the second or third boundary value problem for the same equation is given by:

$$u(x) = - \int_T F(x, y) \, f(y) \, dy + \int_{\partial T} F(x, y) \, \varphi(y) \, d_y \sigma. \qquad (10.5)$$

These formulas suggest a route for arriving at existence theorems for our boundary value problems. It is this: first prove the existence of GREEN's function, then verify that (10.4) or (10.5) actually furnish the solution of the boundary value problem considered.

It will turn out to be useful to this object to observe that the second members of (10.4) and (10.5) have meaning under the hypothesis that f and φ are merely continuous, from which we can expect that the formula gives the solution u of the problem even when u is not of class $C^{(1)}$ in T.

This route is not always the most convenient, and we often succeed more comfortably in proving the existence theorem for a given boundary value problem and its adjoint directly, from which, having also noted the existence of the fundamental solutions of the two equations, the existence of GREEN's function follows, and thus the possibility of writing (10.4) or (10.5).

Thus, for example, if we are treating the DIRICHLET problem for the equation $\mathfrak{M} u = f$, it is possible to write:

$$F(x, y) = L(y, x) + g(x, y),$$

$L(y, x)$ being, as a function of y, a fundamental solution of the equation $\mathfrak{N} v = 0$, and $g(x, y)$ being a regular solution of the adjoint problem:

$$\mathfrak{N}_y g(x, y) = 0, \qquad g = -L(y, x) \quad \text{for} \quad y \in \partial T.$$

It is nevertheless worth pointing out that, even if the aforesaid existence theorems allow the immediate construction of the GREEN's functions, this does not say that formulas like (10.4) and (10.5) can be applied immediately. We must first verify that u as well as F satisfy those regularity conditions which are necessary for the validity of these formulas, summed up in essence by the continuity on ∂T of the first derivatives of u and F.

Postponing such considerations to the future, we wish now to prove the following classical theorem:

10, I. *If the* GREEN'S *functions* $F(x, y)$ *and* $\Phi(x, y)$ *of two adjoint problems exist and admit continuous first derivatives with respect to the* y_i *in* $T - x$, *we have:*

$$F(x, y) = \Phi(y, x). \tag{10.6}$$

Indeed, if we fix two points x and y in T and apply GREEN's formula to the domain $T - \left(I(x, \varrho) \cup I(y, \varrho)\right)$ and to the functions $F(x, z)$ and $\Phi(y, z)$ of the variable point z, we have, regardless of the type of boundary value problem considered:

$$\int_{\partial I(x, \varrho) \cup \partial I(y, \varrho)} \left[F(x, z)\, \mathfrak{P}_z\, \Phi(y, z) - \Phi(y, z)\, \mathfrak{D}_z\, F(x, z)\right] d_z\sigma = 0.$$

From this, passing to the limit as $\varrho \to 0$, (10.6) follows easily, from considerations analogous to those which we used to prove STOKES's formula.

An immediate corollary of theorem 10, I is the following:

10, II. *If a self-adjoint problem admits a* GREEN'S *function, then it is a symmetric function of the two points* x *and* y.

Thus, we note that under the hypotheses of theorem 10, II the function $F(x, y)$ is in fact a LEVI function of the point x, a solution of the equation $\mathfrak{M}_x F = 0$ which for $x \in \partial T$ satisfies the same condition on the boundary, except homogeneous, as for the problem we are trying to solve. It is always proper to consider such a function even when the coefficients of $\mathfrak{M}$ do not have the regularity properties which permit us to write the adjoint equation, and to this, if it exists, we give in every case the name *Green's function*. We can add beforehand that, where an existence theorem holds for the problem we are considering, we can immediately construct a GREEN's function in the latter sense, setting out from a fundamental solution.

If, however, it is not possible to verify at will that this $F(x, y)$ as a function of the point y satisfies (10.1) and (10.2) or (10.3), it is not possible without further effort to apply (10.4) and (10.5). We shall see in the future that in many cases such formulas can be directly verified.

CHAPTER II

Functions represented by integrals

In this chapter we shall occupy ourselves with the study of certain functions represented by either domain or surface integrals, with the purpose of establishing under what conditions such functions turn out to be integrable, or continuous, Hölder continuous, differentiable, etc. Among others, we shall study certain integrals which we can consider a natural extension of the ordinary potential either of a domain or of a single or double layer. This fact shows the importance of the contents of this chapter to the end of a systematic treatment of the theory of elliptic equations.

Finally we shall consider the problem of constructing a function defined in a domain T, satisfying specific regularity conditions in $T - \partial T$ and prescribed boundary conditions on ∂T.

11. Products of composition of two kernels. Let T be a bounded domain which we always suppose, without explicit notice to the contrary, to be of class $A^{(1,\lambda)}$. Let $K(x, y)$ be a function of two variables x and y in T, continuous at least for $x \neq y$. We say that K is a *kernel of class* $N^{(\alpha)}$, with $\alpha < m$, if it satisfies the bound $K = O(r^{\alpha - m})$ uniformly in T, where $r = \overline{xy}$. We say that K is a kernel of class $N^{(m)}$ if it satisfies the bound $K = O(\log 2 R/r)$ uniformly in T, denoting by R the diameter of T. Finally we say K is a kernel of class $N^{(\alpha)}$ with $\alpha > m$ if it is continuous even for $x = y$.

If, then, K is a kernel of class $N^{(\alpha)}$ and if, furthermore, for some $\lambda \leq 1$ and $\leq \alpha$ we have, uniformly for x', x'', y in T:

$$K(x', y) - K(x'', y) = O\left(\overline{x' x''}^{\lambda} \varrho^{\alpha - \lambda - m}\right), \qquad (11.1)$$

designating by ϱ the minimum distance from y to the segment $x' x''$, we say that K is of class $N^{(\alpha, \lambda)}$.

Note that if $\alpha > m$ and $\lambda \leq \alpha - m$, the condition $K \in N^{(\alpha, \lambda)}$ says that K is λ-Hölder continuous with respect to x, uniformly with respect to y. It is also easy to show[1] that for $\mu < \lambda$ it follows that $N^{(\alpha, \lambda)} \subset N^{(\alpha, \mu)}$ and this would hold independently of the condition imposed: $\lambda, \mu \leq 1$

---

[1] See GIRAUD [15], Chapter I, § 1.

or $\lambda, \mu \leq \alpha$. A kernel of class $N^{(\alpha)}$ is then certainly of class $N^{(\alpha, \lambda)}$ if $\partial K/\partial x_i \in N^{(\alpha - 1)}$ or else if:

$$K(x, y) = \big(z(x) - z(y)\big) K_0(x, y),$$

$$z \in C^{(0, \alpha)}, \quad K_0 \in N^{(0)}, \quad \partial K_0/\partial x \in N^{(-1)}.$$

Given two kernels K and K', then, the following integrals are called, respectively, the *product of composition* (or simply the *composition*) on T or ∂T of K by K':

$$U(x, y) = \int_T K(x, t)\, K'(t, y)\, dt,$$

$$V(x, y) = \int_{\partial T} K(x, t)\, K'(t, y)\, d_t\, \sigma.$$

Naturally such integrals are meaningful only under particular hypotheses on K and K'; the following two theorems are classics in this regard:

11, I. *If* $K \in N^{(\alpha)}$, $K' \in N^{(\beta)}$, *with* $\alpha, \beta > 0$, *then* $U \in N^{(\alpha + \beta)}$.

11, II. *If* $K \in N^{(\alpha)}$, $K' \in N^{(\beta)}$, *with* $\alpha, \beta > 1$, *then* $V \in N^{(\alpha + \beta - 1)}$.

Finally, we owe the following further theorems to GIRAUD[1]:

11, III. *If* $K \in N^{(\alpha, \beta)}$, $K' \in N^{(\beta)}$, *with* $\alpha, \beta > 0$, *then* $U \in N^{(\alpha + \beta, \mu)}$ *with* $\mu \leq \lambda$ *and* $\mu < \alpha$ *if* $\alpha + \beta \leq m$, *with* $\mu \leq \lambda$ *and* $\mu < \alpha + \beta - m$ *if* $\alpha + \beta > m$ *and* $\beta \leq m$, *with* $\mu \leq \lambda$ *and* $\mu < \alpha$ *if* $\beta > m$.

11, IV. *If* $K \in N^{(\alpha, \beta)}$, $K' \in N^{(\beta)}$, *with* $\alpha, \beta > 1$, *then* $V \in N^{(\alpha + \beta - 1, \mu)}$ *with* $\mu \leq \lambda$ *and* $\mu < \alpha - 1$ *if* $\alpha + \beta \leq m + 1$, *with* $\mu \leq \lambda$ *and* $\mu < \alpha + \beta - m - 1$ *if* $\alpha + \beta > m + 1$ *and* $\beta \leq m$, *with* $\mu \leq \lambda$ *and* $\mu < \alpha - 1$ *if* $\beta > m$.

12. Functions represented by integrals. In this section we shall occupy ourselves with functions of the type:

$$u(x) = \int_T K(x, y)\, z(y)\, dy, \quad v(x) = \int_{\partial T} K(x, y)\, \zeta(y)\, d_y \sigma,$$

which we intend to study with various hypotheses on the functions z and ζ, supposing these latter to be defined respectively in T and on ∂T. The results of which we shall speak, in the case when z and ζ are bounded, are derived from a classical memoir of E.E. LEVI [1] and from a series of works in which G. GIRAUD [1, 2, 3, 4, 8, 15] is concerned about others of these questions. All that which concerns instead the case in which z and ζ are supposed only integrable can be regarded as a simple extension of results obtained by various authors in particular cases. For these

---

[1] See GIRAUD [15], Chapter I, § 2.

we cite L. AMERIO [3], G. C. EVANS and E. R. C. MILES[1], G. FICHERA [1, 4, 5], K. O. FRIEDRICHS [5].

Meanwhile, it is quite easy to demonstrate that:

12, I. *If $z[\zeta] \in L^\infty$ in T [on ∂T] and if $K \in N^{(\alpha)}$ with $\alpha > 0$ [$\alpha > 1$] then we have in T: $u[v] \in C^{(0)}$.*

We have also as an immediate consequence of theorem 11, III:

12, II. *If $z[\zeta] \in L^\infty$ in T [on ∂T] and if $K \in N^{(\alpha, \lambda)}$ with $\alpha > 0$ [$\alpha > 1$], then we have in T: $u[v] \in C^{(0, \mu)}$ with $\mu \leq \lambda$, $\mu < \alpha$ [$\mu < \alpha - 1$].*

Let us now pass to the study of the properties of u and v when z and ζ are assumed to be only integrable (in T and on ∂T, respectively), instead of bounded.

Because, by virtue of theorem 12, I, the integral $\int_T |K(x, y)| \, dx$ is a continuous function of y (always assuming $K \in N^{(\alpha)}$ with $\alpha > 0$), we have under these hypotheses that the function of y given by $|z(y)| \int_T |K(x, y)| \, dx$ is integrable in T. By a well-known theorem of TONELLI, we have then that $|K(x, y) z(y)|$ is integrable in the domain $T^{(2)} = T \times T$ of R_{2m} defined by the ordered pairs of variables (x, y) in T. By FUBINI's theorem it follows from this that the function $\int_T K(x, y) z(y) \, dy$ is defined for almost all x in T and is integrable in T.

Then the following theorem holds, and the proof can be completed by analogous reasoning:

12, III. *If $z[\zeta] \in L^1$ in T [on ∂T] and if $K \in N^{(\alpha)}$ with $\alpha > 0$, we have $u[v] \in L^1$ in T, the function u being defined almost everywhere in T [v continuous in $T - \partial T$]. If $\alpha > 1$, $u[v]$ is defined even on ∂T up to a set of hypersurface measure zero, and is integrable on ∂T.*

From this theorem and from FUBINI's and TONELLI's theorems it follows immediately that:

12, IV. *If $z \in L^1$ in T, $\zeta \in L^1$ on ∂T, $K \in N^{(\alpha)}$, $K' \in N^{(\beta)}$, $\alpha, \beta > 0$, then we have for almost all x in T:*

$$\int_T K'(x, t) u(t) \, dt = \int_T z(y) \, dy \int_T K'(x, t) K(t, y) \, dt, \quad \text{and for} \quad x \in T - \partial T:$$

$$\int_T K'(x, t) v(t) \, dt = \int_{\partial T} \zeta(y) \, d_y \sigma \int_T K'(x, t) K(t, y) \, dt.$$

[1] EVANS, G. C. and E. R. C. MILES: Potential of general masses in single and double layers. The relative boundary value problems. Amer. J. Math. **53** (1931) 493—516. For other indications of previous works of these authors see also: G. C. EVANS: Complements of potential theory. Amer. J. Math. **54** (1932) 213—234; **55** (1933) 29—49; **57** (1935) 623—626.

The same formulas hold everywhere in T, the first if $z \in L^\infty$, the second if $\zeta \in L^\infty$, $v \in L^\infty$, $\alpha + \beta > 1$. If, moreover, $\alpha > 1$, $\beta > 0$, we also have for $x \in T - \partial T$:

$$\int_{\partial T} K'(x, t) \, u(t) \, d_t\sigma = \int_T z(y) \, dy \int_{\partial T} K'(x, t) \, K(t, y) \, d_t\sigma,$$

$$\int_{\partial T} K'(x, t) \, v(t) \, d_t\sigma = \int_{\partial T} \zeta(y) \, d_y\sigma \int_{\partial T} K'(x, t) \, K(t, y) \, d_t\sigma.$$

If $\alpha, \beta > 1$, these last formulas hold even for $x \in \partial T$, up to a set of hypersurface measure zero, and absolutely everywhere if $z \in L^\infty$, $\zeta \in L^\infty$.

Let us pass now to the establishment of some further properties of the functions u and v under the hypotheses that z and ζ are square integrable. We have for example:

12, V. *If $z [\zeta] \in L^2$ in T [on ∂T] and if $K \in N^{(\alpha)}$ with $\alpha > 0 [\alpha > 1]$, then we have $u [v] \in L^2$ in T [on ∂T]. If furthermore $K \in N^{(\alpha, \lambda)}$ with $\alpha > \dfrac{m}{2} + \lambda \left[\alpha > \dfrac{m+1}{2} + \lambda\right]$ we have $u [v] \in C^{(0, \lambda)}$ in T.*

Let us prove the theorem for u. Putting:

$$K_1(x, y) = \int_T | K(t, x) \, K(t, y) | \, dt, \quad K_p(x, y) = \int_T K_{p-1}(t, x) \, K_{p-1}(t, y) \, dt,$$

and fixing n such that $2^n \alpha > m$, the kernel K_n turns out to be continuous in consequence of theorem 11, I and hence the function $K_n(x, y) | z(x) \, z(y) |$ is integrable in $T^{(2)} = T \times T$. By TONELLI's theorem it now follows that $K_{n-1}(t, x) \, K_{n-1}(t, y) | z(x) || z(y) |$ is integrable in the domain $T^{(3)} = T \times T \times T$ of R_{3m} of ordered triples (t, x, y) of variables t, x, and y in T. From this, by FUBINI's theorem, it follows that the function $\int_T K_{n-1}(t, y) | z(y) | \, dy$ is square integrable in T and therefore that $K_{n-1}(t, y) | z(y) \, z(t) |$ is integrable in $T^{(2)}$. By repetition of the argument we arrive easily at the assertion. The last part of the theorem is a consequence of SCHWARZ's inequality.

After this we prove quickly that:

12, VI. *If $z [\zeta] \in L^2$ in T [on ∂T] and if $K \in N^{(\alpha)}$ with $\alpha > 1 [\alpha > 1/2]$ we have $u [v] \in L^2$ on ∂T [in T]. If $\alpha > 1/2$ we can still assert that $u \in L^1$ on ∂T.*

To prove the theorem, for example for $v(x)$, it suffices to observe that the function $\int_{\partial T} K_1(x, y) | \zeta(y) | \, d_y\sigma$ is square integrable on ∂T by dint of theorem 12, V, since for $\alpha > 1/2$ we have $K_1 \in N^{(\beta)}$ with $\beta = 2\alpha > 1$. From this the customary application of TONELLI's and FUBINI's theorems leads to the assertion.

We have explicitly enunciated theorems 12, III, 12, V, and 12, VI either because they are sufficient for certain applications to boundary value problems or because their proofs are sufficiently simple. Such theorems can, however, be remarkably improved by having recourse to considerations of a more advanced nature introduced in a fundamental memoir by S. L. SOBOLEV [4] and renewed later by the same author in the volumes [5, 6]. Not being able to enter into details, we limit ourselves here to enunciating the following theorem:

12, VII. *If* $z[\zeta] \in L^p$ *in* T *[on* ∂T*] with* $p \geq 1$ *and if* $K \in N^{(\alpha)}$, *we have:*

$$\| u \|_{L^q(T)} = O\big(\| z \|_{L^p(T)}\big) \qquad for \qquad 0 < \alpha < \frac{m}{p} \leq \alpha + \frac{m}{q} \, ,$$

$$\| u \|_{L^r(\partial T)} = O\big(\| z \|_{L^p(T)}\big) \qquad for \qquad \frac{1}{p} < \alpha < \frac{m}{p} \leq \alpha + \frac{m-1}{r} \, ,$$

$$\| v \|_{L^s(T)} = O\big(\| \zeta \|_{L^p(\partial T)}\big) \qquad for \qquad 1 - \frac{1}{p} < \alpha < \frac{m-1}{p} + 1 \leq \alpha + \frac{m}{s} \, ,$$

$$\| v \|_{L^t(\partial T)} = O\big(\| \zeta \|_{L^p(\partial T)}\big) \qquad for \qquad 1 < \alpha < \frac{m-1}{p} + 1 \leq \alpha + \frac{m-1}{t} \, ,$$

the equality sign being able to hold in the bounds concerning q, r, s, t *only for* $p > 1$. *We have moreover if* $K \in N^{(\alpha, \lambda)}$:

$$\| u \|_{C^{(0, \mu)}(T)} = O\big(\| z \|_{L^p(T)}\big) \qquad for \qquad 0 < \mu < \inf \Big[\alpha - \frac{m}{p} \, , \, \lambda\Big] \, ,$$

$$\| v \|_{C^{(0, \mu)}(T)} = O\big(\| \zeta \|_{L^p(\partial T)}\big) \qquad for \qquad 0 < \mu < \inf \Big[\alpha - 1 - \frac{m-1}{p} \, , \, \lambda\Big] \, .$$

Let us terminate this section with the study of some differentiability properties of the functions u and v. Now the following theorem is nearly obvious:

12, VIII. *If* $z \in L^\infty$ *and if* $K \in N^{(\alpha)}$, $\partial K / \partial x_i \in N^{(\alpha-1)}$ *with* $\alpha > 1$, *then* $u \in C^{(1)}$ *and*

$$\frac{\partial u}{\partial x_i} = \int_T \frac{\partial K(x, y)}{\partial x_i} \, z(y) \, dy. \tag{12.1}$$

In fact, if Δx_i is an arbitrary increment of x_i and $\Delta_i u$ the corresponding increment of u, we have:

$$\Delta_i u = \int_{x_i}^{x_i + \Delta x_i} dx_i \int_T \frac{\partial K(x, y)}{\partial x_i} \, z(y) \, dy, \tag{12.2}$$

from which (12.1) follows immediately. But we have further that:

12, IX. *If* $z \in L^p$ *with* $p \geq 1$ *and if* $K \in N^{(\alpha)}$, $\partial K / \partial K_i \in N^{(\alpha-1)}$ *with* $\alpha > 1$, *then* u *is absolutely continuous with respect to* x_i *for almost every*

$(m-1)$-*tuple* $(x_1, \ldots, x_{i-1}, x_{i+1}, \ldots, x_m)$. *Fixing one such* $(m-1)$-*tuple for which* u *is absolutely continuous,* (12.1) *is valid for almost every* x_i, $\partial u / \partial x_i$ *turning out to be integrable in* T.

In fact, under the proposed hypotheses, the function under the simple integral sign in (12.2) is integrable in T, by dint of theorem 12, III. By FUBINI's theorem it now follows that (12.2) is still valid for almost every $(m-1)$-tuple $(x_1, \ldots, x_{i-1}, x_{i+1}, \ldots, x_m)$ from which the theorem, the integrability of $\partial u / \partial x_i$, results, as an obvious consequence of theorem 12, III.

Note that if u were continuous, by theorem 12, IX, u would be absolutely TONELLI continuous. Lacking global continuity for u, one can only assert that it is included in a class of functions amply studied under diverse points of view by C. B. MORREY [2] and by G. STAMPACCHIA [2], and also in classes of type $H^{1,q}$; the value of q can be deduced from theorem 12, VIII.

From the cited memoir by MORREY and from another by C. MIRANDA [6] it would be possible to deduce other results relating to u under the hypothesis that $z \in L^{p,\lambda}[T]$. Here we limit ourselves to enunciating the following theorem, which is obtained by generalizing the results of the preceding authors, relating to Newtonian potentials:

12, X. *If* $z \in L^{1,\lambda}$ *and if* $K \in N^{(\alpha,\lambda)}$ *with* $\alpha > 0$, $\alpha + \lambda - m > 0$, *then we have* $u \in C^{(0, \alpha + \lambda - m)}$. *If further* $\partial K / \partial x_i \in N^{(\alpha-1,\lambda)}$ *with* $\alpha > 1$, $\alpha + \lambda - m > 1$, *then we have* $u \in C^{(1, \alpha + \lambda - m - 1)}$.

Finally, concerning the derivative of $v(x)$, it suffices to observe that in consequence of theorems 12, III and 12, IV we have:

12, XI. *Let* $K \in N^{(\alpha)}$, $\partial K / \partial x_i \in N^{(\alpha-1)}$. *If* $\zeta \in L^1$ *and* $\alpha > 1$, *then we have* $\partial v / \partial x_i \in L^1$ *in* T; *if* $\zeta \in L^2$ *and* $\alpha > 3/2$, *then we have* $\partial v / \partial x_i \in L^2$ *in* T. *In each case* $\partial v / \partial x_i$ *turns out to be continuous in* $T - \partial T$ *and is given by the formula:*

$$\frac{\partial v}{\partial x_i} = \int_{\partial T} \frac{\partial K(x, y)}{\partial x_i} \, \zeta(y) \, d_y \sigma.$$

Note that also for v, theorem 12, VII would permit, more precisely, inclusion in classes of the type $H^{1,q}(T)$.

13. Generalized domain potentials. Let $L(x, y)$ by a LEVI function defined in a region Ω in which the functions a_{ik} are of class $C^{(0,\lambda)}$; if T is a bounded domain of class $A^{(1,\lambda)}$ contained in Ω, then the following function is called the (*generalized*) *domain potential of density* z:

$$u(x) = \int_T L(x, y) z(y) \, dy. \tag{13.1}$$

Because, by virtue of (8.3), (8.4), $L \in N^{(2)}$ and $\partial L / \partial \dot{x}_i \in N^{(1, \mu)}$ with any $\mu < 1$, the properties of u and of its derivatives in T are obtained immediately as a particular case of the theorems of the preceding section, while the analogous properties in $\Omega - T$ are of an elementary character. We have thus, for example, in consequence of theorems 12, VII and 12, IX:

13, I. *If $z \in L^p$ with $p \geq 1$ we have:*

$$\frac{\partial u}{\partial x_i} = \int_T \frac{\partial L(x, y)}{\partial x_i} z(y) \, dy, \qquad (13.2)$$

with $u \in H^{1, q}(T_1)$ in every bounded domain $T_1 \subset \Omega$ with $q < \dfrac{n}{n-1}$ if $p = 1$, and $q = \dfrac{np}{n-p}$ if $1 < p < n$. If $p > n$, then $u \in C^{(1, \mu)}(T_1)$ for any $\mu < 1 - \dfrac{n}{p}$.

We observe also that for $x \in \Omega - T$ we have, if $z \in L^1$:

$$\frac{\partial^2 u}{\partial x_i \, \partial x_k} = \int_T \frac{\partial^2 L(x, y)}{\partial x_i \, \partial x_k} z(y) \, dy. \qquad (13.3)$$

Concerning the second derivatives of u in $T - \partial T$ the following theorem, extending a classical property of ordinary potentials, holds:

13, II. *If $z \in C^{(0, \mu)}$ then $u \in C^{(2)}$ in $T - \partial T$ and we have for $x \in T - \partial T$:*

$$\frac{\partial^2 u}{\partial x_i \, \partial x_k} = -\frac{1}{m} A_{ik}(x) z(x) + \int_T^* \frac{\partial^2 L(x, y)}{\partial x_i \, \partial x_k} z(y) \, dy, \qquad (13.4)$$

where the integral with the asterisk is to be taken as a principal value, i.e., as the limit as $\varrho \to 0$ of the integral over $T - I(x, \varrho)$. Finally, under the hypothesis that the second derivatives of $L - H$ are of class $N^{(\lambda, \mu)}$ with $\mu < \lambda$, we also have that $u \in C^{(2, \lambda)}$ in $T - \partial T$.

Let us begin with the supposition that $u = u_1 + u_2$, having denoted by u_1 and u_2 integrals of the type (13.1) which are obtained by sustituting H for L one time, and $L - H$ for L the other.

Because, by hypothesis:

$$L - H \in N^{(\lambda + 2)}, \quad \frac{\partial(L - H)}{\partial x_i} \in N^{(\lambda + 1)}, \quad \frac{\partial^2(L - H)}{\partial x_i \, \partial x_k} \in N^{(\lambda)},$$

u_2 possesses all the properties which we might wish to demonstrate for u, and its second derivatives are calculated by differentiating twice under the integral sign.

With the meaning given to $I(x, \varrho)$ in § 9 let us put:

$$\varphi_i(x, \varrho) = \int_{T - I(x, \varrho)} \frac{\partial H(x, y)}{\partial x_i} z(y) \, dy.$$

We then have:

$$\frac{\partial \varphi_i}{\partial x_k} = \int_{T-I(x,\varrho)} \frac{\partial^2 H(x,y)}{\partial x_i \partial x_k} z(y) \, dy - \int_{\partial I(x,\varrho)} \frac{\partial H(x,y)}{\partial x_i} z(y) X_k(y) \, d_y\sigma, \quad (13.5)$$

having designated by $X_k(y)$ the direction cosines of the exterior normal to $\partial I(x, \varrho)$. Now the integral over $\partial I(x, \varrho)$ can also be written:

$$\int_{\partial I(x,\varrho)} \left[\frac{\partial H(x,y)}{\partial x_i} z(y) + \frac{\partial H(y,x)}{\partial y_i} z(x) \right] X_k(y) \, d_y\sigma$$

$$- z(x) \int_{\partial I(x,\varrho)} \frac{\partial H(y,x)}{\partial y_i} X_k(x) \, d_y\sigma,$$

and of these two integrals, the first goes to zero with ϱ, uniformly with respect to x in every domain contained in T, because the function integrated is $O(\varrho^{\lambda+1-m})$. The second integral is constant and equals $z(x) A_{ik}(x)/m$.

Because $\lim_{\varrho \to 0} \varphi_i = \partial u/\partial x_i$, we shall have proved (13.4) if we are able also to show that the first integral on the right side of (13.5) converges for $\varrho \to 0$, uniformly with respect to x in every domain contained in T.

But this integral can be written:

$$\int_{T-I(x,\varrho)} \left[\frac{\partial^2 H(x,y)}{\partial x_i \partial x_k} z(y) - \frac{\partial^2 H(y,x)}{\partial y_i \partial y_k} z(x) \right] dy$$

$$+ z(x) \int_{\partial T} \frac{\partial H(y,x)}{\partial y_i} X_k(x) \, d_y\sigma + \frac{1}{m} A_{ik}(x) z(x),$$

and the aforesaid property follows from the fact that in the first integral, the integrand is $O(\overline{xy}^{\lambda-m})$.

From the proof thus developed the following formula follows:

$$\frac{\partial^2 u_1}{\partial x_i \partial x_k} = \int_T \left[\frac{\partial^2 H(x,y)}{\partial x_i \partial x_k} z(y) - \frac{\partial^2 H(y,x)}{\partial y_i \partial y_k} z(x) \right] dy$$

$$+ z(x) \int_{\partial T} \frac{\partial H(y,x)}{\partial y_i} X_i(y) \, d_y\sigma, \qquad (13.6)$$

in which we recognize easily that $u_1 \in C^{(2,\mu)}$ in $T - \partial T$ with $\mu < \lambda$, because the two integrals on the right are both functions of class $C^{(0,\mu)}$ in $T - \partial T$. This is almost immediate for the integral over ∂T, while for the integral over T it suffices to apply theorem 12, II, after having verified that the integrand is a kernel of class $N^{(\lambda,\mu)}$.

With this, our theorem is completely proved. Let us observe that if $L = H$, we can prove the theorem holds even for $\mu = \lambda < 1$; in an important particular case, we can show that $u \in C^{(2,\lambda)}$ throughout T. For this, however, we refer to a memoir [8] of GIRAUD[1]. In the general case we have $u \in C^{(2,\mu)}(T)$ with $\mu < \lambda$ if $a_{ik} \in C^{(1,\mu)}(T)^2$.

An important consequence of (13.3) and (13.4) is the following formula:

$$\mathfrak{M}u = \begin{cases} \int\limits_T \mathfrak{M}_x L(x, y) z(y)\, dy & \text{for} \quad x \in \Omega - T \\ -z(x) + \int\limits_T \mathfrak{M}_x L(x, y) z(y)\, dy & \text{for} \quad x \in T - \partial T, \end{cases} \tag{13.7}$$

which reduces to the well-known formula of POISSON in the case of ordinary potentials.

We note finally that if $\Omega \equiv T - \partial T$, all the properties of the potential of the domain for $x \in T$ hold on condition that the a_{ik} are λ-Hölder continuous in T, the L and $\partial L/\partial x_i$ are continuous in $T - y$ and satisfy the first two of (8.4) uniformly in T, and the $\partial^2 L/\partial x_i\, \partial x_k$ are integrable with respect to y in $T - I(y, \varrho)$.

Let us see now what can be said of the derivatives of the domain potential when the density is supposed neither Hölder continuous nor bounded. Along this line the following theorem holds;

13, III. *If $z \in L^p$ with $p > 1$, the $\partial u/\partial x_k$ given by (14.2) are absolutely continuous with respect to all x_i, for almost every $(m - 1)$-tuple $(x_1, \ldots, x_{i-1}, x_{i+1}, \ldots, x_m)$ and we have $\partial^2 u_i/\partial x_i\, \partial x_k \in L^p$, (13.4) and (13.7) continuing to hold almost everywhere in T; and thus $u \in H^{2,p}(T)$.*

In the case of ordinary potentials, this theorem has been proved for $p = 2$ by L. LICHTENSTEIN [1] for $m = 2$ and by K.O. FRIEDRICHS [5] for $m > 2$. In the general case the theorem can be deduced from a fundamental result of CALDERON and ZYGMUND[3] related to the theory of singular integrals. If we abstract from the verification of (13.4), however, the result can be obtained in a more elementary manner, at least in the case $p = 2$, with the following reasoning.

Let $T_1 \subset \Omega$ be a domain of class $A^{(2)}$ to which T is interior, and let $\{z_n(x)\}$ be a sequence of polynomials which converge in the mean in T_1

[1] Chapter IV, § 2, and Chapter XVI.

[2] See C. MIRANDA [19], § 4. For $L = H$ we can even assume $\mu = \lambda < 1$.

[3] This result and its application to the case of ordinary potentials are contained in CALDERON, A.P. and ZYGMUND, A., On the existence of certain singular integrals, Acta. Math. **88** (1952) 85—139. For application to the general case see for example C. MIRANDA [19] § 4.

to the function which is zero in $T_1 - T$ and equal to z in T. Putting for $x \in T_1$:

$$u_n(x) = \int_{T_1} L(x, y) \, z_n(y) \, dy \,,$$

u_n and the $\partial u_n / \partial x_i$ converge in the mean in T to u and the $\partial u / \partial x_i$. This follows, indeed, from theorems 13, III and 12, VII. From the other part of (13.7) and then from theorem 12, VII follow the existence of a constant M for which:

$$\int_{T_1} [\mathfrak{M} u_n]^2 \, dx \le M \int_T z^2 \, dx \,.$$

By dint of a result of R. CACCIOPPOLI [11] of which we shall speak later (Chapter V, § 37), there now follows a bound of the type:

$$\sum_{i, k = 1}^{m} \int_T \left(\frac{\partial^2 u}{\partial x_i \, \partial x_k} \right)^2 dx \le M' \int_T z^2 \, dx$$

and from this, by a theorem of G. STAMPACCHIA [2], the assertion follows.

For $p \ne 2$, the reasoning developed can be repeated exactly, substituting for the theorem of CACCIOPPOLI an extension obtained by A. I. KOŠELEV [4] and D. GRECO [1]. However, while the proof of CACCIOPPOLI's theorem is elementary, one can not say as much for the theorem of KOŠELEV and GRECO (see § 37).

For more regarding the contents of this section other works can be consulted: the memoirs of LEVI and GIRAUD already cited in § 12 and the memoir [8] of M. GEVREY[1]. For a study of derivatives of higher order see C. MIRANDA [19].

14. Generalized single layer potentials. Preserving the hypotheses of § 13 on $L(x, y)$ and T, a function of the following type is called a (*generalized*) *single layer potential with density* ζ:

$$v(x) = \int_{\partial T} L(x, y) \, \zeta(y) \, d_y \sigma \,. \tag{14.1}$$

If $\zeta \in L^1$, v is continuous with its first and second derivatives in $T - \partial T$ and $\Omega - T$. Then the following theorems hold, where we denote by T_1 a domain contained in Ω and containing or coinciding with T.

14, I. *If* $\zeta \in L^p$ *with* $1 \le p < m - 1$, *we have* $v \in L_{q1}(T_1)$ *with* $q_1 \le \dfrac{p \, m}{m - p - 1}$ *and* $v \in L^q(\partial T)$ *with* $q \le \dfrac{p \, (m - 1)}{m - p - 1}$, *the equality holding only for* $p > 1$. *Moreover, for* x_0 *almost everywhere on* ∂T:

---

[1] See also W. POGORZELSKI [2].

$$\lim_{x \to x_0} v(x) = v(x_0), \tag{14.2}$$

provided that x tends to x_0 along an axis l such that $\cos(l, n) > 0$. For $p > m - 1$ we have $v \in C^{(0, \mu)}(T_1)$ with $\mu < 1 - \dfrac{m-1}{p}$.

The first affirmation is an immediate consequence of theorem 12, VII. In order to prove (14.2) it suffices to observe that, fixing $x_0 \in \partial T$ in such a manner that $L(x_0, y)\,\zeta(y) \in L^1$, a bound of the following type is valid for $x \in l$ and $\overline{xx_0}$ sufficiently small:

$$\left| L(x, y)\,\zeta(y) \right| \le \frac{M}{\cos^{m-2}(l, n)} \left| L(x_0, y)\,\zeta(y) \right|; \tag{14.3}$$

from this arises the possibility of passing to the limit under the integral sign in (14.1). We arrive at (14.3) by observing that for a convenient M':

$$\left| \frac{L(x, y)}{L(x_0, y)} \right| < M' \left(\frac{\overline{x_0 y}}{\overline{x y}} \right)^{m-2},$$

thence taking account of the fact that T is of class $A^{(1, \lambda)}$. The last affirmation of the theorem is then a consequence of theorem 12, VII.

Because we can calculate the $\partial v / \partial x_i$ by differentiating under the integral sign, it follows, once more from theorem 12, VII, that:

14, II. *For $\zeta \in L^p$ with $p \ge 1$ we have $\partial v / \partial x_i \in L^s(T_1)$ with $s \le \dfrac{p\,m}{m-1}$, equality holding only for $p > 1$.*

We note also that as a particular case of theorem 14, I, or directly from theorem 12, II, we have:

14, III. *If $\zeta \in L^\infty$ we have $v \in C^{(0, \mu)}$ in Ω for any $\mu < 1$.*

We now wish to pass to the study of the behavior of the derivatives of v on ∂T. For this, pick any x_0 on ∂T whatsoever, and let us denote by $J(x_0, \varrho)$ a neighborhood of x_0 on ∂T varying with ϱ, whose projection on the hyperplane tangent to ∂T at x_0 is contained in the hypersphere of center x_0 and radius $k\varrho$, and which contains the hypersphere of center x_0 and radius $k'\varrho$ with k and k' constant. We premise the following lemma:

Lemma. *Let $K(x, y)$ be a kernel continuous for $x, y \in \Omega$ and $x \ne y$, and such that for $\alpha, \beta \ge 0$, $K = O\left(\overline{xx_0}^\alpha \cdot \overline{xy}^{-\beta}\right)$ for fixed x_0 on ∂T and a given axis l coming out of x_0 with $\cos(l, n) \ne 0$, $x \in l$ and $y \in \partial T$ being both in a neighborhood of x_0. If $\zeta \in L^\infty$ and if $\alpha > 0$, $\alpha + m > \beta + 1$, then we have:*

$$\lim_{\substack{x \to x_0 \\ x \in l}} \int_{\partial T} K(x, y)\,\zeta(y)\,d_y\,\sigma = \int_{\partial T} K(x_0, y)\,\zeta(y)\,d_y\,\sigma.$$

The result holds if $\zeta \in L^1$, on condition that $\alpha > 0$, $\alpha + m > \beta + 1$, and moreover:

$$\lim_{\varrho \to 0} \varrho^{1-m} \int_{J(x_0, \varrho)} |\zeta(y)| \, d_y\sigma < \infty \,, \tag{14.4}$$

or if $\alpha > 0$, $\alpha + m \geq \beta + 1$ and:

$$\lim_{\varrho \to 0} \varrho^{1-m} \int_{J(x_0, \varrho)} |\zeta(y)| \, d_y\sigma = 0 \,. \tag{14.5}$$

Finally, if $\zeta(x_0) = 0$, $\zeta \in C^{(0, \mu)}$ the result holds even for $\alpha \geq 0$, $\alpha + m + \mu > \beta + 1$.

To demonstrate the lemma it clearly suffices to show that the integral

$$\int_{J(x_0, \varrho)} |K(x, y) \zeta(y)| \, d_y\sigma \tag{14.6}$$

goes to zero as ϱ and $\overline{x x_0}$ tend to zero. But, letting y' be the projection of y on the hyperplane tangent to ∂T at x_0 and putting $\overline{x x_0} = \delta$, $\overline{x_0 y'} = t$, then if T is of class $A^{(1, \lambda)}$ we can verify that for $x \in l$ and $y \in \partial T$ the expression $\overline{xy}/\sqrt{t^2 + \delta^2}$ is bounded from below by a positive number. Now it follows that, if $\zeta \in L^\infty$, the integral (14.6) can be majorized for some constant M, by the quantity

$$M \int_0^{k\varrho} \frac{\delta^\alpha \, t^{m-2} \, dt}{(t^2 + \delta^2)^{\beta/2}} \,,$$

which, if $\alpha + m > \beta + 1$, goes to zero with δ; from which the first case of the lemma is proved. If instead $\zeta \in L^1$, and $S(x_0, \varrho)$ is the portion of ∂T whose projection on the hyperplane tangent to ∂T at x_0 is the hypersphere of center x_0 and radius ϱ, then we put:

$$Z(\varrho) = \int_{S(x_0, \varrho)} |\zeta(y)| \, d\sigma$$

and we assume that for $\varrho < \varrho_\varepsilon$, $Z < \varepsilon \varrho^{m-1}$ with ε a finite quantity in hypothesis (14.4), sufficiently small in hypothesis (14.5). The integral (14.6) can then be majorized by

$$M \int_0^{k\varrho} \frac{\delta^\alpha \, Z'(t) \, dt}{(t^2 + \delta^2)^{\beta/2}}$$

and because by an integration by parts we recognize that this quantity is $O(\varepsilon \, [\delta^\gamma - \delta^\alpha \log \delta])$, γ being the smaller of the two numbers α and

$\alpha + m - \beta - 1$, the lemma stands proved both for the second and the third cases. In the fourth case the proof is analogous, where we keep in mind that $Z = O\left(\varrho^{\mu+m-1}\right)$.

With this behind us, let x_0 be a fixed point on ∂T and l and l_1 be two axes coming out of x_0 and such that $\cos(l, n) \neq 0$, $\cos(l_1, n) \neq 0$, and let us calculate dv/dl for $x \in l_1$; here we intend to recognize that as $x \to x_0$ on l_1, this derivative admits in general two distinct limits according as x tends to x_0 on $l_1 \cap T$ or on $l_1 \cap (\Omega - T)$. These limits, which turn out to be independent of l_1, will be denoted respectively by $\left(\dfrac{dv}{dl}\right)^-$ and $\left(\dfrac{dv}{dl}\right)^+$. If $l \equiv v$ the following theorem is valid:

14, IV. *If $\zeta \in L^1$ we have, for x almost everywhere on ∂T:*

$$\left(\frac{dv}{dv}\right)^{\pm} = \mp \frac{\zeta(x_0)}{2\,a(x_0)} + \int\limits_{\partial T} \frac{dL(x_0, y)}{dv}\,\zeta(y)\,d_y\,\sigma\,, \qquad (14.7)$$

$a(x)$ being the function defined by (4.3). If $\zeta \in C^{(0)}$, (14.7) holds everywhere. Under the first hypothesis the second member of (14.7) is integrable on ∂T, under the second, continuous. The integral on the right side of (14.7) is, moreover, of class $L^q(\partial T)$ with $q \leq \dfrac{p\,(m-1)}{m - \lambda p - 1}$ if $\zeta \in L^p$ with $1 \leq p < \dfrac{m-1}{\lambda}$, equality holding only for $p > 1$, and of class $C^{(0, \mu)}(\partial T)$ with $\mu < \lambda - \dfrac{m-1}{\lambda}$ if $p > \dfrac{m-1}{\lambda}$. In particular, if $\zeta \in C^{(0)}$ this integral is of class $C^{(0, \mu)}(\partial T)$ for every $\mu < \lambda$.

We verify now easily that for $x \in l_1$ and $y \in \partial T$ we can write:

$$\frac{dL(x, y)}{dv} = A(x, y) + B(x, y),$$

with $B(x, y) \in N^{(\lambda+1)}$ and:

$$a(x_0)\,A(x, y) = -\frac{\sum\limits_{i=1}^{m} X_i(y)\,(x_i - y_i)}{\omega_m\,\sqrt{A(x_0)}\left[\sum\limits_{i,k=1}^{m} A_{ik}(x_0)\,(x_i - y_i)\,(x_k - y_k)\right]^{m/2}}\,.$$

To be precise suppose $m > 2$ and consider the function:

$$H_0(x, y) = \frac{1}{(m-2)\,\omega_m\,\sqrt{A(x_0)}}\left[\sum\limits_{i,k=1}^{m} A_{ik}(x_0)\,(x_i - y_i)\,(x_k - y_k)\right]^{\frac{2-m}{2}}\,.$$

As a function of y, H_0 is a fundamental solution of the equation $\mathfrak{M}_0\,u = 0$, where $\mathfrak{M}_0 = \Sigma\,a_{ik}(x_0)\,\partial^2/\partial x_i\,\partial x_k$. Moreover, if v_0 is the conormal at the point y relative to this operator and a_0 the function for

$\mathfrak{M}_0$ analogous to a, we have $a(x_0)\,A(x,y) = -a_0(y)\,dH_0/dv_0$. From STOKES's formula applied to the functions $u = 1$ and $L = H_0$ and GREEN's formula applied to the functions $u(y) = 1$ and $v(y) = H_0(x, y)$ with $x \in \Omega - T$, we have:

$$\int_{\partial T} A(x,y)\,a(x_0)\,d_y\,\sigma = \theta, \qquad (14.8)$$

with $\theta = 1$ for $x \in T - \partial T$ and $\theta = 0$ for $x \in \Omega - T$. Choosing now on the conormal to ∂T at the point x_0 two points x and x' symmetrical with respect to x_0, we recognize that we have:

$$A(x,y) + A(x',y) = A_1(x,y) + A_2(x,y),$$

with $A_1 = O\left(\overline{xx_0^2} \cdot \overline{xy}^{\lambda-1-m}\right)$, $A_2 = O\left(\overline{xy}^{\lambda+1-m}\right)$.

Therefore, we also have $A(x_0, y) = O\left(\overline{x_0 y}^{\lambda+1-m}\right)$, and it follows from the lemma that:

$$\lim_{x \to x_0} \int_{\partial T} [A(x,y) + A(x',y)]\,a(x_0)\,d_y\,\sigma = z \int_{\partial T} A(x_0,y)\,a(x_0)\,d_y\sigma \qquad (14.9)$$

and therefore:

$$\int_{\partial T} A(x_0,y)\,a(x_0)\,d_y\,\sigma = \frac{1}{2}. \qquad (14.10)$$

Hence, for every x_0 for which:

$$\lim_{\varrho \to 0} \varrho^{1-m} \int_{J(x_0,\varrho)} |\zeta(y) - \zeta(x_0)|\,d\sigma = 0, \qquad (14.11)$$

we have, observing that for $x \in l_1$, $A = O\left(\overline{xx_0} \cdot \overline{xy}^{-m} + \overline{xy}^{\lambda+1-m}\right)$, and applying the lemma:

$$\lim_{x \to x_0} \int_{\partial T} A(x,y)\,[\zeta(y) - \zeta(x_0)]\,d_y\,\sigma = \int_{\partial T} A(x_0,y)\,\zeta(y)\,d_y\,\sigma - \frac{\zeta(x_0)}{2\,a(x_0)}.$$

Because (14.11) and

$$\lim_{x \to x_0} \int_{\partial T} B(x,y)\,\zeta(y)\,d_y\,\sigma = \int_{\partial T} B(x_0,y)\,\zeta(y)\,d_y\sigma$$

are satisfied for x_0 almost everywhere on ∂T if $\zeta \in L^1$ and everywhere if $\zeta \in C^{(0)}$, from the identity:

$$\frac{dv}{dv} = \frac{\theta\,\zeta(x_0)}{a(x_0)} + \int_{\partial T} A(x,y)\,[\zeta(y) - \zeta(x_0)]\,d_y\,\sigma + \int_{\partial T} B(x,y)\,\zeta(y)\,d_y\,\sigma$$

(14.7) follows at once. The property of integrability or continuity of $(dv/dv)^{\pm}$ is a consequence of theorem 12, VII after having verified previously that for $y, x_0 \in \partial T$, we have $dL(x_0, y)/dv \in N^{(\lambda+1, \lambda)}$.

Passing now to consider the case of an arbitrary axis l, we have the theorem:

14, V. *If $\zeta \in C^{(0, \mu)}$, we have for every $x_0 \in \partial T$:*

$$\left(\frac{dv}{dl}\right)^{\pm} = \mp \frac{\zeta(x_0)}{2 a^{(l)}(x_0)} + \int_{\partial T}^{*} \frac{dL(x_0, y)}{dl} \zeta(y) \, d_y\sigma , \qquad (14.12)$$

$a^{(l)}$ *being the function defined by (4.7) and the integral with the asterisk being taken in the sense of a principal value, that is, as the limit of the integral over* $\partial T - I(x_0, \varrho)$.

Referring to GIRAUD [15] for the general case, we limit ourselves here to proving the theorem supposing T to be of class $A^{(2)}$ and the direction cosines of l to be of class $C^{(1)}$.

Under these hypotheses, given that a_0 and $a_0^{(l)}$ are the functions defined by (4.3) and (4.7) relative to the operator $\mathfrak{M}_0$, the functions

$$\alpha_{0i} = \alpha_0^{(l)} \cos(l, x_i) - a_0 \cos(v_0, x_i)$$

are of class $C^{(1)}$.

We recognize easily that:

$$\frac{dL(x, y)}{dl} = A^{(l)}(x, y) + B^{(l)}(x, y) ,$$

with $B^{(l)} = O(\overline{xy}^{\lambda+1-m})$ and:

$$a^{(l)}(x_0) A^{(l)}(x, y) = a(x_0) A(x, y) - \sum_{i=1}^{m} \alpha_{0i}(y) \frac{\partial H_0}{\partial y_i} .$$

With a procedure analogous to that which we used to prove (6.7) we then have easily:

$$\int_{\partial T} A^{(l)}(x, y) a^{(l)}(x_0) \, d_y\sigma = \int_{\partial T} A(x, y) a(x_0) \, d_y\sigma + \int_{\partial T} H_0(x y), b_0'(y) \, d_y\sigma$$

$$= \theta + \int_{\partial T} H_0(x, y) b_0'(y) \, d_y\sigma$$

with $b_0' = J^{-1} \Sigma(\partial\beta_{0i}/\partial t_i)$, the β_{0i} being related to the α_{0i} as the β_i to the α_i in § 6.

Analogously, if $I'(x_0, \varrho)$ is the projection of $\partial T \cap I(x_0, \varrho)$ on the hyperplane tangent to ∂T at x_0, we have:

$$\int_{\partial T - I(x_0, \varrho)} A^{(l)}(x_0, y)\, a^{(l)}(x_0)\, d_y\sigma = \int_{\partial T - I(x_0, \varrho)} A(x_0, y)\, a(x_0)\, d_y\sigma$$

$$+ \int_{\partial T - I(x_0, \sigma)} H_0(x_0, y)\, b_0'(y)\, d_y\sigma$$

$$+ \int_{\partial I'(x_0, \varrho)} H_0(x_0, y) \sum_{i=1}^{m-1} \beta_{0i}\, dt_{i+1} \ldots dt_{m-1}\, dt_1 \ldots dt_{i-1}$$

and because the last integral goes to zero with ϱ, H_0 being constant on $\partial I'$, we now deduce that the left side admits a limit as $\varrho \to 0$, namely:

$$\overset{*}{\int_{\partial T}} A^{(l)}(x_0, y)\, a^{(l)}(x_0)\, d_y\sigma = \frac{1}{2} + \int_{\partial T} H_0(x_0, y)\, b_0'(y)\, d_y\sigma\;.$$

After this the proof of the theorem proceeds as for those preceding. There remains only the observation that the principal integral on the right side of (14.12) can certainly be written meaningfully as:

$$\overset{*}{\int_{\partial T}} \frac{dL(x_0, y)}{dl}\, \zeta(y)\, d_y\sigma = \int_{\partial T} A^{(l)}(x_0, y)\, [\zeta(y) - \zeta(x_0)]\, d_y\sigma$$

$$+ \frac{\zeta(x_0)}{a^{(l)}(x_0)} \overset{*}{\int_{\delta T}} A^{(l)}(x_0, y)\, a^{(l)}(x_0)\, d_y\sigma + \int_{\delta T} B^{(l)}(x_0, y)\, \zeta(y)\, d_y\sigma\;,$$

where the first integral on the right side of the preceding formula is certainly meaningful, since the quantity in the brackets which multiplies $A^{(l)}$ is $O(\overline{x_0\, y^\mu})$.

Let us consider now the operator $\mathfrak{P}$ defined by (7.1), and attach to it a meaning even for $x \neq x_0$, $x \in l_1$, putting:

$$\mathfrak{P}\, v(x) = a^{(l)}(x_0)\, \frac{dv(x)}{dl} + \beta(x)\, v(x)\;. \tag{14.13}$$

For $x \to x_0$ on l_1, this operator will admit two distinct limits:

$$(\mathfrak{P}v)^\pm = a^{(l)}(x_0) \left(\frac{dv}{dl}\right)^\pm + \beta(x_0)\, v(x_0)$$

according as $x \to x_0$ on $\Omega - T$ or on $T - \partial T$, and the following theorem obviously holds:

14, VI. *If $\zeta \in C^{(0,\,\mu)}$ we have for every $x_0 \in \partial T$:*

$$(\mathfrak{P}v)^{\pm} = \mp \frac{\zeta(x_0)}{2} + \int\limits_{\partial T}^{*} \mathfrak{P}_x \, L(x_0, y) \, \zeta(y) \, d_y\sigma \,. \qquad (14.14)$$

If $l \equiv v$, (14.14) holds under the sole hypothesis that $\zeta \in C^{(0)}$ and the integral on the right side, which is an ordinary integral, is a continuous function of x_0. Always under the hypothesis $l \equiv v$, and if $\zeta \in L^p$ with $p \geq 1$, then (14.4) holds now for x_0 almost everywhere on ∂T and the integral on the right hand side has the same properties as that which figures in (14.7).

It is worthwhile to note explicitly the following formula which will be useful frequently in what follows:

$$(\mathfrak{P}v)^{+} - (\mathfrak{P}v)^{-} = -\zeta(x_0) \,. \qquad (14.15)$$

Let us complete these results by enunciating the following theorems:

14, VII. *If $\zeta \in C^{(0,\,\lambda)}$, both the derivative of v in $T - \partial T$ and that in $\Omega - T$ can be extended by continuity to T and to $\Omega - (T - \partial T)$ respectively, the functions thus extended being of class $C^{(0,\,\lambda)}$. And if the function β and the direction cosines of l are of class $C^{(0,\,\lambda)}$ then the integral on the right side of (14.14) is also of class $C^{(0,\,\lambda)}$. In the particular case when $l \equiv v$, this integral is of class $C^{(0,\,\mu)}$ with $\mu < \lambda$ under the sole hypotheses that $\beta \in C^{(0,\,\mu)}$, $\zeta \in C^{(0)}$.*

14, VIII. *If $\zeta \in L^p$ with $p \geq 1$ and $l \neq v$, then (14.14) holds for x_0 almost everywhere on ∂T, and if $p > 1$, the integral on the right side is also of class L^p.*

For the proof of theorem 14, VII we refer to Chapter VII, § 5 of GIRAUD's memoir [8]; for that of theorem 14, VIII, which is obtained as a consequence of the theorem of ZYGMUND and CALDERON already cited in § 12, see C. MIRANDA [19]. For more details on all this, it would be wise to consult the accounts in the memoirs of GIRAUD, AMERIO, EVANS and MILES, FICHERA already indicated in § 12[1]. For a study of the derivatives of v of higher order, see C. MIRANDA [19].

15. Generalized double layer potentials. In this section we shall suppose that the functions a_{ik} are of class $C^{(1,\,\lambda)}$ in Ω and that $L(x, y)$ is a Levi function which admits derivatives $\partial L/\partial y_i$ and $\partial^2 L/\partial x_k \, \partial y_i$ which are continuous for $x \neq y$ and satisfy the bounds:

$$\frac{\partial(L - H)}{\partial y_i} = O(r^{2-m}), \quad \frac{\partial^2(L - H)}{\partial x_k \, \partial x_i} = O(r^{1-m}) \,. \qquad (15.1)$$

---

[1] See also W. POGORZELSKI [4].

3*

Moreover, we shall suppose that (8.4) is also satisfied for $\lambda = 1$. The domain T will, moreover, always be assumed to be of class $A^{(1,\lambda)}$ with $\lambda \leq 1$. Given that β is a function continuous on ∂T and that $l(x)$ is an axis coming out of the generic point x of ∂T whose direction cosines are continuous functions of x, such that $\cos(l, n) > 0$, let $\mathfrak{D}$ be the operator defined by (7.2) and let us put:

$$w(x) = \int_{\partial T} \mathfrak{D}_y L(x, y)\, \zeta(y)\, d_y\sigma . \tag{15.2}$$

To this integral we give the name (generalized) double layer potential, of moment ζ.

Since $\mathfrak{D}_y L(x, y) \in N^{(1)}$, from theorems 12, III and 12, VI we have immediately:

15, I. If $\zeta \in L^p$ with $p \geq 1$, we have also that $w \in L^s(T)$ with $s \leq \dfrac{p\,m}{m-1}$, equality holding only for $p > 1$.

We now wish to study the behavior of w when x tends to a point x_0 of ∂T along an axis l_1 coming out of x_0 and such that $\cos(l_1, n) \neq 0$. We shall recognize that w has in general two distinct limits according as x tends to x_0 in $T - \partial T$ or in $\Omega - T$ and these limits, which do not depend on l_1, we shall denote respectively by $w^-(x_0)$ and $w^+(x_0)$.

Beginning with the case $l \equiv \nu$, let us prove that:

15, II. If $l \equiv \nu$, $\zeta \in L^p$ with $p \geq 1$, we have for x_0 almost everywhere on ∂T:

$$w^{\pm}(x) = \pm \frac{\zeta(x_0)}{2} + \int_{\partial T} \mathfrak{D}_y L(x_0, y)\, \zeta(y)\, d_y\sigma , \tag{15.3}$$

the integral on the right of this formula having the same regularity properties as that which figures in (14.7). In particular, if $\zeta \in C^{(0)}$, (15.3) holds everywhere and the aforesaid integral is of class $C^{(0,\mu)}$ for every $\mu < 1$; moreover, the function equal to w in $T - \partial T$ [in $\Omega - T$] and to w^- [w^+] on ∂T is continuous in T [in $\Omega - (T - \partial T)$].

In fact, (15.3) follows from (14.7) where we observe that for $x \in l_1$:

$$\mathfrak{D}_y L(x, y) = -\mathfrak{P}_x L(x, y) + O\left(\overline{xy}^{\lambda + 1 - m}\right) .$$

Analogously we justify the asserted properties of the integral on the right of (15.3). The possibility of continuously extending w in T or in $\Omega - (T - \partial T)$ follows then from the fact that the limiting expression (15.3) holds uniformly as x_0 varies on ∂T if, for example, we assume $l_1 \equiv n$.

If, now, we observe that we can write:

$$w = - \sum_{i=1}^{m} \frac{\partial}{\partial x_i} \int_{\partial T} L(x, y) \cos(\nu, y_i) \, \zeta(y) \, d_y \sigma + \int_{\partial T} K(x, y) \, \zeta(y) \, d_y \sigma$$

with $K \in N^{(2, \lambda)}$, we deduce immediately from theorem 14, VII that:

15, III. *If $\zeta \in C^{(0, \mu)}$ with $\mu \leq \lambda$, then we have also for the function w, continued in T or in $\Omega - (T - \partial T)$ as in the preceding theorem, that $w \in C^{(0, \mu)}$.*

In a perfectly analogous manner we prove that:

15, IV. *(15.3) holds also if $l \neq \nu$ provided that the direction cosines of l and the function ζ are of class $C^{(0, \mu)}$, the integral figuring in (15.3) being taken in the sense of a principal value, that is, as the limit of the integral over $\partial T - I(x_0, \varrho)$. Under the same hypotheses, theorem 15, III also holds for $l \neq \nu$.*

Let us note explicitly that from (15.3) it follows that:

$$w^+(x_0) - w^-(x_0) = \zeta(x_0) . \tag{15.4}$$

Now we wish to specialize the function L for the purpose of establishing new properties of w. If $L(x, y)$ is, as a function of y, a fundamental solution of the equation $\mathfrak{N}_y L = 0$, supposing $l \equiv \nu$, we have, from GREEN's (7.5) and STOKE's (9.3) formulas, applied as in the previous section,

$$\int_{\partial T} \mathfrak{D}_y L(x, y) \, d_y \sigma = \int_{\partial T} L(x, y) \, \beta(y) \, d_y \sigma - \int_T L(x, y) \, c(y) \, dy - \theta, \tag{15.5}$$

where as usual θ is 1 in $T - \partial T$ and 0 in $\Omega - T$.

Naturally, to arrive at this formula we must put hypotheses on the coefficients of $\mathfrak{M}$ so that it is valid to speak of the adjoint operator $\mathfrak{N}$. We shall see, however, in what follows, that there are functions L for which (15.5) holds even when the operator $\mathfrak{N}$ has no significance. Without altering the hypotheses other than those made on the a_{ik}, let us assume as hypothesis that L is a LEVI function satisfying (15.5); if we denote by w_0 the integral on the left side of (15.5) and by v_0 the first integral on the right, we have:

$$(\mathfrak{P} w_0)^+ - (\mathfrak{P} w_0)^- = (\mathfrak{P} v_0)^+ - (\mathfrak{P} v_0)^- + \beta(x_0)$$

and hence by (14.15):

$$(\mathfrak{P} w_0)^+ - (\mathfrak{P} w_0)^- = 0. \tag{15.6}$$

Let us now proceed to prove the following theorem:

15, V. *If L satisfies* (15.5) *and if* $l \equiv v$ *and* $\zeta \in C^{(0,\,\mu)}$ *with* $\mu > 1 - \lambda$, *and if x and x' are two points on the conormal through x_0 symmetric with respect to x_0, we have:*

$$\lim_{x \to x_0} [\mathfrak{P}\, w\,(x) - \mathfrak{P}\, w\,(x')] = 0, \tag{15.7}$$

$\mathfrak{P}$ *being the operator defined by* (14.13). (15.7) *holds even under the sole hypothesis* $\zeta \in C^{(0)}$, *provided T is of class $A^{(2)}$ and* $\partial^2 (L - H)/\partial x_k\, \partial y_i \in N^{(1,\,h)}$. *Under this last hypothesis and if* $\zeta \in L^1$, *then* (15.7) *holds almost everywhere on* ∂T.

We have, in fact:

$$\mathfrak{P}\, w\,(x) = \int_{\partial T} \mathfrak{P}_x\, \mathfrak{D}_x\, L\,(x,\,y)\, [\zeta\,(y) - \zeta\,(x_0)]\, d_y\sigma + \zeta\,(x_0)\, \mathfrak{P}\, w_0\,(x).$$

By virtue of (15.6) everything reduces to showing that:

$$\lim_{x \to x_0} \int_{\partial T} [\mathfrak{P}_x\, \mathfrak{D}_y\, L\,(x,\,y) - \mathfrak{P}_x\, \mathfrak{D}_y\, L\,(x',\,y)]\, [\zeta\,(y) - \zeta\,(x_0)]\, d_y\sigma = 0. \tag{15.8}$$

Now we can verify that for y near x_0 we have:

$$\mathfrak{P}_x\, \mathfrak{D}_y\, L\,(x,\,y) - \mathfrak{P}_x\, \mathfrak{D}_y\, L\,(x',\,y) = \sum_{\alpha=1}^{3} K_\alpha\,(x,\,y)$$

with $K_2 = O(\overline{xy}^{1-m})$ and $K_\alpha = O(\overline{xx_0}^{\alpha} \cdot \overline{xy}^{\lambda-\alpha-m})$ for $\alpha = 1, 3$; and thus if $\zeta \in C^{(0,\,\mu)}$ with $\mu > 1 - \lambda$, (15.7) follows from the lemma of § 14. If now T is of class $A^{(2)}$, in the preceding bounds on the kernels K_α we can assume $\lambda = 1$ and then the same lemma permits us to attain the conclusion even in the case when ζ is of class $C^{(0)}$ or L^1, where we bear in mind that, for the supplementary hypothesis on $L - H$, we have:

$$K_2 = O(\overline{xx_0}^{h} \cdot \overline{xy}^{1-h-m}).$$

Finally, always under the hypothesis that L satisfies (15.5), the following theorems hold:

15, VI. *If L satisfies* (15.5) *and if* $l \equiv v$, $\beta \in C^{(0,\,\lambda)}$, *and* $\zeta \in C^{(0,\,\mu)}$ *with* $\mu > 1 - \lambda$, *then the integral on the right of* (15.3) *is of class* $C^{(1,\,\lambda)}$ *on* ∂T.

15, VII. *If L satisfies* (15.5) *and if* $l \equiv v$, $\beta \in C^{(0,\,\lambda)}$, *and* $\zeta \in C^{(1,\,\lambda)}$, *then the function w extended in T or in $\Omega - (T - \partial T)$ as in theorem* 15, II *is also of class* $C^{(1,\,\lambda)}$.

Referring for the proofs of these theorems to the memoir [8] of GIRAUD, here we shall limit ourselves to the observation, regarding theorem 15, VI, that, if t_i is one of the parameters in a local representation of ∂T in a neighborhood of x_0, we have formally:

$$\frac{\partial}{\partial t_i} \int_{\partial T} \mathfrak{D}_y L(x_0, y) \zeta(y) \, d_y \sigma = \int_{\partial T} \frac{\partial}{\partial t_i} \mathfrak{D}_y L(x_0, y) [\zeta(y) - \zeta(x_0)] \, d_y \sigma$$

$$+ \zeta(x_0) \frac{\partial}{\partial t_i} \int_{\partial T} \mathfrak{D}_y L(x_0, y) \, d_y \sigma,$$

where the first integral on the right is meaningful inasmuch as the integrand is $O(\overline{x_0 y}^{\lambda + \mu - m})$. As for the final term of this formula it suffices to observe that, since we have in consequence of (15.5) that:

$$\int_{\partial T} \mathfrak{D}_y L(x_0, y) \, d_y \sigma = \int_{\partial T} L(x_0, y) \beta(y) \, d_y \sigma - \int_T L(x_0, y) c(y) \, dy - \frac{1}{2},$$

the function on the left is of class $C^{(1, \lambda)}$ on ∂T by dint of theorems 13, I and 14, VII.

Also, for the arguments in this section, the bibliography to consult is that already recorded at the end of § 14.

16. Construction of functions satisfying assigned boundary conditions. The solution of the following preliminary problem often has importance in various questions involving the theory of elliptic equations:

Given a bounded domain T of class $A^{(1)}$ and $s + 1$ functions $u_0, u_1, \ldots, u_s$ defined on ∂T and continuous there, to construct a function $u(x)$ of class $C^{(\infty)}$ in $T - \partial T$ and satisfying on ∂T the boundary condition

$$u = u_0, \quad \frac{d^q u}{d n^q} = u_q \, (q = 1, 2, \ldots, s). \tag{16.1}$$

Moreover, it is interesting to be able to pin down the behavior of $u(x)$ and of its derivatives as x tends toward a point of ∂T.

This problem is naturally susceptible to many different solutions E.E. LEVI in an appendix to the memoir [2] for the case $m = 2$, G. GIRAUD [2, p. 211 and 8, pp. 101—104], and M. GEVREY [8] are all occupied with these. We shall give some idea of the method of GEVREY, which is the one requiring the minimum number of hypotheses but furnishing more precise results. It is worth mentioning nevertheless that E.E. LEVI considered the case in which the u_q present some singularities, which we shall not.

Putting for $\mu > 0$:

$$c_\mu = \pi \frac{m-2}{2} \, \Gamma\left(\frac{\mu}{2}\right) \left[\Gamma\left(\frac{\mu-m-1}{2}\right)\right]^{-1},$$

let us consider the integral:

$$I_\mu(x;v) = \frac{d^\mu}{c_\mu} \int_{\partial T} \frac{v(y)\,dy}{\overline{xy}^{\mu+m-1}},$$

in which $v(y)$ denotes a function continuous on ∂T and d is the distance of the point x from ∂T. Procedures analogous to those adopted in §§ 14 and 15 permit us to prove easily that:

16, I. *If T is of class $A^{(1+h)}$ with $\mu > h \geq 0$, and if $v \in C^{(h)}$, then the function equal to I_μ in $T - \partial T$ and to $v(x)$ on ∂T is of class $C^{(h)}$ in a neighborhood of ∂T. If T is of class $A^{(1+h,\lambda)}$ with $\mu - \lambda > h \geq 0$ and if $v \in C^{(h,\lambda)}$, then the aforesaid function is of class $C^{(h,\lambda)}$ in a neighborhood of ∂T.*

Supposing now that T is of class $A^{(1)}$, fix a point x_0 on ∂T and denote by δ the distance of a generic point x to the hyperplane tangent to ∂T at x_0, assumed positive if x is situated on the interior normal to ∂T at x_0. Let us put:

$$I_\mu^* = \left(\frac{\delta}{d}\right)^\mu I_\mu.$$

It is obvious that for $x \to x_0$, along the interior normal to ∂T at x_0, I_μ^* has a behavior analogous to that of I_μ, namely $\lim I_\mu^* = v(x_0)$; and moreover $I_\mu^* - v(x_0) = O(\overline{xx_0}^\lambda)$, if we make a further hypothesis that T is of class $A^{(1,\lambda)}$ and $v \in C^{(0,\lambda)}$. The integral I_μ^* is moreover of class $C^{(\infty)}$ in $T - \partial T$ and, denoting by $I_\mu^{(k)}$ one of its derivatives of order k calculated at a generic point of the interior normal to ∂T at x_0, we have that this derivative will admit a finite limit for $x \to x_0$ if, supposing T of class $A^{(1+h)}$ and $v \in C^{(h)}$ ($\mu > h$), we have $k \leq h$. If instead $k > h$ we have:

$$I_\mu^{(k)} = o\left(\overline{xx_0}^{h-k}\right).$$

This bound can be replaced by the more precise:

$$I_\mu^{(k)} = O\left(\overline{xx_0}^{\lambda+h-k}\right)$$

if T is of class $A^{(1+h,\lambda)}$, $v \in C^{(h,\lambda)}$, $\mu > h + \lambda$.

Once we have advanced this, the following theorem then holds:

16, II. *Let T be of class $A^{(1+h)}$ $\left[A^{(1+h,\lambda)}\right]$ with $h \geq 0$, $u_0 = u_1, = \ldots = u_{s-1} = 0$, $u_s \in C^{(h)}$ $\left[C^{(h,\lambda)}\right]$. If $\mu > h$ $[\mu > h + \lambda]$, the equation*

$$u(x) = \frac{(-d)^s \, I_\mu(x;u_s)}{s! \, I_{\mu+s}(x;1)} \tag{16.2}$$

defines a function which satisfies (16.1) *and for which we have* $u \in C^{(\infty)}$
in $T - \partial T$ *and* $u \in C^{(s+h)} \left[C^{(s+h, \lambda)} \right]$ *in* T. *Moreover, if* $u^{(k)}$ *is any deriv-*
ative of u *of order* $k > s + h$, *then for this* $u^{(k)}$ *calculated at a generic*
point of the interior normal to ∂T *at any point* x_0 *we have:*

$$u^{(k)} = o\left(\overline{xx_0}^{\,s+h-k}\right) \left[= O\left(\overline{xx_0}^{\,\lambda+s+h-k}\right)\right]. \tag{16.3}$$

All this is easily proved by observing that however we choose x_0
on ∂T, (16.2) can be written:

$$u(x)\, I^*_{\mu+s}(x;\,1) = \frac{(-\delta)^s}{s!}\, I^*_{\mu}(x;\, u_s),$$

a formula from which it is easy to extract the expression for any derivative of order q of u, by means of the derivative of I^* of order not
greater than q and the derivative of the same u of lower order. From
this, what has been premised about the properties of the derivatives
of I^* leads to the assertion.

We can also prove in an analogous manner that:

16, III. *Under the same hypotheses as theorem* 16, II *and if moreover*
$u_s > 0$, *the expression* (16.2) *can be for all practical purposes replaced by*

$$u(x) = \frac{(-d)^s}{s!\, I_s\left(x;\, \dfrac{1}{u_s}\right)}. \tag{16.4}$$

Passing now to consider the general problem posed at the beginning
of this section, we can easily demonstrate that:

16, IV. *Let* T *be of class* $A^{(s+h+1)} \left[A^{(s+h+1, \lambda)} \right]$ *with* $h \geq 0$,
$u_q \in C^{(s-q+h)} \left[C^{(s-q+h, \lambda)} \right]$ *for* $q = 0, 1, \ldots, s$. *Then there exists at least*
one function u, *of class* $C^{(\infty)}$ *in* $T - \partial T$ *and of class* $C^{(s+h)} \left[C^{(s+h, \lambda)} \right]$ *in*
T, *which satisfies* (16.1). *This function also satisfies* (16.3).

In fact, by use of theorem 16, II it is possible to determine functions
$U_0, U_1, \ldots, U_s$ in such a manner that these have all the qualitative
properties demanded by our statement for u, the U_q satisfying, moreover, the boundary conditions:

$$U_q = \frac{d U_q}{dn} = \ldots = \frac{d^{q-1} U_q}{dn^{q-1}} = 0, \qquad \frac{d^q U_q}{dn^q} = u_q - \sum_{i=0}^{q-1} \frac{d^q U_i}{dn^q}.$$

Putting $u = \Sigma\, U_q$ we then obtain the function the existence of which
our theorem affirms.

We should note that the hypotheses made in theorem 16, IV on the
domain T are more restrictive than those considered in theorem 16, II
for the particular case $u_0 = u_1 = \ldots = u_{s-1} = 0$. For this purpose the
following obvious observation is useful:

16, V. *If a function* $v \in C^{(s+h)}$ $[C^{(s+h,\lambda)}]$ *exists in* T *satisfying the first* q *conditions* (16.1), *then theorem* 16, IV, *where we leave out of consideration that* u *is in* $C^{(\infty)}$ *in* $T - \partial T$ *and the validity of* (16.3), *holds even if* T *is only of class* $A^{(s+h-q+1)}$ $[A^{(s+h-q+1,\lambda)}]$.

Based on this observation we wish now to demonstrate that:

16, VI. *Let* T *be of class* $A^{(s+h)}$ $[A^{(s+h,\lambda)}]$ *with* $h \geq 0$, $u_q \in C^{(s-q+h)}$ $[C^{(s-q+h,\lambda)}]$ *for* $q = 0, 1, \ldots, s$. *Then there exists at least one function* u *of class* $C^{(s+h)}$ $[C^{(s+h,\lambda)}]$ *in* T *which satisfies* (16.1).

It will suffice for this to demonstrate that, under the hypotheses posed, there exists at least one function $u(x)$ of class $C^{(s+h)}$ $[C^{(s+h,\lambda)}]$ in T which satisfies the first of (16.1).

To this end, for every point ξ of ∂T let us pick an axis $l(\xi)$ whose direction cosines are of class $C^{(s+h)}$ $[C^{(s+h,\lambda)}]$ and such that $\cos(l, n) > 0$. We see now easily that there exists a neighborhood I of ∂T such that every point x of I can be placed on only one of these axes $l(\xi)$. Denoting by t_i the parameters which identify the point ξ and putting $t_m = \overline{x\xi}$, we shall have for the coordinates of the point x:

$$x_i = x_i(t_1, t_2, \ldots, t_{m-1}) + t_m\, l_i(t_1, t_2, \ldots, t_{m-1})$$

and vice versa:

$$t_r = t_r(x_1, x_2, \ldots, x_m), \quad t_r \in C^{(s+h)}[C^{(s+h,\lambda)}].$$

Therefore, putting $u(x) = u_0(\xi)$ in I, we define a function of class $C^{(s+h)}[C^{(s+h,\lambda)}]$. If now T' is a domain of class $A^{(s+h+1)}[A^{(s+h+1,\lambda)}]$ inside T and such that its boundary is interior to I, $u(x)$ defined as above in $T - T'$ can be extended into T' maintaining continuity [λ-Hölder continuity] of the derivatives of order $s + h$; this follows, in fact, from theorem 16, IV.

u thus defined in all of T agrees with the requisites demanded by our statement.

Referring to some works of H. WHITNEY[1] for other results on the problems just considered, let us finish this paragraph by giving some idea of a different method of stating these questions.

Let us begin by recalling that for every function $u \in H^{1,p}(T)$ it is possible to define a *trace* γu almost everywhere on ∂T, which is a function of class $L^p(\partial T)$, which in the case $u \in C^{(0)}(T)$ is simply the

[1] H. WHITNEY: Analytic extensions of differentiable functions defined in closed sets, Trans. Amer. Math. Soc. **36** (1934), 63—89. Functions differentiable on the boundaries of regions. Ann. of Math. **35** (1934), 482—485.

restriction of u to ∂T. The definition of "trace" can be given in various ways. For example, if $T \in A^{(2)}$ we can define γu by means of the expression:

$$\gamma u (x_0) = \lim_{\substack{x \to x_0 \\ x \in n_{x_0}}} u (x)$$

and by proving that this limit exists for x_0 almost everywhere on ∂T, and is of class L^p there. Another procedure consists of proving that for $u \in C^{(1)}(T)$, a bound of the type $\| \gamma u \|_{L^p(\partial T)} = O(\| u \|_{H^{1,p}(T)})$ holds for the restriction γu of u to ∂T, then that the mapping $u \to \gamma u$ from $C^{(1)}(T)$ to $C^{(1)}(\partial T)$ initially defined can be extended in one and only one way to a mapping from $H^{1,p}(T)$ to $L^p(\partial T)$. Under the hypothesis posed on T, the two definitions are equivalent.

The subspace of $L^p(\partial T)$ constituted of the functions φ to which one can associate a $u \in H^{k,p}$ such that:

$$\gamma u = \varphi \tag{16.5}$$

is denoted by $H_{k - \frac{1}{p}}$; this can be considered as a BANACH space by defining a norm with the expression:

$$\| \varphi \|_{H_{k - \frac{1}{p}}} = \inf \| u \|_{H^{k,p}(T)}, \tag{16.6}$$

where the infimum is taken as u varies over the set of functions for which (16.5) is valid. Then the problem arises of characterizing these spaces intrinsically, of defining for them a norm equivalent to (16.6) which can be calculated by taking into account only the values assumed by φ on ∂T. For example, for $k = 1$ a norm equivalent to the $H_{1 - \frac{1}{p}}$ norm is furnished, according to a theorem of G. GAGLIARDO[1], by the quantity:

$$\| \varphi \|_{L^p(T)} + \left[\int_{\partial T \times \partial T} \int \frac{|\varphi(x) - \varphi(y)|^p}{|x - y|^{p+2}} \, d_x \sigma \, d_y \sigma \right]^{\frac{1}{p}} .$$

Note that for a function $u \in H^{k,p}$ the traces on ∂T of all the derivatives of u up to order $k - 1$ turn out to be defined. It therefore turns out to be evident how one could pose the problem of constructing a function $u \in H^{s+1,p}(T)$ satisfying (16.1) in the generalized sense.

---

[1] E. GAGLIARDO: Caratterizzazione delle traccie sulla frontiera relative ad alcune classi di funzioni in n variabili. Rend. Sem. Mat. Padova **27** (1957), 284—305.

On this argument and on other related ones the literature is quite vast, and we shall not be able to dwell on it further. It is possible to come across various bibliographic indications in this regard, for example in the works of E. MAGENES and G. STAMPACCHIA [1], J.L.LIONS and E. MAGENES [1, III], E. MAGENES [9], and in a memoir by S.M. NIKOL'SKIĬ[1].

---

[1] S.M. NIKOL'SKII: On imbedding, continuation and approximation theorems for differentiable functions of several variables. Usp. Mat. Nauk **16** no. 6, (1961) 63—144 (Russian); transl. in Russian Math. Surveys **16** no. 5., (1961), 55—104.

CHAPTER III

Transformation of the boundary value problems into integral equations

A classical method to attain the proofs of existence theorems for the solutions of various boundary value problems consists of reducing the solution of these problems to the solution of an integral equation or of a system of integral equations. This method originated in the works of FREDHOLM for Δ_2 and the later researches of HILBERT, POINCARÉ, PICARD, LICHTENSTEIN, and others[1], and has been most often applied to the study of particular problems on which we cannot dwell here. Problems under the general conditions in which they will be posed here were treated by this method for the first time by E. E. LEVI in the case $m = 2$. M. GEVREY and G. GIRAUD successively studied the general case under much less restrictive hypotheses than those considered by LEVI. There is not room for doubt that the definitive results along this line are those of GIRAUD, and on these we shall dwell copiously. Nevertheless, at the end of the chapter we shall not fail to illustrate briefly the research of LEVI and of GEVREY also. In § 23 we shall also mention some results of other authors concerning the oblique derivative problem in the non-regular case.

For the convenience of the reader, we have prefaced the exposition of this research with some review of the theory of integral equations.

17. Review of basic knowledge about integral equations. Let T be a domain of class $A^{(1,\lambda)}$, x and y be variable points in T, and ξ and η be variable points on ∂T. The integral equations with which we shall generally be occupied will be of the type:

$$\left. \begin{array}{l} \varphi_1(x) - \mu \int_T K_{11}(x,y)\, \varphi_1(y)\, dy - \mu \int_{\partial T} K_{12}(x,\eta)\, \varphi_2(\eta)\, d_\eta \sigma = f_1(x) \\[2ex] \varphi_2(x) - \mu \int_T K_{21}(\xi,y)\, \varphi_1(y)\, dy - \mu \int_{\partial T} K_{22}(\xi,\eta)\, \varphi_2(\eta)\, d_\eta \sigma = f_2(\xi). \end{array} \right\} \quad (17.1)$$

---

[1] See L. LICHTENSTEIN [5] for items regarding the older literature on the argument.

We treat, therefore, a system of two integral equations of second kind in two unknown functions, of which the first $\varphi_1(x)$ is defined in T and the second $\varphi_2(\xi)$ is defined on ∂T; however, we do not exclude the case in which the kernels K_{i1} or K_{i2} are zero for $i = 1$ or $i = 2$ and the system reduces to a single equation in the single unknown $\varphi_2(\xi)$ or $\varphi_1(x)$.

Supposing that the functions f_1 and f_2 are real and continuous, we shall seek out solutions φ_1 and φ_2 in the class of functions continuous in T and on ∂T, respectively. For the time being we shall assume that the kernels K_{ij} are real and continuous.

We now wish to summarize the principal theorems which will occur in applications, and for this end it will be necessary to change our notation, in order to write the system (17.1) in a more convenient form.

To begin with, we put $x = x_1$, $y = y_1$, $\xi = x_2$, $\eta = y_2$, $T = T_1$, $\partial T = T_2$, $dy = dT_1$, and $d_\eta\sigma = dT_2$, and we agree to denote by φ and f the *vectors* with components (φ_1, φ_2) and (f_1, f_2), respectively, and by K the *kernel matrix* $\| K_{ij}(x_i, y_j) \|$. Next we put:

$$f * \varphi = \sum_{i=1}^{2} \int_{T_i} f_i(y_i)\, \varphi_i(y_i)\, dT_i$$

and we denote by $K * \varphi$ the vector ψ with components:

$$\psi_i(x_i) = \sum_{j=1}^{2} \int_{T_j} K_{ij}(x_i, y_j)\, \varphi_i(y_j)\, dT_j$$

and by $\varphi * K$ the vector χ with components:

$$\chi_i(x_i) = \sum_{j=1}^{2} \int_{T_j} K_{ij}(y_j, x_i)\, \varphi_j(y_j)\, dT_j.$$

We have thus defined the *composition* of two vectors or of a vector and a matrix.

If H is a second kernel matrix of the same type as K, we now define the *composition* $H * K$ of H and K as that matrix G which has as elements the kernels:

$$G_{ij}(x_i, z_j) = \sum_{k=1}^{2} \int_{T_k} H_{ik}(x_i, y_k)\, K_{kj}(y_k, z_j)\, dT_k.$$

It is obvious that the composition of two matrices is not commutative, as already happened in the composition of matrix and vector.

With this notation our system can be written in the form:

$$\varphi - \mu K * \varphi = f, \tag{17.1}$$

and, besides this, it is also necessary to consider the *transposed system*:

$$\psi - \mu \psi * K = g, \tag{17.2}$$

in which g is a new vector and ψ is a new unknown vector.

It is well known that the solution formulas of the system (17.1), (17.2) can be written, respectively, as:

$$\varphi = f + \mu \Gamma * f, \tag{17.3}$$

$$\psi = g + \mu g * \Gamma, \tag{17.4}$$

where Γ is a kernel matrix whose elements $\Gamma_{ij}(x_i, y_j, \mu)$ are of the form:

$$\Gamma_{ij}(x_i, y_j, \mu) = \frac{D_{ij}(x_i, y_j, \mu)}{D(\mu)}, \tag{17.5}$$

$D(\mu)$ and $D_{ij}(x_i, y_j, \mu)$ being entire functions of μ, the second function being continuous in x_i and y_j for every fixed μ. The matrix Γ, called the *resolvent matrix* of the matrix K, is therefore meromorphic in μ; it is, however, regular for $\mu = 0$ and even admits in a neighborhood of $\mu = 0$ the power series expansion:

$$\Gamma = \sum_{n=0}^{\infty} \mu^n K^{(n)}, \tag{17.6}$$

where the matrices $K^{(n)}$ (*iterated matrices*) are defined by the formulas:

$$K^{(0)} = K,$$
$$K^{(n)} = K^{(n-1)} * K. \tag{17.7}$$

Moreover, the resolvent matrix satisfies the relation:

$$\Gamma = K + \mu K * \Gamma = K + \mu \Gamma * K. \tag{17.8}$$

A classical discussion of (17.3) and (17.4) leads to the recognition that for (17.1) and (17.2), only two cases are possible (FREDHOLM'S *Alternative*):

1^0) μ is not a pole of Γ, in which case (17.1) and (17.2) admit one and only one solution, given by (17.3) and (17.4), whatever f and g may be. In particular the homogeneous equations associated with (17.1) and (17.2) admit only the zero solution, $\varphi = 0$ and $\psi = 0$.

2^0) μ is a pole of Γ, in which case the homogeneous equations associated with (17.1) and (17.2) admit an equal (finite) number p of linearly independent solutions, which we call $\varphi^{(1)}$, $\varphi^{(2)}$, ..., $\varphi^{(p)}$ and $\psi^{(1)}$, $\psi^{(2)}$, ..., $\psi^{(p)}$ respectively. A necessary and sufficient condition that (17.1) [(17.2)] possess a solution in this case is that f [g] satisfy the conditions:

$$f * \psi^{(i)} = 0 \; [g * \varphi^{(i)} = 0]. \tag{17.9}$$

If these conditions are satisfied and if φ [ψ] is a solution of (17.1) [(17.2)] then every solution, and all possible solutions, are of the form $\varphi + \Sigma c_i \varphi^{(i)}$ [$\psi + \Sigma c_i \psi^{(i)}$] where the c_i are arbitrary constants.

In case 2^0) we say that μ is an *eigenvalue of multiplicity* p of the matrix K and the vectors $\varphi^{(i)}$ and $\psi^{(i)}$ are called *eigensolutions* of the homogeneous equations associated with (17.1) and (17.2).

About the actual existence of eigenvalues we can say nothing, in general. If, however, the relations $K_{ij}(x_i, y_j) = K_{ji}(y_j, x_i)$ are satisfied, in which case the kernel matrix K is called *symmetric* and the system (17.1) coincides for $f = g$ with the system (17.2), it can be shown that:

a) The kernel K admits at least one eigenvalue.

b) The eigenvalues are all real and are first order poles of Γ.

c) Ordering all the eigenvalues in a sequence $\{\mu_n\}$, in which each one is repeated the same number of times as its multiplicity, we can associate to each μ_n an eigensolution $\varphi^{(n)}$ in such a way that these eigensolutions constitute an orthonormal system.

d) If f is an arbitrary continuous vector, we have:

$$f * (K * f) = \sum_{n-1}^{\infty} \frac{(f * \varphi^{(n)})^2}{\mu_n} .$$

All the results just recorded can be extended in various ways to the case in which the kernels K_{ij} are not continuous. For example, the extension to the case in which these kernels are measurable and the following integrals are meaningful (in the Lebesgue sense) is classical:

$$\int_{T_i} \int_{T_j} [K_{ij}(x_i, y_j)]^2 \, dT_i \, dT_j.$$

In fact, in this case all the preceding results hold on the condition that the unknown functions φ_1 and φ_2 and the functions f_1 and f_2 belong to the classes $L^2 (T_1)$ and $L^2 (T_2)$, respectively. Note also that if the kernels K_{ij} are not square integrable, the FREDHOLM alternative can fail. Nevertheless, there is one case in which this theorem and those relating to symmetric kernels continue to hold even without the kernels K_{ij} being square integrable provided that f_i and φ_i are assumed to be continuous.

For example, according to an observation of GIRAUD [11] we have:

17, I. FREDHOLM'S *alternative and the theorems on equations with symmetric kernel matrices are valid for the system* (17.1) *under the single hypothesis that the kernels* K_{ij} *are of class* $N^{(\alpha_{ij})}$ *with* $\alpha_{11} > 0$, $\alpha_{12} > 1$, $\alpha_{21} > 1$, $\alpha_{22} > 1$.

The possibility of this extension resides substantially in theorems 11, I, 11, II, and 12, IV.

In fact, theorem 12, IV makes it plain on the one hand, that every solution of (17.1) is also a solution of the system:

$$\varphi - \mu^{p+1} K^{(p)} * \varphi = f + \sum_{n=1}^{p} \mu^n K^{(n-1)} * f, \qquad (17.10)$$

and on the other hand, that if the system:

$$\varrho - \mu^{p+1} K^{(p)} * \varrho = f \qquad (17.11)$$

admits a solution ϱ, then the vector $\varphi = \varrho + \sum_{n=1}^{p} \mu^n K^{(n-1)} * \varrho$ is a solution of (17.1). Theorems 11, I and 11, II guarantee that if p is sufficiently large, the matrix $K^{(p)}$ has elements which are continuous even for $x_i = y_j$. From this, from the validity of the alternative theorem for the system (17.10) and (17.11), the validity of the same theorem follows for (17.1); this is proved by a classical argument which is related in every treatise on integral equations.

Also nearly immediate is the extension to the case we considered, of theorems relating to symmetric kernel matrices, except regarding proposition d), for which we refer to G. GIRAUD's note [13].

We also have:

17, II. *The alternative theorem is valid for the system* (17.1) *under the single hypothesis that the kernels* K_{ij} *are of class* $N^{(\alpha_{ij})}$ *with* $\alpha_{11} > 0$, $\alpha_{12} > 1$, $\alpha_{21} + \alpha_{11} > 1$, $\alpha_{22} > 1$.

In fact, if in the second equation of the system we substitute for φ_1 the expression which is obtained from the first equation, we obtain a new equation which with the first of (17.1) constitutes a system equivalent to (17.1). And for this new system the hypotheses of theorem 17, I hold, and therefore the alternative theorem is valid.

17, III. *If the kernel* K_{12} *is such that for* $\zeta \in C^{(0)}$ *on* ∂T *we even have* $\int K_{12} \zeta \, d\sigma \in L^{(\infty)}$ *in* T, *then the alternative theorem is valid for the system* (17.1) *under the further hypothesis that the kernels* K_{ij} *are of class* $N^{(\alpha_{ij})}$ *with* $\alpha_{11} > 0$, $\alpha_{12} + \alpha_{11} > 1$, $\alpha_{21} > 1$, $\alpha_{22} > 1$.

In fact, upon changing the unknown function:

$$\varphi_1(x) = \varphi_1'(x) + \mu \int_{\partial T} K_{12}(x, \eta)\ \varphi_2(\eta)\ d_\eta \sigma\ ,$$

we recognize here an equivalent system which satisfies the hypotheses of theorem 17, I. The particular hypotheses made on the kernel K_{12} are necessary (theorem 12, V) in order to justify the change in the order of two integrations.

It is also useful to enunciate explicitly the following theorems which we shall frequently have to apply:

17, IV. *Under the hypotheses of theorem* 17, I *the elements* Γ_{ij} *of the resolvent* Γ *of the kernel matrix* K *are continuous for* $x_i \neq y_j$ *and they satisfy the same bounds admitted by hypothesis for the* K_{ij}.

17, V. *Under the hypotheses of theorem* 17, I *the resolvent* Γ *is holomorphic for* $|\mu| < \frac{1}{M}$ *where:*

$$M = \sum_{i,j=1}^{2} \max_{T_i} \int_{T_j} |\ K_{ij}(x_i, y_j)\ |\ dT_j\ .$$

Let us conclude by recalling that G. GIRAUD [12, 17] has proved the validity of the alternative theorem even in certain cases in which continuity of the kernels K_{ij} fails for $x_i \neq y_j$ and in others in which the domain T is not bounded.

18. The method of potentials. The way which presents itself most naturally for translating boundary value problems for elliptic equations into integral equations is that of extending in some way the classical treatment of DIRICHLET's and NEUMANN's problems for harmonic functions. The first results along these lines were developed for the case $m = 3$ by W. STERNBERG [1]. Most interesting and perhaps worthy of complete development are the researches of W. FELLER [1] which depart from the observation that every self-adjoint operator $\mathfrak{M}\ u$, with $m > 2$ and $c = 0$, can be considered, at least up to multiplication by a factor, as the second differential parameter of u in a convenient RIEMANN space. This makes it possible to extend various properties of harmonic functions to solutions of the equation $\mathfrak{M}\ u = 0$. Among these properties are those relating to single and double layer potentials, which are basic for the translation of boundary value problems into integral equations. The most general and truly definitive results in this area of research, however, are due to G. GIRAUD, who dedicated to the argument a series of memoirs in which the study of the three boundary value problems and of other related questions is unrolled in the most complete way, and with the indispensable minimum of hypotheses. Referring to the next paragraphs for a detailed exposition of GIRAUD's research, we wish to give in this section a first idea of this author's method, in order to underline the difficulties we must surmount and to establish some preliminary results.

Therefore, let $\mathfrak{M}$ be an elliptic operator whose coefficients are defined in a region Ω and for which the a_{ik} are of class $C^{(1,\lambda)}$ and the b_i and c are of class $C^{(0,\lambda)}$.

Moreover, let T be a domain of class $A^{(1,\lambda)}$, $f(x)$ be a function which is λ-Hölder continuous in T, and $\varphi(x)$ be a continuous function on ∂T. If we consider the DIRICHLET problem:

$$\mathfrak{M}\, u = f \quad \text{for} \quad x \in T - \partial T, \qquad u = \varphi \quad \text{for} \quad x \in \partial T,$$

we can seek to determine a solution which is of the form:

$$u(x) = -\int_T L(x,y)\, z(y)\, dy - 2\int_{\partial T} \mathfrak{D}_\eta L(x,\eta)\, \zeta(\eta)\, d_\eta \sigma, \qquad (18.1)$$

where $L(x,y)$ is a LEVI function satisfying the hypotheses of § 15 and endowed with the derivatives $\partial^3 L/\partial x_h\, \partial x_k\, \partial y_i$, $\mathfrak{D}$ is the operator defined in (7.2) with l coinciding with the conormal ν, β is an arbitrary function, for example, continuous on ∂T, and finally z and ζ are two functions to be determined, the first continuous in T and of class $C^{(0,\lambda)}$ in $T - \partial T$, the second continuous on ∂T. If we demand of these functions that they satisfy the equations of the problem, we discover, by dint of (13.7) and (15.3), the integral equations:

$$\left. \begin{aligned} z(x) - \int_T \mathfrak{M}_x L(x,y)\, z(y)\, dy - 2\int_{\partial T} \mathfrak{M}_x \mathfrak{D}_\eta L(x,\eta)\, \zeta(\eta)\, d_y\sigma &= f(x) \\ \zeta(\xi) - \int_T L(\xi,y)\, z(y)\, dy - 2\int_{\partial T} \mathfrak{D}_\eta L(\xi,\eta)\, \zeta(\eta)\, d_y\sigma &= \varphi(\xi)\,. \end{aligned} \right\} \qquad (18.2)$$

This system is formally of type (17.1), but the order of infinity for $x = \eta$ of the element K_{12} of the kernel matrix is in general too high for us to be able to apply FREDHOLM's theory to this system.

Still, if the system (18.2) admits a solution (z, ζ) and if z and ζ have the prescribed properties, the function u given by (18.1) furnishes a solution for our boundary value problem. If instead system (18.2) does not admit a solution, we can not deduce that the boundary value problem has no solution, because we can not say that a solution of this problem ought necessarily to have the form (18.1). Thus, so far, no single relation can be established in general between the uniqueness or non-uniqueness of solutions for the system (18.2) and for the boundary value problem under consideration. Namely, the system (18.2) could admit a single solution and the boundary value problem could admit, other than the solution (18.1), another of different type. Or the system (18.2) could admit more solutions and the corresponding $u(x)$'s furnished by (18.1) could be coincident among themselves. In order to clear up this inconvenient situation, we dispose of a broad arbitrariness in the choice of the function L; GIRAUD's method consists precisely of determining L in such a way that (18.1) places a one-to-one correspondence between the solutions of (18.2) and those of the boundary value problem being considered. With this particular choice of L, moreover, the alternative theorem turns out to be valid for the system (18.2), and from this an analogous theorem follows for our boundary value problem. All this will be amply considered, as we have already said, in subsequent paragraphs; for the moment, we wish to limit ourselves to a first application of the method, laying down some ideas which will also be useful in what follows.

Let T be a bounded domain of class $A^{(1,\lambda)}$ contained in Ω, and let T_1 be another domain having the same property, to which T is interior. Putting $\mathfrak{M}_x H(x,y) = K(x,y)$, we form the iterated kernels:

$$K^{(0)}(x,y) = K(x,y), \quad K^{(n)}(x,y) = \int_T K(x,t)\, K^{(n-1)}(t,y)\, dt .$$

Even under the sole hypothesis that the a_{ik} are, as the b_i and c, of class $C^{(0,\lambda)}$, we have $K = O(r^{\lambda - m})$ and thus by theorem 11, I:

$$K^{(n)} \in N^{([n+1]\lambda)}, \tag{18.3}$$

it being possible to put $\lambda = 1$ in (18.3) if, as we previously supposed, the a_{ik} are of class $C^{(1,\lambda)}$.

Under this same hypothesis we have, moreover, that the $K^{(n)}$ are differentiable with respect to the y_i, and for $x, y \in T$:

$$\frac{\partial K^{(n)}}{\partial y_i} \in N^{(n)} . \tag{18.4}$$

Once (18.4) is established for $n = 1$, the same formula is proved by recurrence for arbitrary n, taking account of theorems 12, VIII and 11, I from which it follows, among other things, that for $n > 1$:

$$\frac{\partial K^{(n)}}{\partial y_i} = \int_{T_1} K(x,t)\, \frac{\partial K^{(n-1)}(t,y)}{\partial y_i}\, dt .$$

The case $n = 1$ is instead treated separately, because, since $\partial K / \partial y_i = O(r^{-m})$, this presents the same difficulty which we encountered in the calculation of the second derivative of a domain potential. If, however, we observe that:

$$\frac{\partial K(x, y)}{\partial y_i} = \frac{\partial K(y, x)}{\partial x_i} + O(r^{\lambda-m}), \quad K(y, x) \in N^{(1, \mu)}, \quad (18.5)$$

with $\mu \leq 1$, procedures analogous to those which were valid to prove (13.6) permit us to demonstrate that:

$$\frac{\partial K^{(1)}}{\partial y_i} = \int\limits_{T_1} \left[K(x, t) \frac{\partial K(t, y)}{\partial y_i} - K(x, y) \frac{\partial K(y, t)}{\partial t_i} \right] dt$$

$$+ K(x, y) \int\limits_{T_1} K(y, t) X_i(t) \, d_t\sigma \qquad (18.6)$$

and from this, now taking account of (18.5) and theorem 11, I, we easily deduce the validity of (18.4) for $n = 1$.

We shall now consider for $p \geq 1$ the kernel:

$$H^{(p)}(x, y) = H(x, y) + \sum_{n=1}^{p} \int\limits_{T_1} H(x, t) K^{(n-1)}(t, y) \, dt. \qquad (18.7)$$

Under the sole hypothesis that the a_{ik}, b_i, and c are of class $C^{(0, \lambda)}$, we easily prove, bearing in mind (18.3) and theorems on the domain potential and theorem 11, I, that $H^{(p)}(x, y)$ is a LEVI function in $T_1 - \partial T_1$, and that we have:

$$\mathfrak{M}_x H^{(p)}(x, y) = K^{(p)}(x, y). \qquad (18.8)$$

If then the a_{ik} are of class $C^{(1, \lambda)}$, then $H^{(p)}$ is also differentiable with respect to the y and we have:

$$\frac{\partial H^{(p)}}{\partial y_i} = \frac{\partial H}{\partial y_i} + \int\limits_{T_1} \left[H(x, t) \frac{\partial K(t, y)}{\partial y_i} - H(x, y) \frac{\partial K(y, t)}{\partial t_i} \right] dt$$

$$+ H(x, y) \int\limits_{\partial T_1} K(y, t) X_i(t) \, d_t\sigma + \sum_{n=2}^{p} \int\limits_{T_1} H(x, t) \frac{\partial K^{(n-1)}(t, y)}{\partial y_i} \, dt. \qquad (18.9)$$

From (18.9), applying the theorems of § 13, we recognize easily that the $\partial H^{(p)}/\partial y_i$ are differentiable twice with respect to the x_k, and the following results:

$$\frac{\partial H^{(p)}}{\partial y_i} = \frac{\partial H}{\partial y_i} + O(r^{2-m}), \quad \frac{\partial^2 H^{(p)}}{\partial y_i \partial x_k} = \frac{\partial^2 H}{\partial y_i \partial x_k} + O(r^{1-m}), \quad (18.10)$$

$$\mathfrak{M}_x \frac{\partial H^{(p)}}{\partial y_i} = \frac{\partial K^{(p)}}{\partial y_i}. \qquad (18.11)$$

Because of this it is obvious that in (18.1) it is valid to assume $L = H^{(p)}$; with this, for the kernels K_{ij} of the system of integral equations (18.2), to which we thus return, the following bounds are valid:

$$K_{11} = O(r^{1-m}), \qquad K_{12} = O(r^{p-m}),$$
$$K_{21} = O(r^{2-m}), \qquad K_{22} = O(r^{\lambda+1-m}),$$

in which, as likewise in (18.10), $O(r^{2-m})$ is replaced by $O(\log 2R/r)$ for $m = 2$, denoting by the customary R the diameter of T.

For $p \geq 2$, the preceding bounds are sufficient to assure the validity of the alternative theorem for the system (18.2), the solvability of which stands proved if it is possible to show that the associated homogeneous system admits only the zero solution.

With this procedure we can, for example, prove the following theorem of existence *in the small*:

18, I. *If T is a domain of class $A^{(1,\lambda)}$ with a single boundary, and if T_ϱ is the domain we obtain by a contraction of coordinates in the ratio ϱ with respect to a center $x_0 \in T - \partial T$, then the* DIRICHLET *problem in the domain T_ϱ admits a solution for sufficiently small ϱ.*

Because this theorem is included in another more general one (21, III) which is proved later, we will refer to GIRAUD[1] for the details of the direct proof.

Whatever we do for the DIRICHLET problem is easily extended to NEUMANN's problem. For this problem, supposing that the a_{ik}, b_i and c are of class $C^{(0,\lambda)}$ in Ω, we substitute for equation (18.1), the following:

$$u(x) = -\int_T L(x,y)\, z(y)\, dy + 2\int_{\partial T} L(x,\eta)\, \zeta(\eta)\, d_\eta\sigma, \qquad (18.12)$$

with which, if the boundary condition is of the form $\mathfrak{P} u = \varphi$, we attain, by dint of (13.7) and (14.14), the system of integral equations:

$$\left.\begin{aligned} z(x) - \int_T \mathfrak{M}_x L(x,y)\, z(y)\, dy + 2\int_{\partial T} \mathfrak{M}_x L(x,\eta)\, \zeta(\eta)\, d_\eta\sigma = f(x) \\[2mm] \zeta(\xi) - \int_T \mathfrak{P}_\xi L(\xi,y)\, z(y)\, dy + 2\int_{\partial T} \mathfrak{P}_\xi L(\xi,\eta)\, \zeta(\eta)\, d_\eta\sigma = \varphi(\xi), \end{aligned}\right\} \qquad (18.13)$$

for which the same considerations already treated for the system (18.2) are valid. In particular, if we assume $L = H^{(p)}$, we carry through the study of the system (18.13) to establish an existence theorem in the small. More precisely:

---

[1] GIRAUD, G. [2] p. 170.

18, II. *Let T be a domain of class $A^{(1,\lambda)}$, possibly with several boundaries, and let T_ϱ be its contraction with ratio ϱ to a similar domain with respect to a center $x_0 \in T - \partial T$. If ϱ is sufficiently small, NEUMANN's problem with respect to the domain T_ϱ and with the boundary condition*

$$a \frac{du}{dv} + \frac{\beta}{\varrho} u = \varphi \quad (\beta > 0)$$

admits a solution.

This theorem is also contained in a more general one (22, II) which will be proved later. For a direct proof based on our assumptions here, we refer to GIRAUD[1].

19. Existence of fundamental solutions. Unique continuation property. The first proof of the existence of a fundamental solution of a general elliptic equation of the second order $\mathfrak{M} u = 0$ was given by J. HADAMARD in 1904 under the hypothesis that the coefficients of the equation are analytic[2]. Because of this these fundamental solutions also turn out to be analytic and therefore, STOKES's formula holding, we can prove that[3]:

19, I. *Every regular solution of a linear elliptic equation with analytic coefficients is analytic.*

From this theorem it follows that the solutions of an analytic equation possess the following properties:

Weak unique continuation property: Every regular solution in a region Ω vanishing in a neighborhood of a point of Ω vanishes identically in Ω.

Strong unique continuation property: Every regular solution in a region Ω which at a point of Ω has a zero of infinite order, is identically zero in Ω.

Let us also recall that, by the CAUCHY-KOWALEWSKY theorem, the CAUCHY problem for an analytic elliptic equation, which consists of seeking a regular solution in a neighborhood of a hypersurface Σ satisfying on Σ the initial conditions:

$$u = \varphi, \frac{du}{dv} = \psi,$$

admits one and only one solution if Σ, φ, and ψ are analytic.

---

[1] GIRAUD: [3] pp. 379—384 and [8] pp. 21—29. In theorem 18, II the function β is supposed constant with respect to $\overline{x x_0}$.

[2] For bibliographic references we refer to L. LICHTENSTEIN [5] p. 1296 and to J. HADAMARD [1].

[3] For more details see [44].

Now, abandoning the hypothesis of analyticity, we wish to examine either the problem of the existence of a fundamental solution or the possibility of extending the validity of the unique continuation property and the uniqueness theorem for CAUCHY's problem. In this case we do not usually think of an extension of the existence theorem for CAUCHY's problem inasmuch as this problem is not to be considered *well posed* for the elliptic equation in other than the analytic field. HADAMARD in fact observed that the problem does not in general admit a solution, and that when it does, this solution does not depend continuously on the data[1]. We note also that the validity of the unique continuation property carries with it as a consequence the validity of the uniqueness theorem for CAUCHY's problem. In fact, if u is a solution of the homogeneous CAUCHY problem which is regular in a neighborhood I of Σ, we can easily prove that, if Σ divides I into two regions A and B devoid of points in common, then the function v equal to u in A and vanishing in $B \cup \Sigma$ is of class $C^{(2)}$ in I and therefore is identically zero in I, thanks to the unique continuation property.

The study of these questions was initiated in 1933 in a note by T. CARLEMAN [1] in which he proved both the strong continuation property and the validity of the uniqueness theorem for CAUCHY's problem in the case of an equation in two variables with coefficients of class $C^{(2)}$. For an equation in $m > 2$ variables, the first direct proof of the uniqueness theorem for CAUCHY's problem was due to E.M. LANDIS [1], which he obtained in 1955, supposing $a_{ik} \in C^{(2)}$, $b_i \in C^{(1)}$, $c \in C^{(0)}$. Further proofs were given by M.M. LAVRENT'EV [1] and by L. HÖRMANDER [1]. The latter proved that:

19, II. *The uniqueness theorem for* CAUCHY'S *problem holds under the sole hypothesis that the* a_{ik} *are* LIPSCHITZ *continuous, the* b_i *and c are bounded, and* Σ *admits a representation of type* (1.1) *with* ζ *endowed with first derivatives which are* LIPSCHITZ *continuous.*

A final reduction of hypotheses in the case $m > 2$ appears very difficult indeed to obtain, because A. PLIŚ [4] has given an example, for every $\lambda < 1$, of an equation with λ-Hölder continuous coefficients, for which the uniqueness theorem fails. Other proofs obtained not directly, but through the proof of a unique continuation property, are due to C. MÜLLER [4], E. HEINZ [3], P. HARTMAN and A. WINTNER [3], P.D. LAX [2], YA.B. LOPATINSKII [6], R.N. PEDERSON [1], M.H. PROTTER [2], for the case of equations with constant a_{ik}; and to N. ARONSZAJN [4], H.O. CORDES [1], N. ARONSZAJN, A. KRZYWICKI and J. SZARSKI [1], for the general case. In the last of the works cited, and in a work by L. BERS and L. NIRENBERG [2], for the case $m = 2$, it is shown that:

---

[1] See J. HADAMARD [1], I.G. PETROWSKIĬ [3], C. PUCCI [3].

19, III. *The strong unique continuation property holds under the same hypotheses as theorem 19, II. In the case m = 2 this is valid even for an equation with measurable and bounded coefficients.*

In the work cited by P.D. LAX it is also placed in evidence that the existence of the unique continuation property is a necessary and sufficient condition for the validity of the so-called RUNGE *approximation property*. By this expression we mean the possibility of approximating the solutions of a homogeneous elliptic equation defined in an open Ω_1 by means of solutions of the same equation defined in an open $\Omega \supset \Omega_1$ such that $\Omega - \Omega_1$ does not have compact components. The interest of this question, which can be considered in relation to various types of approximations, goes beyond the theory of elliptic equations of the second order, because it can be posed and studied in the compass of the general theory of partial differential equations. On this argument see also B. MALGRANGE [1], H. BECKERT [6], F.E. BROWDER [13], and Chapters III and VIII of HÖRMANDER's volume [2], where other bibliographic references are also indicated.

Passing now to the study of the existence of the fundamental solution outside of the analytic field we shall see that this is also connected with the problem of establishing whether or not an elliptic equation, whatever $f \in C^{(0,\,\lambda)}(T)$ may be, admits at least one solution regular in $T - \partial T$. This study will be conducted by modifying a classical procedure due to E.E. LEVI [1], according to methods indicated by F. JOHN [4], JU.I. LJUBIČ [1, 2], C. MIRANDA [8 (Russian transl.), 10]. Let us begin with the proof of the following theorem:

19, IV. *If the coefficients of the operator $\mathfrak{M}$ are of class $C^{(0,\,\lambda)}$ in the region Ω and if $T \subset \Omega$ is a bounded domain of class $A^{(1,\,\lambda)}$, a necessary and sufficient condition that $\mathfrak{M}\,u = 0$ admit in T a fundamental solution is that, whatever $f \in C^{(0,\,\lambda)}$ may be, the equation $\mathfrak{M}\,u = f$ admit at least one regular solution in $T - \partial T$ of class $C^{(1,\,\lambda)}$ in T.*

That the condition is necessary is evident. Indeed, if a fundamental solution $G(x, y)$ exists, the equation $\mathfrak{M}\,u = f$, by dint of (13.7) and theorems 13, I and 13, II admits at least the solution:

$$u(x) = -\int_T G(x, y)\, f(y)\, dy\,.$$

Let us now prove that the condition is also sufficient. Given $L(x, y)$, a LEVI function, T, a bounded domain of class $A^{(1,\,\lambda)}$ contained in Ω, $Z(x, y)$ a kernel of class $N^{(\lambda,\,\mu)}$ in T, let us put:

$$G(x, y) = L(x, y) + \int_T L(x, t)\, Z(t, y)\, dt\,. \tag{19.1}$$

Recalling the properties of domain potentials (§ 13) as well as theorem 11, I, we see immediately that $G - L$ and its derivatives with respect to the x_i are continuous for $x, y \in T$ and $x \neq y$ and satisfy the following bounds uniformly in T:

$$G - L = O\left(r^{2 + \lambda - m}\right), \qquad \frac{\partial(G - L)}{\partial x_i} = O\left(r^{1 + \lambda - m}\right).$$

From theorem 18, II it now follows that the second derivatives of $G - L$ also exist and are continuous for $x, y \in T - \partial T$ and $x \neq y$, and we have:

$$\frac{\partial^2 (G - L)}{\partial x_i \, \partial x_k} = Z(x, y) \frac{\partial^2}{\partial x_i \, \partial x_k} \int_T L(x, t) \, dt + \int_T \frac{\partial^2 L(x, t)}{\partial x_i \, \partial x_k} \left[Z(t, y) - Z(x, y)\right] dt .$$

From this formula, bearing in mind that $Z \in N^{(\lambda, \mu)}$, we draw the conclusion that in every domain interior to T the following bound holds:

$$\frac{\partial^2 (G - L)}{\partial x_i \, \partial x_k} = O\left(r^{\lambda - m}\right) .$$

G is, therefore, also a LEVI function; and hence it is quite natural to ask whether it is not possible to determine Z in such a way that G is definitely a fundamental solution of the equation $\mathfrak{M} u = 0$. Now, in order that this happen, it is necessary and sufficient, by dint of (18.7), that the following hold:

$$Z(x, y) = \int_T \mathfrak{M}_x L(x, t) Z(t, y) \, dt + \mathfrak{M}_x L(x, y) \qquad (19.2)$$

or, as we recognize by comparing (19.2) with (17.8), that $Z(x, y)$ be the resolvent of the kernel $\mathfrak{M}_x L(x, y)$. Because $\mathfrak{M}_x L(x, y) = O\left(r^{\lambda - m}\right)$, FREDHOLM's theory holds for (19.2) and therefore, if unity is not an eigenvalue of the kernel $\mathfrak{M}_x L(x, y)$, then $Z(x, y)$ is uniquely determined and (theorem 17, IV) is also $O\left(r^{\lambda - m}\right)$.

From theorem 11, III it finally follows that $Z \in N^{(\lambda, \mu)}$ with $\mu < \lambda$ also, if $\mathfrak{M}_x L(x, y) \in N^{(\lambda, k)}$ with $\mu \leq k \leq \lambda$, hypotheses certainly satisfied if, for example, we assume $L = H$.

In the case in which the homogeneous equation associated to (19.2) admits $p \geq \alpha$ linearly independent solutions, the procedure fails. We should then seek $G(x, y)$ in the form:

$$G(x, y) = L(x, y) + \int_T L(x, t) Z(t, y) \, dt + \sum_{i=1}^{p} \alpha_i(x) \beta_i(y) , \qquad (19.3)$$

where the α_i and β_i are determined in such a way that the final known quantity $\mathfrak{M}_x L(x, y) + \Sigma \beta_i(y) \mathfrak{M} \alpha_i(x)$ of the equation which arises upon substitution in (19.2) is, as a function of x, orthogonal to p linearly independent solutions $V_1(x)$, ..., $V_p(x)$ of the homogeneous transposed equation of (19.2); because of this, the integral equation which is to be satisfied by $Z(x, y)$ is solvable, and (19.3) furnishes the required fundamental solution. Now it is evident that in order to apply this procedure it is necessary and sufficient to determine the $\alpha_i(x)$ in such a way that

$$\det \left\| \int\limits_T V_k \, \mathfrak{M} \, \alpha_i \, dx \right\| \neq 0 , \qquad (19.4)$$

and for this it suffices to select p functions $U_i(x)$ in such a way that $\int\limits_T V_k \, U_i \, dx = \delta_{ik}$ and to determine succesively the α_i in such a way that $\mathfrak{M} \, \alpha_i = U_i$. The theorem is thus completely proved.

19, V. *If, under the hypotheses of theorem* 19, IV, *the measure of the domain T is sufficiently small, we obtain a fundamental solution by putting the following in* (19.1):

$$Z(x, y) = \sum_{n=0}^{\infty} K^{(n)}(x, y) ,$$

the $K^{(n)}(x, y)$ being the iterates of the kernel $K(x, y) = \mathfrak{M}_x \, L(x, y)$.

This in fact follows from theorem 17, V as soon as we verify that:

$$\int\limits_T |K(x, y)| \, dy < 1 .$$

But this is obvious because from the hypothesis $K = O(r^{\lambda-m})$ it follows that the aforesaid intergral is $O([\text{mes } T]^{\lambda/m})$.

19, VI. *Under the hypotheses of theorem* 19, IV, *the integral equation:*

$$G(x, y) = L(x, y) + \int\limits_T G(x, t) \, \mathfrak{M}_t \, L(t, y) \, dt \qquad (19.5)$$

admits a unique solution $G(x, y)$, and this $G(x, y)$ is a fundamental solution of the equation $\mathfrak{M} \, u = 0$.

And indeed, if $Z(x, y)$ is the resolvent of the kernel $\mathfrak{M}_x \, L(x, y)$, then the solution of (19.5) is given exactly by (19.1), and therefore is a fundamental solution of the equation $\mathfrak{M} \, u = 0$.

Of considerable importance for what follows is the following theorem, also:

19, VII. *Let the coefficients a_{ik} be of class $C^{(1, \lambda)}$ in Ω and the b_i and c be of class $C^{(0, \lambda)}$; moreover, let $L(x, y)$ be a Levi function such that the $\partial L/\partial y_i$ exist, are differentiable twice with respect to the x_k, and satisfy the bounds (15.1) and the following:*

$$\mathfrak{M}_x \, \frac{\partial L(x, y)}{\partial y_i} = - \mathfrak{M}_y \, \frac{\partial L(y, x)}{\partial x_i} + O(r^{\lambda-m}). \qquad (19.6)$$

If equation (19.2) *admits only one solution, then the fundamental solution* $G(x, y)$ *given by* (19.1) *is differentiable with respect to the* y_i, *and the* $\partial G/\partial y_i$ *are differentiable twice with respect to the* x_k *and also satisfy the bounds* (15.1), *uniformly in every domain interior to* Ω.

Since the hypotheses on $L(x, y)$ are satisfied if $L = H$ let us for brevity limit ourselves here to proving the theorem in this particular case. This has the sole purpose of not obliging us to repeat for the function L some reasoning already unrolled for the function H in § 18.

As in § 18, we put $\mathfrak{M}_x H(x, y) = K(x, y)$ and we construct the iterates $K^{(n)}$ of K, assuming $T_1 = T$.

The differentiability properties of $K^{(n)}$ will consequently hold in every domain interior to T, and as much can be said for the $H^{(p)}$ also, constructed assuming $T_1 = T$. Now we see instantly that, if we put $Z = K + W$ in (19.2), we derive for W the equation:

$$W(x, y) = \int_T K(x, y)\, W(t, y)\, dt + K^{(1)}(x, y),$$

from which we draw the following (recalling that Z is the resolvent of K):

$$W(x, y) = K^{(1)}(x, y) + \int_T Z(x, t)\, K^{(1)}(t, y)\, dt.$$

Now it follows that:

$$G(x, y) = H^{(2)}(x, y) + \int_T \int_T H(x, s)\, Z(s, t)\, K^{(1)}(t, y)\, ds\, dt$$

and the assertion follows easily from the properties of the $\partial H^{(2)}/\partial y_i$ and $\partial K^{(1)}/\partial y_i$ established in § 18.

Let us now return to consider the problem of the existence of a fundamental solution, posing further regularity conditions for the coefficients of the operator. For this purpose we can enunciate the following theorem:

19, VIII. *If in a region* $\Omega: a_{ik} \in C^{(2,\lambda)}$, $b_i \in C^{(1,\lambda)}$, *and* $c \in C^{(0,\lambda)}$, *then the equation* $\mathfrak{M} u = 0$ *admits a fundamental solution in every bounded domain contained in* Ω.

Without loss of generality we can suppose that Ω coincides with R_m and prove the existence of a fundamental solution in every open sphere Γ_ϱ with center at the origin and radius ϱ. From theorem 19, V we already know that for ϱ sufficiently small a fundamental solution defined in $\overline{\Gamma_\varrho}$ exists. Let us denote by r the upper bound of the values of ϱ for which this is satisfied, and let us show that the supposition that r is finite leads to a contradiction. Fixing δ, we apply the procedure indicated in the proof of theorem 19, IV for constructing a fundamental solution in $\overline{\Gamma}_{r+\delta}$; since by hypothesis this construction is impossible, this means that condition (19.4) is never satisfied for any α_i, and this implies that there exist constants c_k such that for the function $V(x) = \sum\limits_{k=1}^{p} c_k V_k(x)$ we have for every $\alpha(x)$:

$$\int\limits_{\overline{\Gamma}_{r+\delta}} V(x) \, \mathfrak{M} \, \alpha(x) \, dx = 0 \, .$$

It follows from this, by a Lemma which we shall see proved later[1], that $V(x)$ is a regular solution of class $C^{(1)}$ in $\overline{\Gamma}_{r+\delta}$ of the equation $\mathfrak{N} v = 0$, which satisfies the boundary condition:

$$V = \frac{dV}{dv} = 0 \quad \text{for} \quad x \in \partial\Gamma_{r+\delta}.$$

By theorem 19, II $V(x)$ vanishes in a neighborhood of $\partial\Gamma_{r+\delta}$, whose size can be fixed independently of δ. Now it follows that for sufficiently small $\delta V(x)$ is of compact support in Γ_r; thus, if the support of $V(x)$ is contained in Γ_ϱ for $\varrho < r$, from STOKES's formula we draw the following, whatever $\alpha \in C^{(2)}$ may be:

$$\int\limits_{\overline{\Gamma}_\varrho} V(x) \, \mathfrak{M} \, \alpha(x) \, dx = 0 \, ,$$

and this proves that in $\overline{\Gamma}_\varrho$ the equation $\mathfrak{M} \, \alpha = V$ is devoid of solutions. But this is absurd, because in $\overline{\Gamma}_\varrho$ there exists a fundamental solution. The theorem is thus completely proved.

We shall finish this section by occupying ourselves with some applications of the results just extablished for which we shall always assume the hypotheses of theorem 19, IV.

We now observe: if T_1 is a domain of class $A^{(1,\lambda)}$ with sufficiently small measure, and if to begin with we have constructed for the domain T_1 and the function $L = H$ the fundamental solution $G(x,y)$ given by theorem 19, II, then as mes T_1 tends to zero we have $G = H(1 + o(1)) > 0$. Consequently the function:

$$\omega(x) = \int\limits_{T_1} G(x,y) \, dy$$

---

[1] We mean the lemma used in the proof of theorem 29, IV. In § 29, this lemma will be proved in the case $m = 2$, but the proof can be extended to the general case.

is also positive in T_1. From (13.7) it follows that $\mathfrak{M}\,\omega = -1$, and therefore, if we perform the change of unknown function $u = v\,\omega$ in the equation $\mathfrak{M}\,u = f$ and divide the same equation by ω, we find an equation for v in which the coefficient of v is equal to $-1/\omega$ and is hence negative.

From this we have immediately the following uniqueness theorem:

19, IX. *If the measure of T is sufficiently small, then* DIRICHLET's *problem for the equation $\mathfrak{M}\,u = f$ admits at most one solution.*

It suffices, in fact, to effect the same change of unknown function as that above, then to apply theorem 5, I to the transformed equation.

We also have:

19, X. *Fixing in any manner a domain T_1 of class $A^{(1,\,\lambda)}$ with sufficiently small measure, we can always determine a number $k > 0$ such that the second and third boundary value problems, for any domain T contained in T_1 and for the boundary condition $a^{(l)}\,du/dl + \beta\,u = \varphi$, admit at most one solution as long as $\beta\cos(n, l) > k$.*

In fact, effecting the change of unknown function $u = v\,\omega$, we work out for v an equation as above and the boundary condition:

$$a^{(l)}\,\frac{dv}{dl} + \left(a^{(l)}\,\frac{d\log\omega}{dl} + \beta\right) v = \frac{\varphi}{\omega}\,.$$

Our assertion follows therefore from theorem 5, II as soon as a k is chosen larger than $a\cos(n, \gamma)\,|\,d\log\omega/dl\,|$.

The same change of unknown function now permits us to prove the following theorem due to GIRAUD[1]:

19, XI. *Under the hypotheses of theorem 19, IV, every function $u(x)$ which is a regular solution in $\Omega - y$ of the equation $\mathfrak{M}\,u = 0$ and for which*

$$u(x) = o(H(x, y)) \qquad as \qquad x \to y, \tag{19.7}$$

is continuous together with its first and second derivatives even for $x = y$.

Indeed, by making use of the aforesaid change of unknown function, where it is necessary, we can always reduce the problem to the case where $c < 0$ in a neighborhood of y.

Now let $G(x, y)$ be the (positive) fundamental solution of the equation $\mathfrak{M}\,u = 0$ constructed in the hypersphere $\Gamma(x, \varrho)$ with the procedure of theorem 19, V; we can verify that if M is sufficiently large, then for every ϱ sufficiently small:

$$\mathfrak{P}\,G = a\,\frac{dG}{dv} + \frac{M}{\varrho}\,G > 0 \qquad for \qquad x \in \partial\Gamma.$$

---

[1] GIRAUD: [9], p. 281. See also GIRAUD [2] p. 194, GEVREY [3] and [10], ASCOLI [1].

By theorem 18, II we can take ϱ so small that the boundary value problem $\mathfrak{M} u_1 = 0$ in $\Gamma - \partial \Gamma$, $\mathfrak{P} u_1 = \mathfrak{P} u$ on $\partial \Gamma$, admits a solution, and from this we have for $\varepsilon > 0$ $[\varepsilon < 0]$:

$$\mathfrak{M} [u - u_1 + \varepsilon G] = 0, \; \underline{\mathfrak{P}} [u - u_1 + \varepsilon G] > 0 \, [< 0].$$

From these conditions it follows that $u - u_1 + \varepsilon G$ is devoid of negative minima [positive maxima] in $\Gamma - y$ and since from (19.7) it follows that in a neighborhood of y, $u - u_1 + \varepsilon G > 0$ $[< 0]$, then this same inequality must be valid in all of $\Gamma - y$. Now it follows that $| u - u_1 | < \varepsilon \, | G |$ and therefore, since ε is arbitrary, $u = u_1$. From this the theorem follows.

A theorem of the same type as that just proved holds even for the solutions of an elliptic equation which are singular not at an isolated point but rather on a $p < m$ dimensional manifold. Postponing this question to § 27, we shall limit ourselves to pointing out the following obvious corollary of theorem 19, XI.

19, XII. *Two fundamental solutions of the equation $\mathfrak{M} u = 0$, defined in the same domain T, differ at most by a function $u(x, y)$ which is a regular solution of $\mathfrak{M}_x u = 0$ for $x \in T - \partial T$.*

20. Principal fundamental solutions. The procedure laid out in the previous section for the construction of fundamental solutions could be utilized for an analogous construction of the GREEN's function, providing it exists, for the first or second boundary value problem. If, for example, we wish to construct GREEN's function for DIRICHLET's problem, we can apply the procedure which we found valid to prove Theorem 19, VIII, starting with a function $L(x, y)$ which is zero for $x \in \partial T$; if (19.2) admits a unique solution, then the $G(x, y)$ given by (19.1) is also zero for $x \in \partial T$ and therefore is the desired GREEN's function.

To the function $L(x, y)$ with which we start it is customary to give the name *quasi-Green's function*. The difficulty of this method, which was developed by M. GEVREY [5], consists of the construction of the $L(x, y)$, which is to be chosen so that the bounds (8.4) and (8.5) hold uniformly for $x, y \in T$. This difficulty did not occur previously since, when we do not impose some boundary condition on L, this function can be considered, as we have done, to be continuous with its derivatives, for $x \neq y$, in a region Ω to which T is interior. Later, GEVREY [8] recognized that this procedure is perfectly equivalent to another in which, rather than GREEN's function, we seek the solution of the boundary value problem directly. We shall postpone this research of GEVREY to § 24; we now wish to see, under the hypothesis that the region Ω of definition of the operator $\mathfrak{M}$ is the entire space R_m, whether or not it is possible to construct a fundamental solution also defined in the entire space and satisfying convenient conditions at infinity. This problem has considerable analogy with that of the construction of a GREEN's function, the condition at infinity taking the place of a boundary condition, but it is simpler in certain aspects, since we do not have to find the function $L(x, y)$ to start from; other complications, however, arise from the fact that the integrations are extended over all space. However, the results obtained by G. GIRAUD in the study of this problem are basic for the translation of the various boundary value problems into integral equations, and we shall now give a quick survey belonging mainly to the procedures followed by GIRAUD in the memoir [9], but taking account also of preceding results obtained by this author in [2], [4], and [8].

We shall therefore say that $G(x, y)$ is a *principal fundamental solution* of the equation $\mathfrak{M} u = 0$ if it is a fundamental solution defined in all of R_m and if there exist two positive constants a and R such that for $r > R$:

$$G = O(e^{-ar}), \quad \frac{\partial G}{\partial x_i} = O(e^{-ar}). \tag{20.1}$$

The theorem which we wish to demonstrate is the following:

20, I. *Let the functions a_{ik}, b_i, c be bounded in R_m and of class $C^{(0, \lambda)}$ there, with the a_{ik} being, moreover, λ-Hölder continuous in R_m. If A admits a positive lower bound and if $c \leq 0$, and moreover, outside a bounded domain we have $c < -g^2$ with $g > 0$, then the equation $\mathfrak{M} u = 0$ admits one and only one principal fundamental solution $G(x, y)$.*

That $G(x, y)$, if it exists, is unique, we prove immediately. Indeed, the difference between two principal fundamental solutions will by theorem 19, XII be a regular solution in all space of $\mathfrak{M} u = 0$, and therefore will turn out to be equal to zero, since it vanishes at infinity and is devoid of positive maxima or negative minima, by the hypothesis that $c \leq 0$.

Let us now indicate briefly the proof of the existence theorem, supposing, to be precise, that $m > 2$.

Let us consider in the first place the equation:

$$\Delta_2 u - k^2 u = 0. \tag{20.2}$$

It is known[1] that (20.2) admits as solutions functions of $r = \overline{xy}$ alone, of the form:

$$F(x, y) = k^{m-2} \varphi(k r), \tag{20.3}$$

---

[1] See WHITTAKER and WATSON, Modern Analysis, Chapters XVI and XVII, and H. G. GARNIR [2].

where $\varphi(t)$ is any solution of BESSEL's equation:

$$\frac{d^2\varphi}{dt^2} + \frac{m-1}{t}\frac{d\varphi}{dt} - \varphi = 0.$$

It is known also that this equation admits a solution, dependent on a single arbitrary constant, which is $O(t^{2-m})$ for $|t| < 1$ and $O(e^{-at})$ with $\alpha < 1$ for $t < 1$. We can therefore determine the aforesaid constant in such a way that $F(x, y)$ is a principal fundamental solution of (20.2).

Now, putting:

$$L(x, y) = \frac{k^{m-2}}{\sqrt{A(y)}} \varphi\left[k\left(\sum_{r,s=1}^{m} A_{rs}(y)(x_r - y_r)(x_s - y_s)\right)^{1/2}\right], \qquad (20.4)$$

we define a LEVI function for the operator $\mathfrak{M}$ which for r sufficiently large satisfies the bounds:

$$L, \frac{\partial L}{\partial x_i}, \frac{\partial^2 L}{\partial x_i \partial x_k} = O(e^{-ar}). \qquad (20.5)$$

This function is also a solution of the equation:

$$\sum_{i,j=1}^{m} a_{ij}(y)\frac{\partial^2 L}{\partial x_i \partial x_j} - k^2 L = 0.$$

If we now propose to determine a fundamental solution $G_k(x, y)$ of the equation $\mathfrak{M}u - k^2 u = 0$, which is defined in all space, we are led by analogy with theorem 19, III, to the integral equation:

$$G_k(x, y) = L(x, y) + \int_{R_m} G_k(x, t) K(t, y)\, dt, \qquad (20.6)$$

with:

$$K(x, y) = \mathfrak{M}_x L - k^2 L = \sum_{i,j=1}^{m} [a_{ij}(x) - a_{ij}(y)]\frac{\partial^2 L}{\partial x_i \partial x_j} + \sum_{i=1}^{m} b_i\frac{\partial L}{\partial x_i} + c L.$$

FREDHOLM's theory is not applicable to (20.6) because the integral is extended over all space. By virtue of (20.5), however, we have $K = O(e^{-ar})$ and this bound now permits the construction of the iterated kernels $K^{(n)}$; now it follows that, if we have:

$$\int_{R_m} |K(x, y)|\, dx < \varrho < 1, \qquad (20.7)$$

(20.6) then admits a solution given by:

$$G_k(x, y) = L(x, y) + \sum_{n=0}^{\infty} \int_{R_m} L(x, t) K^{(n)}(t, y) \, dt. \qquad (20.8)$$

Next, if in the integral on the left of (20.7) we make the change of variables $x_i = y_i + t_i/k$, we easily recognize that this integral is $O(k^{-\lambda})$ uniformly with respect to y. (20.7) is therefore verified for k sufficiently large, and G_k given by (20.8) is the desired fundamental solution.

Note nevertheless that this last conclusion is not as immediate as in the case of theorem 19, VI and results only after laborious verification made necessary by the fact that the integral is not extended over a bounded domain. We also verify that the G_k satisfy (20.1) and it stands proved (at least for sufficiently large k) that the equation $\mathfrak{M} u - k^2 u = 0$ admits a principal fundamental solution.

Now let us take up consideration of the equation $\mathfrak{M} u - (k - \Delta k)^2 u = 0$ with $\Delta k > 0$ and seek the principal fundamental solution $G_{k-\Delta k}$. Applying the usual method, but this time assuming:

$$L = G_k, \quad K = \mathfrak{M}_x L - (k - \Delta k)^2 L = \Delta k (2k - \Delta k) G_k,$$

we are led back to an integral equation of the type (20.6), for which condition (20.7) becomes:

$$\Delta k (2k - \Delta k) \int_{R_m} |G_k(x, y)| \, dx < \varrho < 1. \qquad (20.9)$$

This equation will be solvable for sufficiently small Δk and therefore the equation $\mathfrak{M} u - (k - \Delta k)^2 u = 0$ admits a principal fundamental solution. We shall arrive thus, after a finite number of operations of the type described, at the construction of the principal fundamental solution for the equation $\mathfrak{M} u = 0$, if it is possible to show that a number θ, independent of k, exists so that (20.9) is satisfied for $\Delta k < \theta$ whatever k may be. We are led to this if to the hypotheses listed in the statement we add one which says that $c < 0$. Indeed, in this case it is possible to show that there exists a positive function $\Gamma(x, y)$ majorizing all the G_k and such that the integral $\int_{R_m} \Gamma(t, y) \, dt$ is a bounded function of y.

The theorem is therefore proved under the hypothesis that $c < 0$. To prove it in the general case, we denote by $\chi(x)$ a function of class $C^{(0, \lambda)}$ in R_m, zero outside of a bounded domain T, and such that $c - \chi < 0$ in T. Because of what is premised, the equation $\mathfrak{M} u - \chi u = 0$ admits a principal fundamental solution $G^*(x, y)$ which we can assume as the function L in the construction of G. With this choice of the function L, (19.8) becomes:

$$G(x, y) = G^*(x, y) + \int_T G(x, t)\, \chi(t)\, G * (t, y)\, dt \qquad (20.10)$$

and this equation shows that the values assumed by G for $y \in \complement T$ are determined by those which G assumes for $y \in T$. We can thus limit ourselves to considering (20.10) in the bounded domain T, so that the FREDHOLM theory is applicable. The solvability of (20.10) will thus stand verified if we can show that the homogeneous transposed equation:

$$v(t) = \chi(t) \int_T G * (t, y)\, v(y)\, dy \qquad (20.11)$$

admits only the zero solution. But the integral on the right of (20.11) is a function, say $\omega(t)$, defined in all space, going to zero at infinity and a regular solution of the equation $\mathfrak{M}\,\omega = 0$, and if $c \leq 0$ it follows from theorem 5, VI that $\omega = 0$ and hence $v = 0$.

Our theorem is thus completely proved.

The method adopted in the last part of the proof can be used to construct the principal fundamental solution even when, holding fixed the hypothesis that $c < g^2$ outside a bounded domain, we drop the supposition that $c \leq 0$ throughout R_m. Indeed, choosing χ as previously, we must consider (20.10) anew, for which it is not now possible to prove the solvability. However, the following theorem holds, and we shall limit ourselves to stating it:

20, II. *Retaining all the hypotheses of theorem* 20, I *except* $c \leq 0$, *a necessary and sufficient condition that the equation* $\mathfrak{M}\,u = 0$ *admit a principal fundamental solution is that equation* (20.10) *admit a unique solution. The principal fundamental solution, if it exists, is unique and coincides with the solution* $G(x, y)$ *of* (20.10).

Also important is the following property of principal fundamental solutions:

20, III. *If, under the hypotheses of theorem* 20, II, *the equation* $\mathfrak{M}\,u = 0$ *admits a principal fundamental solution, then for every function* $v(x)$ *of class* $C^{(2,\lambda)}$ *in* R_m *and, with* $\mathfrak{M}\,v$, *approaching zero at infinity, the following formula holds:*

$$v(x) = -\int_{R_m} G(x, y)\, \mathfrak{M}\,v(y)\, dy. \qquad (20.12)$$

If $c \leq 0$, this theorem is proved immediately by observing that the difference $u(x)$ between the two sides of (20.12) is a regular solution of

$\mathfrak{M} u = 0$ approaching zero at infinity and thus is identically zero. In the general case we start from the formula:

$$v(x) = -\int_{R_m} G * (x, y) \left[\mathfrak{M} v(y) - \chi(y) v(y)\right] dy$$

and transforming it, taking account of (20.10), we arrive at (20.12).

We now have:

20, IV. *Theorem* 20, III *is valid even if we replace the hypothesis that v and $\mathfrak{M} v$ go to zero at infinity with another, that v, $\mathfrak{M} v$, and $\partial v/\partial x_i$ are $O(\varrho^p)$, ϱ being the distance of x from the origin and p being an arbitrary positive number.*

In fact, if we write STOKES's formula for a sphere with center x and radius R, the boundary integrals go to zero as $R \to \infty$, and from this (20.12) follows.

We have until now supposed that all the coefficients of the operator are of class $C^{(0,\lambda)}$; now we wish to see what consequences we can draw regarding G by strengthening the aforesaid hypothesis. We have now, by procedures analogous to those which were valid for proving theorem 19, VII:

20, V. *Under the same hypotheses as theorem* 20, I, *and if moreover the $\partial a_{ik}/\partial x_j$ exist and are bounded in R_m and λ-Hölder continuous in every bounded domain, then the $\partial G/\partial y_i$ exist, are endowed for $x \neq y$ with continuous first and second derivatives with respect to the x_i, and satisfy in addition the bounds* (15.1) *for $r < R$ and:*

$$\frac{\partial G}{\partial y_i} = O\left(e^{-ar}\right) \tag{20.13}$$

for $r > R$.

And now:

20, VI. *Retaining the hypotheses of theorem* 20, V *except $c \leq 0$, we suppose moreover that the functions e_i are endowed with first derivatives bounded in R_m and λ-Hölder continuous in every bounded domain. If the equation $\mathfrak{M} u = 0$ admits a principal fundamental solution $G(x, y)$ then the adjoint equation $\mathfrak{N} v = 0$ also admits a unique principal fundamental solution $\Gamma(x, y)$ and we have:*

$$\Gamma(x, y) = G(y, x). \tag{20.14}$$

Let us start by observing that if the adjoint equation admits a principal fundamental solution, (20.14) is proved by procedures analogous to those used for the proof of theorem 10, I. Thus it suffices to apply GREEN's formula to the functions $G(z, x)$ and $\Gamma(z, y)$ of the variable point z, assuming as domain of integration a hypersphere of radius R containing the two points x and y and deprived of two neighborhoods $I(x, \varrho)$ and $I(y, \varrho)$, afterwards causing $R \to \infty$ and $\varrho \to 0$. In the general case we choose $\chi > 0$ so that the two equations $\mathfrak{M} u - \chi u = 0$ and $\mathfrak{N} v - \chi v = 0$ both admit principal fundamental solutions; if this is $G^*(x, y)$ for the first equations, $G^*(y, x)$ will be it for the second. From (20.10) we now easily deduce that $G(x, y)$ is twice differentiable with respect to the y_i and satisfies the equation $\mathfrak{N}_y G(x, y) = 0$. Bearing in mind (20.13) also, this proves that $G(y, x)$ is a principal fundamental solution of the equation $\mathfrak{N} v = 0$. The uniqueness of Γ follows finally from (20.14) and the uniqueness of G (theorem 20, II).

Let us finish this section with the following final theorem:

20, VII. *Under the hypotheses of theorem* 20, V, *the following formula holds for every function* $u(x)$ *of class* $C^{(1)}$ *in* T *and endowed in* $T - \partial T$ *with second derivatives continuous in* $T - \partial T$ *and integrable in* T:

$$\theta \, u(x) = -\int_T G(x, y) \, \mathfrak{M} \, u(y) \, dy$$

$$+ \int_{\partial T} [G(x, y) \, \mathfrak{P} \, u(y) - u(y) \, \mathfrak{D}_y \, G(x, y)] \, d_y \, \sigma, \qquad (20.15)$$

with $\theta = 1$ *for* $x \in T - \partial T$ *and* $\theta = 0$ *for* $x \in \complement \, T$.

(20.15) is an immediate consequence of Green's and Stokes's formulas if the hypotheses of theorem 20, VI are satisfied, which allow us to use the adjoint operator.

In the general case we arrive at (20.15) by approximating the coefficients of $\mathfrak{M}$ with regular functions, writing the formula analogous to (20.15) except for using the approximating operator and its principal fundamental solution, then passing to the limit and bearing in mind that the principal fundamental solutions vary continuously with the variation of the coefficients of $\mathfrak{M}$[1].

21. Transformation of the Dirichlet problem into integral equations.
In this section we shall exhibit the treatment of Dirichlet's problem developed by G. Giraud in the memoirs [2], [8], and [9]. Of these memoirs, the first examines the Dirichlet problem which we posed in § 4, and the other two consider the same problem in a generalized sense, which we shall touch upon in the next chapter (§ 25). The procedure followed in the memoirs [8] and [9], especially those of the memoir

---

[1] See Giraud [8], p. 65, and [9], p. 306.

[8], preserve their validity even if we remain in the compass of the ordinary problem, yielding however, a greater economy of hypotheses than those attained in memoir [2]. In our exposition we shall moreover simplify others of the procedures of the memoirs [8] and [9], with these simplifications and clarifications made possible or necessary by the fact that the basis of our problem is ordinary and not generalized.

Therefore, let us consider the DIRICHLET problem:

$$\mathfrak{M} u = f \quad \text{for} \quad x \in T - \partial T, \qquad u = \varphi \quad \text{for} \quad x \in \partial T, \qquad (21.1)$$

supposing that f is continuous in T and of class $C^{(0,\,\lambda)}$ in $T - \partial T$, that φ is continuous on ∂T and finally that T is a bounded domain of class $A^{(1,\,\lambda)}$. For the time being, we place on the operator $\mathfrak{M}$ all the hypotheses necessary for the validity of theorem 20, V. Under these hypotheses, if we consider a double-layer potential constructed by assuming as LEVI function $L(x, y)$ the principal fundamental solution $G(x, y)$ of the equation $\mathfrak{M} u = 0$, then for this potential all the theorems of § 15 are valid, including, by dint of theorem 20, VII, also theorems 15, VI, VII, and VIII. Analogous observations hold regarding the domain potential constructed by putting $L = G$, and moreover, for the validity of the contingent theorems it would be sufficient that $\mathfrak{M}$ satisfy only the hypotheses of theorem 20, I. With this premised, let us see if a solution of problem (21.1) exists in the form:

$$u(x) = - \int_T G(x, y) \, f(y) \, dy - 2 \int_{\partial T} \mathfrak{D}_\eta(x, \eta) \, \zeta(\eta) \, d_\eta \sigma, \qquad (21.2)$$

the operator $\mathfrak{D}$ being defined by (7.2) by causing l and ν to coincide and assuming for β a λ-Hölder continuous function on ∂T and negative there, and so that moreover $\beta - b < 0$.

From (13.7) and (15.3) it follows that the function $u(x)$ given by (21.2) is a regular solution in $T - \partial T$ of the equation $\mathfrak{M} u = f$, and satisfies the boundary condition of our problem if and only if, for $\xi \in \partial T$:

$$\zeta(\xi) = 2 \int_{\partial T} \mathfrak{D}_\eta \, G(\xi, \eta) \, \zeta(\eta) \, d_\eta \sigma + \int_T G(\xi, y) \, f(y) \, dy + \varphi(\xi). \qquad (21.3)$$

(21.3) is an integral equation in the unknown function $\zeta(\xi)$ to which, since $\mathfrak{D}_\eta \, G(\xi, \eta) = O\big(\overline{\xi\eta}^{\lambda+1-m}\big)$, FREDHOLM's theory is applicable; the solvability of the DIRICHLET problem will thus stand proved if we can show that the homogeneous equation associated with (21.3) admits only the zero solution. Let us now suppose that ζ_0 is a solution of the homogeneous equation associated with (21.3) and that u_0 is the function obtained by substituting ζ_0 in place of ζ and zero in place of f in (21.2). This function is in T a solution of the homogeneous problem associated with problem (21.1) and is therefore identically zero by the uniqueness theorem 5, I. From this $(\mathfrak{P}\, u_0)^- = 0$ and also, therefore, $(\mathfrak{P}\, u_0)^+ = 0$ as follows from (15.7), which is applicable since by theorem 15, II $\zeta_0 \in C^{(0,\,\mu)}$. Now u_0 is a regular solution of $\mathfrak{M}\, u = 0$ in $\mathfrak{C}\, T$, and goes to zero at infinity; from theorem 5, VI it follows that u_0 is identically zero in $\mathfrak{C}\, T$ and therefore continuous in all space. Consequently (15.4) it follows as required that $\zeta_0 = 0$. Taking account of the uniqueness theorem 5, I we can therefore conclude:

21, I. *Let $\mathfrak{M}$ be an elliptic operator whose coefficients in a domain T of class $A^{(1,\,\lambda)}$ satisfy $a_{ik} \in C^{(1,\,\lambda)}$, b_i, $c \in C^{(0,\,\lambda)}$; and moreover let f be a continuous function in T of class $C^{(0,\,\lambda)}$ in $T - \partial T$ and φ be a function continuous on ∂T. If $c \leq 0$, then the* DIRICHLET *problem (21.1) admits one and only one solution given by (21.2) with ζ the solution of (21.3).*

Note that in the statement of this theorem we have not mentioned any hypotheses relating to the behavior of the coefficients of $\mathfrak{M}$ in $\mathfrak{C}\, T$. This is because, once the coefficients of $\mathfrak{M}$ are given in T, it is always possible to extend them into all R_m in such a way that they satisfy the hypotheses of theorem 20, V. Indeed, if Γ is a hypersphere to which T is interior, it will suffice to define the coefficients of $\mathfrak{M}$ in $\mathfrak{C}\, \Gamma$ by making them coincide with those of the equation $\Delta_2 u - u = 0$ outside Γ, then extend the definition into $\Gamma - T$ in such a way that they and the derivatives of the a_{ik} turn out to be λ-Hölder continuous. And, because T is of class $A^{(1,\,\lambda)}$, the possibility of this extension is ensured by theorem 16, VI. The operator $\mathfrak{M}$ thus defined in all space could fail to be elliptic at most in a set I interior to $\Gamma - T$; in order to make it so, it will suffice to add to the a_{ii} in $\Gamma - T$ some non-negative functions of class $C^{(1,\,\lambda)}$, zero with their derivatives on $\partial(\Gamma - T)$ and sufficiently large in I.

Let us now resume the discussion of problem (21.1), maintaining all the hypotheses of theorem 21, I, except without supposing that $c \leq 0$. Also, without loss of generality we can suppose that the operator $\mathfrak{M}$ is defined in all space and satisfies all the hypotheses of theorem 20, V, the only exception being $c \leq 0$. Now, let χ be a function of class $C^{(0,\,\lambda)}$, zero outside a bounded domain and such that $c - \chi \leq 0$ in R_m. Under these conditions the principal fundamental solution $G^*(x, y)$ of the equation $\mathfrak{M}\, u - \chi\, u = 0$ exists, and by theorem 20, I every solution of problem (21.1) can be written in the form:

$$u(x) = -\int_T G^*(x, y)\, z(y)\, dy - 2\int_{\partial T} \mathfrak{D}_\eta G^*(x, \eta)\, \zeta(\eta)\, d_\eta \sigma\,, \qquad (21.4)$$

with:

$$z = f - \chi\, u = \int_T \chi(x)\, G^*(x, y)\, z(y)\, dy + 2\int_{\partial T} \chi(x)\, \mathfrak{D}_\eta G^*(x, \eta)\, \zeta(\eta)\, d_\eta \sigma + f(x) \qquad (21.5)$$

and ζ a solution of the equation:

$$\zeta(\xi) = \int_T G^*(\xi, y) \, z(y) \, dy + 2 \int_{\partial T} \mathfrak{D}_\eta \, G^*(\xi, \eta) \, \zeta(\eta) \, d_\eta \sigma + \varphi(\xi) \, . \qquad (21.6)$$

On the contrary, it follows from (13.7) and (15.3) that, if z and ζ satisfy (21.5) and (21.6), then the function $u(x)$ given by (21.4) is a solution of problem (21.1). The system of integral equations (21.5), (21.6), together with (21.4) is therefore perfectly equivalent to our boundary value problem. On the other hand, this system, as we verify immediately, satisfies the hypotheses of theorem 17, III and therefore the alternative theorem holds for it. Two cases are therefore apparent. Either the system admits a unique solution for arbitrary f and φ and then the same occurs in problem (21.1), or else the associated homogeneous system admits p linearly independent solutions.

In this case, the transposed system:

$$\left.\begin{aligned}
v(x) &= \int_T \chi(y) \, G^*(y, x) \, v(y) \, dy + \int_{\partial T} G^*(\eta, x) \, \omega(\eta) \, d_\mu \sigma \\
\omega(\xi) &= 2 \int_T \chi(y) \, \mathfrak{D}_\xi \, G^*(y, \xi) \, v(y) \, dy + 2 \int_{\partial T} \mathfrak{D}_\xi \, G^*(\eta, \xi) \, \omega(\eta) \, d_\xi \sigma
\end{aligned}\right\} \qquad (21.7)$$

also admits p linearly independent solutions (v_i, ω_i) and the compatibility condition of the system (21.5), (21.6):

$$\int_T f(x) \, v_i(x) \, dx + \int_{\partial T} \varphi(\xi) \, \omega_i(\xi) \, d_\xi \sigma = 0 \qquad (21.8)$$

will also be one for problem (21.1). Moreover, the homogeneous DIRICHLET problem also admits p linearly independent solutions in this case. In fact, by reasoning quite similar to the proof of theorem 21, I we immediately prove that every solution (z_0, ζ_0) of the homogeneous system associated with the system (21.5), (21.6) is zero, and to this (21.4) causes a function $u(x)$ to correspond which vanishes identically in T. And this proves that the homogeneous DIRICHLET problem admits at least p linearly independent solutions, while it is quite obvious that it can not admit more than p. Summing up:

21, II. *Under the same hypotheses as theorem* 21, I, *except for* $c \leq 0$, *the following alternative holds for the* DIRICHLET *problem* (21.1). *Either the associated homogeneous problem admits only the zero solution, and then problem* (21.1) *admits for arbitrary* f *and* φ *one and only one solution given by* (21.4) *with* (z, ζ) *the solution of the system* (21.5), (21.6); *or else the homogeneous problem admits* p *linearly independent solutions* $u_1, u_2, \ldots,$

u_p, *and then the system* (21.7) *also admits* p *linearly independent solutions* (v_i, ω_i) *and problem* (21.1) *is solvable if and only if* (21.8) *is satisfied. If this condition is satisfied, problem* (21.1) *admits infinitely many solutions, and, if* $\bar{u}$ *is one of these, all the others are of the form* $\bar{u} + \Sigma\, c_i\, u_i$, *with the* c_i *constants.*

We note that from this theorem and the uniqueness theorem 19, V an obvious corollary follows:

21, III. *Under the hypotheses of theorem* 21, II *and if, moreover, the measure of* T *is sufficiently small, the* DIRICHLET *problem admits one and only one solution.*

Let us now posit the hypotheses of theorem 20, VI so that along with problem (21.1) it is valid to consider the adjoint problem:

$$\mathfrak{R}\, v = g \quad \text{for} \quad x \in T - \partial T, \quad v = \gamma \quad \text{for} \quad x \in \partial T. \tag{21.9}$$

Even the hypothesis that the coefficients of $\mathfrak{M}$ are defined in all of R_m is not particularly restrictive, since if these are defined only in T, we can always extend them to all space in such a way that the e_i satisfy the desired property of theorem 20, VI. It will suffice for this to extend the $a_{i\,k}$ and e_i first and then deduce the extension of the b_i.

After this premise, let us show that, if (v, ω) is a solution of the system (21.7), v is also a solution of the adjoint homogeneous problem. Meanwhile the right side of the first of (21.7) is meaningful even for $x \in \complement\, T$, through which $v(x)$ can be considered defined in all of space. From (14.14) and from the second of (21.7) if then follows easily that:

$$(\mathfrak{D}v)^- = \omega(\xi), \quad (\mathfrak{D}v)^+ = 0 \quad \text{for} \quad \xi \in \partial T. \tag{21.10}$$

Finally, from theorem 21, VI and from (13.7) we draw immediately:

$$\mathfrak{R}\, v = 0 \quad \text{for} \quad x \in T - \partial T, \quad \mathfrak{R}\, v - \chi\, v = 0 \quad \text{for} \quad x \in \complement\, T. \tag{21.11}$$

Now, if we choose χ so that $c^* - \chi < 0$ also, c^* being the coefficient of v in $\mathfrak{R}\, v$, then from the second of (21.11) and the second of (21.10) it follows by theorem 5, VI that $v = 0$ in $\complement\, T$. Then, from the continuity of v in all of R_m and from the first of (21.11), the assertion follows.

In what we have said is the implicit conclusion that:

21, IV. *Under the same hypotheses as theorem* 21, II *and if moreover* $e_i \in C^{(1,\,\lambda)}$, *then the homogeneous* DIRICHLET *problem and its adjoint have the same number (positive or zero) of linearly independent solutions. Therefore if problem* (21.1) *admits a unique solution for arbitrary known terms, then the same occurs for problem* (21.9). *If the homogeneous problem admits* p

linearly independent solutions, then the compatibility condition for the nonhomogeneous problem can be written in the form:

$$\int_T f(x)\, v_s(x)\, dx + \int_{\partial T} \varphi(\xi)\, a(\xi)\, \frac{dv_i}{d\nu}\, d_\xi\sigma = 0\,, \qquad (21.12)$$

the v_i being solutions of the adjoint homogeneous problem.

In fact, if the homogeneous problem admits p linearly independent solutions and its adjoint admits p', it follows from what we have premised that $p \le p'$. But since we can exchange the places of the two operators, we also have $p' \le p$ and therefore $p = p'$.

From the first of (21.10) and from $v = 0$ on ∂T we now draw the conclusion that $\omega = a\, dv/d\nu$ and therefore (21.8) assumes the form (21.12).

Note that these formulas coincide with conditions (7.11) which we have already recognized necessary for the compatibility of the DIRICHLET problem. Moreover, besides having also established the sufficiency of these conditions, now we have shown their necessity without supposing that u and the v_i are of class $C^{(1)}$ in T. This is important because, while by dint of theorems 13, I and 14, VII, the v_i are certainly of class $C^{(1,\lambda)}$, we can not say as much for u.

The observation that the solutions of the homogeneous problem are of class $C^{(1,\lambda)}$ now permits us to assert that under the hypotheses of theorem 21, IV, the uniqueness theorem 7, III is applicable even to solutions of class $C^{(0)}$ of the DIRICHLET problem. It follows from this that:

21, V. *If, under the hypotheses of theorem 21, IV, $c + c^* \le 0$ also, then the DIRICHLET problem and its adjoint admit one and only one solution.*

Let us complete our exposition now with research on the GREEN's function for the DIRICHLET problem; we shall see then how the results to which we are led will permit a significant extension of the existence theorems just proved.

Let us put forth the hypotheses of theorem 21, IV and, supposing $c \le 0$, let us remember first of all that, according to the first definition we posed in § 10, the GREEN's function $F(x, y)$ of problem (21.1) is a LEVI function characterized by the conditions:

$$\mathfrak{N}_y\, F(x, y) = 0 \quad \text{for} \quad x, y \in T - \partial T\,, \quad F(x, y) = 0 \quad \text{for} \quad y \in \partial T\,.$$

We can therefore put $F(x, y) = G(x, y) + g(x, y)$, g being a regular solution in $T - \partial T$ of the equation $\mathfrak{N}_x\, g(x, y) = 0$ satisfying the boundary condition $g(x, y) = -G(x, y)$ for $y \in \partial T$. g is therefore a solution of an adjoint DIRICHLET problem which certainly admits one and only one solution. This problem can be solved by methods analogous to those followed for the solution of problem (21.1) in the case $c \le 0$, and this leads to writing definitively:

$$F(x, y) = G(x, y) + 2 \int_{\partial T} \mathfrak{P}_\eta G(\eta, y) \, \omega(x, \eta) \, d_\eta \sigma, \qquad (21.13)$$

with $\omega(x, \eta)$ a solution, for every $x \in T - \partial T$, of the integral equation:

$$\omega(x, \xi) = 2 \int_{\partial T} \mathfrak{P}_\eta G(\eta, \xi) \, \omega(x, \eta) \, d_\eta \sigma + G(x, \xi). \qquad (21.14)$$

That this equation is solvable is proved immediately. In fact, if $\omega'(\xi)$ is a solution of the homogeneous transposed equation, then the function:

$$v(x) = \int_T G(x, \xi) \, \omega'(\xi) \, d_\xi \sigma$$

satisfies the equation $\mathfrak{M} v = 0$, $(\mathfrak{P} v)^+ = 0$ in $\complement \, T$, from which it follows that $v = 0$ in $\complement \, T$. From this, by continuity of v in all space, we draw $v = 0$ in T and therefore $(\mathfrak{P} v)^- = \omega' = 0$.

Now from theorems 15, II and 15, VII it easily follows that for every $x \in T - \partial T$, $\omega(x, \xi)$ is of class $C^{(1, \lambda)}$ on ∂T with respect to ξ, and hence from theorem 15, VIII we deduce that $F(x, y)$ is of class $C^{(1, \lambda)}$ in $T - x$ with respect to y. From what we established in § 10 it is valid to conclude that every solution of problem (21.1) of class $C^{(1)}$ in T is given by the formula:

$$u(x) = - \int_T F(x, y) \, f(y) \, dy - \int_{\partial T} \mathfrak{D}_\eta F(x, \eta) \, \varphi(\eta) \, d_\eta \sigma. \qquad (21.15)$$

However, we can verify directly that the function $u(x)$ given by (21.15) satisfies (21.1) in every case. Indeed, this follows from the properties of domain and double layer potentials once we have verified the following properties of $F(x, y)$:

1°) $F(x, y)$ satisfies the equation $\mathfrak{M}_x F = 0$ and is zero for $y \in T - \partial T$, $x \in \partial T$.

2°) F and its first and second derivatives with respect to the x_i are continuous with respect to y in $T - x$, for every fixed x. We have moreover that $F = O(r^{2-m})$ uniformly in T.

3°) For $x \in T$ and $\eta \in \partial T$ we have (uniformly):

$$\mathfrak{D}_\eta F(x, \eta) = 2 \, \mathfrak{D}_\eta G(x, \eta) + O(\overline{x\eta}^{\lambda+1-m}).$$

And of these properties, the first follows obviously from theorem 10, I, the second is an immediate consequence of (21.13) and (21.14), and only the third requires some analytic development, for which we refer to GIRAUD[1].

The fact that (21.13) and (21.14) are fully meaningful even if we drop the hypothesis that the functions e_i are of class $C^{(1,\lambda)}$ is now important. $F(x, y)$ can therefore be defined only under the hypotheses of theorem 21, I and retains properties 1°), 2°), 3°) as a direct proof[2] can verify regarding 1°), nothing being altered regarding 2°) and 3°). Finally, the results hold even if the hypothesis $c \leq 0$ is replaced by another which causes a theorem of uniqueness to hold for problem (21.1). Indeed, in this case we can prove[3], if χ is chosen so that $c - \chi < 0$ and if $F^*(x, y)$ is the GREEN's function for the DIRICHLET problem for the operator $\mathfrak{M} u - \chi u$, that the GREEN's function for problem (21.1) can be obtained by solving one or the other of the integral equations:

$$F(x, y) = F^*(x, y) + \int_T F^*(x, t)\, \chi(t)\, F(t, y)\, dt\,, \qquad (21.16)$$

$$F(x, y) = F^*(x, y) + \int_T F(x, t)\, \chi(t)\, F^*(t, y)\, dt\,. \qquad (21.17)$$

We can therefore conclude that:

21, VI. *If for problem* (21.1) *under the hypotheses of theorem* 21, II *a uniqueness theorem holds, then the* GREEN's *function* $F(x, v)$ *for the problem exists, and every solution of it is given by* (21.15).

We shall now add that when a uniqueness theorem does not hold for problem (21.1), we can still construct a function $F(x, y)$ so that a solution of problem (21.1), when it exists, is still given by (21.15). To this function $F(x, y)$ which on the other hand is not uniquely determined, we generally give the name *Green's function in the broad sense*. With a procedure due to L. LICHTENSTEIN[4] we can, for example, define $F(x, y)$ by imposing on it, other than the usual behavior for $x \to y$, that it satisfy the equations:

$$\mathfrak{M}_x F = \sum_{n=1}^{p} v_n(x)\, v_n(y) \quad \text{for} \quad x \in T - \partial T,$$

$$F(x, y) = 0 \quad \text{for} \quad x \in \partial T, \quad \int_T u_n(x)\, F(x, y)\, dx = 0\,,$$

where the u_n and v_n are respectively solutions vanishing on ∂T of the equation $\mathfrak{M} u = 0$ and its adjoint, forming two systems of orthonormal

[1] GIRAUD: [9] pp. 338—340.
[2] Loc. cit. [1].
[3] See G. GIRAUD [17], p. 90.
[4] LICHTENSTEIN, L.: [6], Chapter II, § 3.

solutions. With this definition of $F(x, y)$, we have, in fact, that the function u defined by (21.15) satisfies the equation:

$$\mathfrak{M} u = f(x) - \sum_{n=1}^{p} v_n(x) \left[\int_T f(y) \, v_n(y) \, dy + \int_{\partial T} \varphi(\eta) \, a(\eta) \, \frac{dv_n}{dv} \, d_\eta \sigma \right]$$

and therefore is a solution of the problem (21.1), if the conditions (21.12) are satisfied. This is also the unique solution of the problem which is orthogonal to all the u_n. For details regarding the proof of the actual existence of $F(x, y)$, we refer to the volume by L. LICHTENSTEIN cited, not to mention G. GIRAUD's memoir [20] in which various other ways of defining $F(x, y)$ are considered.

We shall finish this section by indicating how we can now enlarge the scope of the existence theorems just proved. For this purpose we observe that the procedure for constructing a GREEN's function based, under the hypothesis $c \leq 0$, on the equations (21.13) and 21.14), is always valid even if the a_{ik} are supposed only of class $C^{(0,\lambda)}$. In the light of these hypotheses, this ensures (theorem 20, I) the existence of G, as results from the fact that the procedure described is still applicable even if G possesses derivatives only with respect to the variables x_i. Thus likewise properties 1°) and 2°) are valid for $F(x, y)$, the impossibility of differentiating G with respect to the y_i having as a consequence only the absence of property 3°). However, from the properties still true of F it follows that the function:

$$u(x) = - \int_T F(x, y) \, f(y) \, dy$$

is a solution of problem (21.1) under the hypothesis $\varphi = 0$.

If now, always supposing $\varphi = 0$, we want to solve this problem without the hypothesis $c \leq 0$, we can construct the GREEN's function $F^*(x, y)$ for the operator $\mathfrak{M} u - \chi u$, with $c - \chi \leq 0$ as usual, and observe that the same problem is equivalent to the integral equation:

$$u(x) = \int_T \chi(y) \, F^*(x, y) \, u(y) \, dy - \int_T F^*(x, y) \, f(y) \, dy, \qquad (21.18)$$

to which FREDHOLM's theory is applicable.

From this we get the theorem:

21, VII. *If the coefficients of the operator $\mathfrak{M}$ are all of class $C^{(0,\lambda)}$ in the domain T of class $A^{(1,\lambda)}$, and if moreover f is continuous in T and of class $C^{(0,\lambda)}$ in $T - \partial T$, then the alternative theorem holds for the boundary value problem (21.1) with $\varphi = 0$. If $c \leq 0$ the solution exists and is unique.*

It is now natural to inquire whether a similar result does not hold even for $\varphi \neq 0$. This is tied up with the possibility of proving the differentiability of F^* with respect to the y_i, and GIRAUD [18] indicated a procedure for attaining this end. He was nevertheless limited to giving the essential outlines of the method without entering into the details of a discussion which, at least at first sight, seems somewhat laborious. Since, on the other hand, we can arrive at the desired result by an entirely different procedure, which will be examined in Chapter V (§ 36), we shall not insist further on the argument.

22. Transformation of Neumann's problem into integral equations. In this section we shall discuss the treatment of NEUMANN's problem developed by GIRAUD in the memoirs [4], [8], and [9]. This treatment is quite similar to that of DIRICHLET's problem, and this will allow a more concise exposition than that of the previous section.

Therefore let us consider NEUMANN's problem:

$$\mathfrak{M}\, u = f \quad \text{for} \quad x \in T - \partial T, \qquad \mathfrak{P}\, u = \varphi \quad \text{for} \quad x \in \partial T, \qquad (22.1)$$

and let us suppose that the domain T is of class $A^{(1,\,\lambda)}$, that the coefficients of $\mathfrak{M}$ are of class $C^{(0,\,\lambda)}$ in T, that the function β which occurs in the operator $\mathfrak{P}$ is continuous on ∂T, and finally, that f and φ satisfy the same hypotheses as in § 21.

Let us begin by considering the case in which $c \leq 0$, $\beta \geq 0$ without c and β being simultaneously identically zero. Also, without loss of generality we can allow the coefficients of $\mathfrak{M}$ to be defined in all of R_m and to satisfy the hypotheses of theorem 20, I. Denoting by $G(x, y)$ the principal fundamental solution of the equation $\mathfrak{M}\, u = 0$, we seek a solution of (22.1) of the form:

$$u(x) = - \int_T G(x, y)\, f(y)\, dy + 2 \int_{\partial T} G(x, \eta)\, \zeta(\eta)\, d_\eta \sigma, \qquad (22.2)$$

ζ being a continuous function on ∂T. (22.2) is already a solution of the equation $\mathfrak{M}\, u = f$ and in order that this satisfy the boundary condition, it is necessary and sufficient, by (14.14), that for $\xi \in \partial T$:

$$\zeta(\xi) = - 2 \int_{\partial T} \mathfrak{P}_\xi\, G(\xi, \eta)\, \zeta(\eta)\, d_\eta \sigma + \int_T \mathfrak{P}_\xi\, G(\xi, y)\, f(y)\, dy + \varphi(\xi). \qquad (22.3)$$

And for this integral equation the alternative theorem is valid at once, since its kernel is $O(\overline{\xi\eta}^{\lambda+1-m})$. As in the case of (21.14), only exchanging T and $\complement\, T$, we prove that every solution of the homogeneous equation associated with (22.3) vanishes, and from this it follows that this equation admits one and only one solution.

If we now wish to drop the hypothesis that $c \leq 0$, $\beta \geq 0$, let us consider two functions χ and ψ, the first of class $C^{(0,\lambda)}$ and zero outside a bounded domain, the second continuous on ∂T, so that $c - \chi < 0$, $\beta + \psi > 0$. Denoting by $G^*(x, y)$ the principal fundamental solution of the equation $\mathfrak{M}\, u - \chi\, u = 0$, we have that every solution of (22.1) is necessarily of the form:

$$u(x) = -\int_T G^*(x, y)\, z(y)\, dy + 2 \int_{\partial T} G^*(x, \eta)\, \zeta(\eta)\, d_\eta\sigma\,, \qquad (22.4)$$

with $z = f - \chi\, u$ and ζ determinable on the basis of the condition $\mathfrak{P}\, u + \psi\, u = \varphi + \psi\, u$. On the other hand, a necessary and sufficient condition that (22.4) be a solution of (22.1) is that z and ζ satisfy the system of integral equations:

$$\left.\begin{aligned} z(x) &= \int_T \chi(x)\, G^*(x, y)\, z(y)\, dy - 2 \int_{\partial T} \chi(x)\, G^*(x, \eta)\, \zeta(\eta)\, d_\eta\sigma + f(x)\,, \\[2mm] \zeta(\xi) &= \int_T \mathfrak{P}_\xi\, G^*(\xi, y)\, z(y)\, dy - 2 \int_{\partial T} \mathfrak{P}_\xi\, G^*(\xi, \eta)\, \zeta(\eta)\, d_\eta\sigma + \varphi(\xi)\,. \end{aligned}\right\} \qquad (22.5)$$

This system, with (22.4), is therefore equivalent to problem (22.1) and it is easy to prove that every solution of the homogeneous system associated with (22.5) is zero for which the corresponding $u(x)$ is identically zero. From this equivalence and from the fact that, if the hypotheses of theorem 17, II are satisfied by system (22.5), the alternative theorem holds for this, it follows that an analogous theorem is also valid for the problem (22.1). On the other hand, if the homogeneous problem associated with (22.1) admits p (and no more) linearly independent solutions, then this many (v_i, ω_i) are also solutions of the transposed homogeneous system of (22.5):

$$\left.\begin{aligned} v(x) &= \int_T \chi(y)\, G^*(y, x)\, v(y)\, dy + \int_{\partial T} \mathfrak{P}_\eta G^*(\eta, x)\, \omega(\eta)\, d_\eta\sigma\,, \\[2mm] \omega(\xi) &= -2 \int_T \chi(y)\, G^*(y, x)\, v(y)\, dy - 2 \int_{\partial T} \mathfrak{P}_\eta G^*(\eta, \xi)\, \omega(\eta)\, d_\eta\sigma\,, \end{aligned}\right\} \qquad (22.6)$$

and the compatibility conditions of system (22.5) and therefore of problem (22.1) are written:

$$\int_T f(x)\, v_i(x)\, dx - \int_{\partial T} \varphi(\xi)\, v_i(\xi)\, d_\xi\sigma = 0 \qquad (22.7)$$

given that, as we immediately recognize making use of (15.3), we have on ∂T that $\omega_i = -v_i$.

Summing up and recalling theorem 5, IV also, we can conclude:

22, I. *Let $\mathfrak{M}$ be an elliptic operator whose coefficients are all of class $C^{(0,\lambda)}$ in the domain T of class $A^{(1,\lambda)}$; moreover, let f be a function continuous in T and of class $C^{(0,\lambda)}$ in $T - \partial T$ and φ be a continuous function on ∂T. If the function β which occurs in the operator $\mathfrak{P}$ is also continuous, then the following alternative holds for the* NEUMANN *problem (22.1). Either the associated homogeneous problem admits only the zero solution, and then the problem (22.1) admits for arbitrary f and φ one and only one solution given by (22.4) with (z, ζ) solutions of the system (22.5). Or else the associated homogeneous problem admits p linearly independent solutions u_1, u_2, $\ldots$, u_p, and then the system (22.6) also admits p linearly independent solution (v_i, ω_i) and problem (22.1) is solvable if and only if (22.7) are satisfied. If these conditions are satisfied, then problem (22.1) admits infinitely many solutions, and if $\bar{u}$ is one of these, all the others are of the form $\bar{u} + \Sigma c_i u_i$. If $c = 0$, $\beta = 0$ we have $p = 1$, $u_1 = 1$. If $c \leq 0$, $\beta \geq 0$ without c and β being both identically zero, the existence and uniqueness theorem holds.*

Bearing in mind theorem 19, X, we also have:

22, II. *Under the hypotheses of the preceding theorem, and if, moreover, the measure of T is sufficiently small, we can determine a number k so that the* NEUMANN *problem for an arbitrary domain contained in T admits one and only one solution whenever $\beta > k$.*

We see how this result can be made precise when, if a_{ik}, $e_i \in C^{(1,\lambda)}$, it is valid to consider the adjoint problem:

$$\mathfrak{N} v = g \quad \text{for} \quad x \in T - \partial T, \qquad \mathfrak{Q} v = \gamma \quad \text{for} \quad x \in \partial T. \qquad (22.8)$$

In this case, supposing that the homogeneous problem associated to (22.1) admits p linearly independent solutions, then for the corresponding functions v_i:

$$\mathfrak{N} v_i - \chi v_i = 0 \quad \text{for} \quad x \in \mathfrak{C} \, T, \qquad v_i^+ = 0 \quad \text{on} \quad \partial T.$$

From this follows that the v_i are identically zero in $\mathfrak{C} \, T$, whence we have $(\mathfrak{Q} v_i^-)^+ = 0$, and therefore, by (15.7) applied after exchanging $\mathfrak{P}$ with $\mathfrak{Q}$, $(\mathfrak{Q} v_i)^- = 0$.

Because in $T - \partial T$ we have $\mathfrak{N} v_i = 0$ also, we deduce from this that the v_i are solutions of the adjoint homogeneous problem. From this, as in the case of DIRICHLET's problem, the theorem follows:

22, III. *Under the same hypotheses as theorem 22, I and if, moreover, a_{ik}, $e_i \in C^{(1,\lambda)}$, the homogeneous* NEUMANN *problem and its adjoint have the same number (positive or zero) of linearly independent solutions.*

Consequently, if problem (22.1) *admits a unique solution for arbitrary known terms, then the same occurs for problem* (22.8). *If the homogeneous problem admits p linearly independent solutions, then the compatibility condition of the non-homogeneous problem can be written in the form* (22.7), *with the v_i solutions of the adjoint homogeneous problem.*

As for the DIRICHLET problem we have found again, but under more general hypotheses, the compatibility conditions (22.7) already established in § 7 by means of GREEN's formula. Moreover, analogous considerations to those which we used in the statement of theorem 21, V permit us to prove that:

22, IV. *If, under the hypotheses of theorem* 22, III, *we also have $c + c^* \leq 0$, $2\beta - b \geq 0$, without c and β both being identically zero, then the* NEUMANN *problem and its adjoint admit one and only one solution.*

We can now add, referring to GIRAUD[1] for the proof, that, provided the uniqueness theorem holds for problem (22.1), the solution can be written in the form (10.5), $F(x, y)$ being the GREEN's function for the problem.

If $c \leq 0$, $\beta \geq 0$, we even have:

$$F(x, y) = G(x, y) + 2 \int_{\partial T} G(\eta, y)\, \omega(x, \eta)\, d_\eta \sigma ,$$

ω being a solution of the equation:

$$\omega(x, \xi) = -2 \int_{\partial T} \mathfrak{D}_\xi G(\eta, \xi)\, \omega(x, \eta)\, d_\eta \sigma + G(x, \xi) ,$$

while in the general case $F(x, y)$ is a solution of the integral equations:

$$F(x, y) = F^*(x, y) + \int_T F(x, t)\, \chi(t)\, F^*(t, y)\, dt - \int_{\partial T} F(x, t)\, \psi(t)\, F^*(t, y)\, d_t \sigma ,$$

$$F(x, y) = F^*(x, y) + \int_T F^*(x, t)\, \chi(t)\, F(t, y)\, dt - \int_{\partial T} F^*(x, t)\, \psi(t)\, F(t, y)\, d_t \sigma ,$$

where F^* is the GREEN's function for the problem $\mathfrak{M} u - \chi u = f$, $\mathfrak{P} u + \psi u = \varphi$, χ and ψ being chosen as previously.

Finally, in the case where a uniqueness theorem does not hold, the solution of the problem, when it exists, can always be written in the form (10.5) by introducing a convenient GREEN's function in the broad sense.

---

[1] GIRAUD: [9], p. 324, [17], p. 96, and [20].

23. Transformation of the oblique derivative problem into integral equations. If we seek to treat the third boundary value problem with the same procedure adopted in § 22, we encounter a grave difficulty. If $\mathfrak{P}$ is defined starting with an axis l not coinciding in general with ν, but such that $\cos(l, n) > 0$, we can then seek to solve the problem by assuming for the unknown u an expression of the type (22.2) or (22.4), and we are thus led anew to equation (22.3) or to the system (22.5), but the integral which operates on the unknown ζ in (22.3) and the analogous integral which occurs in the second of (22.5) must be taken as principal values (theorem 14, VI), which are meaningful only if ζ is λ-Hölder continuous.

And for such equations the FREDHOLM theory is no longer valid, in general, because in the iteration procedures theorems 11, II and 12, IV are no longer applicable. Still, GIRAUD proved first [15] for $m = 2$ and $m = 3$ and then [19] in the general case that for the particular equations which come up in the study of our problem, the alternative theorem is still valid, provided we assume that the functions β and φ are λ-Hölder continuous, as well as the direction cosines of the axis l, the domain T always being assumed of class $A^{(1,\lambda)}$.

We now wish to give in summary an idea of GIRAUD's procedure, limiting ourselves to the case (22.3) which we write introducing a parameter μ:

$$\zeta(\xi) = \mu \int_T^{\bullet} K(\xi, \eta)\, \zeta(\eta)\, d_\eta\sigma + \varphi(\xi) . \tag{23.1}$$

The most complicated part of GIRAUD's treatment consists in showing, for the particular kernel K of interest, that it is possible to construct a kernel $H(\xi, \eta, \mu)$ which is $O(\overline{\xi\eta}^{1-m})$, exactly like K, and such that, for arbitrary (real or complex) μ, we have:

$$M(\xi, \eta, \mu) = K(\xi, \eta) - H(\xi, \eta, \mu)$$

$$+ \mu \int_{\partial T}^{\bullet} H(\xi, \tau, \mu)\, K(\tau, \eta)\, d_\tau\sigma = O(\overline{\xi\eta}^{\lambda+1-m}) , \tag{23.2}$$

$$N(\xi, \eta, \mu) = K(\xi, \eta) - H(\xi, \eta, \mu)$$

$$+ \mu \int_{\partial T}^{\bullet} K(\xi, \tau)\, H(\tau, \eta, \mu)\, d_\tau\sigma = O(\overline{\xi\eta}^{\lambda+1-m}) ,$$

the two integrals which occur in these formulas being taken as principal values either with respect to the singularity at the point ξ or with respect to that at the point η. Moreover, there exists a function $\Phi(\xi, \mu)$, real and positive for μ real, such that for any λ-Hölder continuous function ζ:

$$\int\limits_{\partial T}^{*} H(\xi, \tau, \mu) \, d_\tau \sigma \int\limits_{\partial T}^{*} K(\tau, \eta) \, \zeta(\eta) \, d_\eta \sigma = -\zeta(\xi) \, \Phi(\xi, \mu)$$

$$+ \int\limits_{\partial T}^{*} \zeta(\eta) \, d_\eta \sigma \int\limits_{\partial T}^{*} H(\xi, \tau, \mu) \, K(\tau, \eta) \, d_\tau \sigma \, ,$$

the same formula holding with the same Φ if we switch the two kernels H and K.

From this it is easy to verify that every solution of (23.1) is also a solution of the equation:

$$[1 + \mu^2 \, \Phi(\xi, \eta)] \, \zeta(\xi) = \mu \int\limits_{\partial T} M(\xi, \eta, \mu) \, \zeta(\eta) \, d_\eta \sigma + f(\xi, \mu) \, , \quad (23.3)$$

with:

$$f(\xi, \mu) = \varphi(\xi) + \mu \int\limits_{\partial T}^{*} H(\xi, \eta, \mu) \, \varphi(\eta) \, d_\eta \sigma \, .$$

Now, by dint of the first of (23.2), the alternative theorem is valid for (23.3), provided that μ does not belong to the set M of values of η for which there exists at least one point ξ such that $1 + \mu^2 \, \Phi(\xi, \mu) = 0$. This set is distributed on two rays of the imaginary axis emanating from two points symmetric with respect to the origin. Moreover, since the functions H and Φ turn out to be holomorphic with respect to μ, the values of μ not belonging to M for which (23.3) is not solvable coincide with the zeros of a holomorphic function and therefore constitute at most a denumerable set M_1. If μ does not belong to $M \cup M_1$, then (23.1) admits, therefore, at most one solution.

On the other hand, putting:

$$\zeta(\xi) = \varrho(\xi) + \mu \int\limits_{\partial T}^{*} H(\xi, \eta, \mu) \, \varrho(\eta) \, d_\eta \sigma \, , \quad (23.4)$$

(23.3) becomes an equation for ϱ:

$$[1 + \mu^2 \, \Phi(\xi, \mu)] \, \varrho(\xi) = \mu \int\limits_{\partial T} N(\xi, \eta, \mu) \, \varrho(\eta) \, d_\eta \sigma + \varphi(\xi) \, , \quad (23.5)$$

every solution of which furnishes through (23.4) a solution of (23.1). But (23.5) is certainly solvable if μ does not belong to $M \cup M_2$, M_2 being another denumerable set; therefore, if μ does not belong to $M \cup M_1 \cup M_2$, then (23.1) admits one and only one solution. We see also that this solution is given by a formula of the type:

$$\zeta(\xi) = \frac{\varphi(\xi)}{1 + \mu^2\,\Phi(\xi,\mu)} + \mu \int\limits_{\partial T}^{*} \Gamma(\xi,\eta,\mu)\,\varphi(\eta)\,d_\eta\sigma\,,$$

Γ being a meromorphic function of μ in the complex plane outside of M, whose poles do not depend on ξ and η.

A deep study of the resolvent $\Gamma(\xi,\eta,\mu)$ shows finally that corresponding to every pole of Γ, the homogeneous equation associated with (23.1) and its transposed equation have the same (finite) number of linearly independent solutions and that if μ is a pole of Γ, a necessary and sufficient condition for the solvability of (23.1) is the orthogonality of φ to the solutions of the homogeneous transposed equation.

Extending in this way FREDHOLM's theory to integral equations corresponding to the third boundary value problem, the discussion of these equations is identical to that unrolled for the NEUMANN problem, and we attain the following theorems:

23, I. *Under the hypothesis that β, φ, and the direction cosines of l are* HÖLDER *continuous on ∂T, theorem* 22, I *holds even for the third boundary value problem.*

23, II. *Under the further hypothesis that the direction cosines of l are of class $C^{(1,\,\lambda)}$, even theorem* 22, III *holds for the third boundary value problem.*

For the proof of this last theorem, however, some complications arise due to the fact that we established (15.7) only in the case $l \equiv \nu$; but on this we shall not insist. Instead, we shall observe that the treatment developed is simplified considerably in the case $m = 2$, it now being valid in the study of (23.1) to assume $H = K$.

We seek now to examine briefly what difficulties are encountered in the study of the third boundary value problem when we drop the hypothesis $\cos(l, n) > 0$. We agree now to write the boundary condition in the form:

$$\mathfrak{L}\,u = a\cos(n,\nu)\,\frac{du}{dl} + \beta\,u = \varphi\,,$$

with which, if for example we seek a solution of the problem of the form (22.2), we find for ζ the integral equation:

$$\zeta(\xi)\cos(l,n) = -2\int\limits_{\partial T}\mathfrak{L}_\xi G(\xi,\eta)\,\zeta(\eta)\,d_\eta\sigma + \int\limits_{\partial T}\mathfrak{L}_\xi G(\xi,y)\,f(y)\,dy + \varphi(\xi)\,,$$
$$\tag{23.6}$$

which can no longer be said to be of second kind because the coefficient of ζ could vanish at some points or on some portion of ∂T. Still, if it is valid to apply to this equation the procedure indicated for the study of (23.1), then (23.6) is transformed into an equation in which the coefficient of ζ is $[\cos(l, n) + \mu^2 \Phi]$ and which therefore is of second kind if $\cos(l, n)$ and Φ are not simultaneously zero. We can therefore hope to be able to discuss equation (23.6) by this means. In the case $m = 2$ this study was developed by GIRAUD in the memoir [25] taking into account the fact that for an equation of the type (23.6) it can happen that the associated homogeneous equation and its transposed equation do not have the same number of linearly independent solutions; nevertheless, the fact remains that the compatibility condition is expressed by the orthogonality of the known terms to the solutions of the homogeneous transposed equation. This obviously carries the consequence that the existence of solutions does not imply uniqueness, and vice versa.[1]

However, GIRAUD, while he arrived at interesting results related to the theory of integral equations, dedicated only a few lines to the applications of these results to the study of the third boundary value problem. These were indeed limited to asserting that by this procedure it would be possible to extend to boundary value problems for elliptic equations in two variables the results obtained by A. LIÉNARD [1, 2] for the analogous problem relating to harmonic functions, with which the difference between the number of linearly independent solutions of the homogeneous problem and those in the compatibility conditions is put in relation either to the order of connectedness of the domain T, or to the variation which the angle formed by the axis $l(\xi)$ with a fixed direction undergoes as ξ varies on a detached portion of ∂T.

These observations by GIRAUD, though being quite suggestive and essentially well founded, were undoubtedly a bit premature at the time when his memoir was published. In fact, it has been only after later progress in the theory of singular integral equations that it has been possible to conduct to the end a thorough study of the third boundary value problem in the non-regular case and for $m = 2$. We now wish to give some indication of the results obtained in this direction, referring to S.G. MIHLIN's monographs [1, 10] and to the volumes by N.I. MUSHE-LIŠVILI [2] and F.D. GAHOV [1] for items concerning the general theory of singular integral equations and the first applications to boundary value problems for harmonic functions.

The first works related to a general elliptic operator are by B.V. HVEDELIDZE [1] and by I.N. VEKUA [3][2] and concern the equation:

$$\Delta_2 u + b_1 \frac{\partial u}{\partial x} + b_2 \frac{\partial u}{\partial y} + \varepsilon c u = f \tag{23.7}$$

[1] This, however, was already noted by F. NÖTHER [1] many years earlier.

[2] This volume by VEKUA sums up and completes a series of previous researches related to elliptic equations in two variables with analytic coefficients. For a survey of VEKUA's theory see P. HENRICI [1].

with the boundary condition:

$$\alpha \frac{\partial u}{\partial y} - \beta \frac{\partial u}{\partial y} + \varepsilon \gamma u = \varphi(s) , \quad \text{for} \quad x \in \partial T , \tag{23.8}$$

where ε is a real parameter, $\alpha^2 + \beta^2 = 1$, and b_1, b_2, c, f are entire functions of the complex variables x and y, real valued for real x and y. In this research, the transformation into integral equations is obtained by making use of an integral representation of the solutions of (23.7) given by VEKUA in preceding works; α and β are real and the domain T can even be multiply connected.

Then in 1953 VEKUA [6] proposed another procedure for the study of the problem[1], which is applicable under hypotheses of great generality on the coefficients of (23.7). Under these hypotheses, moreover, the supposition that the equation is of type (23.7) is no limitation, since the general equation in two variables can always be reduced to the canonical form (23.7) by a change of variables (see § 54). A detailed treatment of the problem with this procedure was given by VEKUA only some years later in Chapter IV (§ 8) of his volume [9] and is limited to the case of a simply connected domain T. Between VEKUA's first note and the definitive treatment is included G. FICHERA's[2] memoir [21] in which the problem is studied under the more general hypothesis that α and β are complex functions. FICHERA's treatment is interesting because it concerns an elliptic equation of general type and is unrolled parallel to a re-elaboration of the theory of singular integral equations, in which this theory, rather than depending on the theory of holomorphic functions as usual, is based on the preliminary study of the boundary value problem which is obtained by associating the condition (23.8) with $\gamma = 0$ to the equation (6.2) with $e_i = c = 0$.

We shall limit ourselves to recalling briefly VEKUA's result, supposing for simplicity that the functions $b_i, c, f, \alpha, \beta, \gamma, \varphi$ are HÖLDER continuous. We denote by $2n\pi$ the variation which $\arg[\alpha(s) + i\beta(s)]$ undergoes when s describes ∂T in the positive direction. If we leave out of consideration a sequence ε_k of values of ε (eigenvalues), two cases can be presented:

23, III. *If $n \geq 0$, the problem is solvable for arbitrary f and φ and the associated homogeneous problem admits $2n + 2$ linearly independent solutions.*

23, IV. *If $n < 0$, the problem is solvable if and only if f and φ satisfy $-(2n + q)$ compatibility conditions of the type:*

---

[1] This procedure is also applicable in the case $m = 2$ to the study of the DIRICHLET and NEUMANN problems. For this purpose see I. N. VEKUA [6, 9] and V. S. VINOGRADOV [1].

[2] For some partial results see also JOACHIM and IOHANNES NITSCHE [1].

$$\int_{\partial T} \varphi \, g_j \, ds + \int \int_T f \, h_j \, dx \, dy = 0,$$

where q can be either 1 or 2. The associated homogeneous equation admits $2 - q$ *linearly independent solutions.*

We also note that:

23, V. *In every case, denoting by* $\varkappa$ *the difference between the number of linearly independent solutions of the homogeneous problem and the number of compatibility conditions for the non-homogeneous problem, we have:*

$$\varkappa = 2\,n + 2\,. \tag{23.9}$$

To the number $\varkappa$ we give the name *index* of the problem[1]. We have also that:

23, VI. *If* ε *coincides with one of the eigenvalues* ε_k *then the value of* $\varkappa$ *is still given by* (23.9).

These results are no longer valid in the case $m \geq 3$. For example, B. PETTINEO [3, 4, 5] observed that if we seek to solve the problem by representing the solution by a single layer potential, we find an integral equation for which, if $\cos(l, n) = 0$ even at a single point of ∂T, the alternative theorem no longer holds, not even in the generalized form in which it is presented for singular integral equations on a curve. The deep reason for this lies in the fact revealed by L. HÖR-MANDER [2][2], that for $m \geq 3$, the problem, in the non-regular case, is to be considered in a certain sense not well-posed. Referring for some details to § 52, in which analogous considerations are unrolled in general for elliptic equations of arbitrary order, we shall recall that many recent results of notable interest were obtained by R. L. BORELLI [1] and L. HÖRMANDER [4] for harmonic functions[3] with a method which approaches those which, for DIRICHLET's and NEUMANN's problem, are illustrated in §§ 37, 38, 39. In particular, HÖRMANDER indicated some cases in which, if φ satisfies a finite number of compatibility conditions, the problem admits at least one solution; in this case, however, we can not prove that the homogeneous problem admits only a finite number of linearly independent solutions. In other cases, moreover, even the existence theorem is lacking. HÖRMANDER concludes with the affirmation, "It seems rather doubtful that one can ever get a problem of finite index."

Posing anew the hypothesis that $\cos(l, n) > 0$, we recall that G. GIRAUD [23], generalizing a procedure adopted by C. W. OSEEN [1] in

---

[1] VEKUA gives the name "index" to the number n, which by others is instead called the *characteristic*.

[2] See for example 2 of § 10.5 of [2].

[3] See also A. V. BICADZE [2].

the case of Δ_2, also gave another method for establishing the alternative theorem for the third boundary value problem by transforming it into a system of nonsingular integral equations. And of this method we want to give a brief idea, referring to E. MAGENES [7] for an exposition reviewed more extensively.

Supposing $\beta > 0$, let us begin by considering the third boundary value problem for the equation $\mathfrak{M} u - g^2 u = f$, with g constant. We have now that there exists a LEVI function $L_g(x, y)$ such that $\mathfrak{P}_\xi L_g(\xi, \eta)$ is $O(\overline{\xi \eta}^{\lambda + 1 - m})$ for $\xi, \eta \in \partial T$. Assuming for u the expression (18.12) with $L = L_g$ we arrive now at a system of integral equations for which the alternative theorem is valid. Moreover it comes out that, if g is sufficiently large, the number M (from theorem 17, V) is < 1 and from this the solvability of the system follows, and thus follows the existence theorem for the particular boundary problem considered. It follows from this that every possible solution of the third boundary value problem for the equation $\mathfrak{M} u = f$, even in the case when β is not > 0, can always be written in the form (18.12); from this we obtain the possibility of writing a system of integral equations perfectly equivalent to our boundary value problem.

We have in this way the following remarkable theorem:

23, VII. *Theorem* 23, I *holds even under the sole hypothesis that* β *and* φ *are continuous on* ∂T.

Naturally the most complicated part of the proof of this theorem is the construction of the function $L_g(x, y)$.

24. The method of the quasi-Green's functions. In this section we shall occupy ourselves with a method for transforming the various boundary value problems into integral equations, which was applied for the first time by E. E. LEVI [2] in the case $m = 2$. In the general case the method was later found again by M. GEVREY [8]. This is analogous to the so-called *parametrix* method adopted by D. HILBERT [1] in a particular case, the study of which was then resumed by L. SAUER [1] and B. PETTINEO [2].

Let us immediately say that this method has not as yet given results comparable to those obtained by G. GIRAUD by means of the systematic use of the principal fundamental solution. Still, this is interesting exactly because it does not require the prior construction of a fundamental solution. Moreover, this is applicable even to the study of boundary value problems for equations of order $2r$ with $r > 1$ (see § 52).

For equations of the second order, with which we wish to occupy ourselves exclusively, the method is not distinct from that of § 18, other than by the particular choice of the function $L(x, y)$. We shall display it, following LEVI, for the DIRICHLET problem with vanishing boundary data: $\varphi = 0$. Without making precise, for now, the hypotheses on the data of the problem, let us suppose that there exists a LEVI function $L(x, y)$ zero for $x \in \partial T$ and such that the domain potential:

$$u(x) = -\int_T L(x, y) \, z(y) \, dy \qquad (24.1)$$

has all the properties established in § 13, under more restrictive hypotheses, and from which we must now leave out of consideration that L is defined in a region Ω containing T. Such an $L(x, y)$ will be called a *quasi*-GREEN's *function*. If we ask that the function (24.1) satisfy the equation $\mathfrak{M} u = f$, we find for z the integral equation:

$$z(x) - \int_T K(x, y) \, z(y) \, dy = f(x) , \qquad (24.2)$$

with $K(x, y) = \mathfrak{M}_x L(x, y)$. We shall then suppose that the alternative theorem is valid for this equation and that the possible solutions of the associated homogeneous equation and of its transposed equation are of class $C^{(2)}$ in $T - \partial T$ and of class $C^{(1)}$ in T.

Now two cases can be given. Either (24.2) admits, for arbitrary $f(x)$, one and only one solution and then the DIRICHLET problem admits *at least* one solution. However, we can not affirm *a priori* the uniqueness of this solution, since we can not say that all the solutions of the DIRICHLET problem are of type (24.1). Or else the homogeneous equation associated to (24.2) admits p linearly independent solutions $\{z_k\}$ and as many $\{v_k\}$ are admitted by its transposed equation. In this case we substitute for the relation (24.1) the other:

$$u(x) = -\int_T \left[L(x, y) - \sum_{k=1}^{p} u_k(x) \, z_k(y) \right] z(y) \, dy , \qquad (24.3)$$

where the u_k are, at least for now, arbitrary functions zero on ∂T and of class $C^{(2)}$ in T. (24.2) then becomes:

$$z(x) - \int_T K(x, y) \, z(y) \, dy + \sum_{k=1}^{p} \mathfrak{M} u_k(x) \int_T z_k(y) \, z(y) \, dy = f(x), \qquad (24.4)$$

and the solvability of this equation for arbitrary $f(x)$ is certainly assured if the system of linear equations:

$$\sum_{k=1}^{p} c_k \int_T v_i(x) \, \mathfrak{M} u_k(x) \, dx = \int_T f(x) \, v_i(x) \, dx \qquad (24.5)$$

has a determinant different from zero. Indeed, under this hypothesis, denoting by $\{c_k\}$ a solution of (24.5), the integral equation:

$$z(x) - \int_T K(x, y)\, z(y)\, dy = f(x) - \sum_{k=1}^{p} c_k\, \mathfrak{M}\, u_k(x)$$

admits one and only one solution $z(x)$ satisfying the conditions:

$$\int_T z_k(x)\, z(x)\, dx = c_k,$$

and this function is obviously a solution of (24.4). If, therefore, it is possible to determine the u_k in such a way that the determinant of system (24.5) is non-zero, then the DIRICHLET problem admits at least one solution.

Let us suppose now that instead, for arbitrary u_k, the aforesaid determinant is always zero; then it is not difficult to recognize that we should get:

$$\int_T v_i(x)\, \mathfrak{M}\, u(x)\, dx = 0 \quad (i = 1, 2, \ldots, p)$$

for arbitrary $u(x)$ zero on ∂T and of class $C^{(2)}$ in T. The preceding formula can also be written as follows, by applying GREEN's formula:

$$\int_T u(x)\, \mathfrak{M}\, v_i(x)\, dx + \int_T a(x)\, v_i(x)\, \frac{du}{dv}\, d\sigma = 0$$

and from this it follows, by the arbitrariness of u, that the v_i are solutions of the adjoint homogeneous problem.

It thus stands established that, *if the uniqueness theorem holds for the adjoint problem, then the existence theorem holds for the given problem.*

We shall now consider the equation:

$$\mathfrak{M}^{(\lambda)}\, u = \sum_{i,k=1}^{m} \frac{\partial}{\partial x_i}\left(a_{ik}\, \frac{\partial u}{\partial x_k}\right) + \lambda \sum_{i=1}^{m} e_i\, \frac{\partial u}{\partial x_i} + \left(c + \frac{\lambda-1}{2}\sum_{i=1}^{m} \frac{\partial e_i}{\partial x_i}\right) u = f(x),$$

$$(24.6)$$

which reduces to $\mathfrak{M}\, u = f$ for $\lambda = 1$ and to $\mathfrak{N}\, u = f$ for $\lambda = -1$. If the uniqueness theorem holds for the adjoint problem, it is possible, in virtue of what has been premised, to determine $L(x, y)$ in such a way that (24.2) admits a unique solution. Let us now make use of the relation (24.1) to seek a solution of (24.6) vanishing in ∂T. Because the integral equation to which we are led is solvable for $\lambda = 1$, it will be solvable for arbitrary λ with the exception of, at most, a denumerable infinity of values of this parameter. And the boundary value problem considered relative to (24.6) will admit a solution of the form:

$$u(x, \lambda) = \frac{u_1(x, \lambda)}{D(\lambda)} ,$$

u_1 and D being entire functions of λ. It follows from this that if $D(-1) \neq 0$ then the DIRICHLET problem for the equation $\mathfrak{N} u = f$ admits a solution. Let us now show that we are led to this last conclusion even when $D(-1) = 0$. It suffices for this to prove that, if $\lambda = -1$ is a zero of order n for $D(\lambda)$, then u_1 also has a zero of the same order for $\lambda = -1$, by means of which $u(x, \lambda)$ is regular for $\lambda = -1$. And this is immediate. In fact, if $u_1(x, \lambda)$ should have a zero of order $k < n$ for $\lambda = -1$, putting $u_1 = u_2(x, \lambda) (\lambda + 1)^k$, we would have:

$$\mathfrak{M}^{(\lambda)} u_2(x, \lambda) = \frac{f(x) D(\lambda)}{(\lambda + 1)^k} ,$$

and therefore $\mathfrak{N} u_2(x, -1) = 0$, contrary to the hypothesis made that for the adjoint problem the uniqueness theorem holds. It thus stands proved that, *if the uniqueness theorem holds for the adjoint problem, then the existence theorem holds as well for the given problem as for the adjoint, and also, the solution of the given problem is unique.* This last affirmation follows from the fact already perceived in § 7 that when the given homogeneous problem admits nonzero solutions, the nonhomogeneous adjoint problem can not be solved for arbitrary known terms.

The validity of these results is dependent on the construction of a quasi-GREEN's function $L(x, y)$ satisfying all the hypotheses which we have admitted. This problem, which presents the difficulties indicated at the beginning of § 20, was solved by LEVI only in the case $m = 2$, and under the hypotheses that T is of class $A^{(3)}$, that the functions a_{ik} are of class $C^{(3)}$, the b_i of class $C^{(2)}$, c of class $C^{(1)}$. LEVI supposed moreover, that f is also of class $C^{(1)}$, and considered only solutions of the problem which were of class $C^{(2)}$ in T. For the study of NEUMANN's problem, we can follow a similar procedure by assuming as a quasi-GREEN's function a LEVI function which has vanishing conormal derivatives on ∂T.

GEVREY [8], reconsidering this research many years later without knowing of LEVI's memoir, then succeeded in constructing the quasi-GREEN's function for the first two boundary value problems in the case of arbitrary m and under the sole hypothesis that T is of class $A^{(1, \lambda)}$ and the coefficients of $\mathfrak{M}$ are of class $C^{(0, \lambda)}$. Then in the notes [11] [12], [13], and [14] he was occupied with the third boundary value problem. Under the general hypotheses considered by GEVREY, however, the method adopted by LEVI to deduce the alternative theorem for the boundary value problem from the existence of the quasi-GREEN's function fails to work, but GEVREY indicated no other procedures which could be adopted for this purpose[1]. Along this line of ideas, then, the only concrete results are those of LEVI, and these have only quite limited interest, because they occur as particular cases of theorems 21, IV and 22, III, in whose statements we meet a much greater economy of hypotheses.

---

[1] Indeed, GEVREY's memoir does not lack some considerations of this type, which however appear inadequate for this purpose. See, regarding GIRAUD's review, Zbl. Math. **11** (1935) 403—404.

CHAPTER IV

Generalized solutions of the boundary value problems

When we seek to extend the field of validity of the existence theorems proved in the preceding chapter by reducing the hypotheses there admitted for the data of the problem, we are conducted in quite a natural way to consider as solutions of a boundary value problem functions which satisfy only in part the properties specified in § 4, these properties being replaced by others which are less restrictive. The *generalized solutions* of the various boundary value problems, which we thus come to consider, can be defined in different ways, which differ from one another according as the major scope of the hypotheses which we want to follow relate to the coefficients of the equation, or else to the domain, or even to the boundary conditions. In this chapter we shall occupy ourselves with those existence theorems for generalized solutions which can be proved by having recourse to the theory of integral equations or by putting to service other procedures based ultimately on the HAHN-BANACH-ASCOLI theorem. We shall also take into consideration PERRON's method of super- and subfunctions and we shall give some idea of the methods of the minimum. Other types of generalized solutions, the study of which requires different procedures, will be considered instead in Chapter V (§§ 37, 38, 39).

25. Generalized elliptic operators. Let us decompose the quadratic form $\Sigma a_{ik}(x)\,\xi_i\,\xi_k$ into the sum of m squares of linear forms:

$$\sum_{i,k=1}^{m} a_{ik}\,\xi_i\,\xi_k = \sum_{r=1}^{m}\left(\sum_{s=1}^{m} g_{rs}\,\xi_s\right)^2,$$

and denoting by $y^{(r)}$ and $z^{(r)}$ the points with coordinates $\left(x_i + h\,g_{r_i}\right)$ and $\left(x_i - h\,g_{r_i}\right)$, respectively, let us put:

$$\mathfrak{D}\,u = \lim_{h \to 0}\sum_{r=1}^{m}\frac{u\left(y^{(r)}\right) + u\left(z^{(r)}\right) - 2\,u(x)}{h^2}. \qquad (25.1)$$

If $u(x)$ is of class $C^{(2)}$, expanding $u(y^{(r)})$ and $u(z^{(r)})$ by TAYLOR's formula up to terms of the second order, we recognize immediately that:

$$\mathfrak{D}\, u = \sum_{i,k=1}^{m} a_{ik} \frac{\partial^2 u}{\partial x_i \, \partial x_k}\,.$$

The operator $\mathfrak{D}$ can, however, have significance even for functions which are not endowed with second derivatives; a fundamental example is furnished by the following theorem, for the proof of which we refer to GIRAUD [9]:

25, I. *If the functions a_{ik} are of class $C^{(0,\,\lambda)}$ in Ω and if $z(x)$ is a continuous function in a domain T contained in Ω, then, denoting by the usual $L(x, y)$ a LEVI function, for the domain potential:*

$$u(x) = \int_T L(x, y)\, z(y)\, dy,$$

$\mathfrak{D}\, u$ *is meaningful at every point x interior to T, and we have:*

$$\mathfrak{D}\, u = -\, z(x) + \int_T \mathfrak{D}_x L(x, y)\, z(y)\, dy\,. \tag{25.2}$$

From this result arises the idea of studying elliptic equations obtained by defining the operator $\mathfrak{M}$ by means of the formula:

$$\mathfrak{M}\, u = \mathfrak{D}\, u + \sum_{i=1}^{m} b_i \frac{\partial x}{\partial x_i} + c u\,.$$

We can now say that u is a generalized solution of the equation:

$$\mathfrak{M}\, u = f \tag{25.3}$$

in a region Ω if u is endowed at every point of Ω with continuous first derivatives and with a continuous $\mathfrak{D}\, u$ und if it satisfies (25.3).

It can be proved that the maximum and minimum principles continue to hold for generalized solutions of (25.3), and consequently so do the uniqueness theorems enunciated in §§ 3 and 5 for regular solutions.

Moreover, if in the development of § 20, we suppose the coefficients b_i and c to be merely continuous and no longer HÖLDER continuous, we then can arrive at a proof of the existence of a principal fundamental solution, which is now no longer a regular, but a generalized, solution of $\mathfrak{M}\, u = 0$.

After this, while we limit ourselves to seeking generalized solutions of (25.3), a great part of what is said in §§ 21, 22, 23 concerning the various boundary value problems for (25.3) holds unaltered, solely under the hypotheses that the a_{ik} are λ-Hölder continuous and the b_i, c, f continuous.

The only real difficulty is in the fact that under the new hypotheses we can no longer affirm that the principal fundamental solutions are differentiable with respect to the y_i, which, while this does not alter the treatment of the second or third boundary value problem, does limit the study of DIRICHLET's problem to the case of zero boundary data, making possible only the application of the method based on the integral equation (21.18) and not the one based on (21.3), (21.5), and (21.6).

All the preceding was established by GIRAUD in the memoir [9] for the DIRICHLET and NEUMANN problems and in the memoirs [15] and [19] for the third boundary value problem.

It is useful to take up the fact that (25.1) is not the only way to generalize the operator $\sum a_{ik} \, \partial^2 u / \partial x_i \, \partial x_k$.

Various other definitions of $\mathfrak{D}u$, for which theorem I still holds, are found in the works [7, 8] by GEVREY, in which he even shows that it is valid to eliminate the hypothesis of HÖLDER continuity on the functions b_i, c, and f.

If instead we want to eliminate the hypothesis of HÖLDER continuity only on the functions c and f while preserving it on the b_i, supposing the a_{ik} to be of class $C^{(1, \lambda)}$, we can achieve this purpose with another procedure, one which was followed by GIRAUD in the memoir [8]. If we denote by χ a function such that $c - \chi$ is negative and HÖLDER continuous, and so that the equation $\mathfrak{M}u - \chi u = 0$ satisfies the other hypotheses of theorem 20, V, then a principal fundamental solution $G^*(x, y)$ exists for this equation. If T' is an arbitrary domain interior to T and if $x \in T' - \partial T'$, then (theorem 20, VII) STOKES's formula holds for every solution u of the equation $\mathfrak{M} u = f$:

$$u(x) = - \int_{T'} G^*(x, y) \, [f(y) - \chi(y) \, u(y)] \, dy$$

$$+ \int_{\partial T'} [G^*(x, y) \, \mathfrak{P} \, u(y) - u(x) \, \mathfrak{Q}_y \, G^*(x, y)] \, d_y \sigma . \qquad (25.4)$$

We can now assume as a generalized solution of (25.3) every function of class $C^{(1)}$ which satisfies (25.4) in every T' interior to T. Such a generalized solution in general does not possess second derivatives, but if these do exist and are continuous, then this generalized solution is also a regular solution of (25.3).

The study of the first and second boundary value problems can now be conducted with the same methods used in §§ 21 and 22, limiting ourselves, however, to the investigation of these generalized solutions. We are thus led to the same integral equations already considered in the previous sections, but the discussion of them is naturally more laborious. The result is that under the hypotheses recently detailed for the coefficients and for the known terms of (25.3), theorems 21, I, II, III and 22, I and II hold without alteration. Even theorems 21, IV and 22, III continue to hold, leaving out the hypothesis that the e_i are of class $C^{(1,\lambda)}$, but on the understanding that we consider as a generalized solution of the adjoint equation $\mathfrak{N} v = g$ every function $v(x)$, for which we have for $x \in T' - \partial T'$ and any arbitrary T' interior to T:

$$v(x) = - \int_{T'} G^*(y, x)\, [g(y) - \chi(y)\, v(y)]\, dy$$

$$+ \int_{\partial T'} [G^*(y, x)\, \mathfrak{D}\, v(y) - v(y)\, \mathfrak{P}_y\, G^*(y, x)]\, d_y\sigma\,.$$

26. Equations with singular coefficients and known terms.
In this section we wish to consider some of G. GIRAUD's research concerning the case in which the coefficients b_i, c and the known term f become singular on certain manifolds of dimension less than m.

We shall denote for this purpose by $\mathfrak{U}_p$ a manifold of $m - p$ dimensions of R_m enjoying the following properties:

1. $\mathfrak{U}_p$ can be covered by a finite number of manifolds A_{pk} each possessing a parametric representation of the type:

$$x_i = x_i\,(t_1, t_2, \ldots, t_{m-p})\,,$$

the x_i being finite and continuous functions together with their derivatives in a bounded domain D of R_{m-p} and moreover such that their jacobian matrix has rank $m - p$. For $p = 1$ the derivatives of the x_i are assumed λ-Hölder continuous in D.

2. Every $\mathfrak{U}_p$ constitutes a bounded set in R_m.
3. Every point of $\mathfrak{U}_p$ is interior to at least one of the manifolds A_{pk}.
4. $\mathfrak{U}_m$ consists of a finite number of points.

The manifolds thus defined are not necessarily either orientable or connected and can even admit multiple points; by virtue of property 3, these are, however, closed, that is, devoid of a boundary. In what follows we shall denote by $\mathfrak{V}_p$ an arbitrary portion, possibly endowed with a boundary, of $\mathfrak{U}_p$, and by $r_p(x)$ the distance from a generic point x of R_m to $\mathfrak{V}_p$.

Let us now consider an elliptic operator $\mathfrak{M}$ and a function f satisfying the following hypotheses:

a) The a_{ik} are bounded and uniformly λ-Hölder continuous in all space.

b) The b_i are continuous and $O(r_1^{\lambda-1})$ in $R_m - \mathfrak{V}_1$, and bounded outside a bounded domain containing $\mathfrak{V}_1$.

c) c is continuous and $O(r_1^{\lambda-1} + r_2^{\lambda-1})$ in $R_m - (\mathfrak{V}_1 \cup \mathfrak{V}_2)$, bounded and less than $-g^2$ outside of a bounded domain containing $\mathfrak{V}_1 \cup \mathfrak{V}_2$.

d) f is continuous and $O\left(\sum_{p=1}^{m} r_p^{\lambda-1}\right)$ in $T - \bigcup_{p=1}^{m} \mathfrak{V}_p$.

Under these hypotheses it is possible to develop for the equation $\mathfrak{M}\,u = f$ a theory analogous to that which holds in the absence of the manifolds $\mathfrak{B}_p$; in this case, however, we ought to understand by a *solution* of the equation a function u which enjoys the following properties:

α) u is continuous and $O\left(\sum\limits_{p=3}^{m} r_p^{\lambda+2-p}\right)$ in $T - \bigcup\limits_{p=3}^{m} \mathfrak{B}_p$.

β) The first derivatives of u are continuous in $T - \bigcup\limits_{p=2}^{m} \mathfrak{B}_p$.

γ) The equation is satisfied in a generalized sense[1] in $T - \bigcup\limits_{p=1}^{m} \mathfrak{B}_p$.

As for the boundary conditions for the various problems, these are posed in the following way:

In the case of DIRICHLET's problem the function φ is assumed to be continuous and $O\left(\sum\limits_{p=3}^{m} r_p^{\lambda+2-p}\right)$ on $\partial T - \bigcup\limits_{p=3}^{m} \mathfrak{B}_p$, while we require that the condition $u = \varphi$ is satisfied on $\partial T - \bigcup\limits_{p=3}^{m} \mathfrak{B}_p$. In the case of the second and third boundary value problems the function β is assumed to be continuous and $O\left(r_2^{\lambda-1}\right)$ on $T - \mathfrak{B}_2$ and φ to be continuous and $O\left(\sum\limits_{p=1}^{m} r_p^{\lambda+1-p}\right)$ on $\partial T - \bigcup\limits_{p=2}^{m} \mathfrak{B}_p$, while we require that the condition $\mathfrak{P}\,u = \varphi$ be satisfied only on $T - \bigcup\limits_{p=2}^{m} \mathfrak{B}_p$.

Having thus specified the different problems, we shall be satisfied with the customary method of transforming them into integral equations and establishing the alternative theorem.

When we want to consider the adjoint equation $\mathfrak{N}\,v = g$, the coefficients of $\mathfrak{N}$ are to satisfy the same hypotheses as those of $\mathfrak{M}$.

In every domain contained in T in which b_i, c, and f are λ-Hölder continuous, u will be a regular solution of the equation.

In the case $c \leq 0$ for DIRICHLET's problem, and $c \leq 0$, $\beta \geq 0$ with c and β not together zero for the second and third problem, the solution exists and is unique. We also have uniqueness for quite a vast class of functions if we substitute for hypothesis α) one which requires that u be continuous and $o\left(\log \dfrac{2R}{r_2} + \sum\limits_{p=3}^{m} r_p^{2-p}\right)$ for $\inf r_p \to 0$, denoting by R the diameter of T.

---

[1] In the sense of § 25.

This uniqueness is proved, however, by a procedure quite different from that adopted in § 5; we shall give an idea of it at least for the DIRICHLET problem.

Choosing $\chi \leq 0$, in such a way that $c - \chi$ is continuous and ≤ 0 in R_m and less than $-g^2$ outside of a bounded domain, one proves that the equation $\mathfrak{M}\,u - \chi\,u = 0$ admits a principal fundamental solution $G^*(x, y)$ which is positive and continuous in all of space for $x \neq y$. If we denote by $d\sigma^{(p)}$ the element of measure on $\mathfrak{U}_p$, the function:

$$w(x) = \sum_{p=2}^{m} \int_{\mathfrak{U}_p} G^*(x, y)\, d_y \sigma^{(p)} \tag{26.1}$$

turns out to be positive and continuous in $T - \bigcup\limits_{p=2}^{m} \mathfrak{U}_p$. Moreover, there exist two positive constants a and b such that:

$$a < \dfrac{w(x)}{\log \dfrac{2R}{r_2} + \sum\limits_{p=3}^{m} r_p^{2-p}} < b, \tag{26.2}$$

R being the diameter of T. Now if ε is a positive constant and u is a solution of the homogeneous DIRICHLET problem, $u + \varepsilon\,w$ will be positive on $\partial T - \bigcup\limits_{p=2}^{m} \mathfrak{U}_p$ and in a neighborhood of $\bigcup\limits_{p=2}^{m} \mathfrak{U}_p$, and being a solution of the equation $\mathfrak{M}(u + \varepsilon\,w) = \varepsilon\chi w \leq 0$, will be positive (theorem 3, II). But in an analogous way we see that $u - \varepsilon\,w < 0$, and this proves, since ε is arbitrary, that $u = 0$.

A detailed analysis of this research is beyond the scope of our exposition and for it we refer to GIRAUD's original works. These are the memoirs [12] and [17] for the first two boundary problems and the memoirs [19] and [21] for the third boundary value problem. Analogous results under slightly more general hypotheses are indicated in the notes [12], [16], and [24]. We recall also that G. PRODI [2] then proved the existence of the solution of DIRICHLET's problem, assuming that the a_{ik}, b_i, and c were of class $C^{(0)}$ in T and $C^{(0,\lambda)}$ in $T - \partial T$, and that $f(x) = O(r^{2-\mu})$ with $\mu > 0$, r being the distance of x from ∂T.[1]

Also for a general equation, but from a different point of view, we refer to a work of M. KRZYŽAŃSKI [3], in which, under the assumption that the points of discontinuity of f and φ constitute a closed set situated on the boundary of a region Ω, the DIRICHLET problem for $\overline{\Omega}$ is studied. The further hypotheses which render this study possible are inspired by those of some uniqueness theorems for the equation $\varDelta_2\,u - c\,u = 0$

---

[1] For this purpose see also C. MIRANDA [12].

established by M. PICONE [11, 12] and perfected by E. LEJA [1] and B. PINI [5, 6].

Let us now pass on to give notice of various results obtained in the study of particular equations but under hypotheses much less restrictive than those considered by GIRAUD. We shall consider in the first place the equation in two variables:

$$\Delta u + \frac{b}{y}\frac{\partial u}{\partial y} + \frac{c}{y^2} u = f \tag{26.3}$$

and pose for this DIRICHLET's problem in a domain T of the half-plane $y \geq 0$, whose boundary consists of a segment Σ of the x axis and of a curve Γ in the half-plane $y > 0$. The study of problems of this type for (26.3), begun between 1932 and 1934 by G. BOULIGAND [1, 2, 3] and J. CAPOULADE [1, 2, 3], was renewed after about twenty years by P. BROUSSE [1, 2, 3] and P. BROUSSE and H. PONCIN [1] for equations with constant coefficients and by M.B. KAPILEVIČ [1, 2] and G. TALENTI [3] for the case of variable coefficients. The new fact which becomes apparent in the study of DIRICHLET's problem is that in certain cases a solution of (26.3) is uniquely determined by the boundary condition assigned on Γ, it being instead impossible to assign the boundary values of u on Σ. For example, P. BROUSSE [4] proved that for the equation:

$$\Delta u - \frac{k}{y}\frac{\partial u}{\partial y} = 0, \tag{26.4}$$

with k constant, it is possible to assign the values of u on $\Sigma \cup \Gamma$ only if $k > -1$. Instead, if $k \leq -1$, we can assign u arbitrarily on Γ, while on Σ the only admissible boundary condition is $u = 0$. We note that for $k = -1$, (26.4) is the equation, already studied by BELTRAMI[1], which must be satisfied by harmonic functions of three variables endowed with axial symmetry. Consequently the study of (26.4) has taken the name of *generalized axially symmetric potential theory*. To this theory we can also assign the equation considered by G. TALENTI [3] which arises from the study of axially symmetrical solutions of an elliptic equation with variable coefficients. The study of (26.4) received considerable impulse from the initial research of A. WEINSTEIN. For more ample material on this subject we refer to the works [1, 3] of this author, in which his research and that of others relating to the general theory of solutions of (26.4) is summarized. Other notes can be found in an exposition by A. HUBER [2]. More particularly, regarding boundary value problems for (26.4), other than the work [4] of P. BROUSSE, we mention those of P.I. LIZORKIN [2] and of A.A. VASARIN and P.I. LIZORKIN [1].

---

[1] E. BELTRAMI, Opere Matematiche Hoepli Milano (1911) vol. 3, pp. 349—382.

7*

For the equation:

$$\Delta u + \frac{k}{x_m} \frac{\partial u}{\partial x_m} = 0,\tag{26.5}$$

which can be considered a generalization of (26.4) to m variables, see M.N. OOLEVSKIĬ [1], J.B. DIAZ and A. WEINSTEIN [1], P.I. LIZORKIN [1], A. HUBER [1, 2]. (26.5) is also included in a more general equation studied by M. SCHECHTER [9].

Another equation with singular coefficients which has been studied only recently is the equation:

$$\Delta u + \sum_{i=1}^{m} \frac{a_i}{r} \frac{\partial u}{\partial x_i} + \frac{b}{r^2} u = 0,\tag{26.6}$$

in which r denotes the distance of x from the origin. Research related to this equation is due to A. DŽUAREV [3] and L.G. MIHAĬLOV [6] for the case $m = 2$, and to MIHAĬLOV [4, 5] and G. NAZIROV [1] for the general case.

We also note that M. BRELOT [2, ..., 7, 11, 12] and P. HARTMAN and A. WINTNER [5] studied the equation $\Delta u + c\, u = 0$ with $c \le 0$ singular at one point. For an equation a bit more general see H. BREMER-KAMP [1].

Finally, we should bear in mind that many questions studied in this section can be considered from a different point of view, in the compass of the theory of elliptic-parabolic equations. A propos of this see § 56.

27. Local properties of the solutions of elliptic equations. In this section we propose to display some properties of the solutions of elliptic equations considered in a subset of their region Ω of definition. This subset can be a neighborhood of a point of Ω or of an isolated point, or of a component of $\partial \Omega$. Some examples of properties of this type have already been considered in the previous sections. Examples of this are the theorems of § 3 and theorems III, XI, and XII of § 19, and it is precisely for these theorems that we now wish to consider different extensions.

Recalling theorem 19, III which asserts the impossibility of nonzero solutions of a homogeneous equation admitting a zero of infinite order, it is natural to ask what might be the behavior of a solution of the equation $\mathfrak{M}\, u = 0$ in a neighborhood of a point x_0 which is a zero of order N of this solution. This question was studied first in the case $m = 2$ by L. BERS [2, 9, 15] and I.N. VEKUA [5]; P. HARTMAN and A. WINTNER [3], other than the case $m = 2$, considered the case $m = 3$ for an operator $\mathfrak{M}$ with the a_{ik} constants. In the general case we owe the following theorem to L. BERS [16]:

27, I. *Let $\mathfrak{M}$ be an elliptic operator with coefficients of class $C^{(0,\lambda)}$. If $u(x)$ is a solution of the equation $\mathfrak{M}u = 0$ in a region Ω, and for a point $x_0 \in \Omega$ we have $u(x) = O(|x - x_0|^N)$ with N a positive integer, then there exists a homogeneous polynomial $p(x)$ of degree N, a solution of the equation:*

$$\sum_{i,k=1}^{m} a_{ik}(x_0) \frac{\partial^2 p}{\partial x_i \, \partial x_k} = 0,$$

for which we have $u(x) = p(x - x_0) + O(|x - x_0|^{N+\lambda})$. This equation can be differentiated $n + 2$ times if the coefficients of $\mathfrak{M}$ are of class $C^{(n,\lambda)}$.

In the case $m = 2$ this theorem is valid even under considerably weaker hypotheses on $\mathfrak{M}$; regarding this see L. BERS [18] and also L. BERS and L. NIRENBERG [1]. Another result of great importance for its different applications and which can be considered a completion of the theorems of § 3 is the extension to the solutions of elliptic equations of the well-known theorem of HARNACK for harmonic functions. We shall state this theorem in the following form[1] due to J. SERRIN [1]:

27, II. *Let $\mathfrak{M}$ be an elliptic operator for which $c \leq 0$, $a_{ik} \in C^{(0,\lambda)}(T)$, b_i, $c \in L^{\infty}(T)$. Then there exists a constant K such that for every solution of the equation $\mathfrak{M}u = 0$ which is positive in $\Gamma(x_0, \varrho) \subset T$:*

$$u(x) \geq K\, u(x_0) \quad for \quad |x - x_0| \leq \frac{\varrho}{3}.$$

If $m = 2$, the theorem holds under the sole hypothesis for the a_{ik} that $\mathfrak{M}$ is uniformly elliptic in T.

Note that, at least till now, we have not succeeded in the case $m > 2$ in eliminating the hypothesis on HÖLDER continuity on the a_{ik}; this hypothesis can nevertheless be considerably weakened, as SERRIN himself has indicated.

Theorem 27, II is the basis of an interesting memoir by D. GILBARG and J. SERRIN [1] dedicated to the study of the behavior of a solution of the equation $\mathfrak{M}u = 0$ in a neighborhood of an isolated singular point y. For this purpose we have already proved theorem 19, XI, which is valid, however, only under the hypothesis that the coefficients of $\mathfrak{M}$ are of class $C^{(0,\lambda)}$. It is therefore interesting to see what can be said when this hypothesis is replaced by a less restrictive one. Along these lines we shall recall some of the results of GILBARG and SERRIN[2], limiting ourselves for simplicity to the case in which $c = 0$. We shall suppose, moreover, that the coefficients a_{ik} and b_i are bounded and that the operator $\mathfrak{M}$ is uniformly elliptic. Under these hypotheses the following

---

[1] For the case $m = 2$ see also L. BERS and L. NIRENBERG [2].
[2] See also M. GEVREY [3].

theorems hold, for the second of which it is useful also to consult C. Pucci [11].

27, III. *Under the hypotheses indicated, if* $m = 2$, *then every function* $u(x)$ *which is a regular solution in* $\Gamma(y, \varrho) - y$ *of the equation* $\mathfrak{M} u = 0$ *and which is bounded on one side admits a limit (possibly infinite) as* $x \to y$.

27, IV. *Let the coefficients* a_{ik} *be Dini-continuous[1] at the point* y. *For every non-constant function* $u(x)$ *which is a regular solution in* $\Gamma(y, \varrho) - y$ *of the equation* $\mathfrak{M} u = 0$ *and which for* $x \to y$ *satisfies (19.7), we have in* $\Gamma - (y \cup \partial \Gamma)$:

$$\min_{\partial \Gamma} u < u < \max_{\partial \Gamma} u. \tag{27.1}$$

For $m > 2$, *and if* $u = O(|x - y|^{2-m+\delta})$ *with* $\delta > 0$, (27.1) *holds even under the sole hypothesis that the* a_{ik} *are continuous at the point* y; *in this case* u *is convergent as* $x \to y$.

Also of interest is the following theorem which contains theorem 19, XII as a particular case and which in the case $m = 2$ has been generalized even more by GILBARG and SERRIN.

27, V. *Let the coefficients of the operator* $\mathfrak{M}$ *be of class* $C^{(0,\lambda)}$ *and let* $c \leq 0$. *Then every positive function* $u(x)$ *which is a regular solution in* $\Gamma(y, \varrho) - y$ *of the equation* $\mathfrak{M} u = 0$ *has the form:*

$$u = h G(x, y) + v(x),$$

where h *is a constant,* $G(x, y)$ *is a fundamental solution, and* $v(x)$ *is of class* $C^{(2)}$ *in* $\Gamma(y, \varrho)$.

The theorems now stated, which it is not possible to prove here, constitute only a part of the results of GILBARG and SERRIN. N. MEYERS and J. SERRIN [1] have added some complements to this. For a summary see D. GILBARG [2]. For other properties, among them the extension of HADAMARD's three circle theorem, see E. M. LANDIS [2, 3] and A. A. NOVRUZOV [1, 2].

Another interesting question is the study of the behavior of a solution of the equation $\mathfrak{M} u = 0$ which has at a point y a singularity of higher order than that of the fundamental solution. For equations with analytic coefficients, F. JOHN [2] has proved that every solution which has a polar singularity, which is, namely, $O(|x - y|^{2-m-k})$ with k a positive integer, is a linear combination of derivatives of a fundamental

---

[1] $f(x)$ is Dini-continuous at the point y if $f(x) - f(y) = O(g(|x - y|))$ with $g(r) \, r^{-1} \in L^1([0, a])$.

solution. The case of essential singularities was recently studied by M. WACHMAN [1]. An extension of JOHN's results to the non-analytic case was given by M. MARCUS [1]. Outside the analytic field, quite complete results were obtained in the case $m = 2$ also by L. BERS [2, 9, 14, 18, 19] and by L. BERS and L. NIRENBERG [1].

Let us now pass to the study of removable singularities of the solutions of the equation $\mathfrak{M} \, u = 0$, proving the following important extension of theorem 19, XI:

27, VI. *Let Ω be a region of diameter R which contains all the manifolds $\mathfrak{U}_p$ of the preceding section, and let $\mathfrak{M}$ be an elliptic operator with coefficients of class $C^{(0,\lambda)}$ in Ω. Then every function $u(x)$ which is a regular solution in $\Omega - \bigcup\limits_{p=2}^{m} \mathfrak{U}_p$ of the equation $\mathfrak{M} \, u = 0$ and for which:*

$$u(x) = o \left(\log \frac{2\,R}{r_2} + \sum_{p=3}^{m} r_p^{2-p} \right) \tag{27.2}$$

is continuous with its first and second derivatives throughout Ω.

As in the proof of theorem 19, XI, without loss of generality we can assume that $c < 0$ in a neighborhood I of $\bigcup \mathfrak{U}_p$. If the measure of I is sufficiently small, it is possible by theorem 22, II to determine a $k > 0$ such that however we may fix a domain $T \subset I$ of class $A^{(1,\lambda)}$ and a constant $\beta > k$, NEUMANN's problem for the domain T for the boundary operator $\mathfrak{P} = a \, d/d\nu + \beta$ admits a unique solution. Having chosen T in such a way that $\bigcup \mathfrak{U}_p$ is interior to T, we denote by $\bar{u}$ the solution of the problem:

$$\mathfrak{M} \, \bar{u} = 0 \quad \text{for} \quad x \in T - \partial T, \qquad \mathfrak{P} \, \bar{u} = \mathfrak{P} \, u \quad \text{for} \quad x \in \partial T.$$

As has already been observed a propos of the proof of theorem 19, IX, there certainly exists a fundamental solution $G(x, y)$ of the equation $\mathfrak{M} \, u = 0$ which is positive in T. Proceeding from this G, we construct the function w given by (26.1) and choose β so large that $\mathfrak{P} \, w > 0$. For $\varepsilon > 0$ we then have

$$\mathfrak{M} \, (u - \bar{u} + \varepsilon \, w) = 0 \quad \text{for} \quad x \in T - (\partial T \cup \bigcup \mathfrak{U}_p),$$

$$\mathfrak{P} \, (u - \bar{u} + \varepsilon \, w) > 0 \quad \text{for} \quad x \in \partial T.$$

These relations, by dint of theorems 3, II and 3, IV, assure us that $u - \bar{u} + \varepsilon \, w$ is devoid of negative minima in $T - \bigcup \mathfrak{U}_p$ and because this function is, by (26.2) and (27.1), positive in a neighborhood of $\bigcup \mathfrak{U}_p$, we can conclude that it is non-negative in $T - \bigcup \mathfrak{U}_p$. But analogously we prove that for $\varepsilon < 0$, $u - \bar{u} + \varepsilon \, w \leq 0$, so that we have as a result

$|u - \overline{u}| \leq |\varepsilon| w$ and from this, since ε is arbitrary, it follows that $u = \overline{u}$, which obviously proves the theorem.

For other theorems of this type for the equation $\Delta_2 u + c u = 0$, see the works of M. BRELOT [1, 2, 3, 4, 10], M. PICONE [11, 12], E. LEJA [1], B. PINI [5, 6]; for equations of more general type, those of G. ASCOLI [2] and M. KRZYŻAŃSKI [3]. Recently J. SERRIN [6, I] has proved other results similar to theorem 27, VI. For example we have

27, VII. *If a_{ik}, b_i, $c \in C^{(0, \lambda)}$ and if Ω contains a single manifold $\mathfrak{U}_p$ with $p > 2$, then every function $u(x)$, which is a regular solution in $\Omega - \mathfrak{V}_p$ of the equation $\mathfrak{M} u = 0$ for which $u \in L^q(\Omega)$ with $q = p/(p - 2)$, is continuous with its first and second derivatives in Ω. The result continues to hold under the sole hypothesis that $\mathfrak{V}_p$ is a compact set in Ω having finite $(m - p)$ dimensional HAUSDORFF measure.*[1]

28. Generalized solutions according to Wiener of Dirichlet's problem.

Let $\mathfrak{M}$ be an elliptic operator defined in all space and let Ω be a bounded region. It is well known, for harmonic functions, that DIRICHLET's problem:

$$\mathfrak{M} u = 0 \quad \text{for} \quad x \in \Omega, \qquad u = \varphi \quad \text{for} \quad x \in \partial\Omega, \qquad (28.1)$$

can not admit a solution. As in the case of harmonic functions, we can, however, always associate to every function φ continuous on $\partial\Omega$ a function $u_\varphi(x)$ satisfying for $x \in \Omega$ the equation $\mathfrak{M} u = 0$, which coincides with the solution of the aforesaid DIRICHLET problem whenever this solution exists. The function u_φ is called a *generalized solution according to* WIENER of the DIRICHLET problem (28.1). A point x_0 is then called *regular with respect to the operator* $\mathfrak{M}$ if for every φ we have $\lim_{x \to x_0} u_\varphi(x) = \varphi(x_0)$, *irregular* in the contrary case.

The study of these generalized solutions was initiated, in particular cases, by M. BRELOT [1, 8, 9, 10]. W. PÜSCHEL [1] considered the case of a self-adjoint equation in three variables, and G. TAUTZ [1] that of the equation in two variables:

$$\Delta_2 u + a p_1 + b p_2 + c u = 0. \qquad (28.2)$$

TAUTZ also then extended his results, considering instead of solutions of (28.2) those functions of class $C^{(1)}$ in Ω which for every domain $T \subset \Omega$, simply connected and of class $A^{(2)}$, satisfy the equation:

$$\int_{\partial T} \frac{du}{dn}\, ds + \int_T \frac{\partial u}{\partial x_1}\, dA + \frac{\partial u}{\partial x_2}\, dB + u\, dC = 0,$$

---

[1] SERRIN assumes $c \leq 0$, but as we have seen in the proof of the previous theorem, this hypothesis is superfluous.

where A, B, and C are completely additive set functions satisfying some further restrictions.

Finally, the case of an equation in m variables of quite general type was exhaustively treated almost contemporaneously by O.A. OLEJNIK [1] and by G. TAUTZ [3].

The fundamental result of all this research, which can also be found in PÜSCHEL's work, for the particular equations considered there, is that *every point of $\partial\Omega$ which is regular with respect to an operator $\mathfrak{M}$ is regular with respect to any other operator*. All the different regularity conditions, either necessary and sufficient or only sufficient, established by various authors in the theory of harmonic functions, are therefore valid unaltered regarding boundary value problem (28.1). Thus, for example, every point of $\partial\Omega$ which is on a hypersphere having no other points in common with $\overline{\Omega}$ will be regular with respect to an arbitrary operator $\mathfrak{M}$.

Naturally, the fundamental result which we have enunciated above is valid for operators $\mathfrak{M}$ which have sufficiently regular coefficients. In OLEJNIK's treatment, which substantially adheres to that of PÜSCHEL, we suppose for example that $a_{ik} \in C^{(3,\lambda)}$, $b_i \in C^{(2,\lambda)}$, $c \in C^{(1)}$. In that of TAUTZ, who obtains the theorem as a particular case of a more general one concerning a boundary value problem defined abstractly, it suffices to suppose that a fundamental solution exists and that DIRICHLET's problem for any hypersphere Γ contained in Ω is always solvable and the solution is devoid in $\Gamma - \partial\Gamma$ of positive relative maxima. And these conditions are certainly satisfied if $a_{ik} \in C^{(1,\lambda)}$, b_i, $c \in C^{(0,\lambda)}$, $c \le 0$.

Referring to the original memoirs for the proof of the fundamental theorem, we limit ourselves here to indicating how one can construct the function u_φ.

Let v be an arbitrary continuous function in Ω. Let $u[v; \Gamma]$ be the solution of the equation $\mathfrak{M} u = 0$ which coincides with v on $\partial\Gamma$. If for arbitrary Γ we have $v \ge u[v; \Gamma]$ $[v \le u[v; \Gamma]]$ in $\Gamma - \partial\Gamma$, then we say that v is $\mathfrak{M}$-convex [$\mathfrak{M}$-concave]. Let us now consider the family of all functions v which are $\mathfrak{M}$-convex in Ω, continuous in $\overline{\Omega}$, and less than or equal to φ on $\partial\Omega$. This family is not empty because it contains at least the non-positive constants less than or equal to the minimum of φ. For every point x of Ω, the upper bound $u'_\varphi(x)$ of the values assumed at that point by functions of the family is defined, and one can prove that this function satisfies the equation $\mathfrak{M} u = 0$ in Ω. Now this same equation is also satisfied by the function $u''_\varphi(x)$ equal at every point x to the lower bound of the $\mathfrak{M}$-concave functions in Ω which are continuous in $\overline{\Omega}$ and greater than or equal to φ on $\partial\Omega$; we have, moreover, $u'_\varphi \le u''_\varphi$. Now, if a solution u of the DIRICHLET problem (28.1) exists, then, being simultaneously $\mathfrak{M}$-convex and $\mathfrak{M}$-concave, this must be both $\leq u'_\varphi$ and $\geq u''_\varphi$ and from this it follows that $u'_\varphi = u''_\varphi$ $= u$. Both u'_φ and u''_φ can therefore be considered generalized solutions according to WIENER of problem (28.1). This procedure, which is an obvious extension of the method of super- and subfunctions devised by O. PERRON [1] for the study of DIRICHLET's problem for harmonic functions, was followed by TAUTZ. PÜSCHEL instead considered a sequence of domains of class $A^{(2)}$: T_1, T_2, ..., T_n, ... such that $T_n \subset T_{n+1}$, $\lim T_n = \Omega$, and, denoting by Φ an arbitrary continuous function in $\overline{\Omega}$ equal to φ on $\partial\Omega$, he put $u_\varphi = \lim u_n(x)$, u_n being the solution in T_n of DIRICHLET's problem $\mathfrak{M} u_n = 0$, $u_n = \Phi$ on ∂T_n. This function, which turns out to be independent of the choice of the domains T_n and of the extension of φ, is a solution of the equation $\mathfrak{M} u = 0$ and can be considered a generalized solution of problem (28.1).

Along these same lines, we can cite the works of V.I. KARABEGOV [1], S.I. MOGILEVSKIĬ [1], K. AKÔ [2], S. SIMODA [6]; and we recall that PERRON's method has also been used by G. PRODI in the work [2] cited in § 26.

Another procedure to make the DIRICHLET problem meaningful without introducing regularity hypotheses for ∂T has been indicated by A.D. MYŠKIS [1].

To all these works we finally add a note by M.G. ŠUR [1] in which the construction of the so-called "Martin boundary" is extended to a general elliptic equation. For harmonic functions this important notion was given by R.S. MARTIN in 1941 and was repeated and used to advantage later by M. BRELOT [13, 14, 17], to whom we refer for the formalities and the details which reasons of brevity prevent us from developing. In a recent memoir by E.B. DYNKIN [1] applications of it are made, always for harmonic functions, to the non-regular oblique derivative problem.

29. Generalized boundary conditions. In this section we wish to see how we could arrive at an existence theorem for the DIRICHLET problem when we no longer assume continuity of the boundary values assigned to the unknown function on ∂T. We could already obtain a first answer to this question with the procedures of § 28, if the function φ were assumed no longer continuous, but just bounded. In this case, in fact, it is easy to construct a generalized solution of the DIRICHLET problem according to WIENER, imposing as boundary condition that for every $x_0 \in \partial T$, u admits as $x \to x_0$ in T the same bounds of indeter-

---

[1] See O. PERRON, Eine neue Behandlung der ersten Randwertaufgabe für $\Delta u = 0$. Math. Z. **18** (1923) 42—54 and M. BRELOT, Familles de Perron et problème de Dirichlet. Acta Szeged **9** (1939) 133—153. On the properties of $\mathfrak{M}$-convex functions see also F.F. BONSALL [1].

minateness admitted by φ as $x \to x_0$ on ∂T. And this condition turns out to be effectively satisfied by the generalized solution at all regular points of ∂T.

However, this method does not appear to be applicable when for the hypothesis that $\varphi(x)$ is bounded we wish to substitute that of integrability of φ or of one of its powers. The question posed in these terms has been studied in the case $m = 2$ by G. Cimmino [4] extending a procedure given by R. Caccioppoli [10] in the proof of existence theorems for Abelian integrals on a Riemann surface. This procedure does not presuppose knowledge of the fundamental solutions of the equation, nor does it make use of the theory of integral equations, being based only on some simple considerations from linear functional analysis. This is susceptible to various other applications, of which we shall give an idea at the end of this section.

Therefore, let T be a plane domain of class $A^{(3)}$ which, for preciseness, we assume to have a single boundary. Denote by s the arc length $(0 \le s \le L)$ and by $x_1 = \psi_1(s)$, $x_2 = \psi_2(s)$ the parametric equations of ∂T, and let us suppose that s is chosen in such a way that the outer normal to ∂T has direction cosines ψ_2' and $-\psi_1'(s)$. For every fixed positive t which is sufficiently small, the curve with parametric equations:

$$x_1 = x_1(s, t) = \psi_1(s) - t\,\psi_2'(s), \quad x_2 = x_2(s, t) = \psi_2(s) + t\,\psi_1'(s) \quad (29.1)$$

then delimits a domain $T(t)$ of class $A^{(2)}$ interior to T. We can even specify that this happens for $t \le t_0 < \max|\varrho(s)|$, where $1/\varrho(s)$ is the algebraic curvature of ∂T. This comes from the fact that:

$$\frac{\partial(x_1, x_2)}{\partial(s, t)} = \sqrt{\left(\frac{\partial x_1}{\partial s}\right)^2 + \left(\frac{\partial x_2}{\partial s}\right)^2} = 1 - \frac{t}{\varrho}. \qquad (29.2)$$

From (29.2) it now follows that (29.1) defines s and t implicitly as functions of x_1, x_2 at least in $T - T(t_0)$; and we shall now suppose as is valid by theorem 16, VI, that the function $t = t(x_1, x_2)$ thus defined in $T - T(t_0)$, has been extended into all of T so that it is of class $C^{(2)}$.

If, now, on ∂T, $\varphi(s) \in L^2$, we shall say that a function $u(x_1, x_2)$ which is continuous in $T - \partial T$ *assumes the value $\varphi(s)$ in the mean on ∂T*, if, for a given weight function $P(s, t)$ continuous and positive for $0 \le s \le L$, $0 \le t \le t_0$, the following holds:

$$\lim_{t \to 0} \int_0^L P(s, t)\,[u(x_1(s, t), x_2(s, t)) - \varphi(s)]^2\,ds = 0. \qquad (29.3)$$

The problem considered by Cimmino consists of seeking a solution of the equation $\mathfrak{M}\,u = f$ which is regular in $T - \partial T$ and assumes prescribed values $\varphi(s)$ in the mean on ∂T.

Let us suppose that the coefficients of $\mathfrak{M}$ and of the adjoint operator $\mathfrak{N}$ are of class $C^{(2)}$, and let us put:

$$P(s, t) = \frac{a_{11}\,\psi_2'^2 + 2\,a_{12}\,\psi_1'\,\psi_2' + a_{22}\,\psi_1'^2}{1 - \dfrac{t}{\varrho}}.$$

This last equation is not essential for the validity of CIMMINO's reasoning, but it permits considerable simplification of the calculations.

The following uniqueness theorem, which shows that the problem considered is well posed, now holds:

29, I. *If $c \leq 0$ then the boundary value problem considered above admits at most one solution.*

For, the following indentity holds:

$$\frac{d}{dt} \int_0^L P u^2\, ds = \int_0^L (\mathfrak{N}\, t - c\, t)\left(1 - \frac{t}{\varrho}\right) u^2\, ds$$

$$+ \int_{T(t)} \left[c u^2 - u\, \mathfrak{M}\, u - \left(a_{11}\left(\frac{\partial u}{\partial x_1}\right)^2 + 2\, a_{12}\, \frac{\partial u}{\partial x_1}\, \frac{\partial u}{\partial x_2} + a_{22}\left(\frac{\partial u}{\partial x_2}\right)^2 \right) \right] dx ,$$

$$(29.4)$$

which is arrived at by elementary considerations related to GREEN's formula. From this, if we had $\mathfrak{M}\, u = 0$ and if u were zero in the mean on ∂T without being identically zero, then the right side of (29.4) would be negative for sufficiently small t and therefore the integral of $P u^2$ would have to tend to zero from below, which is absurd because it is positive.

Let us now pass to the study of a property of the mean characteristic of the solutions of the equation $\mathfrak{M}\, u = f$, the utilization of which is essential for the proof of the existence theorem. This property of the mean constitutes one of the infinite possible extensions of the GAUSS property for harmonic functions. In our exposition we shall bring to CIMMINO's treatment some simplifications suggested in part by some works by B. PINI[1].

We begin by putting:

$$H_\varrho(x, y)\begin{cases} = H(x, y)\left(1 - \dfrac{\overline{xy}^5}{\varrho^5}\right)^5 & \text{for} \quad \overline{xy} \leq \varrho , \\[2mm] = 0 & \text{for} \quad \overline{xy} > \varrho . \end{cases} \qquad (29.5)$$

---

[1] See B. PINI [1, 4, 9] and G. CIMMINO [6]. For the extension to the case of arbitrary m, in the original direction of CIMMINO, see G. ZWIRNER [2], and for other properties of the mean W. FELLER [1], H. PORITSKY [1], S. G. MIHLIN [3], F. TRICOMI [1], Z. V. REĬTBLATT [1].

H_ϱ, which, after an extension of the coefficients of $\mathfrak{M}$, can be considered defined in all space, is obviously continuous for $x \neq y$ with all its derivatives with respect to x_1 and x_2 up to those of fourth order. Moreover, the same derivatives of the function $H - H_\varrho$ are continuous even for $x = y$. This proves that H_ϱ is a LEVI function for which, if u is a regular solution of the equation $\mathfrak{M} u = f$ in $T - \partial T$, we have, from STOKES's formula:

$$u(y) = \int_T [u(x)\, \mathfrak{N}_x\, H_\varrho(x, y) - f(x)\, H_\varrho(x, y)]\, dx , \qquad (29.6)$$

with ϱ always less than the distance $d(y)$ of y from ∂T. In (29.6) the integration is in fact only over the hypersphere $\varGamma(y, \varrho)$ and therefore this formula can be considered as a property of the mean of the solutions of the equation $\mathfrak{M} u = f$. We now want to show that this property of the mean is characteristic of solutions, that is:

29, II. *Every function $u \in L^p$ with $p > 1$ which, for y almost everywhere in $T - \partial T$ and for $\varrho < d(y)$ satisfies the property of the mean (29.6), in which we suppose that $f \in C^{(0,\, \lambda)}$, coincides almost everywhere in T with a function of class $C^{(2)}$ in $T - \partial T$ which is a solution of the equation $\mathfrak{M} u = f$.*

Let $K_\varrho^{(n)}(x, y)$ be the nth iterate over the domain T of the kernel $K_\varrho(x, y) = \mathfrak{N}_x\, H_\varrho(x, y)$, and put:

$$H_\varrho^{(p)}(x, y) = H_\varrho(x, y) + \sum_{n=1}^{p} \int_T H_\varrho(x, t)\, K_\varrho^{(n-1)}(t, y)\, dt .$$

If $D \subset T - \partial T$ is a domain which has distance δ from ∂T, iterating (29.6) three times, we have, for $y \in D$ and $\varrho < \delta/3$:

$$u(y) = \int_T u(x)\, K_\varrho^{(3)}(x, y)\, dx - \int_T f(x)\, H_\varrho^{(3)}(x, y)\, dx . \qquad (29.7)$$

Now the procedures of § 18 permit the proof that $K_\varrho^{(3)}$ is continuous with its first derivatives with respect to x_1 and x_2 and admits for $x \neq y$ second derivatives which are $O\left(\log \dfrac{2\varrho}{xy}\right)$. By theorems 12, VIII and 12, IX, the first integral on the right side of (29.7) can be differentiated twice under the integral sign, where an immediate generalization of theorem 12, I makes it possible to see, by dint of the hypothesis that $p > 1$, that the second derivativev of the aforesaid integral are continuous. The second integral on the right side of (29.7) is then a domain potential and as such is of class $C^{(2,\, \lambda)}$ in $T - \partial T$. It thus stands proved that u coincides almost everywhere with a function of class $C^{(2)}$ in $T - \partial T$.

Now, applying STOKES's formula to u on the hypersphere $\Gamma(y, \varrho)$ and comparing with (29.6) we have for $y \in D$ and for arbitrary, sufficiently small ϱ:

$$\int\limits_{\Gamma(y, \varrho)} H_\varrho(x, y) \, [\mathfrak{M} \, u(x) - f(x)] \, dx = 0 \, ,$$

and from this, since ϱ is arbitrary and $H_\varrho > 0$, we draw the conclusion that $\mathfrak{M} \, u = f$.

The theorem just proved permits us to prove easily the following compactness criterion which occurs in the proof of the existence theorem:

29, III. *Let $\{u_n\}$ be a sequence of functions of class L^2 in T and $C^{(2)}$ in $T - \partial T$ which assume the values $\varphi_n(s)$ in the mean on ∂T, and suppose that the sequence $\{\varphi_n\}$ converges in the mean on ∂T to a square integrable function $\varphi(s)$. If the sequence $\{\mathfrak{M} \, u_n\}$ also converges in the mean in T to a function $f(x)$ which is square integrable in T and of class $C^{(0, \lambda)}$ in $T - \partial T$ and if $\int\limits_T u_n^2 \, dx < M$, then it is possible to extract from the sequence of the $u_n' s$ a subsequence which converges uniformly in every domain interior to T to a function $u_0(x)$ which assumes the values $\varphi(s)$ in the mean on ∂T and which is a regular solution of the equation $\mathfrak{M} \, u = f$. The aforesaid subsequence also converges in the mean in T to $u_0(x)$.*

Moreover, the property of the mean (29.6), which is satisfied by all the u_n with $f = \mathfrak{M} \, u_n$, and the assumption that the integrals of u_n^2 and $(\mathfrak{M} \, u_n)^2$ are bounded, ensure that the $u_n' s$ are equicontinuous in every domain interior to T; this in fact follows from (29.7) and from theorem 18, III. From $\{u_n\}$ it will now be possible to extract a subsequence which converges uniformly in every domain interior to T to a function $u_0(x)$ continuous in $T - \partial T$ and satisfying in addition (29.6). By the preceding theorem, $u_0(x)$ will be a solution of the equation $\mathfrak{M} \, u = f$. Continuing to call the subsequence $\{u_n\}$, let us prove that u_0 assumes the values $\varphi(s)$ in the mean on ∂T. For this purpose we begin with the observation that, if $u(x)$ is a function satisfying the same hypotheses as the $u_n's$, we draw easily from (29.4) the following, by integrating with respect to t:

$$\int\limits_0^L P(s, t) \, u^2(s, t) \, ds \le \int\limits_0^L P(s, 0) \, u^2(s, 0) \, ds$$

$$+ K \int\limits_0^t d\tau \int\limits_0^L P(s, \tau) \, u^2(s, \tau) \, ds + K \int\limits_0^t d\tau \int\limits_{T(\tau)} [u^2(x) + (\mathfrak{M} \, u(x))^2] \, dx \, ,$$

$$(29.8)$$

where K is a constant independent of u and $u(s, t) = u(x_1(s, t), \, x_2(s, t))$.

And from (29.8), for $Kt < 1$, it follows easily that:

$$(1 - Kt) \int_0^L P(s, t)\, u^2(s, t)\, ds \leq \int_0^L P(s, 0)\, u^2(s, 0)\, ds$$

$$+ K \int_0^t d\tau \int_{T(\tau)} [u^2(x) + (\mathfrak{M}\, u\,(x))^2]\, dx.$$

Now, if we put $u = u_n - u_k$ in this formula, it is certainly possible by the assumed convergence in the mean of the $u_n(s, 0) = \varphi_n(s)$ and by the bounds admitted by the integrals of u_n^2 and $(\mathfrak{M}\, u_n)^2$, to determine n_ε and t_ε in such a way that for $n, k > n_\varepsilon$ and $t < t_\varepsilon$, the right side is less than $\varepsilon\,(1 - Kt)$; from this, passing to the limit as $n \to \infty$, we deduce that for $k > n_\varepsilon$ and $t < t_\varepsilon$ we have:

$$\int_0^L P(s, t)\, [u(s, t) - u_k(s, t)]^2\, ds \leq \varepsilon. \tag{29.9}$$

Now, fixing $k > k_\varepsilon$ such that we also have:

$$\int_0^L P(s, t)\, [\varphi_k(s) - \varphi(s)]^2\, ds \leq \varepsilon,$$

we have for $t < t_\varepsilon$:

$$\int_0^L P(s, t)\, [u(s, t) - \varphi(s)]^2\, ds \leq 2\,\varepsilon + \int_0^L P(s, t)\, [u_k(s, t) - \varphi_k(s)]^2\, ds,$$

from which the assertion obviously follows.

From (29.9), integrating with respect to t and bearing in mind that $dx = (1 - t/\varrho)\, ds\, dt$, we draw a bound of the type:

$$\int_{T(t_\varepsilon)} [u(x) - u_k(x)]^2\, dx \leq K\varepsilon,$$

from which, by dint also of the uniform convergence of the u_k's in $T - T(t_\varepsilon)$, it follows that the u_k's converge in the mean in T to u.

Now let us pass to the proof of the following existence theorem:

29, IV. *If $f \in C^{(0,\,\lambda)}$ in T and $\varphi \in L^2$ on ∂T, then a necessary and sufficient condition for the existence of a solution of the boundary value problem being considered is that f and φ satisfy*:

$$\int_T f(x)\, v(x)\, dx + \int_{\partial T} a(x)\, \varphi(x)\, \frac{dv}{dv}\, ds = 0 \tag{29.10}$$

*for every solution $v(x)$ of the adjoint homogeneous problem which is conti-
nuous in T. In particular, if the adjoint homogeneous problem admits only
the zero solution, then the boundary value problem being considered is
solvable for arbitrary f and φ.*

Let us denote by Σ the BANACH space of pairs $\{f(x),\ \varphi(s)\}$ with
$f \in L^2$ in T and $\varphi \in L^2$ on ∂T, assuming as *norm* of the pair $\{f,\ \varphi\}$:

$$\| \{f,\ \varphi\} \| = \left[\int_T f^2(x)\, dx \right]^{1/2} + \left[\int_0^L P(s,\ 0)\ \varphi^2(s)\, ds \right]^{1/2}.$$

Let Σ_1 be the linear manifold of Σ constituting the pairs $\{\mathfrak{M}u(x),$
$u(x_1(s,\ 0),\ x_2(s,\ 0))\}$ with $u(x)$ a function of class $C^{(2)}$ in T, and let $\overline{\Sigma}_1$
be the closure of Σ_1. If $\overline{\Sigma}_1$ does not exhaust Σ then there will exist, by the
HAHN-BANACH-ASCOLI theorem[1], a linear functional zero on $\overline{\Sigma}_1$ and
not identically zero on Σ. And this means, in other words, that, due to the
fact that Σ is a HILBERT space, there exists a pair $\{v,\ \omega\}$ with nonzero
norms such that for any pair $\{f,\ \varphi\}$ of Σ_1 we have:

$$\int_T vf\, dx + \int_0^L P(s,\ 0)\ \omega\ \varphi\ ds = 0. \tag{29.11}$$

Then the following holds:

Lemma. *If $\{v,\ \omega\}$ satisfies (29.11) for every $\{f,\ \varphi\} \in \Sigma_1$, then v is,
almost everywhere in T, equal to a function of class $C^{(2)}$ in $T - \partial T$ and of
class $C^{(1)}$ in T for which:*

$$\mathfrak{N}v = 0 \quad for \quad x \in T - \partial T, \quad v = 0 \quad for \quad x \in \partial T. \tag{29.12}$$

Moreover, we have almost everywhere on ∂T:

$$P(s,\ 0)\ \omega = a\ \frac{dv}{dv}. \tag{29.13}$$

Fixing $y \in T - \partial T$ and assuming $\varrho < d(y)$, we put:

$$u_n(x) = \frac{\gamma_n(\tau)}{2\ \pi\ \sqrt{A(y)}} \left(1 - \frac{\overline{xy}^5}{\varrho^5} \right)^5 \quad for \quad \overline{xy} \leq \varrho,\ u_n(x) = 0 \quad for \quad \overline{xy} > \varrho,$$

---

[1] See H. HAHN: Über lineare Gleichungssysteme in linearen Räumen.
J. für Math. **157** (1927) 214—229. S. BANACH: Sur les fonctionnelles linéares.
Studia Math. **1** (1929) 211—216 and 223—239. G. ASCOLI: Sugli spazi
lineari matrici e le loro varietà lineari. Ann. Mat. Pura appl. **10** (1932)
33—81 and 203—232. In the applications of this theorem to the study of
boundary value problems, the spaces which are considered are always
separable. Under this hypothesis the theorem is proved without the use of
the transfinite.

where:

$$\gamma_n(\tau) = \frac{1}{2} \int_{\tau^2}^{1} [1 - (1-t)^n] \frac{dt}{t}, \qquad \tau = \left[\sum_{r,s=1}^{2} A_{rs}(y) (x_r - y_r) (x_s - y_s) \right]^{1/2}.$$

Because $u_n(x)$ is of class $C^{(2)}$ we have from (29.11): $\int_T v \, \mathfrak{M} \, u_n \, dx = 0$ and, therefore, applying GREEN's formula:

$$\int_T \mathfrak{M} \, u_n(x) \, [v(x) - v(y)] \, dx = -v(y) \int_T \mathfrak{M} \, u_n(x) = -v(y) \int_T u_n(x) \, \mathfrak{N} \, 1 \, dx.$$

$$(29.14)$$

Because $\lim_{n \to \infty} u_n(x) = H_\varrho(x, y)$, if it is valid to pass to the limit under the integral sign in the preceding formula, we have:

$$\int_T \mathfrak{M}_x H_\varrho(x, y) \, [v(x) - v(y)] \, dx = -v(y) \int_T H_\varrho(x, y) \, \mathfrak{N} \, 1 \, dx ,$$

from which, transforming the second integral with STOKES's formula, we draw:

$$v(y) = \int_T v(x) \, \mathfrak{M}_x H_\varrho(x, y) \, dx . \qquad (29.15)$$

This formula is analogous, except for the replacement of $\mathfrak{N}$ by $\mathfrak{M}$, to (29.6) by which its truth, also almost everywhere in T, proves that v coincides (at least up to a set of measure zero) with a regular solution of the equation $\mathfrak{N} v = 0$. It now remains to justify the passage to the limit under the integral sign in (29.14) and this is immediately done as far as the right side is concerned because $0 < u_n < H_\varrho$.

For the left side, because $\lim_{n \to \infty} \mathfrak{M} \, u_n = \mathfrak{M} \, H_\varrho$ uniformly in every domain which excludes the point y, it suffices to show that, if we denote by $I(y, \delta)$ the neighborhood of y defined by the bound $\tau \le \delta$, then we have uniformly in n:

$$\lim_{\delta \to 0} \int_{I(y, \delta)} \mathfrak{M} \, u_n \, [v(x) - v(y)] \, dx = 0 . \qquad (29.16)$$

But this relation holds for y almost everywhere in $T - \partial T$ and precisely for every y for which:

$$\lim_{\delta \to 0} \frac{1}{\delta^2} \int_{I(y, \delta)} |v(x) - v(y)| \, dx = 0 . \qquad (29.17)$$

Indeed, we easily recognize that we can write:

$$\mathfrak{M}\, u_n = \alpha_0 \left(\gamma_n'' + \frac{1}{\tau}\, \gamma_n' \right) + \alpha_1\, \tau\, \gamma_n' + \alpha_2\, \gamma_n\,,$$

with α_0, α_2, and $\tau\, \alpha_1$ bounded functions, whence, putting:

$$\beta_i(\tau) = \int\limits_{I(y,\,\tau)} \alpha_i\, [v(x) - v(y)]\, dx\,,$$

we have:

$$\int\limits_{I(y,\,\delta)} \mathfrak{M}\, u_n\, [v(x) - v(y)]\, dx = \int\limits_0^\delta \left(\gamma_n'' + \frac{1}{\tau}\, \gamma_n' \right) \frac{d\beta_0}{d\tau}\, d\tau$$

$$+ \int\limits_0^\delta \tau\, \gamma_n'\, \frac{d\beta_1}{d\tau}\, d\tau + \int\limits_0^\delta \gamma_n\, \frac{d\beta_2}{d\tau}\, d\tau\,. \qquad (29.18)$$

Now if y is chosen so that (29.17) holds, we can, given ε, determine a δ_ε such that for $\tau < \delta_\varepsilon$ we have $|\beta_0|$, $|\beta_2| < \varepsilon\tau^2$, $|\beta_1| < \varepsilon\tau$. For $\delta < \delta_\varepsilon$ we then have, by integrating (29.18) by parts and bearing in mind that $\gamma_n'' + \gamma_n'/\tau$, $\tau\, \gamma_n'$ and γ are monotone functions:

$$\left| \int\limits_{I(y,\,\delta)} \mathfrak{M}\, u_n\, [v(x) - v(y)]\, dx \right| \le \varepsilon \left[\delta^2 \left| \gamma_n''(\delta) + \frac{1}{\delta}\, \gamma_n'(\delta) \right| \right.$$

$$+ \delta^2\, |\gamma_n'(\delta)| + \delta^2\, |\gamma_n(\delta)|$$

$$\left. + \int\limits_0^\delta \tau^2\, \frac{d}{d\tau} \left(\gamma_n'' + \frac{1}{\tau}\, \gamma_n' \right) d\tau - \int\limits_0^\delta \tau\, \frac{d}{d\tau}\, (\tau\, \gamma_n')\, d\tau - \int\limits_0^\delta \tau^2\, \gamma_n'\, d\tau \right],$$

from which (29.16) follows without further ado, since the quantity in brackets can be bounded independently of δ and n.

To prove now that v vanishes on ∂T, it suffices to apply (29.11) anew, putting $u = u_n$ but assuming ϱ so large that the domain T is interior to the hypersphere $\Gamma(y, \varrho)$ and choosing y first in $T - \partial T$ and then in $\complement\, T$. We thus have the two formulas:

$$\int\limits_T v(x)\, \mathfrak{M}_x\, H_\varrho(x, y)\, dx + \int\limits_{\partial T} H_\varrho(x, y)\, P(s, 0)\, \omega\, d_x s = \begin{cases} v(y) & \text{for } y \in T - \partial T \\ 0 & \text{for } y \in \complement\, T\,, \end{cases}$$

from the first of which it follows that v is continuous in T (theorems 12, V and 14, I), while the second ensures by comparison that $v = 0$ on ∂T.

But because every solution of problem (29.12) is (see § 22) of class $C^{(1)}$ in T, associated with (29.11) GREEN's formula holds:

$$\int\limits_T vf\,dx + \int\limits_{\partial T} a\,\frac{dv}{dv}\,\varphi\,ds = 0\,,$$

and from this, since φ is arbitrary, (29.13) follows.

Our lemma is thus completely proved, and we can resume the proof of the existence theorem. We note first of all that, by what we see in § 22 there exists at most a finite number (say p) of solutions of problem (29.12); let these be v_1, v_2, ... v_p and let $\{f,\varphi\}$ be a point of Σ satisfying (29.10) for $v = v_i$ ($i = 1, 2, \ldots, p$).

Evidently $\{f,\varphi\}$ belongs to $\overline{\Sigma}_1$, otherwise by repeating our reasoning we could prove the existence of another solution of problem (29.12) linearly independent of the v_i. Consequently there exists a sequence $\{u_n\}$ of functions of class $C^{(2)}$ in T such that the u_n converge in the mean on ∂T to φ, and $\mathfrak{M}\,u_n$ converges in the mean in T to f. Supposing f of class $C^{(0,\lambda)}$ two cases can be presented. Either the integrals $\int\limits_T u_n^2\,dx$ are uniformly bounded, and then from theorem 29, III the existence of a solution to the boundary value problem does follow. Or else the aforesaid integrals are not uniformly bounded. Putting:

$$u_n' = u_n\left[\int\limits_T u_n^2\,dx\right]^{-1/2},$$

it is then possible, always by theorem 29, III, to extract from the sequence $\{u_n'\}$ a subsequence which converges to a solution of the equation $\mathfrak{M}\,u = 0$ which is not identically zero but which assumes the value zero in the mean on ∂T. By theorem 29, III, the set of such solutions of the homogeneous problem is compact, and therefore of these there exist at most a finite number which are linearly independent; they are $u^{(1)}$, $u^{(2)}$, ..., $u^{(q)}$. We can now easily prove that it is possible to determine the constants c_{ni} so that the integrals:

$$\int\limits_T\left[u_n + \sum_{i=1}^{q} c_{ni}\,u^{(i)}\right]^2 dx$$

are uniformly bounded, from which, now applying theorem 29, III to the sequence $\left\{u_n + \sum\limits_{i=1}^{q} c_{ni}\,u^{(i)}\right\}$, we arrive anew at the proof of the existence of a solution of the boundary value problem.

8*

The existence theorem 29, IV is thus completely proved; if, however, we want to give this the form customary for the alternative theorems, we are led to prove that:

29, V. *The number of solutions of the homogeneous problem associated with the problem considered is equal to the number of compatibility conditions* (20.10).

Let χ be a positive number greater than c and greater than the coefficient c^* of v in the operator $\mathfrak{N}$. By theorems 29, I and IV, there then exists, for each f of class $C^{(0,\lambda)}$ in T, one and only one solution of the equation $\mathfrak{M} u - \chi u = f$, which assumes in the mean on ∂T the value zero. For this solution u we could thus write $u = \mathfrak{T}(f)$, $\mathfrak{T}$ being the symbol for a linear functional transformation. Now the reasoning adopted for the proof of theorem 29, III makes it possible to show easily that the transformation $\mathfrak{T}$, initially defined only for f of class $C^{(0,\lambda)}$, can be extended to a transformation of the space $L^2(T)$ into the subspace of $L^2(T)$ consisting of all the functions in $L^2(T)$ which assume the value zero in the mean on ∂T and which are of class $C^{(0,\lambda)}$ in $T - \partial T$. Moreover, the transformation turns out to be *completely continuous*, in the sense that, if the functions of the sequence $\{f_n\}$ have uniformly bounded norms, then from the sequence $\{\mathfrak{T}(f_n)\}$ we can extract a subsequence which is convergent in the mean. Thus the problem of determining a solution of the equation $\mathfrak{M} u = f$ which assumes the value zero in the mean on ∂T is equivalent to the solution of the functional equation $u + \chi \mathfrak{T}(u) = \mathfrak{T}(f)$. And for this equation a theorem due to F. RIESZ[1] ensures precisely that the number of compatibility conditions which f must satisfy equals the number of linearly independent solutions of the associated homogeneous equation. Our theorem thus stands completely proved. From this it immediately follows that:

29, VI. *Every solution of the equation $\mathfrak{M} u = f$ which assumes the values φ in the mean on ∂T is continuous in T if φ is continuous on ∂T.*

Let us begin by proving the assertion in the homogeneous case. Then, if the equation $\mathfrak{M} u = 0$ admits q solutions continuous in T and zero on ∂T and q' solutions zero in the mean on ∂T, and if p and p' are the analogous numbers for the equation $\mathfrak{N} v = 0$, then we have by theorem 29, V: $q' = p$, $p' = q$. And because we also have (theorem 22, I) $p = q$, we conclude as desired $q = q'$. From this we are led to the assertion even in the non-homogeneous case by using theorem 21, IV.

We can moreover prove that with the methods developed in this section we are also led to the existence theorem for the problem posed in

---

[1] See e.g. S. BANACH, Théorie des opérations linéaires. Monografie Mat. Warszawa (1932) Chapter X.

ordinary form; we can even add that when we take into consideration the problem in ordinary form exclusively, various noteworthy simplifications are possible[1]. On the other hand, it is also valid to substitute into the study of generalized problems convergence in the mean of order $p > 1$ for the convergence in the mean of order 2. For the case $p = 1$, which requires some cleverness, we refer to B. PINI [12].

The method appears to be usable both in the case $m > 2$ and for the study of NEUMANN's problem. It would also be interesting to try this procedure in the case in which T is a domain of general type, already considered from another point of view in § 28. For some indications on all these questions relative to the particular case $\mathfrak{M} = \varDelta_2$, one can usefully consult CIMMINO's works [4] and [5]. Other applications of the method will be indicated in Chapter VII (§§ 48, 49, 52, 55).

30. Weak solutions of the boundary value problems. If we consider in a region Ω an elliptic operator written in the form (6.2) with $a_{ik} \in C^{(1)}$, e_i, $c \in C^{(0,\lambda)}$ and denote by $u(x)$ a regular solution in Ω of the equation $\mathfrak{M} u = f$, then it is immediately seen that for every $v \in H_0^{1,2}(\Omega)$ we have:

$$\int_{\Omega}\left[-\sum_{i,k=1}^{m} a_{ik} \frac{\partial u}{\partial x_i} \frac{\partial v}{\partial x_k} + \sum_{i=1}^{m} e_i \frac{\partial u}{\partial x_i} v + c\, u\, v\right] dx = \int_{\Omega} f v\, dx. \qquad (30.1)$$

On the contrary, if $u \in C^{(2)}(\Omega)$ satisfies (30.1) for every $v \in H_0^{1,2}(\Omega)$, we have after an integration by parts $\int_{\Omega} v\, (\mathfrak{M} u - f)\, dx = 0$ and therefore $\mathfrak{M} u = f$. Consequently we shall agree to call every function $u \in H^{1,2}(\Omega)$ which satisfies (30.1) for every $v \in H_0^{1,2}(\Omega)$ a *weak solution of the equation* $\mathfrak{M} u = f$, and we can assert that *under the hypotheses indicated, every weak solution of the equation* $\mathfrak{M} u = f$ *which is of class $C^{(2)}$ is also a regular solution of this equation.*

This observation suggests for the solution of the boundary value problem for the equation $\mathfrak{M} u = f$ in a domain T a procedure which we articulate in two successive stages:

I) To determine a *weak solution of the problem*, i.e., a function $u(x)$ which is a weak solution of the equation in the region $\Omega = T - \partial T$ and which satisfies in some way, possibly generalized, the boundary condition.

II) To proceed to the *regularization* of $u(x)$, proving that it is of class $C^{(2)}(\Omega)$ and satisfies the boundary condition in the ordinary sense.

---

[1] See P. PINI [1] and, for the case $\mathfrak{M} = \varDelta_2$, C. MIRANDA [3].

More generally, we shall be able to speak of regularization of $u(x)$ any time we have established qualitative properties for it which are stronger than those normally demanded of a weak solution.

Note that point I), at least if the boundary condition being considered is homogeneous, does not in general present many difficulties, and the determination of the weak solution of the problem can be done under conditions of great generality both on the domain T and on the operator $\mathfrak{M}$ and the function f. It is only at point II) that the hypotheses on the data of the problem must be specified in such a way as to make the regularization possible. But this can be done if we bear in mind the purpose which we wish to attain, and therefore with the maximum economy.

This procedure was adopted for the first time by H. WEYL [1] in the study of boundary value problems for harmonic functions, under the name, tied to certain particular methods of application, of the *method of orthogonal projections*. This procedure leads to a different treatment of the generalized problem for which we use it here, but is quite similar in the analytic methods it employs to that adopted some years earlier by R. CACCIOPPOLI and G. CIMMINO of which we have exposed the essential elements in the preceding section. Indeed, at the basis of WEYL's method are, on one hand, the same Lemma[1] which we used in the proof of theorem 29, IV, and on the other, as in the method of CACCIOPPOLI-CIMMINO, the theorem of HAHN-BANACH-ASCOLI, not applied directly, however, but through some of its corollaries. The first applications of the procedure to general elliptic equations we owe to K. KODAIRA [1] and to M.I. VIŠIK [1]. Many other works follow these, which we shall illustrate briefly after having better illustrated the means of application of the method. For this purpose we consider DIRICHLET's problem for the equation $\mathfrak{M} u = f$ posed in a domain T and with the homogeneous boundary condition:

$$u(x) = 0 \quad \text{for} \quad x \in \partial T . \tag{30.2}$$

First we shall treat this problem under the hypothesis that $\mathfrak{M}$ is self-adjoint. Putting $\Omega = T - \partial T$, we shall mean by a *weak solution of the*

---

[1] In the particular case $\mathfrak{M} = \Delta_2$, $\omega = 0$ this Lemma, which is obviously to be considered a true and proper regularization theorem, is in general called "WEYL's Lemma". This nomenclature is inappropriate because the same proposition can be found in a more general form in CACCIOPPOLI's memoir [10]. And even the extension to the case of arbitrary $\mathfrak{M}$ given by G. CIMMINO [3, 4], precedes WEYL's memoir by some years. To be accurate, CACCIOPPOLI and CIMMINO considered the case $m = 2$, WEYL the case $m = 3$. For further information, some elements of the theory of orthogonal projections can be found in research by S. ZAREMBA [1], repeated later by O. NIKODYM [2], and in a memoir by S.L. SOBOLEV [3]. On the argument see also the note [2] by L. GÅRDING.

problem a function $u \in H_0^{1,\,2}(\Omega)$ which, for every $v \in H_0^{1,\,2}(\Omega)$, satisfies the equation:

$$\int_{\Omega} \left[\sum_{i,\,k=1}^{m} a_{ik} \frac{\partial u}{\partial x_i} \frac{\partial v}{\partial x_k} - c\,u\,v \right] dx = - \int_{\Omega} f\,v\,dx. \tag{30.3}$$

In order to prove the existence of u, it will be sufficient to assume that T is bounded and measurable and that the a_{ik} and c are of class $L^{\infty}(\Omega)$ and such that (2.1) is satisfied and $c < 0$. Under these hypotheses we can consider $H^{1,\,2}(\Omega)$ as a real HILBERT space Σ, defining the left side $a(u,v)$ of (30.3) as the scalar product of two elements u and v of Σ. It will also be convenient to denote by Σ_0 the subspace of Σ consisting of the elements which belong to $H_0^{1,\,2}$. Because the norms of an element u in $H_0^{1,\,2}$ and in Σ_0 are equivalent and $H_0^{1,\,2}$ is complete, Σ_0 is also complete; now, because the right side of (30.3) is a linear continuous functional of the element v of Σ_0, by a well-known theorem there certainly exists an element u of Σ_0 for which we have $a(u,\,v) = - \int_{\Omega} f\,v\,dx$. This element u is the required weak solution of the problem.

In order to treat the general case, we must now assume some notions from linear functional analysis. We shall limit ourselves here to what is strictly necessary for the purposes of this paragraph. For more ample reports, also from different points of view, see M. I. VIŠIK and O. A. LADYŽENSKAJA [1], E. MAGENES and G. STAMPACCHIA [1], J. L. LIONS [6, 7], G. FICHERA [22, 23], F. E. BROWDER [9].

Let S be a linear space over the real or complex field, Σ_1 and Σ_2 two BANACH spaces, Σ_1^* and Σ_2^* their duals, M_1 and M_2 two linear mappings of S into Σ_1 and Σ_2 respectively. Picking $\Phi \in \Sigma_1^*$ we propose to see if $\Psi \in \Sigma_2^*$ exists such that we have:

$$< \Psi,\, M_2(v) > \; = \; < \Phi,\, M_1(v) > \quad \forall\, v \in S. \tag{30.4}$$

For this purpose the following theorem due to G. FICHERA [16][1] holds:

30, I. *A necessary and sufficient condition that there exist a solution Ψ of (30.4) for arbitrary Φ is that there exist a constant k such that:*

$$\|\, M_1(v)\, \|_{\Sigma_1} \leq k\, \|\, M_2(v)\, \|_{\Sigma_2}.$$

Among the solutions of (30.4) there exists further one for which we have (with the same k):

$$\|\, \Psi\, \|_{\Sigma_1^*} \leq k\, \|\, \Phi\, \|_{\Sigma_2^*}. \tag{30.5}$$

---

[1] This theorem can also be used for the direct study of boundary value problems understood in the ordinary sense. For this use see G. FICHERA [21] and for a generalization S. FAEDO [2].

Referring to the work already cited for the proof of the necessity of the condition, we shall relate the proof of the sufficiency[1]. Let $M_2(S)$ be the range of M_2; for every $w_2 \in M_2(S)$ we consider a v such that $M_2(v) = w_2$. By dint of (30.5), $M_1(v)$ depends only on w_2 and furthermore, putting:

$$< \psi, \, w_2 > \; = \; < \Phi, \, M_1(v) > \quad \forall \, w_2 \in M_2(S),$$

we define on $M_2(S)$ a linear functional ψ which, since $\| \psi(w_2) \| \leq k \, \| \Phi \|_{\Sigma_1 *}, \; \| w_2 \|_{\Sigma_2 *}$, is even continuous. By the Hahn-Banach-Asoli theorem, ψ can be extended to a functional Ψ defined on all of Σ_2 for which (30.5) is satisfied.

From theorem 30, I we can now deduce[2] the following theorem of Lax and Milgram[3] :

30, II. *Let Σ be a real or complex Hilbert space and $B(u, v)$ be a bilinear functional defined for every ordinate pair $u, v \in \Sigma$, for which there exist two constants h and k such that:*

$$| B(u, v) | \leq h \, \| u \| \, \| v \|, \; \| u \|^2 \leq k \, | B(u, v) |. \qquad (30.6)$$

Then however we assign the functional $\Phi \in \Sigma^*$ there exists one and only one $\Psi \in \Sigma$ such that:

$$< \Phi, u > \; = B(u, \Psi), \| \Psi \|_{\Sigma} \leq k \, \| \Phi \|_{\Sigma *}. \qquad (30.7)$$

We observe that, for every fixed u, $B(u, v)$ is a linear and continuous function of v and therefore there exists an element $T(u)$ of Σ such that:

$$B(u, v) = (T(u), v).$$

We obviously have:

$$\| u \|^2 \leq k \, | B(u, u) | \leq k \, \| T u \| \, \| u \|$$

and thus:

$$\| u \| \leq k \, \| T u \|.$$

Then from the preceding theorem, assuming

$$S = \Sigma, \quad M_1(u) = u, \quad M_2(u) = T(u)$$

---

[1] G. Fichera [22, 23].

[2] G. Fichera [22].

[3] P. D. Lax and A. M. Milgram: Parabolic equations. Annals of Math. Studies **33** (1954), 167—190.

it follows that there exists a Ψ such that for every u:

$$< \Phi, u > \, = (T(u), \Psi),$$

from which (30.6) follows. The uniqueness of Ψ follows from the fact that if Ψ_1 and Ψ_2 satisfy (30.7), we have $B(u, \Psi_1 - \Psi_2) = 0$ and therefore for $u = \Psi_1 - \Psi_2$ it follows from (30.6) that $\| \Psi_1 - \Psi_2 \| = 0$.

Having premised these two theorems, let us resume our examination of the question of the existence of a weak solution for DIRICHLET's problem, taking into consideration instead of (30.1) the more general equation in the unknown $u \in H_0^{1,2}(T)$:

$$a(u, v) = -\int_\Omega \left(f v + \sum_{i=1}^m f_i \frac{\partial v}{\partial x_i} \right) dx \quad \forall\, v \in H_0^{1,2}(\Omega), \qquad (30.8)$$

having put:

$$a(u, v) = \int_\Omega \left[\sum_{i,k=1}^m a_{ik} \frac{\partial u}{\partial x_i} \frac{\partial v}{\partial x_k} - \sum_{i=1}^m \left(d_i u \frac{\partial v}{\partial x_i} + e_i v \frac{\partial u}{\partial x_i} \right) - c\, u\, v \right] dx.$$

(30.8) is the transformation into the weak form of DIRICHLET's problem for the equation:

$$\mathfrak{M}\, u = \sum_{i,k=1}^m \frac{\partial}{\partial x_i} \left(a_{ik} \frac{\partial u}{\partial x_k} \right) + \sum_{i=1}^m e_i \frac{\partial u}{\partial x_i}$$

$$- \sum_{i=1}^m \frac{\partial}{\partial x_i} (d_i u) + c\, u = f - \sum_{i=1}^m \frac{\partial f_i}{\partial x_i}, \qquad (30.9)$$

while the equation in the unknown $v \in H_0^{1,2}(\Omega)$:

$$a(u, v) = 0 \quad \forall\, u \in H_0^{1,2}(\Omega) \qquad\qquad (30.10)$$

is the transformation of the homogeneous adjoint problem:

$$\mathfrak{N}\, v = 0, \qquad v(x) = 0 \text{ for } x \in \partial T.$$

We immediately see that the bilinear form $a(u, v)$ is *coercive* on $\Sigma = H_0^{1,2}(\Omega)$ in the sense of ARONSZAJN [3], which means that there exists a real number λ_0 such that for $\lambda \geq \lambda_0$ the bilinear functional $B(u, v) = a(u, v) + \lambda(u, v)$ satisfies (30.6). From this, because the right side of (30.8) is a linear and continuous functional of v in Σ, from theorem 30, II the existence and uniqueness for $\lambda \geq \lambda_0$ of a solution $u \in H_0^{1,2}$ of the following equation follows:

$$a(u, v) + \lambda(u, v) = -\int_\Omega \left(f v + \sum_{i=1}^m f_i \frac{\partial v}{\partial x_i} \right) dx \quad \forall\, v \in H_0^{1,2}(\Omega) \qquad (30.11)$$

and for this u:

$$\| u \|_{H_0^{1,2}} \leq k \left(\sum_{i=0}^{m} \| f_i \|_{L^2} \right)^{1/2}.$$

Obviously we can write:

$$u = \sum_{i=0}^{m} G_i(f_i, \lambda),$$

where the $G_i(\cdot, \lambda)$ are, for every $\lambda \geq \lambda_0$ linear and continuous mappings of L^2 into $H_0^{1,2}$. From this, if we want to solve (30.11) for an arbitrary value of λ, in particular for $\lambda = 0$, it will suffice to observe that this equation is equivalent to:

$$u = (\lambda - \lambda_0) \, G_0(u, \lambda_0) + \sum_{i=0}^{m} G_i(f_i, \lambda_0). \tag{30.12}$$

Since the inclusion of $H_0^{1,2}$ in L^2 is compact, by a theorem of RELLICH[1], (30.12) can be considered as a transformation of RIESZ type in L^2, and from this follows:

30, III. *Under hypotheses* (2.1) *and* (30.9) *the homogeneous equation associated with* (30.8) *and its adjoint* (30.10) *have the same finite number of linearly independent solutions. A necessary and sufficient condition that* (30.8) *be solvable is that:*

$$\int_{\Omega} \left(f_0 v + \sum_{i=1}^{m} f_i \frac{\partial v}{\partial x_i} \right) dx = 0 \tag{30.13}$$

for every v which is a solution of (30.10). *There exists a λ_0 such that for $\lambda \geq \lambda_0$* (30.11) *admits one and only one solution.*

Having thus established the existence, under condition (30.13), of a solution of the DIRICHLET problem, let us pass to the study of its regularization. It is obvious that, in order to make this regularization possible, it will be necessary to reinforce the hypotheses on the data of the problem. We can suppose, for example, that:

$$a_{ik}, e_i, d_i, f_i \in C^{(1, \lambda)}, c, f_0 \in C^{(0, \lambda)}, \tag{30.14}$$

hypotheses which are a bit more restrictive than those strictly necessary for (30.9) to be meaningful, and under which the operator $\mathfrak{M}$ is endowed with an adjoint. Moreover we shall assume $T \in A^{(3)}$. Under these hypo-

---

[1] See S.L. SOBOLEV [5].

theses, assuming $v \in H^{2,2} \cap H_0^{1,2}$, we draw from (30.8) using integration by parts:

$$\int_\Omega u \, \mathfrak{N} \, v \, dx = \int_\Omega v \left(f_0 - \sum_{i=1}^m f_i \frac{\partial v}{\partial x_i} \right) dx .$$

From this, by reasoning analogous to that applied in the proof of the Lemma of § 29, and which preserves its validity even for $m > 2$, we draw, for y almost everywhere in $T - \partial T$:

$$u(y) = \int_\Omega \left[u(x) \, \mathfrak{N}_x \, H_\varrho(x, y) - \left(f(x) - \sum_{i=1}^m \frac{\partial f_i}{\partial x_i} \right) H_\varrho(x, y) \right] dx ,$$

which suffices, by theorem 29, II which can also be extended to the case $m > 2$, to establish that u coincides almost everywhere in $T - \partial T$ with a regular solution of (30.9).

In order to complete the regularization of u it now remains to prove that $u \in C^{(0)}(T)$ in such a way that the boundary condition (30.2) would come to be satisfied in the ordinary sense. This proof, on the other hand, can be developed under considerably less restrictive hypotheses than (30.14), namely, by assuming in T together with (2.1) the following conditions:

$$e_i \in L^m, \quad d_i \in L^{m+\sigma}, \quad c \in L^{\frac{m}{2}+\sigma}, \quad \sigma > 0, \qquad (30.15)$$

$$f_i \in L^r \ (i = 1, 2, \ldots, m), \quad f_0 \in L^{r/2}, \qquad (30.16)$$

(30.16) for $r > m$.

We note that, since by S. L. SOBOLEV's inclusion theorems [5], we have $H_0^{1,2}, \in L^{\frac{2m}{m-2}}$, the hypotheses on the coefficients e_i, d_i, and c ensure that (30.8) is meaningful, while the same equation would lose its sense if (30.15) held only for $\sigma < 0$.

The study of weak solutions of (30.9) and of the related boundary value problems under hypotheses of such great generality was produced and considerably deepened between 1938 and 1943 by C. B. MORREY [1, 2] in the case $m = 2$, but only many years later, from 1956 to 1957, did some fundamental works of E. DE GIORGI [2] and J. NASH [1] give a start at all to a series of researches on the general case. These authors, considering (30.9) in the case $e_i = d_i = c = f_0 = f_i = 0$, proved with two quite distinct procedures that every weak solution of this equation coincides almost everywhere with a function of class $C^{(0,\alpha)}(T - \partial T)$, α being a convenient real number which depends only on T and on the ellipticity constant which figures in (2.1). Shortly afterwards many authors, inspired above all by DE GIORGI's method, extended his results in various directions, addressing themselves at first toward the study of equations more general than those considered by DE GIORGI and toward research on HÖLDER conditions for weak solutions, no longer of the equation alone, but of the boundary value problem for it, valid not only in $T - \partial T$ but in all of T. The first works along this line, appearing almost simultaneously, are by C. B. MORREY [6], G. STAMPACCHIA [8, 9], O. A. LADYŽENSKAJA and N. N. URAL'CEVA [1]. A little earlier came another work by G. STAMPACCHIA [7], in which he studied the integrability properties of solutions of the boundary value problem as they depend on similar properties of the f_i. A little later, a note by J. MOSER [1] contained a new simplified proof of DE GIORGI's theorem. Many other works followed these first ones, from among which, for details about the questions on regularization, we cite those of G. STAMPACCHIA [10, 12, 15], J. MOSER [3], V. G. MAZ'JA [2], H. F. WEINBERGER [1], O. A. LADYŽENSKAJA and N. N. URAL'CEVA [5, 6, 7], S. N. KRUŽKOV [2], N. MEYERS [1], C. MIRANDA [16], J. SERRIN [6]. Systematic expositions of the theory are now contained in the monographs [14] and [16] by G. STAMPACCHIA, in Chapters III and IX of the volume [8] by O. A. LADYŽENSKAJA and N. N. URAL'CEVA, and in Chapters I and V of the volume [9] by C. B. MORREY. Taken together these works have borne the theory to quite an advanced state. Let us sum up the results obtained with the following theorem, treated in G. STAMPACCHIA's work [15], which in our opinion concludes this research in a certain sense, at least concerning DIRICHLET's problem:

30, IV. *Let hypotheses* (2.1), (30.15), (30.16) *be satisfied and let D be a domain interior to T. Then there exist three numbers* $K, K_1,$ *and* α, *depending only D, T, r and on the coefficients of the equation, such that for every weak solution of* (30.9) *we have*

$$
\left.\begin{array}{l}
if \ \dfrac{1}{r} - \dfrac{1}{m} < 0 : \| u \|_{C^{(0,\,a)}(D)} \\[2mm]
if \ \dfrac{1}{s} = \dfrac{1}{r} - \dfrac{1}{m} > 0 : \| u \|_{L^s(D)}
\end{array}\right\} \leq K \left[\| f_0 \|_{L^{r/2}(T)} + \sum_{i=1}^{m} \| f_i \|_{L^r(T)} \right] + K_1 \| u \|_{L^1(T)}.
$$

The same bounds hold even for $D = T$ *if u is a solution of* DIRICHLET's *problem with boundary condition* (30.2). *If, now:*

$$
\frac{\partial d_i}{\partial x_i} \in L^{m/2+\sigma}, \qquad c - \sum_{i=1}^{m} \frac{\partial d_i}{\partial x_i} \leq c_0 < 0, \tag{30.17}
$$

with c_0 *constant and* $\sigma > 0$, *then the solution of* DIRICHLET's *problem is unique and in the preceding formula we can put* $K_1 = 0$.

For the regularity properties of u in the case $r = m$ see G. STAM-
PACCHIA [12]. Naturally, by posing hypotheses intermediate between
those of this theorem and (30.14) we can obtain intermediate results.
For example:

30, V. *Let* $T \in A^{(2,\lambda)}$, $a_{ik} \in C^{(0)}$, d_i, e_i, $c \in L^\infty$, $f_i \in L^{(2,\mu)}$ *with* $\mu > m - 2$;
then for every weak solution $u(x)$ *of* DIRICHLET'S *problem with boundary
condition* (30.2), *we have* $u \in H^{1,2,\mu}$. *If*, *now*, a_{ik}, d_i, $f_i \in C^{(0,\lambda)}$ *we have*
$u \in C^{(1,\lambda)}$.

This theorem, although not explicitly enunciated, can be deduced
from some works by C. MIRANDA [6] for the case $e_i = d_i = c = 0$ and
by C.B. MORREY [3, 4] for the general case. In the case $e_i = d_i = c = 0$
this has now been reobtained and finally generalized by S. CAMPANATO
[7]. For other results see also C.B. MORREY [9, Chapter 5].

A query which we now raise spontaneously is whether we can
establish whether convenient integrability hypotheses on f_i will ensure
integrability with an exponent exceeding 2 for the derivatives of u.
For this purpose N. MEYERS [1] has proved that:

30, VI. *If* $d_i = e_i = c = 0$, *then there exists a constant* $Q > 2$ *depending on* T *and on the ellipticity constant* a, *such that if* $f_i \in L^p$ *with* $p < Q$
we have also $u \in H^{1,p}(D)$ *for every weak solution* $u \in H^{1,2}(T)$ *of* (30.9)
and for every open D *of compact support in* $T - \partial T$. *For* $a \to 0$ *we have*
$Q \to 2$.

In the case $m = 2$ a similar result was established previously by
B.V.BOYARSKIĬ [2, 4].

We now add that analogous (even if less precise) results to those
of theorem 30, IV hold for NEUMANN'S problem, where by a weak solution of this problem we mean a function $u \in H^{1,2}(T)$ which satisfies
(30.8) for arbitrary $v \in H^{1,2}(T)$. We note that this time we do not
impose some boundary condition on u *a priori*. It is easy, however,
to verify that, under conditions of sufficient regularity on the data of
the problem and on the weak solution u, it automatically satisfies
the condition:

$$\sum_{i=1}^m X_i \left[\sum_{k=1}^m a_{ik} \frac{\partial u}{\partial x_k} + d_i u + f_i \right] = 0,$$

which is precisely of NEUMANN type and which therefore takes the name
natural boundary condition for NEUMANN'S problem in the weak sense.
For this we can consult the works of M.I.VIŠIK [4] and of E. MAGENES
and G. STAMPACCHIA [1] for the general foundation, and the works
[7, 9, 10, 14] of G. STAMPACCHIA for the regularization. Finally, for the
extension even to the oblique derivative problem, see M.I.VIŠIK [7],
J.L.LIONS [4], E. MAGENES and G. STAMPACCHIA [1], R. FIORENZA [3].

Now, let us point out briefly various results obtained by G. STAM-PACCHIA [15] and of others, concerning the extension to weak solutions of the theorems of §§ 3, 27, 28. Regarding the extension of the theorems of § 3, we can for example prove that under hypothesis (30.17), if $a(u, v) \leq 0$ for every non-negative function $v \in C_0^\infty (T - \partial T)$, we have:

$$\max_T u \leq \max(0, \max_{\partial T} u).$$

We can also prove that under hypothesis (30.17) the solution of DIRICHLET's problem in the case $f_i = 0$ for $i = 1, 2, \ldots, m$ can be put in the form:

$$u(x) = - \int_T G(x, y) \, f_0(y) \, dy,$$

where $G(x, y)$, which is called the GREEN's function of the problem, is as a function of x, a weak solution in $T - (\partial T \cup y)$ of the equation $\mathfrak{M} u = 0$ and satisfies in a neighborhood of y a bound of the type:

$$k^{-1} \, | \, x - y \, |^{2-m} \leq G(x, y) \leq k \, | \, x - y \, |^{2-m}.$$

These results were proved by W. LITTMAN, G. STAMPACCHIA and H. F. WEINBERGER [1] for the case $e_i = d_i = c = 0$ and by G. STAMPACCHIA [15] in the general case[1].

In the memoir [15] G. STAMPACCHIA also gave, for weak solutions, a complete extension of HARNACK's theorem (see § 27), already obtained in particular cases by J. MOSER [2] and by S.N. KRUŽKOV and L.P. KOPKOV [1]. Another extension of HARNACK's theorem, which also concerns the solutions of nonhomogeneous equations, was given by J. SERRIN [6] which is useful [6, 7, 8] to prove the following generalization of theorems 27, V and 19, XI:

30, VII. *Let the hypotheses of theorem 30, IV be satisfied with $r > m$. If* (30.17) *is satisfied, and y is a point of $T - \partial T$, then every function $u(x)$ which is a weak solution in $T - (\partial T \cup y)$ of* (30.9) *is necessarily of the form:*

$$u(x) = h \, G(x, y) + v(x),$$

where h is a constant and $v(x)$ is a weak solution of (30.9) *in $T - \partial T$. Therefore, if $u(x) = o(| \, x - y \, |^{2-m})$ then the point y is a removable singularity of u, which will then turn out to be a weak solution in $T - \partial T$ of* (30.9).

Other theorems of this type are found in SERRIN's works already cited, and also in J. SERRIN and H.F. WEINBERGER [1]. We owe the following extension of theorem 27, VII to J. SERRIN [6][2]:

---

[1] For the first of the results noted see also W. LITTMAN [1, 2] and J. KADLEC [2], for the second H. L. ROYDEN [1].

[2] For more particular extensions see also J. SERRIN [5. I] and Y. KATO [1].

30, VIII. *Let the hypotheses of theorem* 30, IV *be satisfied with* $r > m$ *and let* Q *be a compact subset of* $T - \partial T$ *with vanishing s-capacity for* $2 \leq s \leq m$. *Every function* $u \in L^q$ *with* $q > \dfrac{s}{s-2}$ *which is a weak solution of* (30.9) *in* $T - (Q \cup \partial T)$ *can be extended to* Q *by continuity, and the resulting function is a weak solution of* (30.9) *in* $T - \partial T$. *In particular, if* Q *has a vanishing ordinary capacity, then the singularity on* Q *is removable for every bounded weak solution.*

In this regard we recall that we define s-capacity of a bounded set Q as the infimum of

$$\int_{R_m} \Big[\sum_{|\alpha|=1} |D^\alpha \varphi|^2 \Big]^{s/2} dx,$$

where φ varies over the set of functions of class $C_0^1(R_m)$ which are ≥ 1 on Q; moreover, every bounded set Q having finite p-dimensional HAUSDORFF measure with $0 \leq p < m - 1$ also has vanishing s-capacity for $s < m - p$.

The DIRICHLET problem with which we have been occupied up till now is for the homogeneous boundary condition (30.2). We want now to see what can be said if in place of (30.2) we consider a non-homogeneous condition:

$$u(x) = \varphi \quad \text{for} \quad x \in \partial T. \tag{30.18}$$

It is obvious that if $\varphi \in H_{1-1/p}$ with $p \geq 2$, and if $u_0 \in H^{1,p}$ is a function such that $\gamma u_0 = \varphi$, then we can go back immediately to the case with the boundary condition (30.2) by a change of unknown function of the type $u = u_0 - z$. Regarding this, it is instructive to observe that, if we consider in particular the self-adjoint equation (30.3) under the hypothesis $f = 0$, $c \leq 0$ and remember the definitions of the HILBERT spaces Σ and Σ_0, solving the equation with boundary condition (30.18) means finding two functions z and u, the first belonging to Σ_0 and the second orthogonal to Σ_0, such that $u_0 = u + z$. The functions z and u thus come to be the *projections* of u_0 on Σ_0 and on its complementary space in Σ, respectively. From this the name *method of orthogonal projections* was initially given by WEYL to this procedure. On the other hand, it is necessary to warn the reader that this way of treating the problem is applicable only in the case of self-adjoint problems.

When the hypothesis $\varphi \in H_{1-1/p}$ is not satisfied, the procedures followed are modified by resorting to function spaces different from $H^{1,2}$. Referring to M. I. VIŠIK [11] and to J. NEČAS [1, 4, 5, 7] for the details, we will instead examine the question from a different point of view similar to that of § 28.

We want to solve, under hypothesis (30.17), the weak DIRICHLET problem with $f_i = 0$ and with boundary condition (30.18), assuming $\varphi \in C^{(0)}(\partial T)$; in this case we approximate φ by means of a sequence of functions $\varphi_n \in C^{(0)}(\partial T) \cap H_{1/2}$, and let u_n be the weak solution of DIRICHLET's problem for the boundary condition $u_n = \varphi_n$ on ∂T; we prove that the sequence of the u_n's converges uniformly in T to a function u which is a weak solution of the equation $\mathfrak{M}\,u = 0$ in every domain interior to T, and which does not depend on the sequence φ_n chosen to approximate φ. Every point of ∂T at which, for arbitrary φ, the corresponding u turns out to be continuous and to satisfy the boundary condition, is called *regular for the operator* $\mathfrak{M}$; one can prove that *every point is regular for* $\mathfrak{M}$ *if and only if it is regular for* $\varDelta_2$. For this theorem, then, see G. STAMPACCHIA [15], and for some preceding results for more particular equations, W. LITTMAN, G. STAMPACCHIA and H. F. WEINBERGFR [1], R. M. HERVÉ [2], V. G. MAZ'JA [4].

To conclude, we wish to observe that the procedure described is not the only one which allows a weak basis for boundary value problems. For example, C. B. MORREY [3, 4] considers as a weak solution of (30.9) every function $u \in H^{1,2}$ which for almost all cells $R \in T - \partial T$ satisfies the equation:

$$\int_{\partial R} \sum_{i=1}^{m} X_i \left[\sum_{k=1}^{m} a_{ik} \frac{\partial u}{\partial x_k} + d_i u + f_i \right] d\sigma + \int_{R} \left[\sum_{i=1}^{m} e_i \frac{\partial u}{\partial x_i} + c\,u - f_0 \right] dx = 0.$$

Along these lines see also LIANG SHI-TING [1] and for the case $m = 2$ G. TAUTZ [2]. Other possible weak bases are bound up with the notions of weak and strong extension of the operator $\mathfrak{M}$ introduced by K. O. FRIEDRICHS [2, 3] which we propose to recall briefly here.

Let $\mathfrak{M}$ be an elliptic operator with coefficients of class $C^{(0)}(T)$ and let us consider the mapping of $C^{(2)}(T)$ into $C^{(0)}(T)$ which to every $u \in C^{(2)}(T)$ causes the function $f = \mathfrak{M}\,u$ to correspond. Denoting by $\mathfrak{H}$ the closure of $C^{(2)}(T)$ with respect to the norm:

$$\| u \|_{\mathfrak{H}} = \| u \|_{L^2(T)} + \| \mathfrak{M}\,u \|_{L^2(T)} \tag{30.19}$$

$f = \mathfrak{M}\,u$ can obviously be extended to a mapping of $\mathfrak{H}$ into L^2, which takes the name "strong extension of the operator $\mathfrak{M}$".

The weak extension of the operator $\mathfrak{M}$ is obtained instead by putting $f = \mathfrak{M}\,u$ every time it is possible to associate to the function $u \in L^2$ a function $f \in L^2$ such that:

$$\int_{T} (u\,\mathfrak{N}\,v - f\,v)\, dx = 0, \qquad \forall\, v \in C_0^2(T - \partial T). \tag{30.20}$$

If, as we have assumed, T is bounded and $\mathfrak{M}$ is endowed with an adjoint, M.S. Narasimhan [2] proved that the weak and strong extensions are identical[1]. Still, for the purpose of the applications, it is necessary to consider them separately. While the notion of strong extension is associated, even if this does not appear openly, with the method of Cimmino discussed in § 29, the notion of weak extension suggests the definition of every function $u \in L^2$ which satisfies the equation:

$$\int_T [u \, \mathfrak{N} \, v - f \, v] \, dx = 0 \qquad \forall \, v \in H^{2,\,2}(T) \cap H_0^{1,\,2}(T) \qquad (30.21)$$

as *a weak solution of* Dirichlet's *problem* for the equation $\mathfrak{M} \, u = f$ with boundary condition (30.2). Indeed, having imposed by (30.21) that the equation is satisfied for every v in a larger class than that considered in (30.20) bears as a consequence that (30.2) presents itself as a *natural condition* on solutions of (30.21), namely as a condition which is automatically satisfied under hypotheses of sufficient regularity for u. In the case $\mathfrak{M} = \varDelta_2$ the study of Dirichlet's problem on this basis and an analogous study of Neumann's problem were amply developed by H.G. Garnir [2]. A final generalization consists in substituting for $f \, dx$ in equation (30.21) $d\mu$, μ being a measure. On this way of defining weak problems see the works [19, 22] of G. Fichera, in which theorem 30, I was used in an essential way, and also the memoir [15] of G. Stampacchia. For some extensions along these lines of the theorems of § 3 see W.L. Littman [1, 2] and K. Hayashida [2] and, for an application to the second boundary value problem, M.P. Colautti [1]. Finally in connection with the Dirichlet problem with an inhomogeneous boundary condition, a treatment in the weak sense can be based on the following technique. If the condition is written in the form (30.18) with $\varphi \in L^2(\partial T)$, then every function $u \in L^2(T)$ such that for every $v \in H^{2,\,2}(T) \cap H_0^{1,\,2}(T)$:

$$\int_T u \, \mathfrak{N} \, v \, dx = \int_T f \, v \, dx + \int_{\partial T} \varphi \, \mathfrak{D} \, v \, d\sigma$$

is called a weak solution of the problem.

A more general basis, then, for the same problem can now be given in the compass of the theory of distributions. Precisely, denoting by $H^{-k,\,2}(R_m)$ the space (of distributions) dual to $H^{k,\,2}(R_m)$ and given a distribution S of $H^{-k,\,2}(R_m)$ with compact support in T, we can pose the problem of determing a distribution U of $H^{2-k,\,2}(R_m)$, also with compact support in T, such that we have for every $v \in H^{k,\,2} \cap H_0^{1,\,2}(T)$:

$$< U, \mathfrak{N} \, v > \; = \; < S, v > .$$

---

[1] A propos of this see also K.O. Friedrichs [3].

For more precision in this regard and for information concerning the results which can be obtained on this basis we refer to M.I. Višik and S.L. Sobolev [1], J.L. Lions [5, 7], E. Magenes and G. Stampacchia [1].

We add, finally, that all the procedures studied in this section can also be applied to the study of boundary value problems more general than those just considered. To this we shall return in § 50.

31. The method of Fischer-Riesz equations. In this section we shall occupy ourselves with a method for solving boundary value problems proposed by M. Picone [8] and succesively studied with important results by L. Amerio [3, 5, 7] and by G. Fichera [2, 4, 7]. This method on the one hand offers the advantage of furnishing a procedure for the numerical calculation of the solutions of boundary value problems, and on the other hand is interesting also for questions related to existence problems; it is, moreover, also capable of various applications to problems different from those which interest us here. Some of these latter applications will be indicated in Chapter VII (§§ 50, 52, 55); for others we refer to the memoir [1] by M. Picone and G. Fichera, in which the method is exposed in an abstract form which is totally general; we also point out the memoir [11] by G. Fichera in which the close bonds which run between this method and that of R. Caccioppoli and G. Cimmino discussed in § 29 are put in evidence.

Let us begin by recalling some classical notions from the theory of linear approximations which we must immediately employ. Let us now consider two vectors $\alpha \equiv (a_1, a_2, \ldots, a_p)$ and $\alpha' \equiv (a'_1, a'_2, \ldots, a'_p)$ whose k^{th} components a_k and a'_k are real functions defined and square integrable on a manifold I_k of R_m; we shall call the quantity:

$$(\alpha, \alpha') = \sum_{k=1}^{p} \int_{I_k} a_k \, a'_k \, dI_k$$

the *scalar product* of the two vectors α and α'.

Now, given a system of linearly independent vectors $\{\alpha^{(n)}\}$ and a sequence of real constants $\{c_n\}$, we propose to determine a vector α such that for every n, $(\alpha, \alpha^{(n)}) = c_n$. If we denote by a_k the components of α and by $a_k^{(n)}$ those of $\alpha^{(n)}$, then these equations are written:

$$\sum_{k=1}^{p} \int_{I_k} a_k \, a_k^{(n)} \, dI_k = c_n \tag{31.1}$$

and to this it is customary to give the name *system of integral equations of* Fischer-Riesz *type*. Now, by the Schmidt orthogonalization process it is possible, in a unique way, to determine the constants A_{ni} in such a manner that the vectors $\{\beta^{(n)}\}$ given by $\beta^{(n)} = \sum_{i=1}^{n} A_{ni} \alpha^{(i)}$ are orthonormal. Denoting by $b_k^{(n)}$ the components of $\beta^{(n)}$, the system (31.1) is now equivalent to:

$$\sum_{k=1}^{p} \int_{I_k} a_k \, b_k^{(n)} \, dI_k = \sum_{i=1}^{n} A_{ni} \, c_i, \qquad (31.2)$$

which, manifestly, is solvable if and only if:

$$\sum_{n=1}^{\infty} \left(\sum_{i=1}^{n} A_{ni} \, c_i \right)^2 < \infty . \qquad (31.3)$$

If condition (31.3) is satisfied then the most general solution of the system (31.1) is obtained by putting:

$$a_k = \bar{a}_k + \sum_{n=1}^{\infty} b_k^{(n)} \sum_{i=1}^{n} A_{ni} \, c_i, \qquad (31.4)$$

where $\bar{\alpha} = (\bar{a}_1, \bar{a}_2, \ldots, \bar{a}_p)$ is an arbitrary vector orthogonal to all the $\alpha^{(n)}$ and the series on the right side is considered in the sense of convergence in the mean on I_k. In particular, if the sequence $\{\alpha^{(n)}\}$ is *closed*, namely if every vector which is orthogonal to all the $\alpha^{(n)}$ has components zero almost everywhere on their manifolds of definition, then the solution of the system (31.1) if it exists, is unique and is given by (31.4) with $\bar{a}_k = 0$.

Having premised all of this, let us pass to the exposition of PICONE's method and the results which it permits us to reach. Without preoccupying ourselves for now with specifying the hypotheses which ought to be satisfied by the functions being considered, we denote by u a solution in T of the equation $\mathfrak{M} \, u = f$ and by $\{w^{(n)}\}$ a sequence of functions also defined in T. From GREEN's formula we obtain for arbitrary n:

$$\int_{T} \left(w^{(n)} f - u \, \mathfrak{N} \, w^{(n)} \right) dx - \int_{\partial T} \left(w^{(n)} \, \mathfrak{P} \, u - u \, \mathfrak{Q} \, w^{(n)} \right) d\sigma = 0 , \qquad (31.5)$$

where $\mathfrak{P}$ and $\mathfrak{Q}$ are defined by (7.1) and (7.2) assuming $l \equiv \nu$. PICONE's method consists of deriving from (31.5) a system of FISCHER-RIESZ equations in a convenient unknown vector the solution of which permits us to attain the calculation of the solution of the given boundary value problem.

This can be done in various ways, which we shall describe, limiting ourselves for now only to consideration of DIRICHLET's problem with the boundary condition $u = \varphi$.

9*

Method 1°. Let $\{z^{(n)}\}$ be a sequence of functions which are solutions of the equation $\mathfrak{N} z^{(n)} = 0$; putting $w^{(n)} = z^{(n)}$ in (31.5), this equation can be written in the form:

$$\int_{\partial T} z^{(n)} \, \mathfrak{P} \, u \, d\sigma = \int_{T} z^{(n)} \, f \, dx + \int_{\partial T} \varphi \, \mathfrak{Q} \, z^{(n)} \, d\sigma \,. \tag{31.6}$$

And (31.6) can be considered a system of FISCHER-REISZ equations in the unknown function (a vector with a single component) $\mathfrak{P} \, u$. If this equation allows calculation of $\mathfrak{P} \, u$ and if a fundamental solution $G(x, y)$ is known, then STOKES's formula:

$$u(x) = - \int_{T} G(x, y) \, f(y) \, dy + \int_{\partial T} [G(x, y) - \mathfrak{P} \, u(y) - \varphi(y) \, \mathfrak{Q}_y \, G(x, y)] \, d_y \sigma$$
$$\tag{31.7}$$

now furnishes the expression for u.

Method 2°. Having calculated $\mathfrak{P} \, u$ as above, and denoting by $\{w^{(n)}\}$ a sequence of functions which are *not* solutions of the equation $\mathfrak{N} \, w^{(n)} = 0$, (31.5) written in the form:

$$\int_{T} u \, \mathfrak{N} \, w^{(n)} \, dx = \int_{T} w^{(n)} \, f \, dx - \int_{\partial T} \left(w^{(n)} \, \mathfrak{P} \, u - \varphi \mathfrak{Q} \, w^{(n)} \right) d\sigma \tag{31.8}$$

constitutes a system of FISCHER-RIESZ equations by means of which we can attempt to calculate u directly, avoiding any recourse to (31.7).

Method 3°. (31.5) written in the form:

$$\int_{T} u \, \mathfrak{N} \, w^{(n)} \, dx + \int_{\partial T} w^{(n)} \, \mathfrak{P} \, u \, d\sigma = \int_{T} w^{(n)} \, f \, dx + \int_{\partial T} \varphi \, \mathfrak{Q} \, w^{(n)} \, d\sigma \tag{31.9}$$

can be considered a system of FISCHER-RIESZ equations in the unknown vector which has components u in T and $\mathfrak{P} \, u$ on ∂T. And from the solution of this system we can obtain the simultaneous calculation of u and $\mathfrak{P} \, u$.

Similar considerations are obviously valid for NEUMANN's problem; the procedure can then be adapted even to the solution of the mixed problem, upon which, on the other hand, we shall not dwell here; we shall wait to give some indications of the argument in Chapter VII (§ 50).

Let us now observe that even if it is true that every solution of the boundary value problem being considered is also a solution of the system of FISCHER-RIESZ equations indicated above, the converse does not hold for an arbitrarily choice of the functions $z^{(n)}$ or $w^{(n)}$. And in his note [8] on the argument, PICONE proposes precisely the question of seeing how we ought to choose these functions to make sure that the systems of FISCHER-RIESZ equations have as solutions those and only those solutions of the boundary value problem which we wish to solve. And a first answer to this question, for the particular case of equations with constant coefficients, is already found in the cited note by PICONE and in the memoir [3] by L. AMERIO. In a succeeding memoir [6] AMERIO then attacked the study of the question in all its generality.

Let, therefore, $\mathfrak{M}$ be an elliptic operator defined in a region Ω and suppose that the a_{ik} are of class $C^{(2, \lambda)}$ in Ω, the b_i are of class $C^{(1, \lambda)}$, and c of class $C^{(0, \lambda)}$. Then let $T \subset \Omega$ be a bounded domain which, for simplicity[1], we assume of class $A^{(2)}$, and let T_1 be another domain also of class $A^{(2)}$ to which T is interior. Under these hypotheses the equation $\mathfrak{M} u = 0$ admits at least one fundamental solution $G(x, y)$ defined in T_1 and which we shall intend to have fixed once and for all. Denoting now by $\mathfrak{P}$ and $\mathfrak{Q}$ the operators defined by (7.1) and (7.2) with $l \equiv v$ and $\beta \in C^{(0)}$, let us denote by $\Gamma^{(1)}$ the class of functions $u(x)$ defined in $T - \partial T$ and satisfying the following conditions there:

a) u is of class $C^{(2)}$ in $T - \partial T$ and $\mathfrak{M} u$ is bounded and of class $C^{(0, \lambda)}$ in $T - \partial T$.

b) for x_0 almost everywhere on ∂T and for $x \to x_0$ along the conormal, the following limits exist:

$$\lim_{x \to x_0} u(x) = \varphi(x_0), \qquad \lim_{x \to x_0} \mathfrak{P} u = \psi(x_0), \qquad (31.10)$$

the functions $\varphi(x_0)$ and $\psi(x_0)$ turning out to be integrable on ∂T[2].

c) We have:

$$\theta u(x) = -\int_T G(x, y)\, \mathfrak{M} u(y)\, dy + \int_{\partial T} [\psi(y)\, G(x, y) - \varphi(y)\, \mathfrak{Q}_y\, G(x, y)]\, d_y\sigma,$$
$$(31.11)$$

with $\theta = 1$ for $x \in T - \partial T$ and $\theta = 0$ for $x \in T_1 - T$.

We shall denote now by $\Gamma^{(2)}$ the subclass of $\Gamma^{(1)}$ consisting of all the functions $u(x)$ for which $\varphi, \psi \in L^2$ and by $\Gamma^{(C)}$ the other subclass consisting of the functions for which $\varphi, \psi \in C^{(0)}$ and (31.10) is satisfied almost everywhere on ∂T.

---

[1] In AMERIO's treatment he considers a domain T of very general type, whose boundary can even admit, under certain conditions, singular points constituting a set of hypersurface measure zero.

[2] For the significance of $\mathfrak{P} u$ for $x \neq x_0$ see (14.13).

Fundamental for the sequel is the following theorem due to AMERIO:

31, I. *If the functions f, assumed bounded and of class $C^{(0,\lambda)}$ in $T - \partial T$, and $\varphi, \psi \in L^1 (\partial T)$ satisfy the following relation for $x \in T_1 - T$:*

$$-\int_T G(x, y)\, f(y)\, dy + \int_{\partial T} [\psi(y)\, G(x, y) - \varphi(y)\, \mathfrak{D}_y\, G(x, y)]\, d_y\sigma = 0 \quad (31.12)$$

then the function $u(x)$ defined in $T - \partial T$ by:

$$u(x) = -\int_T G(x, y)\, f(y)\, dy + \int_{\partial T} [\psi(y)\, G(x, y) - \varphi(y)\, \mathfrak{D}_y\, G(x, y)]\, d_y\sigma$$
$$(31.13)$$

belongs to the class $\Gamma^{(1)}$, and for it (31.10) holds and we have $\mathfrak{M}\, u = f$. If, then, $\varphi, \psi \in C^{(0)}$, we have $u \in \Gamma^{(C)}$.

Because the relation $\mathfrak{M}\, u = f$ is obvious, it evidently suffices to prove (31.10). If we mean by $u(x)$ the right side of (31.13) even for $x \in T_1 - T$, we have by (31.12): $u(x) = 0$ in $T_1 - T$. Denoting by x_0 a point of ∂T and by $x \in T$, $x' \in T_1 - T$ two points on the conormal through x_0 symmetrical with respect to x_0, we can write:

$$\lim_{x \to x_0} u(x) = \lim_{x \to x_0} (u(x) - u(x')), \quad \lim_{x \to x_0} \mathfrak{P}\, u(x) = \lim_{x \to x_0} (\mathfrak{P}\, u(x) - \mathfrak{P}\, u(x'))$$

and after this the first of (31.10) follows from theorems 13, I, 14, II, and 15, II, and the second from theorems 13, I, 14, VI, and 15, VI[1].

Now if $\{v^{(n)}\}$ is a sequence of functions defined and continuous in $T_1 - (T - \partial T)$, then by multiplying (31.12) by $v^{(n)}$ and integrating in T we obtain the equations:

$$\int_{\partial T} z^{(n)}\, \psi\, d\sigma = \int_T z^{(n)}\, f\, dx + \int_{\partial T} \varphi\, \mathfrak{D}\, z^{(n)}\, d\sigma, \quad (31.14)$$

with:

$$z^{(n)}(y) = \int_{T_1 - T} G(x, y)\, v^{(n)}(x)\, dx. \quad (31.15)$$

Whence the theorem:

31, II. *If the sequence $\{v^{(n)}\}$ is closed in $T_1 - (T - \partial T)$ and if $\varphi, \psi \in L^2$, then the truth of (31.14) with the $z^{(n)}$ given by (31.15) is a necessary*

---

[1] The idea of premising for the proof of this theorem an exhaustive theory of generalized potentials is from G. FICHERA [2], who completely developed it in the case $\mathfrak{M} = \Delta_2$, $m = 3$. It is still to be observed that some elements of this theory are contained in L. AMERIO's memoir [6] in which this author gave a direct proof of theorem 31, I.

and sufficient condition that (31.12) *hold. The result is valid even if* φ, $\psi \in L^1$ *provided that every function integrable in* $(T_1 - (T - \partial T))$ *and orthogonal to all the* $v^{(n)}$ *is zero almost everywhere.*

Indeed, it suffices to observe that if $\varphi, \psi \in L^p$ with $p = 1, 2$ then the left side of (31.12) is p-integrable in $T_1 - (T - \partial T)$, by dint of theorems 14, I and 15, I.

Consequently, if we can solve the generalized boundary value problem consisting of the search for a solution in class $\Gamma^{(2)}$ of the equation $\mathfrak{M} u = f$ which satisfies one or the other of (31.10), then by means of (31.14), considered as a system of FISCHER-RIESZ equations, we can determine the one of the functions φ or ψ which is unknown, then apply (31.13). Theorems 31, I and 31, II ensure, in fact, that with this procedure we obtain all solutions, and only solutions, of the boundary value problem being considered. Because (31.14) does not differ from (31.6) other than by the formal substitution of ψ for $\mathfrak{P}u$, and on the other hand the $z^{(n)}$ given by (31.14) are obviously regular solutions in $T - \partial T$ of the equation $\mathfrak{N} z^{(n)} = 0$, we recognize in what we have said an application of method 1° of PICONE to the solution of the generalized DIRICHLET and NEUMANN problems.

However, at the end of the numerical calculation, this method presents the grave inconvenience of requiring a knowledge of the fundamental solution. We shall now see how method 3° is applicable also, without the advance knowledge of $G(x, y)$.

We begin by observing that:

31, III. *If* u *is a solution in class* $\Gamma^{(1)}$ *of the equation* $\mathfrak{M} u = f$, *and if* v *is a function of class* $C^{(2)}$ *in* T, *then* GREEN's *formula holds:*

$$\int_T [u \, \mathfrak{N} v - f v] \, dx = \int_{\partial T} [\varphi \, \mathfrak{Q} v - v \, \psi] \, d\sigma . \tag{31.16}$$

Indeed, because it is valid (theorem 16, VI) to assume v defined and of class $C^{(2)}$ in T_1, (13.16) is immediately satisfied by substituting for v its expression given by STOKES's formula for the domain T_1, then inverting the orders of integration and taking account of (31.12) and (31.13).

After this we can prove that:

31, IV. *If* $\varphi, \psi \in L^1$ *and if* f *is continuous in* T *and of class* $C^{(0, \lambda)}$ *in* $T - \partial T$, *then a necessary and sufficient condition that a function* $u(x) \in L^1(T)$ *coincide almost everywhere in* $T - \partial T$ *with a function of class* $\Gamma^{(1)}$ *satisfying* (31.12) *and* (31.13) *is that* (31.16) *be satisfied by all the functions* $v = w^{(n)}$, $\{w^{(n)}\}$ *being the sequence of monomials* $x_1^{r_1} x_2^{r_2} \ldots x_m^{r_m}$, $r_1, r_2, \ldots r_m$ *being nonnegative integers.*

That the condition is necessary follows from the preceding theorem. Therefore we shall prove that the condition is sufficient. We observe for this purpose that if (31.16) is satisfied for $v = w^{(n)}$, we shall have even for $x \in T_1$:

$$\int_T \left[u(x)\, \mathfrak{N}_y\, P_n(x, y) - f(y)\, P_n(x, y) \right] dy$$

$$= \int_{\partial T} \left[\varphi(y)\, \mathfrak{D}_y\, P_n(x, y) - \psi(y)\, P_n(x, y) \right] d_y \sigma , \qquad (31.17)$$

where:

$$P_n(x, y) = \left(\frac{n}{\pi R^2} \right)^{m/2} \int_{T_1} G(x, t) \prod_{i=1}^m \left[1 - \frac{(t_i - y_i)^2}{R^2} \right]^n dt$$

is the nth STIELTJES polynomial of $G(x, y)$, R denoting the diameter of T_1. It is well known that these polynomials and their first and second derivatives with respect to the y_i converge uniformly to $G(x, y)$ and its same derivatives in every domain interior to T_1 which excludes the point x. Passing to the limit in (31.17) for $n \to \infty$ and $x \in T_1 - T$, we therefore obtain (31.12) immediately.

The same passage to the limit applied in the case when $x \in T - \partial T$ permits the proof that (31.13) holds on the condition that:

$$\lim_{\varrho \to 0} \varrho^{-m} \int_{\Gamma(x, \varrho)} | u(y) - u(x) | \, dy = 0 ,$$

namely for x almost everywhere in T. The proof of this is attained by reasoning similar to that developed in § 29 to justify the passage to the limit which leads from (29.14) to (29.15). In this reasoning the fact, proved by AMERIO, is essential, namely that the $P_n(x, y)$ and the $\mathfrak{N}_y P_n(x, y)$ are $O(r^{2-m})$ and $O(r^{1-m})$ respectively, uniformly as x and y very in T as well as for n.

From what we have now proved it falls out that, if we assume as functions $w^{(n)}$ those of theorem 31, IV, the system of FISCHER-RIESZ equations considered in PICONE's method 3° admit as solutions all those vectors $(u, \mathfrak{P} u)$, and only those, for which u is a solution of the generalized DIRICHLET problem; and the same system can be used for the solution of NEUMANN's problem, assuming as unknown the vector (u, φ). Finally, the same $w^{(n)}$ can be used for the application of the second method. From theorem 31, IV it in fact follows immediately that:

31, V. *If $\{w^{(n)}\}$ is the sequence of monomials of theorem 31, IV, then the sequence $\{\mathfrak{N}\, w^{(n)}\}$ is closed.*

AMERIO also proved that analogous results are obtained by assuming as functions $w^{(n)}$, as well as the monomials $x_1^{r_1} x_2^{r_2} \ldots x_m^{r_m}$, functions of the type $exp\,(2\,\pi\,i\,\Sigma r_k\,x_k/T_k)$. In this way this procedure shows its unity with that of the LAPLACE transform with a finite interval. Along these lines are some memoirs by A. GHIZZETTI [1, 3, 6] for the equation $\Delta_2 u - \lambda^2 u = f$, as well as the first results of PICONE [8] and AMERIO [3] on equations with constant coefficients.

With AMERIO's research, of which we have until now exposed the results, the solution of the generalized boundary value problems is recognized as equivalent to that of a system of FISCHER-RIESZ equations. And consequently every existence theorem for a generalized boundary value problem is transmuted into an analogous theorem for a system of FISCHER-RIESZ equations. In particular, every uniqueness theorem gives occasion for a closure theorem for a system of functions or of vectors. In the case instead where a uniqueness theorem does not hold, the search for solutions of the homogeneous problem is changed into that for the functions (or the vectors) orthogonal to the functions (or the vectors) of a given system. We shall not occupy ourselves with this latter question, for which we refer to AMERIO's note [7]; we wish instead to establish under what conditions theorems of existence and uniqueness are valid for generalized problems. What we shall say constitutes a simple extension of results obtained along this line by G. FICHERA [2] in the particular case $\mathfrak{M} = \Delta_2$, $m = 3$.

We now have:

31, VI. *Whenever the uniqueness theorem holds for the ordinary* DIRICHLET *boundary value problem, the same theorem also holds for the generalized boundary value problem.*

Let us suppose, for example, that u is a solution of the generalized homogeneous DIRICHLET problem. From (31.12) written with $f = \varphi = 0$ we deduce, bearing in mind theorem 14, VI, that for x almost everywhere on ∂T:

$$\frac{1}{2}\,\psi\,(x) = \int\limits_{\partial T} \psi\,(y)\,\mathfrak{P}_x\,G\,(x, y)\,d_y\sigma\,.$$

From this, iterating, we find that ψ satisfies an integral equation with a continuous kernel and is therefore continuous; u is then (theorem 31, I) of class $\Gamma^{(C)}$ and therefore zero as a solution of the ordinary DIRICHLET problem.

Passing now to examining the question of existence, a difficulty presents itself for the generalized DIRICHLET problem. In general, in fact, a solution of $\mathfrak{M} u = f$ which satisfies the first of (31.10) is not endowed with a conormal derivative and therefore does not belong to the class $\Gamma^{(1)}$; this can happen even if φ is assumed continuous, or better, HÖLDER continuous. Instead, it is known (see Chapter V, § 35 and G. GIRAUD [12]) that the conormal derivatives exists if φ is of class $C^{(1,\lambda)}$ on ∂T, but then u turns out to be of class $C^{(1,\lambda)}$ in T. Therefore, this is a case in which PICONE's method of calculation is surely applicable; for other cases see L. AMERIO [8] and D. GRECO [3].

For the generalized NEUMANN problem we prove the following theorem, referring to S. ALBERTONI [1, 2], G. PRODI [1], L. AMERIO [8] for other results:

31, VII. *If $\psi \in L^2$, then a necessary and sufficient condition that the generalized NEUMANN problem admit a solution is that f and ψ satisfy the same compatibility condition required for the ordinary problem. In particular, if the associated homogeneous problem admits only the zero solution, then the generalized NEUMANN problem is solvable for arbitrary f and ψ.*

Indeed, if we seek a solution of the equation $\mathfrak{M} u = f$ which satisfies the second of (31.10), we can, as in § 22, assume for u the expression:

$$u = - \int_T G^*(x, y)\, z(y)\, dy + 2 \int_{\partial T} G^*(x, y)\, \zeta(y)\, d_y\sigma\,, \qquad (31.18)$$

where G^* is the principal fundamental solution of the equation $\mathfrak{M} u - \chi u = 0$ with $c - \chi < 0$. z should this time be assumed of class $C^{(0,\lambda)}$ in $T - \partial T$ and L^2 in T, ζ of class L^2 on ∂T. Taking account as usual of (13.7) and (14.4), apart from theorem 13, V, we thus find again the system of integral equations (22.5) with the substitution of ψ in place of φ. This system need be satisfied only almost everywhere, but this does not imply any alteration of the argument[1]. Our theorem therefore stands completely proved if we can show that the functions f and ψ and the function φ equal almost everywhere on ∂T to the right side of (31.18) satisfy (31.12), because with this we shall have proved the inclusion of u in the class $\Gamma^{(2)}$. Now from STOKES's formula we easily draw the identity:

$$G(x, y) = \int_T G(x, t)\, \chi(t)\, G^*(t, y)\, dt$$

$$+ \int_{\partial T} [G^*(t, y)\, \mathfrak{D}_t\, G(x, t) - G(x, t)\, \mathfrak{P}_t\, G^*(t, y)]\, d_t\sigma\,,$$

---

[1] Iterating (22.5) in a convenient manner, we in fact reach a system with continuous kernel matrix, and known square integrable terms, whose solution (z, ζ) enjoys the prescribed properties for such functions.

valid for $x \in T_1 - T$ and $y \in T$. Now it follows for $x \in T_1 - T$, $y \in \partial T$:

$$\frac{1}{2} G(x, y) = \int_T G(x, t)\, \chi(t)\, G^*(t, y)\, dt$$

$$+ \int_{\partial T} [G^*(t, y)\, \mathfrak{Q}_t\, G(x, t) - G(x, t)\, \mathfrak{P}_t\, G^*(t, y)]\, d_t\sigma .$$

Integrating these formulas with respect to y, the first over T after having multiplied it by $z(y)$ and the second over ∂T after having multiplied it by $2\,\zeta(y)$, then subtracting term for term and taking account of (22.5) and of the expression for φ, we easily arrive at (31.12).

In complement to the theorem now proved we observe that, because we can give the form (31.18) to every function of class $\Gamma^{(2)}$, it follows from theorems 13, III and 14, II that:

31, VIII. *Every function of class $\Gamma^{(2)}$ has first derivatives which are square integrable in T.*

And we can now add that:

31, IX. *For every function u of class $\Gamma^{(2)}$ in T, the following formula holds:*

$$\int_T \left[u\, \mathfrak{M}\, u + \sum_{i,\,k=1}^{m} a_{ik} \frac{\partial u}{\partial x_i} \frac{\partial u}{\partial x_k} \right] dx$$

$$= \frac{1}{2} \int_T (c + c^*)\, u^2\, dx + \int_{\partial T} \varphi \left[\psi - \left(\beta - \frac{b}{2} \right) \varphi \right] d\sigma . \tag{31.19}$$

In fact, (31.19) is reduced to (7.13) if $u \in C^{(1)}$. In the general case there is no loss of generality in supposing $c = 0$, $\beta > 0$. In (31.18) it is now valid to put $G = G^*$, $z = f$; as for ζ, we approximate it in the mean with a sequence $\{\zeta_n\}$ of Hölder continuous functions and denote by u_n the function which we obtain from (31.18) substituting ζ_n in place of ζ. By theorem 12, VII the derivatives of u_n converge in the mean in T to the derivatives of u, and the sequences $\{u_n\}$ and $\{\mathfrak{P}\, u_n\}$ converge in the mean on ∂T to φ and ψ. Because for u_n (31.19) holds, a passage to the limit proves the assertion.

We shall finish by recalling that E. Magenes [3, 7] proved the possibility of also applying the methods of this section to regular oblique derivative problems.

32. The method of the minimum. It is well known that every self-adjoint elliptic equation:

$$\sum_{i,\,k=1}^{m} \frac{\partial}{\partial x_i} \left(a_{ik} \frac{\partial u}{\partial x_k} \right) + c\, u = f \tag{32.1}$$

can be identified with the EULER equation of the multiple integral:

$$\int_T \left[\sum_{i,k=1}^m a_{ik} \frac{\partial u}{\partial x_i} \frac{\partial u}{\partial x_k} - c\,u^2 + 2\,u\,f \right] dx. \tag{32.2}$$

Moreover, the weak form (30.3) of (32.1) coincides with the equation which is obtained by equating to zero the first variation of the integral (32.2).

Because the ellipticity of the equation (32.1) implies that the integral (32.2) is *positive regular* in the sense of the calculus of variations, it is obvious how the search for the solutions of (32.1) which satisfy the boundary condition:

$$u = \varphi \quad \text{for} \quad x \in \partial T, \tag{32.3}$$

is connected with the search for the minimum of the integral (32.2) in a class of functions satisfying (32.3). This affinity between the two problems suggests another procedure for establishing the existence theorem for DIRICHLET's problem, consisting of proving first that the integral (32.2) is endowed with a minimum in an appropriate class of functions satisfying (32.3) and then in showing that the minimizing function is a solution of (32.1).

The first applications of this method originated with GAUSS, LORD KELVIN, and RIEMANN, and are well known in connection with the critical objection of WEIERSTRASS.

At the beginning of this century, a memoir by D. HILBERT on the argument gave impulse to new research regarding, above all, DIRICHLET's problem for harmonic functions. Among these are to be remembered those by B. LEVI, G. FUBINI, S. ZAREMBA, and above all the classical memoir by H. LEBESGUE. Then the problem was considered in all its generality by R. COURANT between 1912 and 1931 in a series of memoirs. COURANT then summarized the results of this research in Chapter VII of volume II of his well known treatise[1]; hence it would be superfluous to expatiate in a detailed scrutiny of his works. Therefore we shall limit ourselves here to recalling that the existence theorem which he obtains for the first boundary value problem is valid under the hypothesis that $\varphi \in H_{1/2}$, (32.3) being satisfied in the sense that, for every $v \in H^{1,2}$ such that $\gamma\,v = \varphi$, we have $u - v \in H_0^{1,2}$. Other theorems concern the second boundary value problem, also considered in the generalized sense. COURANT's entire treatment concerns the case $m = 2$. For the extension to the case $m > 2$, see S. L. SOBOLEV [5] for what concerns harmonic functions, and S. G. MIHLIN [2, 3, 5, 6] for the general

---

[1] R. COURANT and D. HILBERT [1].

case. In particular, MIHLIN's volumes [2, 5] contain an ample exposition of the state of the theory around 1952.

We note now that the solution of a boundary value problem being also a solution of a minimum problem can suggest various procedures for the numerical calculation of this solution. Such procedures are based on the construction of appropriate sequences of functions, minimizing the integral (32.2), chosen in such a way that they turn out to converge to the function which renders the integral a minimum. This is, for example, the basis of the well known method of RITZ as well as other similar methods based on the principle of least squares.

Concerning the researches on this argument we shall limit ourselves to recalling those, in order of time, of M. PICONE [3], E. TREFFTZ [1], L. KANTOROVIČ [1], S. FAEDO [1], PH. COOPERMAN [1].

To the study of linear elliptic equations which are related to the problems of the calculus of variations the so-called algorithm of the "Kernel Function" is applicable. A propos of this we refer to the volumes by S. BERGMAN [2] and S. BERGMAN and M. SCHIFFER [2] dedicated precisely to this theory. On the argument see also G. FICHERA's note [12]. We recall also that this theory can be incorporated into that of the "reproducing kernels" of N. ARONSZAJN [1, 2].

CHAPTER V

A priori majorization of the solutions of the boundary value problems

In the scope of the study of boundary value problems for nonlinear elliptic equations, it is essential to deepen the study of the solutions of linear equations for the purpose of establishing certain majorization formulas for them, which are refined as much moɪe as possible. Along this line of thought, the first problem which presents itself is that of majorizing, by functions of the data of the problem, the maximum moduli and the HÖLDER coefficients of the solution and its derivatives and of establishing which regularity pɪoperties hold for this solution as consequences of appropriate regularity properties on the data of the problem. The importance of this research was recognized by G. GIRAUD since his first works [2, 4] in which, departing from theorems proved by himself on existence and on the representation of the solution by means of potentials, he succeeded in bringing some contribution to the study of the question.

However, to E. HOPF, J. SCHAUDER, and R. CACCIOPPOLI belongs the merit of having studied a procedure with which it is possible to establish "a priori" majorization formulas foɪ the solutions of DIRICHLET's problem, namely, formulas which are applicable to every solution of the problem *possibly* existing, yet being valid under hypotheses *not* implying by themselves the existence of the solution; such formulas can even (SCHAUDER, CACCIOPPOLI) be put at the basis of a new method for the proof of the existence theorems, which has the value of not requiring the prior construction of a fundamental solution and of utilizing, in place of the theory of integral equations, some sɪmple considerations from functional analysis.

The first work along this line is by E. HOPF [3], who succeeded in proving the λ-Hölder continuous character of the second derivatives of the solutions of elliptic equations in the interior of their region of definition under the hypothesis that the coefficients of the equation are HÖLDER continuous with exponent λ. More generally, if the coefficients of the equation are of class $C^{(k, \lambda)}$ then the solutions are of class $C^{(k+2, \lambda)}$ internally to their region of definition. A little later, J. SCHAUDER [9] proved that if the a_{ik} are HÖLDER continuous with exponent $h > \lambda$ then every regular solution of the equation, defined in a domain $T \in A^{(2, \lambda)}$ and of class $C^{(2, \lambda)}$ on ∂T, is of class $C^{(2, \lambda)}$ in all of T.

A bit later, but independently of SCHAUDER's works, R. CACCIOPPOLI [7] arrived at the same result under the less restrictive hypothesis that $h = \lambda$; moreover, CACCIOPPOLI succeeded in making quite precise the dependence of the constants which occur in the majorization formulas on the data of the problem.

Finally, SCHAUDER [12] returned to the question, ultimately perfecting CACCIOPPOLI's results. Some of CACCIOPPOLI's results were established in the same period of time by G. GIRAUD [12] also, using, however, an existence theorem in the small for the solution of NEUMANN's problem.

In our exposition, though inspired directly by the works of SCHAUDER and CACCIOPPOLI, we shall give a treatment a bit redeveloped in detail, for the purpose of better specifying some complementary results which the authors have stressed only in passing.

This treatment will be developed in §§ 33, 34, 35, 36 of this chapter, while in §§ 37, 38, and 39 we shall occupy ourselves with bounds of integral type and with existence theorems for a new type of generalized solutions of an elliptic equation, namely, for those solutions which, belonging to class of the type $H^{k, p}$, satisfy the equation almost everywhere. The beginning of this research, which has had a notable development in the last fifteen years, is found in two memoirs in 1951 by R. CACCIOPPOLI [11] and O. A. LADYŽENSKAJA [1], concerning the majorization in $H^{2, 2}$. In fact, in precedence, the unique contributions to be remembered are, for the case $m = 2$, a formula of S. BERNSTEIN [2] which goes back to 1910[1], the memoir by L. LICHTENSTEIN [1] on the logarithmic potential which is from 1912, and a note by J. SCHAUDER [10] from 1934; and, for the case $m > 2$, the extension of LICHTENSTEIN's theorem obtained by FRIEDRICHS [5] in 1947. All the contents of §§ 37 and 38 are with reference to the results obtained by various authors in the study of DIRICHLET's problem, while in § 39 we shall make note of some results for the second and the third boundary value problem.

33. Orders of magnitude of the successive derivatives of a function and of their Hölder coefficients. In this section, considering a function defined in a bounded domain T and of class $C^{(n, \lambda)}$ there, we propose to establish some dependence relationships between the quantities U_k and $U_{k, \lambda}$ of different index defined in § 1. These relations, even if of elementary character, are fundamental for the study which we propose to set forth in the following sections.

Assuming T of class $A^{(1, \lambda)}$ and of diameter R, the following proposition then holds, of which we shall for brevity omit the proof:

---

[1] See also S. BERNSTEIN [9].

A) *There exist two numbers L and θ, depending only on T, such that:*
1°) two arbitrary points x and y of T can be joined by a regular curve
contained in T of length less than $L\,\overline{xy}$; 2°) however we choose a hyper-
sphere $\Gamma(x, \delta)$ with its center at $x \in T$ and radius $\delta \leq R$, there exists
a segment $x^{(i)} y^{(i)}$ in $T \cap \Gamma(x, \delta)$ parallel to the axis x_i and of length $\theta\,\delta$,
for every i.

Having premised this, let us prove that:

33, I. *For all functions $u(x)$ of class $C^{(1)}$ in T, the following bound*
holds uniformly:

$$U_{0,\lambda} = O\left(U_1^\lambda\, U_0^{1-\lambda}\right). \tag{33.1}$$

And indeed, having fixed some δ in the interval $(0, R)$ and denoting
by ω the oscillation of u in T, we have:

$$\frac{|\,u(x) - u(y)\,|}{\overline{xy}^\lambda} \begin{cases} \leq L\,U_1\,\delta^{1-\lambda} & \text{for} \quad \overline{xy} \leq \delta \\ \leq \omega\,\delta^{-\lambda} & \text{for} \quad \overline{xy} \geq \delta. \end{cases} \tag{33.2}$$

Now it follows, denoting by $\varphi(\delta)$ the largest of the quantities on the
right side of (33.2), that we have $U_{0,\lambda} \leq \min \varphi(\delta)$.

But, taking account of the fact that $\omega \leq R\,L\,U_1$, we recognize
easily that $\min \varphi(\delta) = L^\lambda\,U_1^\lambda\,\omega^{1-\lambda}$ and from this (33.1) follows without
further ado, since $\omega \leq 2\,U_0$.

33, II. *For all the functions of class $C^{(1,\lambda)}$ in T, the following bound*
holds uniformly:

$$U_1 = O\left(U_{1,\lambda}^{\frac{1}{1+\lambda}}\, U_0^{\frac{\lambda}{1+\lambda}} + U_0\,R^{-1}\right). \tag{33.3}$$

Indeed, let x_0 be the point at which $|\,p_i\,|$ assumes its maximum P_i,
and, to be precise let $P_i = p_i(x_0)$. For $\overline{xx_0} \leq \delta$ we have $P_i - \delta^\lambda\,U_{1,\lambda}$
$\leq p_i(x)$. Now either $P_i < \delta^\lambda\,U_{1,\lambda}$ or else for the two points $x^{(i)}$ and $y^{(i)}$
of observation A) we have:

$$\frac{|\,u(x)^{(i)} - u(y)^{(i)})\,|}{\theta\,\delta} \geq P_i - \delta^\lambda\,U_{1,\lambda}.$$

In the one case as in the other it now follows that:

$$U_1 \leq m\,\delta^\lambda\,U^{1,\lambda} + \frac{2\,m}{\theta\,\delta}\,U_0.$$

From this, determining δ in such a way as to render the right side of
this inequality a minimum, and then majorizing this minimum, we
arrive easily at (33.3).

Before proceeding further, we summarize in the following statement some obvious observations, which, however, must be continuously recalled:

B) *If $\alpha, \beta > \varepsilon > 0$, $a, b > 0$ then we have:*

$$(a + b)^\alpha = O(a^\alpha + b^\alpha), \qquad a^\alpha b^\beta = O(a^{\alpha - \varepsilon} b^{\beta + \varepsilon} + a^{\alpha + \varepsilon} b^{\beta - \varepsilon}).$$

If, with $0 < \alpha_i < 1$, $a, b, a_i > 0$:

$$a = O\left(\sum a_i a^{\alpha_i} + b\right),$$

then we also have:

$$a = O\left(\sum a_i^{\frac{1}{1-\alpha_i}} + b\right).$$

Having premised this, we have that:

33, III. *For all functions of class $C^{(n)}$ in T, the following bound holds uniformly and for every $k < n$:*

$$U_k = O\left(U_n^{\frac{k}{n}} U_0^{\frac{n-k}{n}} + U_0 R^{-k}\right), \tag{33.4}$$

$$U_{k,\lambda} = O\left(U_n^{\frac{k+\lambda}{n}} U_0^{\frac{n-k-\lambda}{n}} + U_0 R^{-(k+\lambda)}\right). \tag{33.5}$$

With the same procedure adopted for proving (33.3), we now establish that (33.4) is valid for $n = 2$ and $k = 1$, and from this it follows that for every p we have:

$$U_p = O\left(U_{p+1}^{\frac{1}{2}} U_{p-1}^{\frac{1}{2}} + U_{p-1} R^{-1}\right). \tag{33.6}$$

To prove (33.4) in general, we now suppose that this formula holds for $n = p$, $k < p$, and we prove that this is also true for $n = p + 1$, $k < p + 1$. Now from (33.6) and from (33.4) written for $n = p$, $k = p - 1$, we obtain, by dint of B):

$$U_p = O\left(U_{p+1}^{\frac{1}{2}} U_p^{\frac{p-1}{2p}} U_0^{\frac{1}{2p}} + U_{p+1}^{\frac{1}{2}} U_0^{\frac{1}{2}} R^{\frac{1-p}{2}} + U_p^{\frac{p-1}{p}} U_0^{\frac{1}{p}} R^{-1} + U_0 R^{-p}\right)$$

and from this, with several applications of B), it follows easily that:

$$U_p = O\left(U_{p+1}^{\frac{p}{p+1}} U_0^{\frac{1}{p+1}} + U_0 R^{-p}\right). \tag{33.7}$$

(33.4) is thus proved for $n = p + 1$, $k = p$; finally, writing (33.4) for $n = p$, $k < p$ and majorizing U_p, which appears in this formula, with (33.7), we succeed in establishing (33.4) even in the case $n = p + 1$, $k < p$. From (33.1) it now follows that:

$$U_{k,\lambda} = O\left(U_{k+1}^{\lambda} U_{k-1}^{1-\lambda}\right)$$

and from this formula, taking (33.4) into account, we arrive easily at (33.5).

33, IV. *For all the functions of class* $C^{(n,\lambda)}$ *in* T, *the following bounds hold uniformly:*

$$U_k = O\left(U_{n,\lambda}^{\frac{k}{n+\lambda}} U_0^{\frac{n+\lambda-k}{n+\lambda}} + U_0 R^{-k}\right), \qquad k \le n, \qquad (33.8)$$

$$U_{k,\lambda} = O\left(U_{n,\lambda}^{\frac{k+\lambda}{n+\lambda}} U_0^{\frac{n-k}{n+\lambda}} + U_0 R^{-(k+\lambda)}\right), \qquad k < n. \qquad (33.9)$$

From (33.4) written for $k = n - 1$, taking into account that by (33.3)

$$U_n = O\left(U_{n,\lambda}^{\frac{1}{1+\lambda}} U_{n-1}^{\frac{\lambda}{1+\lambda}} + \mathfrak{U}_{n-1} R^{-1}\right), \qquad (33.10)$$

we obtain:

$$U_{n-1} = O\left(U_{n,\lambda}^{\frac{n-1}{n(1+\lambda)}} U_{n-1}^{\frac{\lambda(n-1)}{n(1+\lambda)}} U_0^{\frac{1}{n}} + U_{n-1}^{\frac{n-1}{n}} U_0^{\frac{1}{n}} R^{\frac{1-n}{n}} + U_0 R^{1-n}\right)$$

and from this, by dint of B), (33.8) follows for $k = n - 1$.

Once U_{n-1} is majorized, (33.8) follows from (33.10) and for $k = n$; substituting the value found for U_n in (33.4), we arrive at (33.8) also for $k < n - 1$. (33.9) follows finally from (33.5) and from (33.8) written for $k = n$.

Finally, it is obvious that:

33, V. *The bounds of the preceding theorem hold uniformly upon variation of the domain* T *in every family of domains of class* $A^{(1,\lambda)}$ *in which the quantities* L *and* $1/\theta$ *are held bounded.*

Now let N be a closed set contained in ∂T and let $4\, d(x)$ be the distance of a generic point x of $T - N$ from N. For every fixed ν which is positive and less than 4, the family of domains $T \cap \Gamma(x_0, \nu\, d(x_0))$, as x_0 varies in $T - N$, enjoys precisely those properties required by theorem 33, V. Therefore, if we denote by $U_k^{(\nu)}$ and $U_{k,\lambda}^{(\nu)}$ the values of U_k and $U_{k,\lambda}$ for the domain $T \cap \Gamma(x_0, \nu\, d(x_0))$, these quantities satisfy all the formulas previously established, uniformly as x_0 varies. It follows easily from this that:

33, VI. *The suprema* $\mathfrak{U}_k^{(\nu)}$ *and* $\mathfrak{U}_{k,\lambda}^{(\nu)}$ *of the quantities* $(\nu\,d)^k\,U_k^{(\nu)}$ *and* $(\nu\,d)^{k+\lambda}\,U_{k,\lambda}^{(\nu)}$, *respectively, satisfy the bounds:*

$$\mathfrak{U}_{0,\lambda}^{(\nu)} = O\big([\mathfrak{U}_1^{(\nu)}]^\lambda\,U_0^{1-\lambda}\big), \tag{33.11}$$

$$\mathfrak{U}_k^{(\nu)} = O\Big([\mathfrak{U}_n^{(\nu)}]^{\frac{k}{n}}\,U_0^{\frac{n-k}{n}} + U_0\Big), \quad \mathfrak{U}_{k,\lambda}^{(\nu)} = O\Big([\mathfrak{U}_n^{(\nu)}]^{\frac{k+\lambda}{n}}\,U_0^{\frac{n-k-\lambda}{n}} + U_0\Big), \tag{33.12}$$

$$\mathfrak{U}_k^{(\nu)} = O\Big([\mathfrak{U}_{n,\lambda}^{(\nu)}]^{\frac{k}{n+\lambda}}\,U_0^{\frac{n+\lambda-k}{n+\lambda}} + U_0\Big), \quad \mathfrak{U}_{k,\lambda}^{(\nu)} = O\Big([\mathfrak{U}_{n,\lambda}^{(\nu)}]^{\frac{k+\lambda}{n+\lambda}}\,\mathfrak{u}^{\frac{n-k}{n+\lambda}} + U_0\Big). \tag{33.13}$$

Moreover, we have easily that:

33, VII. *For every domain* $D \subset T - N$ *which has distance* δ *from* N, *we have, for* $\nu < 4$:

$$U_k(D) \le (\nu\,\delta)^{-k}\,\mathfrak{U}_k^{(\nu)}, \tag{33.14}$$

$$U_{k,\lambda}(D) \le (\nu\,\delta)^{-(k+\lambda)}\,[\mathfrak{U}_{k,\lambda}^{(\nu)} + 2\,\mathfrak{U}_k^{(\nu)}], \tag{33.15}$$

and from this theorem it follows immediately that:

33, VIII. *If* $\nu \le \mu < 4$ *we have, uniformly in* ν:

$$\mathfrak{U}_k^{(\mu)} = O\Big[\Big(\frac{\mu}{\nu}\Big)^k\,\mathfrak{U}_k^{(\nu)}\Big], \quad \mathfrak{U}_{k,\lambda}^{(\mu)} = O\Big[\Big(\frac{\mu}{\nu}\Big)^{k+\lambda}\,\big(\mathfrak{U}_{k,\lambda}^{(\nu)} + \mathfrak{U}_k^{(\nu)}\big)\Big]. \tag{33.16}$$

Finally, it is obvious that:

33, IX. *If* N *is empty, then the preceding results hold, where we assume* $d(x)$ *constant. In particular,* (33.14) *and* (33.15) *hold even for* $D \equiv T$, $\delta = d$.

34. Majorization in $C^{(n,\lambda)}$ of the solutions of equations with constant coefficients.

Let T be a bounded domain of class $A^{(n,\lambda)}$ with $n \ge 2$ and let Ξ be a portion of ∂T admitting a cartesian representation of the type (1.1). For every function $\varphi(x)$ defined in Ξ and of class $C^{(n,\lambda)}$ there, we can, with the meanings of the preceding section and by considering φ as a function of the $m-1$ variables $\xi_1, \xi_2, \ldots, \xi_{m-1}$, define for $k \le n$ the symbols $\Phi_k(\Xi)$ and $\Phi_{k,\lambda}(\Xi)$.

Now let $\Xi_1, \Xi_2, \ldots, \Xi_p$ be portions of ∂T of the preceding type such that every point of ∂T is interior to at least one of them; if X is a set of points of ∂T, open on ∂T, and $\varphi(x)$ is a function of class $C^{(n,\lambda)}$ on the closure $\overline{X}$ of X, we shall put:

$$\Theta_n[\varphi; \overline{X}] = \sum_{i=1}^{p} \Phi_k[\overline{X} \cap \Xi_i], \quad \Theta_{k,\lambda}[\varphi; \overline{X}] = \sum_{i=1}^{p} \Phi_{k,\lambda}[\overline{X} \cap \Xi_i].$$

In what follows the set X is understood to be fixed once and for all, and by N we shall mean the closure of $\partial T - \overline{X}$; $4\,d(x)$ will designate as before the distance of a generic point x of $T - N$ from N.

Now let $\mathfrak{M}$ be an elliptic operator with constant coefficients, containing only the second derivatives, and let $f(x)$ be a function of class $C^{(n,\,\lambda)}$ in $T - N$. Fixing arbitrarily a point $x_0 \in T - N$ and choosing a positive number $\sigma < \sigma_0 < 1$, we propose to establish some majorization formulas for the functions $u(x)$ which satisfy the equations:

$$\mathfrak{M}\,u = f \quad \text{for} \quad x \in T \cap \Gamma\big(x_0,\, 2\,\sigma\,d(x_0)\big)$$

$$u = \varphi \quad \text{for} \quad x \in \overline{X} \cap \Gamma\big(x_0,\, 2\,\sigma\,d(x_0)\big). \tag{34.1}$$

σ_0 being chosen so that $\overline{X} \cap \Gamma\big(x_0,\, 2\,\sigma\,d(x_0)\big)$ is connected and entirely contained in at least one of the regions $\varXi_i$, the following theorem holds:

34, I. *For every solution $u(x)$ of (34.1), the following majorization formula is valid for $n \geq 2$:*

$$U^{(\sigma)}_{n,\,\lambda} = O\!\left(\varrho^{2-n-\lambda}\,F_0^{(3\,\sigma)} + F_{n-2,\,\lambda}^{(3\,\sigma)} + \varrho^{-\lambda}\,U_n^{(2\,\sigma)} + \sum_{k=1}^{n-1} U_k^{(2\,\sigma)} + \Theta_{n,\,\lambda}\big[\varphi;\,\overline{X}\big]\right),$$
$$\tag{34.2}$$

where we have put $\varrho = \sigma\,d(x_0)$. (34.2) holds uniformly as $\mathfrak{M}$ varies in every family of elliptic operators in which the maximum moduli of the a_{ik} and $1/a$ are held bounded, a being the ellipticity constant of $\mathfrak{M}$.

We observe beforehand that it suffices to prove (34.2) assuming $\mathfrak{M} = \varDelta_2$, because we can always be brought back to this case with a change of variables. We assume now in the first place that $f = 0$. It is then well known that, if $\Gamma(x_0,\, 3\,\varrho/2)$ is interior to T, we have $U^{(\sigma)}_{(n+1)} = O\big(\varrho^{-(n+1)}\,U_0^{(2\,\sigma)}\big)$ and from this it easily follows that:

$$U^{(\sigma)}_{n,\,\lambda} = O\big(\varrho^{-(n+\lambda)}\,U_0^{(2\,\sigma)}\big). \tag{34.3}$$

If instead $\overline{X}$ has points in common with $\Gamma(x_0,\, 3\,\varrho/2)$, we denote by $v(x')$ the function which is obtained from $u(x)$ by executing the change of variables:

$$x_i' = \frac{x_i - x_{0i}}{\varrho}. \tag{34.4}$$

By this change of variables, $\Gamma(x_0,\, \varrho)$ and $\Gamma(x_0,\, 2\,\varrho)$ are changed into two hyperspheres Γ_1' and Γ_2' of radii 1 and 2 respectively; $\overline{X} \cap \Gamma(x_0,\, 2\,\varrho)$ is changed into a hypersurface X' with equation:

$$\xi_m' = \frac{1}{\varrho}\,\zeta\big(\varrho\,\xi_1',\, \ldots,\, \varrho\,\xi_{m-1}'\big) = \zeta'\big(\xi_1',\, \xi_2',\, \ldots,\, \xi_{m-1}'\big),$$

and the derivatives of various orders of ζ' can be bounded independently of ϱ; finally, φ is changed into a ψ which on X' can be considered a function of $\xi_1', \xi_2', \ldots, \xi_{m-1}'$.

By dint of some results of O.D. KELLOGG [1] on the behavior on the boundary of the derivatives of harmonic functions[1], for v a bound of the following type then holds:

$$V_{n,\lambda}(\Gamma_1') = O\left(V_0(\Gamma_2') + \sum_{k=1}^{n} [\Psi_k\,(X') + \Psi_{k,\lambda}(X')]\right).$$

On the other hand, because X' has a distance less than $3/2$ from the center of Γ_2', its projection on the hyperplane $\xi_m' = 0$ will contain at least a hypersphere of radius τ, where τ can be bounded from below by a function of the second derivatives of ζ', that is, of the curvatures of ∂T. By what we saw in §33, all the Ψ_k and $\Psi_{k,\lambda}$ can therefore be bounded by functions of V_0 and of $\Psi_{n,\lambda}$ alone, whence for the preceding bound we can substitute the simpler:

$$V_{n,\lambda}(\Gamma_1') = O(V_0(\Gamma_2') + \Psi_{n,\lambda}(X')),$$

from which, returning to the variables x_i, we draw:

$$U_{n,\lambda}^{(\sigma)} = O\left(\varrho^{-(n+\lambda)}\, U_0^{(2\sigma)} + \Phi_{n,\lambda}^{(2\sigma)}\right), \tag{34.5}$$

having put for simplicity:

$$\Phi_{n,\lambda}^{(2\sigma)} = \Phi_{n,\lambda}\left(\overline{X} \cap \Gamma(x_0, 2\,\varrho)\right).$$

And it is obvious that this formula is valid in every case, and that it can take the place of (34.3) even in the case in which that formula would be applicable.

We denote now by $p(x)$ the harmonic polynomial of degree $n-1$, the sum of the terms of degree less than n of the TAYLOR series for the function $u(x)$ near the point x_0.

Applying (34.5) to the function $z = u - p$, whose values on $\overline{X} \cap \Gamma(x_0, 2\,\varrho)$ we denote by ω, and taking into account that

$$Z_{n,\lambda}^{(\sigma)} = U_{n,\lambda}^{(\sigma)}, \; Z_0^{(2\sigma)} = O\left(\varrho^n\, U_n^{(2\sigma)}\right), \; \Omega_{n,\lambda}^{(2\sigma)} = O\left(\Phi_{n,\lambda}^{(2\sigma)} + \sum_{k=1}^{n-1} U_k^{(2\sigma)}\right),$$

we arrive at the formula:

$$U_{n,\lambda}^{(\sigma)} = O\left(\varrho^{-\lambda}\, U_n^{(2\sigma)} + \sum_{k=1}^{n-1} U_k^{(2\sigma)} + \Phi_{n,\lambda}^{(2\sigma)}\right), \tag{34.6}$$

---

[1] See also J. SCHAUDER [5, 6].

from which (34.2) drops out easily. Having thus proved the theorem in the case of harmonic functions, we can examine the case in which u is instead a solution of the inhomogeneous equation $\varDelta_2 u = f$.

In this case (assuming for simplicity $m > 2$), putting:

$$w(x) = -\frac{1}{(m-2)\,\omega_m} \int\limits_{T \cap \Gamma(x_0,\,3\,\varrho)} f(y)\,\overline{xy}^{\,2-m}\,dy, \qquad \overline{\varphi} = \varphi - w \quad \text{for} \quad x \in \overline{X},$$

we could write $u = w + \overline{u}$, where $\overline{u}$ is a harmonic function in $T \cap \Gamma(x_0,\,2\,\varrho)$ equal to $\overline{\varphi}$ on $\overline{X}$. Now, resorting to the change of variables (34.4) and to the known properties of the newtonian potential, we easily establish the following bounds for $k \leq n$:

$$W_k^{(2\,\sigma)} = O\big(\varrho^{2-k}\,F_0^{(3\,\sigma)} + k(k-1)\,\varrho^{n-k+\lambda}\,F_{n-2,\,\lambda}^{(3\,\sigma)}\big),$$

$$W_{k,\,\lambda}^{(2\,\sigma)} = O\big(\varrho^{2-k-\lambda}\,F_0^{(3\,\sigma)} + k(k-1)\,\varrho^{n-k}\,F_{n-2,\,\lambda}^{(3\,\sigma)}\big),$$

while u can be majorized by means of (34.6). If we take into account that:

$$\overline{U}_k^{(2\,\sigma)} = O\big(U_k^{(2\,\sigma)} + W_k^{(2\,\sigma)}\big), \qquad \overline{\varPhi}_{n,\,\lambda}^{(2\,\sigma)} = O\bigg(\varPhi_{n,\,\lambda}^{(2\,\sigma)} + \sum_{k=1}^{n} W_{k,\,\lambda}^{(2\,\sigma)}\bigg),$$

we easily arrive at (34.2), which thus stands completely proved. The result obtained can be completed by the following obvious observation:

34, II. (34.2) *also holds if* $\overline{X}$ *coincides with* ∂T, *and we can now assume* $d(x)$ *constant. If instead* $\overline{X}$ *is empty and if* $4\,d(x_0)$ *denotes the distance of* ∂T *from a generic point* x_0 *of* $T - \partial T$, *then the following formula holds:*

$$U_{n,\,\lambda}^{(\sigma)} = O\big(\varrho^{2-n-\lambda}\,F^{(3\,\sigma)} + F_{n-2,\,\lambda}^{(3\,\sigma)} + \varrho^{-\lambda}\,U_n^{(2\,\sigma)}\big). \tag{34.7}$$

35. General majorization formulas in $C^{(n,\,\lambda)}$. In this section, following the ideas of the works by E. Hopf [3], J. Schauder [9, 12], and R. Caccioppoli [7] already referred to in the introduction[1], we propose to establish various majorization formulas for the solutions of Dirichlet's problem:

$$\mathfrak{M}\,u = f \quad \text{for} \quad x \in T - \partial T, \qquad u = \varphi \quad \text{for} \quad x \in \partial T, \tag{35.1}$$

under the hypothesis that $\mathfrak{M}$ is an elliptic operator of general type. For the moment we shall take into consideration exclusively those solutions of the problem which are of class $C^{(n,\,\lambda)}$ in T with $n \geq 2$, assuming

---

[1] For other summaries of these results, see L. M. Graves [1] and R. B. Barrar [1].

that the coefficients and the known terms of the equation are of class $C^{(n-2,\lambda)}$ in T. φ will be assumed to be of class $C^{(n,\lambda)}$ on ∂T, and the domain T, of class $A^{(n,\lambda)}$. Regarding the functions u, f, φ and the coefficient c of $\mathfrak{M}$, the symbols introduced in the previous sections will continue to be valid, such as $U_k = U_k(T)$, $U_k^{(\sigma)}$, $\mathfrak{u}_k^{(\sigma)}$ etc. We shall now denote by $A_k(T)$, $A_{k,\lambda}(T)$ the sums of the maximum moduli and of the HÖLDER coefficients in T of all the derivatives of order $k(\geq 0)$ of the coefficients a_{ij}, while B_k, $B_{k,\lambda}$ will have the same significance for the coefficients b_i; at least for now, we shall write much more simply A_k instead of $A_k(T)$, $A_{k,\lambda}$ instead of $A_{k,\lambda}(T)$, etc. Finally, we shall continue to denote the ellipticity constant of $\mathfrak{M}$ by a.

Now let us prove the following theorem:

35, I. *Under the hypotheses previously indicated, the following bound holds for every possible solution of the problem* (35.1) *for $n \geq 2$:*

$$
U_{n,\lambda} = O\Bigg[(A_{n-2,\lambda} + 1)(F_0 + U_2 + B_0 U_1 + C_0 U_0)
$$
$$
+ (B_{n-2,\lambda} + B_0^{n-1+\lambda} + 1)U_1 + \Bigg(C_{n-2,\lambda} + C_0^{\frac{n+\lambda}{2}}\Bigg)U_0 + F_{n-2,\lambda} + \Theta_{n,\lambda}[\varphi;\partial T]\Bigg],
$$
$$
(35.2)
$$

uniformly as $\mathfrak{M}$ varies in every family of elliptic operators in which A_0 and $1/a$ are held bounded.

And indeed, having chosen an arbitrary x_0 in $T - \partial T$, we write the equation $\mathfrak{M}u = f$ in the form[1]:

$$
\sum_{i,j=1}^{m} a_{ij}(x_0)\,\frac{\partial^2 u}{\partial x_i\,\partial x_j}
$$
$$
= f(x) - \sum_{i,j=1}^{m}(a_{ij}(x) - a_{ij}(x_0))\frac{\partial^2 u}{\partial x_i\,\partial x_j} - \sum_{i=1}^{m} b_i\,\frac{\partial u}{\partial x_i} - c\,u. \qquad (35.3)
$$

Putting $4\,d(x_0) = 1$ we can, by dint of theorem 34, II, majorize $U_{n,\lambda}^{(\sigma)}$ by means of (34.2) on the condition that we substitute the right side of (35.3) for the known terms of (34.1) and put $\overline{X} = \partial T$. We thus succeed in establishing the existence of a constant K, depending only on n, λ, T and on an upper bound for A_0 and $1/a$, for which we have:

---

[1] This artifice, adopted constantly both by HOPF and SCHAUDER and CACCIOPPOLI, figured for the first time in a memoir by A. KORN published in 1914 in the volume „Schwarz-Festschrift".

$$U_{n,\lambda}^{(\sigma)} \le K\left[\varrho^{2-n-\lambda}\left(F_0^{(3\sigma)} + A_{0,\lambda}\,\varrho^\lambda\,U_2^{(3\sigma)} + B_0\,U_1^{(3\sigma)} + C_0\,U_0^{(3\sigma)}\right) + F_{n-2,\lambda}^{(3\sigma)}\right.$$

$$+ A_{0,\lambda}\left(\varrho^\lambda\,U_{n,\lambda}^{(3\sigma)} + U_n^{(3\sigma)}\right) + \sum_{k=2}^{n-1}\left(A_{n-k,\lambda}\,U_k^{(3\sigma)} + A_{n-k}\,U_{k,\lambda}^{(3\sigma)}\right)$$

$$+ \sum_{k=1}^{n-1}\left(B_{n-k-1,\lambda}\,U_k^{(3\sigma)} + B_{n-k-1}\,U_{k,\lambda}^{(3\sigma)}\right)$$

$$+ \sum_{k=1}^{n-2}\left(C_{n-k-2,\lambda}\,U_k^{(3\sigma)} + C_{n-k-2}\,U_{k,\lambda}^{(3\sigma)}\right)$$

$$+ \varrho^{-\lambda}\,U_n^{(2\sigma)} + \sum_{k=1}^{n-1}U_k^{(2\sigma)} + \Theta_{n,\lambda}[\varphi;\partial T]\bigg]. \tag{35.4}$$

Now by theorem 33, VIII we have:

$$\varrho^{n+\lambda}\,U_{n,\lambda}^{(3\sigma)} \le K_1\left[\mathfrak{U}_{n,\lambda}^{(\sigma)} + \varrho_n\,U_n(T)\right],$$

because of which (35.4), multiplied by $\varrho^{n+\lambda}$, can be written in abbreviated form:

$$\varrho^{n+\lambda}\,U_{n,\lambda}^{(3\sigma)} \le K\,K_1\,A_{0,\lambda}\,\varrho^\lambda\,\mathfrak{U}_{n,\lambda}^{(\sigma)} + \varrho^{n+\lambda}\,H(\sigma).$$

Denoting by σ_1 the upper bound of the values of σ which are less than σ_0 and for which $K\,K_1\,A_{0,\lambda}\,\varrho^\lambda < 1/4$, and choosing x_0 so that $(\sigma_1\,d)^{n+\lambda}\,U_{n,\lambda}^{(\sigma_1)} > \mathfrak{U}_{n,\lambda}^{(\sigma_1)}/2$, it follows from the above that $\mathfrak{U}_{n,\lambda}^{(\sigma_1)} \le 2\,(\sigma_1\,d)^{n+\lambda}\,H(\sigma_1)$ and therefore by theorems 33, VII and 33, IX we also have:

$$U_{n,\lambda}(T) = O\left(H(\sigma_1) + (\sigma_1\,d)^{-\lambda}\,U_n(T)\right).$$

From this, if we take into account that for $\sigma = \sigma_1$ not only $A_{0,\lambda}\,\varrho^\lambda = O(1)$ but also $\varrho^{-\lambda} = O(A_{0,\lambda} + 1)$ and that the quantities $U_k^{(3\sigma)}$, $U_{k,\lambda}^{(3\sigma)}$ can be majorized by the quantities $U_k = U_k(T)$ and $U_{k,\lambda} = U_{k,\lambda}(T)$, we easily draw the formula:

$$U_{n,\lambda} = O\left[\left(A_{0,\lambda}^{\frac{n-2+\lambda}{\lambda}} + 1\right)(F_0 + U_2 + B_0\,U_1 + C_0\,U_0) + F_{n-2,\lambda}\right.$$

$$+ \left(A_{0,\lambda} + 1\right)U_n + \sum_{k=2}^{n-1}\left(A_{n-k,\lambda}\,U_k + A_{n-k}\,U_{k,\lambda}\right)$$

$$+ \sum_{k=1}^{n-1}\left[\left(B_{n-k-1,\lambda} + 1\right)U_k + B_{n-k-1}\,U_{k,\lambda}\right]$$

$$+ \sum_{k=0}^{n-2}\left(C_{n-k-2,\lambda}\,U_k + C_{n-k-2}\,U_{k,\lambda}\right) + \Theta_{n,\lambda}[\varphi;\partial T]\bigg]. \tag{35.5}$$

To arrive at (35.2) it now remains to eliminate from the right side of (35.5) the U_k, A_{k-2}, B_{k-2}, C_{k-2}, $A_{n-k,\lambda}$, $B_{n-k,\lambda}$ and $C_{n-k,\lambda}$ with $k > 2$, and all the $U_{k,\lambda}$.

At this we arrive by the following artifice. We want, for example, to eliminate U_n and $A_{0,\lambda}$; from theorem 33, IV it follows that in (35.5) the term $(A_{0,\lambda} + 1) U_n$ can be replaced by:

$$(A_{0,\lambda} + 1) \left(U_{n,\lambda}^{\frac{n-2}{n-2+\lambda}} U_2^{\frac{\lambda}{n-2+\lambda}} + U_2 \right)$$

and this in turn, by dint of proposition B) of § 33, by:

$$\left(A_{0,\lambda}^{\frac{n-2+\lambda}{\lambda}} + 1 \right) U_2,$$

a quantity which, again by theorem 33, IV, is $O\left[(A_{n-2,\lambda} + 1) U_2 \right]$.

In a similar way we transform all the terms of the first summation, while for those of the second and third sums we shall proceed in the same way except to replace U_2 with U_1 and U_0 respectively. Our theorem is thus completely proved. Moreover, the artifice adopted to pass from (35.5) to (35.2) can be utilized to eliminate from the right side of (35.2) even those terms containing U_2 and U_1.

We thus easily arrive at the following theorem:

35, II. *Under the hypotheses of theorem 35, I, and assuming A_0 and $1/a$ are bounded, then for every possible solution of the problem* (35.1) *and for $n \geq 2$, the following bounds hold:*

$$U_{n,\lambda} = O\big[(A_{n-2,\lambda} + 1)(F_0 + B_0 U_1 + C_0 U_0)$$

$$+ \left(A_{n-2,\lambda}^{\frac{n-1+\lambda}{n-2+\lambda}} + B_{n-2,\lambda} + B_0^{n-1+\lambda} + 1 \right) U_1 + \left(C_{n-2,\lambda} + C_0^{\frac{n+\lambda}{2}} \right) U_0$$

$$+ F_{n-2\,\lambda} + \Theta_{n,\lambda}[\varphi; \partial T] \big], \tag{35.6}$$

$$U_{n,\lambda} = O\bigg[(A_{n-2,\lambda} + 1) F_0 + F_{n-2,\lambda} + \Theta_{n,\lambda}[\varphi; \partial T]$$

$$+ \bigg(\left[(A_{n-2,\lambda} + 1) B_0 \right]^{\frac{n+\lambda}{n-1+\lambda}}$$

$$+ (A_{n-2,\lambda} + 1) C_0 + A_{n-2,\lambda}^{\frac{n+\lambda}{n-2+\lambda}} + B_{n-2,\lambda}^{\frac{n+\lambda}{n-1+\lambda}}$$

$$+ B_0^{n+\lambda} + C_{n-2,\lambda} + C_0^{\frac{n+\lambda}{2}} + 1 \bigg) U_0 \bigg]. \tag{35.7}$$

By dint of theorem 33, IV we deduce from this easily that:

35, III. *Under the hypotheses of the preceding theorem and if, moreover, the a_{ij} are of class $C^{(n-1,\lambda)}$, we have for $n \geq 2$:*

$$U_{n,\lambda} = O\left[(A_{n-2,\lambda} + 1)(F_0 + B_0 U_1 + C_0 U_0)\right.$$

$$+ (A_{n-1,\lambda} + B_{n-2,\lambda} + B_0^{n-1+\lambda} + 1) U_1 + \left(C_{n-2,\lambda} + C_0^{\frac{n+\lambda}{2}}\right) U_0$$

$$\left. + F_{n-2,\lambda} + \Theta_{n,\lambda}[\varphi; \partial T]\right]. \tag{35.8}$$

If in addition the a_{ij} are of class $C^{n,\lambda}$ and the b_i of class $C^{(n-1,\lambda)}$ we have for $n \geq 2$:

$$U_{n,\lambda} = O\left[(A_{n-2,\lambda} + 1) F_0 + F_{n-2,\lambda} + \Theta_{n,\lambda}[\varphi; \partial T]\right.$$

$$+ \left((A_{n,\lambda} + 1)\left(B_0^{\frac{n+\lambda}{n-1+\lambda}} + 1\right) + (A_{n-2,\lambda} + 1) C_0\right.$$

$$\left. + B_{n-1,\lambda} + B_0^{n+\lambda} + C_{n-2,\lambda} + C_0^{\frac{n+\lambda}{2}}\right) U_0\right]. \tag{35.9}$$

So far we have considered solutions of problem (35.1) which are of class $C^{(n,\lambda)}$ in T. Denoting now by X an open set on ∂T, by $\overline{X}$ its closure, and putting $N = \partial T - \overline{X}$, let us consider a solution $u(x)$ of (35.1) which is continuous in T and of class $C^{(n,\lambda)}$ in $T - N$. For such a function a formula similar to (35.4) holds; in it, however, $\Theta_{n,\lambda}[\varphi; \overline{X}]$ takes the place of $\Theta_{n,\lambda}[\varphi; \partial T]$, while $4 d(x_0)$ will represent the distance of x_0 from N. Multiplying this formula by $(\sigma d)^{n+\lambda}$ and reasoning as in the proofs of theorems 35, I and II, we succeed in establishing a series of majorization formulas similar to (35.2), (35.6), and (35.7), in which, however, $\mathfrak{U}_k^{(\sigma)}$ and $\mathfrak{U}_{k,\lambda}^{(\sigma)}$ replace U_k and $U_{k,\lambda}$, with σ appropriately small. Bearing in mind theorem 33, VII also, we can thus enunciate the theorem:

35, IV. *Let D be a domain contained in $T - N$ and having distance $\delta > 0$ from N. Assuming the quantities A_0, B_0, C_0, $A_{n-2,\lambda}$, $B_{n-2,\lambda}$, $C_{n-2,\lambda}$ and $1/a$ are bounded, then for every solution $u(x)$ of the problem (35.1) continuous in T and of class $C^{(n-\lambda)}$ in $T - N$, the following bound holds:*

$$\delta^{n+\lambda} U_{n,\lambda}(D) = O\left(F_0 + F_{n-2,\lambda} + U_0(T) + \Theta_{n,\lambda}[\varphi; \overline{X}]\right). \tag{35.10}$$

For every solution $u(x)$ of the problem (35.1) continuous in T and of class $C^{(n,\lambda)}$ in $T - \partial T$ we also have, for $n \geq 2$:

$$\delta^{n+\lambda} U_{n,\lambda}(D) = O\left(F_0 + F_{n-2,\lambda} + U_0(T)\right), \tag{35.11}$$

denoting by δ, this time, the (positive) distance of D from ∂T.

Other majorization formulas could now be established by assuming that the coefficients and the known terms of the equation are of class $C^{(n-2,\lambda)}$ only in $T-\partial T$ and do not have too high an order of infinity as x tends toward ∂T. Along these lines, on the other hand, we shall limit ourselves to examining a case which is sufficiently simple but of particular interest for the study of nonlinear equations.

Denoting by the usual $4\,d\,(x_0)$ the distance from N of a generic point x_0 of $T-N$, we denote by $\mathfrak{R}_\mu^{(\sigma)}[u]$ and $\mathfrak{H}_\mu^{(\sigma)}[u]$, or more simply by $\mathfrak{R}_\mu^{(\sigma)}$ and $\mathfrak{H}_\mu^{(\sigma)}$, the upper bounds of the sets of numbers described respectively by the quantities $(\sigma\,d)^\mu\,U_2^{(\sigma)}$ and $(\sigma\,d)^{\mu+\lambda}\,U_{2,\lambda}^{(\sigma)}$ as x varies over $T-N$, μ being a non-negative number not exceeding one. We then denote by $\mathfrak{B}_0^{(\sigma)}$ and $\mathfrak{B}_{0,\lambda}^{(\sigma)}$ the upper bounds of $(\sigma\,d)^\mu\,B_0^{(\sigma)}$, $(\sigma\,d)^{\mu+\lambda}\,B_{0,\lambda}^{(\sigma)}$, where $B_0^{(\sigma)} = B_0\big(T\cap\Gamma(x_0,\sigma\,d)\big)$, $B_{0,\lambda}^{(\sigma)} = B_{0,\lambda}\big(T\cap\Gamma(x_0,\sigma\,d)\big)$, and similarly we define $\mathfrak{C}_0^{(\sigma)}$, $\mathfrak{C}_{0,\lambda}^{(\sigma)}$, $\mathfrak{F}_0^{(\sigma)}$, $\mathfrak{F}_{0,\lambda}^{(\sigma)}$; we denote instead by $\mathfrak{A}_{k,\lambda}^{(\sigma)}$ the upper bound of $(\sigma\,d)^{k+\lambda}\,A_{k,\lambda}^{(\sigma)}$. The following theorem holds:

35, V. *If the coefficients of the operator $\mathfrak{M}$ and the function f are of class $C^{(0,\lambda)}$ in T, then for every solution of class $C^{(2,\lambda)}$ of the equation $\mathfrak{M}\,u = f$ the following bound is valid:*

$$\mathfrak{H}_\mu^{(3)} = O\Big[\big(\mathfrak{A}_{0,\lambda}^{(3)}+1\big)\big(\mathfrak{F}_0^{(3)}+\mathfrak{R}_\mu^{(3)}+\mathfrak{B}_0^{(3)}\,U_1+\mathfrak{C}_0^{(3)}\,U_0\big)$$
$$+\mathfrak{B}_0^{(3)}\big(\mathfrak{R}_\mu^{(3)}\big)^\lambda\,U_1^{1-\lambda}+\mathfrak{F}_{0,\lambda}^{(3)}+\big(\mathfrak{B}_{0,\lambda}^{(3)}+\mathfrak{C}_0^{(3)}+1\big)\,U_1 \qquad (35.12)$$
$$+\big(\mathfrak{C}_{0,\lambda}^{(3)}+1\big)\,U_0+\Theta_{2,\lambda}\big[\varphi;\overline{X}\big]\Big]:$$

uniformly as $\mathfrak{M}$ varies in every family of elliptic operators in which A_0 and $1/a$ are held bounded. In particular, for $\mu = 0$:

$$\mathfrak{H}_0^{(3)} = O\Big[\big(\mathfrak{A}_{0,\lambda}^{(3)}+1\big)\,(F_0+U_2+B_0\,U_1+C_0\,U_0)$$
$$+B_0\,U_2^\lambda\,U_1^{1-\lambda}+\mathfrak{F}_{0,\lambda}^{(3)}+\big(\mathfrak{B}_{0,\lambda}^{(3)}+C_0+1\big)\,U_1 \qquad (35.13)$$
$$+\big(\mathfrak{C}_{0,\lambda}^{(3)}+1\big)\,U_0+\Theta_{1,\lambda}\big[\varphi;\overline{X}\big]\Big];$$

and if the a_{ik} are of class $C^{(1,\lambda)}$ we have also, for $\mu = 1$:

$$\mathfrak{H}_1^{(3)} = O\Big[\big(\mathfrak{A}_{0,\lambda}^{(3)}+1\big)\big(\mathfrak{F}_0^{(3)}+\mathfrak{B}_0^{(3)}\,U_1+\mathfrak{C}_0^{(3)}\,U_0\big)+\mathfrak{F}_{0,\lambda}^{(3)}$$
$$+\big(\mathfrak{A}_{1,\lambda}^{(3)}+\mathfrak{B}_{0,\lambda}^{(3)}+[\mathfrak{B}_0^{(3)}]^{1+\lambda}+\mathfrak{C}_0^{(3)}+1\big)\,U_1 \qquad (35.14)$$
$$+\big(\mathfrak{C}_{0,\lambda}^{(3)}+1\big)\,U_0+\Theta_{2,\lambda}\big[\varphi;\overline{X}\big]\Big].$$

The proof is similar to that of the preceding theorems. We depart from a formula similar to (35.4), written, however, for $n = 2$. Moreover, in this formula $U_{1,\lambda}^{(3\,\sigma)}$ and $U_{0,\lambda}^{(3\,\sigma)}$ are majorized (theorem 33, I) by $\left(U_2^{(3)}\right)^\lambda U_1^{1-\lambda}$ and $\sigma^{1-\lambda} U_1$, respectively. Multiplying this formula by $(\sigma\,d)^{\mu+\lambda}$ and majorizing, we easily obtain:

$$(\sigma\,d)^{\mu+\lambda}\,U_{2,\lambda}^{(\sigma)} = O\left[\sigma^\lambda\,\mathfrak{A}_{0,\lambda}^{(3)}\,\mathfrak{H}_\mu^{(\sigma)} + \sigma^\mu\left(\mathfrak{F}_0^{(3)} + \mathfrak{R}_\mu^{(3)} + \mathfrak{B}_0^{(3)}\,U_1 + \mathfrak{C}_0^{(3)}\,U\right)\right.$$

$$+ \sigma^{\mu+\lambda}\left(\mathfrak{A}_{0,\lambda}^{(3)}\,\mathfrak{R}_\mu^{(3)} + \mathfrak{B}_0^{(3)}\left(\mathfrak{R}_\mu^{(3)}\right)^\lambda U_1^{1-\lambda} + \mathfrak{F}_{0,\lambda}^{(3)}\right.$$

$$\left.\left. + \left(\mathfrak{B}_{0,\lambda}^{(3)} + \mathfrak{C}_0^{(3)} + 1\right)U_1 + \left(\mathfrak{C}_{0,\lambda}^{(3)} + 1\right)U_0 + \Theta_{2,\lambda}[\varphi;\overline{X}]\right)\right].$$

From this, for a sufficiently small σ but such that in every case $\sigma^\lambda\,\mathfrak{A}_{0,\lambda}^{(3)} = O(1)$, $\sigma^{-\lambda} = O(\mathfrak{A}_{0,\lambda}^{(3)} + 1)$, we obtain a bound for $\mathfrak{H}_\mu^{(\sigma)}$, from which we finally draw (35.12), taking into account that:

$$\mathfrak{H}_\mu^{(3)} = O\left[\sigma^{-(\mu+\lambda)}\,\mathfrak{H}_\mu^{(\sigma)} + \sigma^{-\lambda}\,\mathfrak{R}_\mu^{(3)}\right].$$

(35.13) is now obtained from (35.12) for $\mu = 0$, because in this case $\mathfrak{R}_0^{(3)} \leq U_2$, $\mathfrak{B}_0^{(3)} \leq B_0$, etc. Finally, (35.14) is obtained from (35.12) by putting $\mu = 1$, observing that:

$$\mathfrak{R}_1^{(3)} = O\left(\left[\mathfrak{H}_1^{(3)}\right]^{\frac{1}{1+\lambda}} U_1^{\frac{\lambda}{1+\lambda}} + U_1\right),$$

then applying proposition B) of § 33 and taking into account that:

$$\mathfrak{A}_{0,\lambda}^{(3)} = O\left(\left[\mathfrak{A}_{1,\lambda}^{(3)}\right]^{\frac{\lambda}{\lambda+1}} + 1\right).$$

Turning now to the consideration of (35.9), (35.10), and (35.11), it is useful to perceive that if U_0 is majorized, then from these formulas we can deduce a bound for all the U_k and $U_{k,\lambda}$ in T or D with subscripts $k \leq n$. Indeed, once U_0 and $U_{n,\lambda}$ are majorized, in order to bound U_n and the U_k and $U_{k,\lambda}$ with $k < n$ we will be able to resort to theorems 33, IV, 33, V, 33, VI. On the other hand, it is interesting to deepen this research in view of a more refined bound for $U_{0,\lambda}$ and $U_{1,\lambda}$. It can in fact be presumed, as indeed happens for equations with constant coefficients devoid of terms in u and $\partial u/\partial x_i$, that the majorization of $U_{k,\lambda}$ for $k \leq 1$ could be reached in terms of functions only of U_0, F_0, $\Theta_{k,\lambda}[\varphi;\partial T]$. Referring to § 37 for better precision in this regard, we limit ourselves here to mentioning that we can give an affirmative reply to the question, by making use of the existence theorem for NEUMANN's problem and the representation of the solution by means of potentials. In this regard see G. GIRAUD [12].

However, at this point we shall call attention to the following theorems which concern the behavior of the first derivatives and of the HÖLDER coefficients on the boundary. Of these theorems, the first two can be obtained by referring to a classical memoir by S. BERNSTEIN [2], and the third was recently proved by C. PUCCI [8, 10]. All of these theorems can be proved in an elementary manner under the sole hypothesis that the coefficients of the equation are bounded and $\mathfrak{M}$ is uniformly elliptic. This last hypothesis is even superfluous for the second of these theorems.

35, VI. *Let $u(x)$ be a solution of the problem* (35.1), *regular in $T - \partial T$ and of class $C^{(1)}(T) \cap C^{(2)}(\partial T)$. If $T \in A^{(2)}$, then we have for $x \in \partial T$:*

$$\frac{du}{dn} = O(F_0 + U_0 + \Theta_2[\varphi; \partial T]), \tag{35.15}$$

uniformly with respect to x as $\mathfrak{M}$ varies in every family of elliptic operators in which A_0, B_0, C_0, and $1/a$ are held bounded.

35, VII. *Let $b_i = c = 0$, $f \geq 0$ $[f \leq 0]$ and let the other hypotheses of the preceding theorem be satisfied. If $T \in A^{(2)}$ is uniformly convex, then there exists a constant K, depending only on T, such that we have:*

$$\frac{du}{dn} \geq K \| \varphi \|_{C^{(2}(\partial T)} \quad [\leq K \| \varphi \|_{C^{(2)}(\partial T)}]. \tag{35.16}$$

35, VIII. *Let $u(x)$ be a solution of the equation $\mathfrak{M} u = f$ regular in $T - \partial T$, continuous in T, and of class $C^{(0,\lambda)}$ on ∂T. If $T \in A^{(1)}$, then we have:*

$$u(x) - u(x_0) = O\left[(F_0 + U_0 + U_{0,\lambda}(\partial T)) \mid x - x_0 \mid^\lambda\right], \tag{35.17}$$

uniformly for $x \in T$, $x_0 \in \partial T$ and as $\mathfrak{M}$ varies in every family of elliptic operators in which A_0, B_0, C_0, and $1/a$ are held bounded.

The third theorem holds even under considerably reduced hypotheses for the domain T and for the coefficients of the equation; a propos of this, other than the work [10] by PUCCI, see also A. OSSICINI and F. ROSATI [1].

In the second theorem, by a domain $T \in A^{(2)}$ being *uniformly convex* we mean a domain $T \in A^{(2)}$ such that for every representation (1.1) of a portion of ∂T, the quadratic form $\sum\limits_{i, k = 1}^{n} \zeta_{\xi_i \xi_k} \lambda_i \lambda_k$ is definite.

For brevity, we shall limit ourselves here to proving theorems 35, VI and VII. For the first of these, we observe that without loss of generality we can assume $c = 0$, $\varphi = 0$. Regarding c, this is obvious, since we can always move the term cu to the right side and absorb it into the known term; regarding φ, it suffices to observe that by theorem 16, VI, if $\varphi \neq 0$, we can always extend φ into T in such a way that:

$$\| \varphi \|_{C^{(2)}(T)} = O\left(\| \varphi \|_{C^{(2)}(\partial T)} \right)$$

and after this, if the theorem has already been proved in the case $\varphi = 0$, we will establish it in general by applying the bound discovered to the function $u - \varphi$.

Now, assuming $c = 0$, $\varphi = 0$ and fixing an arbitrary $x_0 \in \partial T$, let $\Gamma(x_1, \varrho)$ be a sphere of radius ϱ with center x_1 on the exterior normal n to T at the point x_0, having in common with T only the point x_0. We verify easily that for $k > 0$ sufficiently large the function

$$v = \varrho^{-k} - | x - x_1 |^{-k}$$

satisfies the conditions:

$$v(x_0) = 0, \quad v(x) > 0 \quad \forall x \in T - x_0, \quad \mathfrak{M}v < -k\beta < 0 \quad \forall x \in T,$$

where β is a constant which we can fix depending on an upper bound for A_0, B_0, $1/a$. It follows from this that for the function $w = u + \dfrac{F_0}{k\beta} v$ the following holds in T: $\mathfrak{M}w < 0$. Because $w(x) > 0$ for $x \in \partial T$ it follows that $w(x)$ is nonnegative in T and this implies, since $w(x_0) = 0$, that the point x_0 gives a minimum for $w(x)$. It follows from this that at this point $dw/dn \leq 0$ and consequently:

$$\frac{du}{dn} \leq \frac{F_0}{\beta} \varrho^{-k-1} .$$

In a similar way we establish a lower bound for $d\,u/d\,n$ and consequently the theorem.

We shall now indicate the proof of theorem 35, VII, assuming without loss of generality that $\| \varphi \|_{C^{(2)}(\partial T)} = 1$. The hypothesis that T is uniformly convex together with the hypotheses of regularity of ∂T and φ permits the proof that the boundary manifold Γ with equation $u = \varphi(x)$, $x \in \partial T$ satisfies the so-called *bounded slope condition*. Namely, there exist for every point P_0 of Γ two planes π^+ and π^- passing through P the slopes of which are bounded independently of P such that Γ is below π^+ and above π^-. It follows from this that, if $f \geq 0$ $[\leq 0]$, then the graph of $u(x)$ is, for every P_0, below π^+ [above π^-]. In fact, if for example $f \geq 0$, and $u = c_0 + \sum\limits_{i=1}^{m} c_i x_i$ is the equation of π^+, then it can certainly not turn out at any point of T that:

$$v(x) = u(x) - c_0 - \sum_{i=1}^{m} c_i x_i > 0 ,$$

for otherwise, $v(x)$ being < 0 on ∂T, there would exist a domain $D \subset T$ such that:

$$v(x) = 0 \quad \text{for} \quad x \in \partial D, \quad v(x) > 0 \quad \text{for} \quad x \in D - \partial D,$$

which is absurd because, since $\mathfrak{M}v = f \geq 0$, v is devoid of relative maxima in D. Once the aforesaid property of the graph of $u(x)$ has been proved, a lower bound for du/dn on ∂T follows from it, and thus the theorem.

The question of establishing the minimal regularity hypotheses for φ and ∂T under which the bounded slope condition holds was detailed by P. HARTMAN [6], who proved among other things that a necessary and sufficient condition that φ satisfy the bounded slope condition is that it satisfy the $(m + 1)$-point condition, namely, that there exist a constant K such that for every hyperplane $u = c_0 + \sum_{i=1}^{m} c_i x_i$ passing through $m + 1$ points of Γ, we have $\sum_{i=1}^{m} c_i^2 \leq K$. In this regard let us recall that many years previously J. SCHAUDER [8] proved that, in the case $m = 2$, if $T \in A^{(2)}$ is uniformly convex and $\varphi \in C^{(2)}(\partial T)$, then the 3-point condition is valid.

Finally, regarding the a priori majorization of U_0, we can make the following observations:

35, IX. *If there exists a positive function $\omega(x)$ of class $C^{(2)}$ in T such that $\mathfrak{M}\,\omega < 0$, then we have, for every solution $u(x)$ of the problem (35.1) which is continuous in T and of class $C^{(2)}$ in $T - \partial T$, that:*

$$U_0 = O(F_0 + \Phi_0), \tag{35.18}$$

denoting by Φ_0 the maximum modulus of φ on ∂T. (35.18) holds in particular if $c \leq 0$, in which case this holds uniformly as $\mathfrak{M}$ varies in every family of elliptic operators in which A_0, B_0, C_0, and $1/a$ are held bounded.

The result is immediate in the case $c < 0$, since now at every point of $T - \partial T$ at which u assumes a positive maximum [negative minimum] we have $cu \geq f$ [$cu \leq f$] and therefore $|u| \leq F_0/\min|c|$. In the general case it suffices to put $u = v\omega$, an equation in v thus being obtained upon which we can reason as previously, because in it the coefficient analogous to c equals $\mathfrak{M}\omega$ and is therefore negative. Finally, in the case $c \leq 0$ the result holds because, if $\Gamma(x_0, \varrho)$ is a hypersphere containing T in its interior with its center at a point x_0 exterior to T, putting $r = \overline{x\,x_0}$, we can assume $\omega = e^{k\varrho^2} - e^{kr^2}$.

This ω is indeed positive in T and, moreover, we easily recognize (see § 3) that, A_0, B_0, C_0, and $1/a$ being bounded, we can determine a k_0 such that for $k > k_0$ we have $\mathfrak{M}\,\omega < -1$.

This result is in part due to S. BERNSTEIN [2]. However, it has been made considerably more precise, from the quantitative point of view, also by M. PICONE [1, 6][1].

We also have:

35, X. *If the coefficients of the operator $\mathfrak{M}$ are of class $C^{(0,\lambda)}$, then* (35.18) *holds under the sole hypothesis that for problem* (35.1) *the uniqueness theorem is valid in the class* $C^{(0)}(T) \cap C^{(2)}(T-\partial T)$.

Under the further hypothesis that the a_{ik} are of class $C^{(1,\lambda)}$, the theorem could be proved by resorting to the transformation of the problem into integral equations (see § 21). Under the hypotheses stated, an elementary and direct proof was given by N. BOBOC and P. MUSTAŢĂ [1]. We note next that:

35, XI. *Let $\mathfrak{M}$ be an elliptic operator defined in a region Ω, uniformly elliptic and with bounded coefficients there. Then there exists a number ϱ such that* (35.18) *holds uniformly as T varies in the family of domains contained in Ω with diameter less than ϱ.*

Indeed, denoting by x_0 an arbitrary point of T and putting $\omega = 2\,\varrho^2 - \overline{x\,x_0^2}$, we have $\omega > 0$ and:

$$\mathfrak{M}\,\omega = -2\sum_{i=1}^{m} a_{ii} + O\,(\varrho\,B_0 + \varrho^2\,C_0)\,,$$

from which it follows for ϱ sufficiently small that $\mathfrak{M}\,\omega < 0$. From this the assertion follows, by theorem 35, IX.

We shall finish by observing that from the majorization formulas proved above we can deduce various convergence and compactness theorems for sequences of solutions of an elliptic equation. We shall limit ourselves here in this regard to stating the following theorem which is an immediate consequence of theorem 35, IV:

35, XII. *Let $\mathfrak{M}$ be an elliptic operator with coefficients of class $C^{(0,\lambda)}$, and let $\{u_k\}$ be a sequence of solutions of the equation $\mathfrak{M}u = f$ with $f \in C^{(0,\lambda)}$* (T). *If $\{u_k\}$ is uniformly convergent [uniformly bounded] in T, then the sequence of the restrictions of the u_k's to an arbitrary domain $D \subset T - \partial T$ is convergent [compact] in* $C^{(2,\lambda)}(D)$.

36. Method of continuation for the proof of the existence theorem for Dirichlet's problem. In this section, following J. SCHAUDER [9] and R. CACCIOPPOLI [7], we shall give a new proof of the existence theorem for DIRICHLET's problem, which is based, on the one hand, on the majorization formulas established in § 35, and on the other,

---

[1] See also M. PICONE [7], Chapter VII no. 5, C. PUCCI [1], and L. E. PAYNE [1].

on some procedures which can be considered the extension into the functional area of the methods of analytic continuation.

Let Σ be the totality of operators $\mathfrak{M}$ with coefficients of class $C^{(n-2,\lambda)}$ in a domain T of class $A^{(n,\lambda)}$. If we define the norm of an element $\mathfrak{M}$ by putting:

$$\| \mathfrak{M} \| = A_0 + B_0 + C_0 + A_{n-2,\lambda} + B_{n-2,\lambda} + C_{n-2,\lambda}, \qquad (36.1)$$

then Σ can be considered as a BANACH space. The totality of elliptic operators will then be a convex subset of Σ which we shall call E; finally, we denote by $E_0 \subset E$ the other convex subset of Σ consisting of elliptic operators for which $c \leq 0$. Now we shall denote by $\Sigma_{n,\lambda}$ the space of pairs $\{f, \varphi\}$ with $f \in C^{(n-2,\lambda)}(T)$, $\varphi \in C^{(n,\lambda)}(\partial T)$, the norm of an element $\{f, \varphi\}$ of $\Sigma_{n,\lambda}$ being defined by the expression:

$$\| \{f, \varphi\} \| = \| f \|_{C^{(n-2,\lambda)}(T)} + \| \varphi \|_{C^{(n,\lambda)}(\partial T)}. \qquad (36.2)$$

We shall also for simplicity put for every $u \in C^{(n,\lambda)}(T)$:

$$\| u \| = \| u \|_{C^{(n,\lambda)}(T)}. \qquad (36.3)$$

Let E_0^* be the set of elements of E_0 for which the problem (35.1) admits one (and only one) solution $u \in C^{(n,\lambda)}(T)$ for every $\{f, \varphi\} \in \Sigma_{n,\lambda}$. Because E_0^* contains at least the element $\mathfrak{M} = \Delta_2$, E_0 is certainly not empty; moreover, if we can show that E_0^* is simultaneously closed and open in E_0, we shall have proved that $E_0^* = E_0$ and consequently that the DIRICHLET problem is solvable for arbitrary elliptic operators $\mathfrak{M}$ with $c \leq 0$.

For every $\mathfrak{M} \in E_0^*$ we denote by:

$$u = \mathfrak{T}\left[\{f, \varphi\}; \mathfrak{M}\right] \qquad (36.4)$$

the solution of problem (35.1) and we observe by theorems 33, IV, and 35, II, and 35, IX that there exists a constant $K(\mathfrak{M})$, which can be bounded by a function of an upper bound of $\| \mathfrak{M} \|$ and of $1/a$, for which we have:

$$\| \mathfrak{T}\left[\{f, \varphi\}; \mathfrak{M}\right] \| \leq K(\mathfrak{M}) \| \{f, \varphi\} \|. \qquad (36.5)$$

It follows from this that if $\{\mathfrak{M}_k\}$ is a sequence of points of E_0^* which tends to a point $\mathfrak{M}$ of E_0, then the functions $\mathfrak{T}\left[\{f, \varphi\}; \mathfrak{M}_k\right]$ are equicontinuous with their derivatives up to order n and these latter have uniformly bounded HÖLDER coefficients; from the sequence of these functions we can consequently extract another which converges uniformly together with the sequence of its derivatives of order $k \leq n$ to a function $u \in C^{(n,\lambda)}(T)$ and its derivatives, respectively. And because this u

obviously turns out to be a solution of problem (35.1), it stands proved that $\mathfrak{M} \in E_0^*$ and therefore that E_0^* is closed in E_0. Let us now prove that E_0^* is open, i.e., show that if $\mathfrak{M} \in E_0^*$ then $\mathfrak{M} + \mathfrak{M}' \in E_0^*$ also, on condition that $\| \mathfrak{M}' \|$ is sufficiently small. Now it is evident that the solution of DIRICHLET's problem for the equation $\mathfrak{M}u + \mathfrak{M}'u = f$ is equivalent to that for the functional equation:

$$u = \mathfrak{T} \left[\{f, \varphi\}; \mathfrak{M} \right] + \mathfrak{S} \left[u \right], \tag{36.6}$$

where we have put:

$$\mathfrak{S} \left[u \right] = - \mathfrak{T} \left[\{ \mathfrak{M}' u, 0 \}; \mathfrak{M} \right].$$

But by dint of (36.5):

$$\| \mathfrak{S} \left[u \right] \| = O (\| \mathfrak{M}' \| \cdot \| u \|),$$

and therefore for $\| \mathfrak{M}' \|$ sufficiently small:

$$\| \mathfrak{S} \left[u \right] \| < h \| u \|$$

with $h < 1$. The functional correspondence $v = \mathfrak{S}[u]$ is therefore a *contraction* and from this it follows easily that the functional equation (36.6) is solvable by the procedure of successive approximation[1]. We have thus completely proved the theorem:

36, I. *Let $\mathfrak{M}$ be an elliptic operator with coefficients of class $C^{(n-2,\lambda)}$ in T, with $n \geq 2$. If also $f \in C^{(n-2,\lambda)}$ and $\varphi \in C^{(n,\lambda)}$ and if $c \leq 0$, then DIRICHLET's problem admits one (and only one) solution of class $C^{(n,\lambda)}$ in T.*

But in addition we have that:

36, II. *When $c \leq 0$ does not hold, then the alternative theorem holds for the DIRICHLET problem considered in theorem 36, I.*

Indeed, putting $\mathfrak{M}_0 u = \mathfrak{M}u - cu$, the solution of the DIRICHLET problem (35.1) is immediately brought back to that for the linear functional equation:

$$u = \mathfrak{T} \left[\{f, \varphi\}; \mathfrak{M}_0 \right] - \mathfrak{T} \left[\{cu, 0\}; \mathfrak{M}_0 \right], \tag{36.7}$$

which, considered as u varies in $C^{(n-2,\lambda)}(T)$, is of RIESZ type, since the transformation:

$$u = \mathfrak{T} \left[\{ cu, 0 \}; \mathfrak{M}_0 \right]$$

is completely continuous in $C^{(n-2,\lambda)}(T)$, in consequence of the bound:

$$V_0 + V_{n,\lambda} = O (U_0 + U_{n-2,\lambda}),$$

which we immediately deduce from (36.5).

---

[1] See R. CACCIOPPOLI [1] and C. MIRANDA: Problemi di esistenza in analisi funzionale. Quaderni Sc. N. Sup. Pisa (1949) Cap. II n. 6.

The theorem of the alternative follows, consequently, from a result due to F. Riesz already mentioned previously (§ 29).

We can now add that:

36, III. *Under the hypotheses of theorem 36, II, every solution of the homogeneous* Dirichlet *problem which is of class* $C^{(2,\lambda)}$ *is necessarily of class* $C^{(n,\lambda)}$.

Indeed, let us suppose that the homogeneous problem admits p solutions of class $C^{(2,\lambda)}$ and that of these only $q < p$ are of class $C^{(n,\lambda)}$. Under this hypothesis the existence of p linearly independent linear functionals $X_i [\{f, \varphi\}]$ defined for $\{f, \varphi\} \in \Sigma_{2,\lambda}$ and of q linear functionals $Y_j [\{f, \varphi\}]$ defined for $\{f, \varphi\} \in \Sigma_{n,\lambda}$ would follow, such that for $\{f, \varphi\} \in \Sigma_{2,\lambda}$, the problem (35.1) admits solutions when and only when we have:

$$X_i[\{f, \varphi\}] = 0, \quad i = 1, 2, ..., p, \tag{36.8}$$

and for $\{f, \varphi\} \in \Sigma_{n,\lambda}$ the solution exists when and only when we have:

$$Y_j[\{f, \varphi\}] = 0, \quad j = 1, 2, ..., q. \tag{36.9}$$

For $\{f, \varphi\} \in \Sigma_{n,\lambda}$ (36.8) would now be a consequence of (36.9), which is absurd if $p > q$, because then the X_i would have to be linearly dependent in $\Sigma_{n,\lambda}$, and therefore also in $\Sigma_{2,\lambda}$, contrary to the hypothesis, since $\Sigma_{n,\lambda}$ is dense in $\Sigma_{2,\lambda}$.

Let us now show that:

36, IV. *Under the hypotheses of theorem 36, II, and if, moreover, the uniqueness theorem for problem* (35.1) *holds in the class of solutions* $u \in C^0(T) \cap C^{(2)}(T - \partial T)$, *in particular if the hypotheses of theorem 35, IX are valid, then this problem admits one and only one solution belonging to the aforesaid class, even if φ is assumed only continuous on ∂T and of class* $C^{(n,\lambda)}$ *in an open region X of ∂T. This solution is, in this case, continuous in T and of class $C^{(n,\lambda)}$ in $(T - \partial T) \cup X$. The result holds even if X is empty.*

Indeed, under the hypotheses posed, every solution in class $C^{(n,\lambda)}(T)$ of the associated homogeneous problem is zero, and therefore the problem (35.1), by theorem 36, II, admits one and only one solution in class $C^{(n,\lambda)}(T)$ if $\varphi \in C^{(n,\lambda)}(\partial T)$. For every φ which instead satisfies the hypotheses of the statement, it is possible to construct an approximating sequence $\{\varphi_k\}$ of functions of class $C^{(n,\lambda)}$, such that for every region X_1 of compact support in X, the quantity $\Theta_{n,\lambda}[\varphi_k; \overline{X}_1]$ is held bounded; let u_k be the solution of the equation $\mathfrak{M} u = f$ equal to φ_k on ∂T. Because, by theorem 35, X, (35.18) holds, the sequence of the u_ks converges uniformly in T to a function u which is equal to φ on ∂T. On the other hand, by dint of (35.18) and theorem 35, IV, the nth derivatives of the u_ks have uniformly bounded HÖLDER coefficients in every domain $D \subset (T - \partial T) \cap X$. This proves that the sequence of u_ks is compact with respect to convergence in $C^{(n)}(D)$. And from this it finally follows that the aforesaid function u is of class $C^{(n, \lambda)}$ in $(T - \partial T) \cup X$ and satisfies the equation $\mathfrak{M} u = f$.

We now wish to conclude by observing explicitly that from theorem 36, IV, the following theorem of E. HOPF [4] follows immediately:

36, V. *Let $\mathfrak{M}$ be an elliptic operator with coefficients of class $C^{(n-2, \lambda)}$ in a bounded domain T, with $n \geq 2$, and let $f \in C^{(n-2, \lambda)}$ also. Then every solution of the equation $\mathfrak{M} u = f$ of class $C^{(2)}$ in $T - \partial T$ is also of class $C^{(n, \lambda)}$ in $T - \partial T$.*

And indeed, for every $x_0 \in T - \partial T$, it is possible (theorem 35, XI) to determine a hypersphere Γ with center x_0 such that for every solution continuous in Γ, (35.18) holds relative to Γ. By the preceding theorem, there now exists in Γ a solution of the equation which coincides with u on $\partial \Gamma$ and which is of class $C^{(n, \lambda)}$ in $\Gamma - \partial \Gamma$. And this solution, by dint of (35.18), must coincide with u throughout Γ. This proves the theorem.

In a more or less analogous way we have also that:

36, VI. *Let $\mathfrak{M}$ be an elliptic operator with coefficients of class $C^{(n-2, \lambda)}(T)$ with $n \geq 2$ and let $T \in A^{(n, \lambda)}$. If $f \in C^{(n-2, \lambda)}(T)$, then every solution of the equation $\mathfrak{M} u = f$ which is regular in $T - \partial T$ and of class $C^{(n, \lambda)}$ in an open region X of ∂T is also of class $C^{(n, \lambda)}$ in $(T - \partial T) \cap X$.*

In finishing, it will be useful to set up a comparison between these results and those obtained by GIRAUD with the method of integral equations. We observe beforehand that the advantages of the method described in this section are a greater simplicity and the possibility of ensuring the inclusion of the solution in class $C^{(n, \lambda)}$ for $n > 2$, under appropriate hypotheses on the data of the problem. It nevertheless should be perceived that we could also arrive at the same results with the method of integral equations, but at the cost of considerable complications. Another advantage of the method of continuation is that it allows the proof of an existence theorem for DIRICHLET's problem with continuous boundary data even in the case in which the coefficients of $\mathfrak{M}$ are assumed merely HÖLDER continuous. In fact, under these general hypotheses, GIRAUD was successful in exhaustively treating only the case in which the boundary condition was homogeneous[1]. On the contrary, it is to be perceived that with the method of integral equations we succeed in developing the entire treatment under the hypothesis that T is of

---

[1] For the treatment, with GIRAUD's method, of the case $\varphi \neq 0$, see the reference given at the end of § 21; in GIRAUD's memoir [12] the case in which $\varphi \in C^{(1, \lambda)}$ is also studied.

class $A^{(1,\lambda)}$, while the method of continuation requires T to be at least of class $A^{(2,\lambda)}$. Moreover, with the method of integral equations we succeed in specifying the nature of the compatibility conditions which must be satisfied by f and φ when the homogeneous problem admits nonzero solutions, whereas with the method of continuation we can only assert that these conditions are of the type (36.8), without being able to say anything about the form of the functionals X_i.

37. General majorization formulas in $H^{k,p}$. Besides the majorization formulas of § 35, some other formulas can turn out to be useful in many questions, formulas in which the functions which we consider enter by means of certain of their integral norms. Some examples of these formulas have already been considered in § 30 for weak solutions of elliptic equations. In this section we wish to occupy ourselves with formulas valid for regular solutions of the equation, or, more generally, for those *generalized solutions* which, being functions of class $H^{2,p}$, satisfy the equation almost everywhere. We recall (§ 16) that such a function admits a trace γu on ∂T for which the boundary condition for DIRICHLET's problem is understood in the sense that:

$$\gamma u = \varphi(x) \qquad \text{for} \quad x \text{ a.e. on } \partial T. \tag{37.1}$$

The following theorem holds:

37, I. *If $T \in A^{(k+2)}$, a_{ij}, b_i, $c \in C^{(k)}(T)$ with $k \geq 0$, and if X is an open portion of ∂T, then for every solution $u \in H^{k+2,p}(T)$ with $p > 1$ of the equation $\mathfrak{M} u = f$ satisfying (37.1), and for every domain $D \subset (T - \partial T) \cup X$ we have:*

$$\| u \|_{H^{k+2,p}(D)} = O\left(\| f \|_{H^{k,p}(T)} + \| u \|_{L^2(T)} + \| \varphi \|_{H_{k+2-\frac{1}{p}}(X)} \right), \tag{37.2}$$

uniformly as $\mathfrak{M}$ varies in every family of elliptic operators in which the a_{ij} are held equicontinuous and the quantities A_k, B_k, C_k, $1/a$ bounded. In particular, (37.2) holds even if X is empty, in which case the hypothesis on T is superfluous, and also for $X = \partial T$ and $D = T$. For $k = 0$, (37.2) holds also under the sole hypothesis that $a_{ij} \in C^{(0)}$, $b_i \in L^m$, $c \in L^{\frac{m}{2}}$, uniformly as $\mathfrak{M}$ varies in every family of elliptic operators in which the a_{ij} are held equicontinuous, the quantities A_0 and $1/a$ uniformly bounded, and the integrals of $|b_i|^m$ and $|c|^{\frac{m}{2}}$ absolutely equicontinuous.

In this theorem, $H_{s-\frac{1}{p}}(X)$ is the space of functions φ defined on X which for every domain $D \subset (T - \partial T) \cup X$ are traces on $D \cap X$ of a function $u \in H^{s,p}(D)$ and for which the norm:

$$\| \varphi \|_{H_{s-\frac{1}{p}}(X)} = \sup_D \inf_u \| u \|_{H^{s,p}(D)}$$

is finite.

If we leave out of consideration the result of S. BERNSTEIN recalled in the introduction and of which we shall speak later, (37.2) was proved for the first time in 1934 by J. SCHAUDER [10] in the case $m = 2$, $p = 2$, $k = 0$, being based on a similar result established many years previously by L. LICHTENSTEIN [1] for equations with constant coefficients, and on KORN's artifice, used in § 35. In 1947 the theorem of LICHTENSTEIN was extended by K.O. FRIEDRICHS [5] to the case $m > 2$ and after this it was possible to apply SCHAUDER's procedure to the general case. On the other hand, (37.2) was proved in 1951 for $m \geq 2$, $p = 2$, $k = 0$ by R. CACCIOP-POLI [12] and by O.A. LADYŽENSKAJA [1, 2, 3] with direct procedures which, at least under slightly more restrictive hypotheses on the coefficients, do not require some previous study of equations with constant coefficients. SCHAUDER's procedure can now still be used to improve the results obtained. In 1951 SCHAUDER's results were rediscovered by S.G. MIHLIN [4]. Later A.I. KOŠELEV [1, 3] proved (37.2) in the cases $m = 2$, $p > 1$, $k = 0$ and $m \geq 2$, $p > m$, $k = 0$. Finally, in the general form of the statement, but with a slightly stronger hypothesis on T and in the case $k = 0$, the result was established almost simultaneously by A.I. KOŠELEV [4] and D. GRECO [1]. Both of these authors first treat the case of equations with constant coefficients, GRECO making use of the theorem of CALDERON and ZYGMUND referred to in § 13, KOŠELEV using a theorem of J. MARCINKIEWICZ on FOURIER series[1]. The proof of (32.2) for $k > 0$ was then given by O.A. LADYŽENSKAJA [5] for $p = 2$ and by A.A. MILJUTIN [1] in the general case. Some complementary results were obtained by V.I. JUDOVIČ [1, 2]. The possibility of reducing the hypotheses on the domain T in the case when $a_{ij} \in C^{(0, 1)}$ was studied by J.K. KADLEC [1, 3]. For a self-adjoint operator, other integral majorization formulas were established by S. CAMPANATO [7], making use, however, of different function spaces than those we have considered. From CAMPANATO's results we can obtain as particular cases both theorems of type 37, I and theorems similar to those of § 35. In particular, making use of these results, S. CAMPANATO and G. STAMPACCHIA [1] have indicated how one could obtain (37.2) without recourse either to the CAL-DERON-ZYGMUND theorem or to that of MARCINKIEWICZ, but by depending on a theorem on functional interpolation due to STAMPACCHIA himself[2].

---

[1] J. MARCINKIEWICZ: Sur les multiplicateurs des séries de Fourier. Studia Math. **8** (1939) 78—91.

[2] G. STAMPACCHIA: The spaces $\mathscr{L}^{(p, \lambda)}$, $N^{(p, \lambda)}$ and interpolation. Ann. Sc. N. Sup. Pisa **19** (1965) 443—462.

Other results could be deduced, finally, from similar results established in general for equations of higher than second order (see § 52). In particular see A. P. CALDERON and A. ZYGMUND [2, 3] for results which can approach those of CAMPANATO.

It is impossible to enter here into details in the proof of theorem 37, I which is technically quite elaborate; the exposition can be found in the volumes by O. A. LADYŽENSKAJA and N. N. URAL'CEVA [8] and by C. B. MORREY [9], together with that for other theorems of the same type in which the hypotheses on the coefficients are minimized. However, we shall not fail to show how we could prove (37.2) under the hypothesis $\mathfrak{M} = \varDelta_2$, $\varphi = 0$, $k = 0$, $p = 2$, the case which for $p = 2$ is the point of departure for the general case. If we depart from the identity:

$$\sum_{i,k=1}^{m} p_{ik}^2 = (\varDelta_2 u)^2 + \sum_{i,k=1}^{m} \left(p_{ik}^2 - p_{ii} p_{kk} \right)$$

$$= (\varDelta_2 u)^2 + \sum_{i,k=1}^{m} \left[\frac{\partial}{\partial x_i} \left(p_k \, p_{ik} \right) - \frac{\partial}{\partial x_k} \left(p_k \, p_{ii} \right) \right], \quad (37.3)$$

we easily obtain by integrating[1]:

$$\int_T \sum_{i,k=1}^{m} p_{ik}^2 \, dx = \int_T f^2 \, dx - \int_{\partial T} H \left(\frac{du}{dn} \right)^2 d\sigma , \quad (37.4)$$

where H is the mean curvature of ∂T, having chosen as positive orientation that of the exterior normal. Consequently if $H \geq 0$, which is satisfied if for example T is convex, we have the inequality perceived by G. TALENTI [1] and C. MIRANDA [18]:

$$\int_T \sum_{i,k=1}^{m} p_{ik}^2 \, dx \leq \int_T f^2 (x) \, dx .$$

From this, taking account of the fact that we have, in an elementary way:

$$\int_T \sum_{i=1}^{m} p_i^2 \, dx \leq \int_T f^2 (x) \, dx$$

and that for $u \in H_0^{1,2}$ POINCARE's inequality holds:

---

[1] Proceeding formally, it would be necessary to assume $u \in H^{3,2}$; however, here we easily free ourselves of this hypothesis by an approximation procedure.

$$\int_T u^2\, dx = O\left(\int_T \sum_{i=1}^m p_i^2\, dx\right), \tag{37.5}$$

we deduce immediately that:

$$\| u \|_{H^{2,2}(T)} = O\left(\| f \|_{L^2(T)}\right).$$

We arrive at the same result even in the case when $H \geq 0$ does not hold, by taking into account that for $u \in H^{2,2}$ an inequality of the following type holds:

$$\int_{\partial T} \left(\frac{d u}{d n}\right)^2 d\sigma = O\left[\int_T \sum_{i=1}^m p_i^2\, dx + \left(\int_T \sum_{i=1}^m p_i^2\, dx \int_T \sum_{i,k=1}^m p_{ik}^2\, dx\right)^{1/2}\right]. \tag{37.6}$$

Finally, we observe that theorem 37, I permits an answer to the question posed in § 35 on the possibility of majorizing $U_{0,\lambda}$, U_1, $U_{1,\lambda}$ without $F_{0,\lambda}$ taking part. From (37.2) and from known theorems of S. L. SOBOLEV [5], it in fact follows that:

37, II. *Assuming the hypotheses of theorem* 37, I *are satisfied, and putting:*

$$K = \| f \|_{L^p(T)} + \| u \|_{L^2(T)} + \| \varphi \|_{H_2 - \frac{1}{p}}(X).$$

we have, for every solution $u \in H^{2,p}(T)$ *of the equation* $\mathfrak{M}u = f$ *satisfying* (37.1) *and for every domain* $D \subset (T - \partial T) \cap X$, *that:*

$$\| u \|_{L^{\frac{np}{n-2p}}(D)} + \| u \|_{H^{1,\frac{np}{n-p}}(D)} = O(K) \quad \text{for} \quad p < \frac{n}{2}, \tag{37.7}$$

$$U_{0,\lambda}(D) + \| u \|_{H^{1,\frac{np}{n-p}}} = O(K) \quad \text{for} \quad 0 < \lambda = 2 - \frac{n}{p} < 1, \tag{37.8}$$

$$U_1(D) + U_{1,\lambda}(D) = O(K) \quad \text{for} \quad \lambda = 1 - \frac{n}{p} > 0. \tag{37.9}$$

The uniformity of these bounds is ensured as in theorem 37, I.

In the two theorems which we have just discussed, the hypothesis that $a_{ij} \in C^{(0)}$ takes part in an essential way. We wish now to see what could be said of this when we reduce this hypothesis or suppress it completely. The following result regards exclusively the case of two variables:

37, III. *Let* $m = 2$, $T \in A^{(2)}$, a_{ij}, b_i, $c \in L^\infty(T)$. *If* $\mathfrak{M}$ *is uniformly elliptic, then there exist two constants* K *and* K_1, *depending only on* A_0,

B_0, C_0, and $1/a$, such that for every solution $u \in H^{2,2}(T)$ of the equation $\mathfrak{M} u = f$ satisfying (37.1) we have:

$$\| u \|_{H^{2,2}} \leq K \left(\| f \|_{L^2(T)} + \| \varphi \|_{H_{\frac{3}{2}}(\partial T)} \right) + K_1 \| u \|_{L^2(T)}. \qquad (37.10)$$

If $c \leq 0$, then we can assume $K_1 = 0$ in (37.10)[1].

It will suffice to prove (37.10) in the case $\varphi = 0$ and for this purpose we suppose in the first place that $b_i = c = 0$. Applying an artifice due to S. BERNSTEIN [2, 9], we observe that by departing from the identity:

$$(a_{11} + a_{22})^{-1} \left[\sum_{i, j, k = 1}^{2} a_{ij} p_{ki} p_{kj} - \Delta_2 u \, \mathfrak{M} u \right] = p_{11} p_{22} - p_{12}^2$$

we easily have a bound of the type:

$$\sum_{i, k = 1}^{2} p_{ik}^2 = O \left(\varepsilon (\Delta_2 u)^2 + \frac{1}{\varepsilon} f^2 + (p_{11} p_{22} - p_{12}^2) \right).$$

From this, by integrating, the following formula similar to (37.4) emerges:

$$\int_T \sum_{i, k = 1}^{2} p_{ik}^2 \, dx = O \left[\int_T f^2 \, dx - \int_{\partial T} \left(\frac{du}{dn} \right)^2 \frac{ds}{\varrho} \right],$$

where ϱ is the radius of curvature of ∂T. If T is convex, then (37.10) follows immediately with $K_1 = 0$. In the general case, we arrive equally well at (37.10) by employing (37.6) and the inequality of EHRLING[2]:

$$\int_T p_i^2 \, dx \leq \varepsilon \int_T \sum_{i, k = 1}^{2} p_{ik}^2 + k(\varepsilon) \int_T u^2 \, dx,$$

valid for arbitrary $\varepsilon > 0$. Now, if $b_i = c = 0$ is not true, it will suffice to apply the result to the equation written in the form:

$$\sum_{i, k = 1}^{2} a_{ik} p_{ik} = f - \sum_{i=1}^{2} b_i p_i - c u$$

and then to use EHRLING's inequality to eliminate the norms in L^2 of the p_i.

---

[1] For some bounds on $\| u \|_{C^1(D)}$ for every domain $D \subset T - \delta T$ see C. PUCCI [6] and M. BATTEZZATI [1]. On the majorization of u in $H^{2,p}$ with $p < 2$ see C. PUCCI [14].

[2] See for example L. NIRENBERG [4].

It remains to prove that we can assume $K_1 = 0$ if $c \leq 0$. This we can deduce from the following theorem due to C. PUCCI [12] valid for arbitrary $m \geq 2$:

37, IV. *Let $T \in A^{(1)}$, a_{ij}, b_i, $c \in L^\infty$, $c \leq 0$. If $\mathfrak{M}$ is uniformly elliptic, then there exists a constant K, depending only on T and on A_0, B_0, C_0, $1/a$, such that for every solution $u \in H^{2,m}(T)$ of the equation $\mathfrak{M} u = f$, satisfying (37.1), the following holds:*

$$sup_T |u| \leq K \left(\| f \|_{L^m(T)} + \| \varphi \|_{H_{m-\frac{1}{2}}(\partial T)} \right). \tag{37.11}$$

PUCCI arrived at this theorem at the conclusion of a group of researches [7, 11] relating to the extremal elliptic operators[1] $m_\alpha(u)$, $M_\alpha(u)$ defined for every function $u \in H^{2,m}(T)$ by the relations:

$$m_\alpha(u) = \inf_{\mathfrak{M} \in \mathfrak{M}_\alpha} \mathfrak{M} u, \qquad M_\alpha(u) = \sup_{\mathfrak{M} \in \mathfrak{M}_\alpha} \mathfrak{M} u,$$

where $\mathfrak{M}_\alpha$ is the class of elliptic operators $\mathfrak{M}$ with bounded coefficients which have ellipticity constants not less than α. The study of these operators also proves useful, among other things, for the construction of counterexamples[2] which allow the proofs that certain a priori bounds or other properties of the solutions of elliptic equations are not valid because of the lack of a sufficient regularity of the coefficients or that the solutions do not belong to a particular class of functions. An example of this type shows in particular that (37.11) is not valid if we assume only that $u \in H^{2,k}$ with $k < m$.

PUCCI's theorem is included in other more general ones due to A. D. ALEKSANDROV [9, 10], who in other works [6, 7, 8] had already obtained various results along these lines. The results of these authors and others due to I. YA. BAKEL'MAN [9], however, concern only the estimation of u, while we know no estimation of the derivatives of u when $m > 2$ and the coefficients of $\mathfrak{M}$ are assumed only measurable and bounded. Still, some results have been obtained under more restrictive hypotheses on the coefficients which do not imply their continuity. For example, the following theorems proved by C. MIRANDA [13] and G. TALENTI [1], respectively, hold; for simplicity we state them while limiting our consideration to the equation:

$$\mathfrak{M} u = \sum_{i,k=1}^{m} a_{ik} p_{ik} = f(x) \tag{37.12}$$

---

[1] See also M. FRASCA [1].
[2] For other counterexamples see J. SERRIN [4] and K. MILLER [1].

37, V. *If $T \in A^{(3)}$, $a_{ik} \in L^\infty \cap H^{1,m}$, and if $\mathfrak{M}$ is uniformly elliptic, then there exists a constant K depending only on T and on the quantities A_0, $1/a$, $\| a_{ik} \|_{H^{1,m}(T)}$, such that for every solution $u \in H^{2,2}(T)$ of (37.12) satisfying (37.1) we have:*

$$\| u \|_{H^{2,2}(T)} \leq K \left[\| f \|_{L^2(T)} + \| \varphi \|_{H_{\frac{3}{2}}(T)} \right]. \tag{37.13}$$

37, VI. *Let $T \in A^{(3)}$ be a bounded domain whose boundary, oriented positively toward the exterior of T, has positive mean curvature. If $a_{ik} \in L^\infty$ and if, moreover:*

$$\sum_{i,k=1}^{m} a_{ik}^2 \leq \frac{1}{m-1+\varepsilon} \left(\sum_{i=1}^{m} a_{ii} \right)^2, \quad 0 < \varepsilon < 1, \tag{37.14}$$

then there exists a constant K depending only on T, A_0, ε, such that for every solution $u \in H^{2,2}(T)$ of (37.12) satisfying (37.1), (37.13) holds.

We note that in theorem 37, V the hypothesis on the a_{ik} does not imply continuity. A simple example shows that if this hypothesis is replaced by one of the type $a_{ik} \in L^\infty \cap H^{1,p}$ with $p < m$, then (37.13) can not hold. As far as hypothesis (37.14) is concerned, which is equivalent to one introduced earlier by H.O. CORDES [2, 3, 4] for another purpose (see § 38), we observe that it is satisfied for $\mathfrak{M} = \Delta_2$, and expresses essentially that $\mathfrak{M}$ does not differ too much from Δ_2. In this case we can also show with an example that if this condition is not satisfied then (37.13) can not hold.

All the integral bounds which we have just discussed are theoretical bounds understood to prepare the way for the proof of existence or regularity theorems. However, those formulas in which the constants can be numerically determined in a simple way also are interesting in that they furnish an effective first evaluation of the solution of the problem. Along these lines interesting results relating to the evaluation of u and its first derivatives are due to L.E. PAYNE and H.WEINBERGER [4] and to J.H. BRAMBLE and L.E. PAYNE [2, 3, 4, 5]. Other noteworthy formulas were obtained by G. FICHERA [15, 16, 17], by exploiting among other things, theorem 30, I. Finally, other bounds were obtained by J.L.SYNGE [1] with the so-called method of the hypercircle.

38. Existence and regularization of theorems in $H^{k,p}$. The
theorems of the preceding section in general affirm the possibility of majorizing, by functions of the data, various elements relating to the solutions of the DIRICHLET problem, of which we already presuppose the existence. We now wish to display some regularization theorems, that is, some theorems which affirm that a solution assumed to be of

class $H^{2,2}$ belongs in reality to a more restrictive class, and which therefore has properties of greater regularity, as soon as the data of the problem satisfy appropriate hypotheses. Some results of this type have already been given in § 36 for regular solutions of DIRICHLET's problem, and in § 30 for weak solutions. These latter are applicable even to the solutions of class $H^{2,2}$ which we are now considering, on the condition that these can be considered weak solutions. For this, however, it will be necessary that at least $a_{ik} \in H^{1,m}$, which hypothesis will be admitted only in some of the theorems which we wish now to consider.

A first theorem, due to C.B. MORREY [4], is the following:

38, I. *If* $a_{ij} \in C^{(0)}(T)$, b_i, $c \in L^{\infty}(T)$, $f \in L^{2,\mu}(T)$, *then for every solution* $u \in H^{2,2}(T)$ *of the equation* $\mathfrak{M} u = f$ *and for every domain* $D \subset T - \partial T$ *we have:*

$$|| u ||_{H^{2,2,\mu}(D)} = O(|| f ||_{L^2,\mu(T)} + || u ||_{L^1(T)}), \qquad (38.1)$$

uniformly as $\mathfrak{M}$ *varies in every family of elliptic operators in which the* a_{ij} *are held equicontinuous and the quantities* A_0, B_0, C_0, $1/a$ *bounded.*

We note that for $\lambda = (\mu - m + 2)/2 > 0$, (38.1) implies, by dint of a lemma of MORREY [2], that:

$$U_1(D) + U_{1,\lambda}(D) = O(|| f ||_{L^2,\mu(T)} + || u ||_{L^1(T)}). \qquad (38.2)$$

For the case of two variables, we also have:

38, II. *Assuming* $m = 2$, *if* a_{ij}, b_i, $c \in L^{\infty}(T)$, $f \in L^p(T)$ *with* $p > 2$, *and if* $\mathfrak{M}$ *is uniformly elliptic, then there exists a number* α *depending only on* A_0, a, *and* p, *such that, for every solution* $u \in H^{2,2}(T)$ *of the equation* $\mathfrak{M} u = f$ *and for every domain* $D \subset T - \partial T$ *we have:*

$$|| u ||_{C^{(1,\alpha)}(D)} = O(|| f ||_{L^p(T)} + || u ||_{L^1(T)}), \qquad (38.3)$$

uniformly as $\mathfrak{M}$ *varies in every family of elliptic operators in which the quantities* A_0, B_0, C_0, $1/a$ *are held bounded. If* u *is zero on* ∂T *and* $T \in A^{(2)}$, *then* (38.3) *holds even for* $D = T$.

This theorem can be incorporated in other more general ones regarding quasi-conformal mappings (see § 53) also studied by various authors in relation to the determination of α. We cite in order of time: C.B. MORREY [1, 2], M. SHIFFMAN [1], L. NIRENBERG [1], L. BERS and L. NIRENBERG [2], P. HARTMAN [1], R. FINN and J. SERRIN [1], A. ZITAROSA [2], G. TALENTI [2]. For the proof we observe that by dint of MORREY's lemma just mentioned the assertion will be proved once we have established a bound for $|| u ||_{H^{2,2,2\alpha}}$. Now, because for $m = 2$ and

$p > 2$ we have $L^p \subset L^{2,\mu}$ with $\mu = 2(p-2)/p$, if the a_{ij} were continuous the conclusion would derive from (38.1) with $\alpha = (p-2)/p$. Having assumed the a_{ij} only measurable and bounded, (38.1) can be established only for $\mu = 2\alpha$ appropriately small, by making use of artifices similar to those which were valid for the proof of theorem 37, III.

Other results for the case $m \geq 2$ can be joined to theorems 37, V and VI. We shall state them, referring for simplicity to equation (37.12).

38, III. *Under the hypotheses of theorem* 38, V *and if, moreover,* $f \in L^p$ *and* $\varphi \in H_{2-\frac{1}{p}}$ *with* $p \geq 2$, *then for every solution* $u \in H^{2,2}(T)$ *of* (37.12) *satisfying* (37.1), (37.7) *and* (37.8) *hold with:*

$$K = \| f \|_{L^p(T)} + \| \varphi \|_{H_{2-\frac{1}{p}}(\partial T)}.$$

38, IV. *Assuming* $m \geq 3$, *let* $T \in A^{(2)}$ *be a convex domain. If* $a_{ik} \in L^\infty(T)$, $f \in L^{2,\mu}(T)$ *with* $\mu > m - 3$ *and if, moreover,* (37.14) *holds, then there exists a constant* K *depending only on* T, ε, A_0, $1/a$ *such that for every solution* $u \in H^{2,2}(T)$ *of* (37.12) *zero on* ∂T *we have:*

$$\| u \|_{C^{(0,1/2)}(T)} \leq K \| f \|_{L^{2,\mu}(T)}. \tag{38.4}$$

Under the same hypotheses and if, moreover, we have:

$$\sum_{i,k=1}^{m} a_{ik}^2 \leq \frac{1}{m}\left(1 + \frac{(1-\varepsilon)(m+1)}{2m^2 - 2m - 1}\right) \sum_{i=1}^{m} a_{ii}^2, \tag{38.5}$$

then we can determine two numbers λ *and* K_1, *depending only on* T, ε, A_0, $1/a$, μ *such that:*

$$\| u \|_{C^{(1,\lambda)}(T)} \leq K_1 \| f \|_{L^2(T)}. \tag{38.6}$$

This theorem, except for some slight variation, is due to H. O. CORDES [1, 2, 3]; theorem 38, III, on the other hand, was proved by C. MIRANDA [13].

Let us now see how we can deduce existence and, in certain cases, uniqueness for the problems being considered from the majorizations obtained in § 37. We have[1]

38, V. *Under the hypotheses of theorem* 37, III *with* $c \leq 0$ *or, also under the hypotheses of one of theorems* 37, V *and* 37, VI, *the* DIRICHLET *problem being considered admits one and only one solution in class* $H^{2,2}$.

Indeed, uniqueness follows immediately from the same theorems mentioned in the statement. To prove the existence, it suffices to approx-

---

[1] For some partial results along this line see G. STAMPACCHIA [1, 2] and E. GAGLIARDO [1].

imate the coefficients of $\mathfrak{M}$, f, and φ with sequences of regular functions; we have thus a sequence of problems approximating the given problem, each one of which admits a solution. Because, by dint of the majorization established, the sequence $\{u_n\}$ of these solutions is bounded in $H^{2,2}$, the sequence will be compact in $H^{1,2}$ and weakly compact in $H^{2,2}$, by known theorems. Then, having extracted from the $\{u_n\}$ a sequence convergent in $H^{1,2}$ and weakly convergent in $H^{2,2}$, it is easy to prove that the limit of this sequence is a solution of the problem we set out from.

Analogously we prove that:

38, VI. *Under the hypotheses of theorem* 37, I *and if, moreover,* $c \leq 0$, $f \in H^{k,p}$, $\varphi \in H_{k+2-\frac{1}{p}}$, *then the* DIRICHLET *problem admits at least one solution of class* $H^{k+2,p}$. *If* $k > 0$, *then this solution is unique in class* $H^{2,q}$ *for arbitrary* $q \geq \dfrac{2\,m}{m+2}$. *For* $k = 0$, *the uniqueness theorem holds in class* $H^{2,2}$ *where we assume that* $a_{ij} \in H^{1,m}$, $b_i \in L^m$, $c \in L^{\frac{m}{2}+\sigma}$ *with* $\sigma > 0$ *and moreover, that* $c \leq c_0 < 0$ *with* c_0 *constant, or else that* b_i, $c \in L^\infty$, $c \leq 0$.

In fact, we prove the existence as previously. As for the uniqueness, we observe first of all that every function of class $H^{2,q}$ is also, by a theorem of S. L. SOBOLEV [5], of class $H^{1,r}$ with $r = \dfrac{m\,q}{m-q}$. Therefore if $q \geq \dfrac{2\,m}{m+2}$, then every solution of class $H^{2,q}$ of the homogeneous DIRICHLET problem is also of class $H^{1,2}$, and from this it is easy to verify that it is also a weak solution of the same problem. If $c \leq c_0 < 0$, then uniqueness follows this time from theorem 30, IV. The case now in which $c \leq 0$ and $k > 0$ or else $k = 0$ with b_i, $c \in L^\infty$ can be brought back to the preceding by a change of unknown function $u = v\,\omega$, where $\omega > 0$, $\mathfrak{M}\,\omega < 0$.

Let us conclude this section with the following regularization theorem:

38, VII. *Let* $f \in H^{k,p}$, $\varphi \in H_{k+2-1/p}$. *If the hypotheses of theorem* 37, I *are satisfied and if in the case* $k = 0$ *we also have* $c \in L^{r/2}$ *with* $r > m$, *then every solution* $u \in H^{2,q}$ *with* $q \geq \dfrac{2\,m}{m+2}$ *of* DIRICHLET's *problem belongs to class* $H^{k+2,p}$.

For brevity we shall limit ourselves to proving the theorem in the case $k = 0$. We recall that by a theorem of S. L. SOBOLEV [5], if $u \in H^{2,q}$, then we also have $u \in L^{\frac{m\,q}{m-2\,q}}$. Thus, writing the equation in the form

$$\sum_{i,k=1}^{m} a_{ik}\,p_{ik} + \sum_{i=1}^{m} b_i\,p_i - u = f - (1 + c)\,u,$$

we have that the known term is of class L^s with $s = \inf \left(p, \dfrac{2\,r\,m\,q}{2\,q\,(m-2)+r\,m} \right)$; by the preceding theorem we now have $u \in H^{2,s}$, and, because $s \geq \inf$ $(p, 2\,q)$, a repeated application of the artifice leads to the assertion.

For all the theorems of this section, the hypothesis that T is sufficiently regular seems to be essential. On the difficulties which arise in the presence of singular points for ∂T, see M. Š. BIRMAN and G. E. SKVORKOV [1] and V. G. MAZ'JA [5].

39. A priori bounds for the solutions of the second and third boundary value problem. Many of the results of the preceding sections can be extended to the case of the second and third boundary value problems. A propos of this we shall limit ourselves here to giving some bibliographic references, giving up stating the results with precision.

Let us recall to begin with, that many of the results of § 35, and in particular theorems 35, II and III, were extended to the case of the second and third boundary value problem by R. FIORENZA [1, 2], which naturally allows also the extension of the existence and regularity theorems of § 36. Regarding the majorization of U_0 (theorem 35, IX) we can then consult M. PICONE [1, 6, 7] and C. PUCCI [1].

Regarding the integral bounds of § 37, analogous results to theorem 37, I were established for the NEUMANN problem by O. A. LADYŽEN-SKAJA [5] for $p = 2$ and by A. A. MILJUTIN [1] in the general case. For other notes on the argument of which we have just spoken, however, it will be useful to consult the volume [8] by O. A. LADYŽENSKAJA and N. N. URAL'CEVA. We also point out that in the work [13] by C. MIRANDA, the theorems analogous to 37, V and 38, III are proved for the NEU-MANN problem. A theorem analogous to theorem 38, II for the regular oblique derivative problem could now be deduced from a more general result due to R. FIORENZA [4]. For the difficulties which are encountered in the presence of singular points for ∂T see G. M. VERZBINSKIĬ [1].

Finally, regarding the proofs of numerically noteworthy majorization formulas, see J. H. BRAMBLE and L. E. PAYNE [3, 4], G. FICHERA [16, 17], and J. L. SYNGE [1].

All the preceding references concern the NEUMANN problem or the regular oblique derivative problem. For non-regular oblique derivative problems the works of R. L. BORRELLI [1] and L. HÖRMANDER [4], already cited in § 24, are worth remembering.

CHAPTER VI

Nonlinear equations

If we leave out of consideration some works by H. POINCARÉ, E. PICARD, and E. LE ROY, relating to equations of particular type, then the date of initiation of the modern theory of nonlinear elliptic equations of the second order can be fixed at 1900. In that year, in fact, at the International Congress of Paris, D. HILBERT stated his conjecture that every solution of an analytic elliptic equation is analytic.

HILBERT's conjecture caused considerable research by several authors, among which, a few years later, that of S. BERNSTEIN assumed considerable importance; in it he succeeded [1] not only in proving HILBERT's theorem for equations in two variables, but also in leading from this research to a profound study of DIRICHLET's problem for nonlinear equations [2, 3, 4]. BERNSTEIN's fundamental idea, which has dominated this theory until now, is that the existence theorem for this problem is ensured when it is possible to establish a priori certain appropriate majorization formulas for all its possible solutions.

BERNSTEIN arrived at these results with a procedure resulting from a combination of the methods of successive approximations and of analytic continuation which requires an extremely laborious development and which can not lead to the conclusion except at the price of useless and unnatural qualitative restrictions. This accounts for the reason why BERNSTEIN's research did not have an immediate sequel, to such an extent that in 1924 L. LICHTENSTEIN [5] could still write, „Es wäre sehr zu begrüßen, wenn diese wichtigen Untersuchungen vereinfacht und übersichtlicher dargestellt werden könnten." And years still had to pass before J. SCHAUDER, J. LERAY, and R. CACCIOPPOLI, between 1932 and 1937, succeeded in explaining the basic raison d'etre of BERNSTEIN's results, in simplifying the treatment substantially, and in indicating the natural field of validity. Essential instruments for the research of these authors were, on the one hand, a general theory of functional equations in abstract spaces, and on the other, a new series of majorization formulas. The results, with the exception of some works by CACCIOPPOLI, are almost exclusively with regard to the case of equations in two variables.

The new regularization and a priori majorization procedures introduced more recently for the study of linear equations (see §§ 30, 37, 38, 39) were also applied in these latest years by various authors to the study of nonlinear equations in the case $m \geq 2$. In this way important results have been obtained, with which this theory has reached a nearly definitive systemization. In this chapter we shall give a report of these results in summary form, without entering into the details of the proofs, which are technically quite complicated. And because new developments of the theory absorb the results of SCHAUDER, LERAY and CACCIOPPOLI completely, we shall not even report the reworked treatment which we gave of these in the first edition of this monograph. The exposition will be centered above all on the DIRICHLET problem, but we shall also indicate some results concerning the general properties of the solutions, the questions of analyticity, equations in parametric form, the NEUMANN and oblique derivative problems and some equations of particular type. We shall also point out some new procedures from functional analysis (see § 41) which, formulated for the study of nonlinear problems of higher order, find applications also in the study of equations of second order.

40. General properties of the solutions. Let F be a function of the point x in R_m and of the variables p_{ik}, p_i, u, defined and continuous with its first derivatives for $x \in T$ and for arbitrary values of the other variables on which it depends. Every solution $u(x)$ of the equation:

$$F(p_{ik}, p_i, u, x) = 0, \tag{40.1}$$

such that the quadratic form $\Sigma F_{p_{ik}} \left(p_{ik}(x), p_i(x), u(x), x \right) \lambda_i \lambda_k$ is definite, for example positive, for arbitrary x in T, will be called *elliptic in T*. If, further, the quadratic form $\Sigma F_{p_{ik}} \lambda_i \lambda_k$ is (positive) definite for arbitrary values of the variables p_{ik}, p_i, u and for arbitrary x in T, we shall say that equation (40.1) is itself *elliptic*. In our exposition, without explicit notice to the contrary, we shall always assume that the latter hypothesis is satisfied. Moreover, we shall often have to consider equations of the type:

$$\sum_{i,k=1}^{m} a_{ik} p_{ik} + \sum_{i=1}^{m} b_i p_i = f, \tag{40.2}$$

the a_{ik}, b_i, f being functions of the variables x, u, p_j. Equations of type (40.2) will be called *quasi-linear*. If further, the a_{ik} depend only on x, then equation (40.2) will be called *semilinear*.

The solutions of nonlinear elliptic equations enjoy various maximum and minimum properties similar to those of the solutions of linear equations. A piopos of this one can consult the works of E. Hopf [1], A.D.Aleksandrov [2], R.M. Redheffer [2, 3, 4], T.Kusano [1], R.Výborný [1, 2]. To related arguments, namely, to the proof of comparison theorems, are dedicated also the works of S. Simoda [3], R.M. Redheffer [1], L.M. Kuks [1], Y.Hirasawa [1], K.Akô [1], A.McNabb [1]. We shall limit ourselves here to proving the following theorem:

40, I. *If $F_u \leq 0$, then the difference $u - \bar{u}$ between two solutions of* (40.1) *can not have a positive maximum or a negative minimum at a point of $T - \partial T$ unless it is identically constant in a neighborhood of the point.*

The proof is immediate. Indeed, putting:

$$G(p_{ik}, p_i, u, x, t) = F\left(\bar{p}_{ik} + t\left(p_{ik} - \bar{p}_{ik}\right), \bar{p}_i + t\left(p_i - \bar{p}_i\right), \bar{u} + t\left(u - \bar{u}\right), x\right),$$

the function $v = u - \bar{u}$ is a solution of the equation:

$$\sum_{i,k=k}^{m} \frac{\partial^2 v}{\partial x_i \partial x_k} \int_0^1 G_{p_{ik}} \frac{dt}{t} + \sum_{i=1}^{m} \frac{\partial v}{\partial x_i} \int_0^1 G_{p_i} \frac{dt}{t} + v \int_0^1 G_u \frac{dt}{t} = 0, \quad (40.3)$$

whence our assertion is an immediate consequence of theorem 3, II.

Also of noteworthy importance is the following theorem:

40, II. *Let F be continuous with all its derivatives of order not greater than $n (\geq 1)$, and let its nth derivatives be Lipschitz continuous with respect to the variables p_{ik} and λ-Hölder continuous with respect to the p_i, u, x. Then every solution $u(x)$ of (40.1) of class $C^{(2,\lambda)}$ in $T - \partial T$ is also of class $C^{(n+2,\lambda)}$ in $T - \partial T$. Moreover, if $T \in A^{(n+2,\lambda)}$ and $u(x)$ is of class $C^{(n+2,\lambda)}$ on an open portion X of ∂T and of class $C^{(2,\lambda)}$ in $(T - \partial T) \cup X$, then $u(x)$ is also of class $C^{(n+2,\lambda)}$ in $(T - \partial T) \cup X$.*

We suppose in the first place that X is empty, in which case the theorem is due to E. Hopf [4].

Having taken a domain $D \subset T$ which has a positive distance δ from ∂T, we put, for $x \in D$ for a fixed j, and for $h < \delta/2$:

$$x_j^{(h)} \equiv (x_1, \ldots, x_{j-1}, x_j + h, x_{j+1}, \ldots, x_m),$$

$$v_j(x, h) = u\left(x_j^{(h)}\right) - u(x), \quad q_i = \frac{\partial v_j}{\partial x_i}, \quad q_{ik} = \frac{\partial^2 v_j}{\partial x_i \partial x_k},$$

$$G(p_{ik}, p_i, u, x, t, h) = F(p_{ik} + t q_{ik}, p_i + t q_i, u + t v_j, x^{(th)}).$$

It is obvious that the function $w_j(x, h) = v_j(x, h)/h$ satisfies the equation:

$$\sum_{i,k=1}^{m} \frac{\partial^2 w_j}{\partial x_i \partial x_k} \int_0^1 G_{p_{ik}} dt + \sum_{i=1}^{m} \frac{\partial w_j}{\partial x_i} \int_0^1 G_{p_i} dt + w_j \int_0^1 G_u dt + \int_0^1 G_{x_j} dt = 0$$

$$(40.4)$$

and from this the theorem follows immediately. Indeed, it suffices to prove that u has HÖLDER continuous $(k + 1)^{st}$ derivatives if it is of class $C^{(k,\lambda)}$ with $2 \leq k \leq n + 1$. Now under this latter hypothesis, the coefficients of (40.4) are of class $C^{(k-2,\lambda)}$ and the HÖLDER coefficients of their $(k + 2)^{nd}$ derivatives can be bounded independently of h. By theorem 35, IV we then have that in every domain interior to D the HÖLDER coefficients of the kth derivatives of $w_j(x, h)$ are also susceptible to an analogous bound. It follows from this that:

$$\frac{\partial u}{\partial x_j} = \lim_{h \to 0} w_j(x, h)$$

is also of class $C^{(k,\lambda)}$, whence the theorem.

If X is not empty, it will suffice to prove that it is possible to find a neighborhood Γ of every point $x \in X$ such that $u(x)$ is in class $C^{(n+2,\lambda)}(\bar{\Gamma} \cap T)$. Then let Γ be restricted so that with a change of variables of class $C^{(n+2,\lambda)}$ it is possible to transform $\Gamma \cap X$ into a region of the hyperplane $x_m = 0$. The same change of variables transforms (40.1) into an equation of the same type satisfying the same hypotheses; moreover, the $||w_j||_{C^{(n+1,\lambda)}(\bar{\Gamma} \cap X)}$ for $j = 1, 2, \ldots, m - 1$ can be bounded independently of h. Then, applying the preceding procedure, we can prove, always making use of theorem 35, IV, that if $u \in C^{(k,\lambda)}(\bar{\Gamma} \cap T)$ with $2 \leq k \leq n + 1$, then the $\partial u / \partial x_j$ with $j = 1, 2, \ldots, m - 1$ are also in class $C^{(k,\lambda)}(\bar{\Gamma} \cap T)$. On the other hand, because by dint of the ellipticity it is possible to extract $\partial^2 u / \partial x_m^2$ from the equation as a function of x, u, p_i, and of the other second derivatives, we have that $\partial^2 u / \partial x_m^2$ is in class $C^{(k-1,\lambda)}(\bar{\Gamma} \cap T)$ and consequently also that $\partial u / \partial x_m$ is in class $C^{(k,\lambda)}(\bar{\Gamma} \cap T)$. With this the theorem is completely proved.

Then, in the case of quasi-linear equations, we see immediately by applying theorems 36, V and VI that:

40, III. *If the functions a_{ik}, b_i, f are continuous with all their derivatives of order not greater than $n (\geq 0)$ and if their n^{th} derivatives are λ-Hölder continuous with respect to the variables p_i, u, x_j, then every solution $u(x)$ of (40.2) which is of class $C^{(2)}$ in $T - \partial T$ is also of class $C^{(n+2,\lambda)}$ in $T - \partial T$. If, moreover, $T \in A^{(n+2,\lambda)}$ and $u(x)$ is of class $C^{(n+2,\lambda)}$ on an open portion X of ∂T and of class $C^{(2)}$ in $(T - \partial T) \cup X$, then $u(x)$ is also of class $C^{(n+2,\lambda)}$ in $(T - \partial T) \cup X$.*

An analogous though more concealed result holds also for the general equation (40.1). In fact, we have that:

40, IV. *The conclusions of theorem 40, II hold even if, leaving all the other hypotheses fixed, we assume only that $u(x)$ is of class $C^{(2)}$ in $T - \partial T$ for the first assertion and in $(T - \partial T) \cup X$ for the second.*

12*

For $m = 2$, $n \geq 4$ and X empty, the theorem was proved in 1935 by R. CACCIOPPOLI [8][1]. In the general case the proof was given in 1953 by L. NIRENBERG [1] for $m = 2$ and by C. B. MORREY [3, 4, 9][2] for $m \geq 2$. This theorem is an immediate consequence of theorem 40, II once we have proved the following preliminary theorem:

40, V. *Under the sole hypothesis that F is continuous with its first derivatives every solution* $u(x)$ *of* (40.1) *which is of class* $C^{(2)}$ *in* $T - \partial T$ *is also of class* $C^{(2,\lambda)}$ *for arbitrary* $\lambda < 1$. *If* $T \in A^{(2)}$ *and if* $u(x)$ *is of class* $C^{(2,\lambda)}$ *on an open portion X of* ∂T *and of class* $C^{(2)}$ *in* $(T - \partial T) \cup X$, *then* $u(x)$ *is also of class* $C^{(2,\lambda)}$ *in* $(T - \partial T) \cup X$.

And this theorem can be established with methods analogous to those followed for the proof of theorem 40, II, by making use of theorem 37, II and of (37.9) to obtain a bound independent of h for $\| w_j(x, h) \|_{C^{(1,\lambda)}}$.

We observe now that:

40, VI. *Even if* (40.1) *is not of elliptic type, theorems* 40, II, III, IV, V *hold equally well for every elliptic solution of the same equation.*

In fact it suffices to observe that the proof of the cited theorems is based exclusively on the ellipticity of (40.3). And this equation is certainly elliptic under the sole hypothesis that $u(x)$ is an elliptic solution of (40.1) and that h is sufficiently small.

We shall finish this section by occupying ourselves with the extension to nonlinear equations of some results from §§ 19, 27, 30.

Concerning the uniqueness theorem for the CAUCHY problem, continuing to denote by Σ the hypersurface on which the initial conditions are assigned, we have that:

40, VII. *If* Σ *can be represented in the form* (1.1) *with* ζ *of class* $C^{(1,1)}$, *then the uniqueness theorem for the* CAUCHY *problem for* (40.1) *holds under the sole hypothesis that F and the* $F_{p_{i,k}}$ *are* LIPSCHITZ *continuous*[3].

In fact, if the problem admits two solutions u and $\bar{u}$, then by applying theorem 19, II to the equation (40.3) satisfied by $v = u - \bar{u}$, we recognize under the hypotheses indicated that $v = 0$.

Analogously:

40, VIII. *If F satisfies the same hypotheses as in theorem* 40, VII, *and if, moreover,* $F(0, 0, 0, x) = 0$, *then, for the solutions of class* $C^{(2)}$ *of* (40.1), *the strong unique continuation property holds.*

---

[1] CACCIOPPOLI's proof with some slight modification is related in § 44 of the first edition of this monograph.

[2] See also L. NIRENBERG [3].

[3] We recall that the first proof of a uniqueness theorem for the CAUCHY problem outside the analytic field was given in 1931 by H. LEWY [3]. In this work F was assumed to be analytic, ζ of class $C^{(2)}$, and the uniqueness was proved for solutions of class $C^{(4)}$.

In fact, in this case every solution of (40.1) also satisfies the linear equation which is obtained from (40.3) by putting $\bar{u} = 0$; and to this equation, under the hypotheses indicated, theorem 19, III is applicable. In particular this theorem is applicable to the quasi-linear equation:

$$\sum_{i,k=1}^{m} a_{ik}(x, u, p_j)\, p_{ik} = f(x, u, p_j) \qquad (40.5)$$

under the hypothesis that the a_{ik} and f are LIPSCHITZ continuous and that, moreover:

$$f(x, u, p_j) = O\left(|u| + \sum_{j=1}^{m} |p_j|\right). \qquad (40.6)$$

Thus the following more general theorem holds:

40, IV. *For solutions of class $C^{(2)}$ of (40.5), the strong unique continuation property holds under the sole hypothesis that the a_{ik} are LIPSCHITZ continuous and that (40.6) is valid.*

This theorem was proved by N. ARONSZAJN, A. KRZYWICKI and J. SZARSKI [1]; analogous theorems except for stronger hypotheses on a_{ik} and f are also found in almost all the works cited in connection with theorem 19, III.

Let us now pass on to occupy ourselves with the extension of some of the theorems of § 30 to equations of the type:

$$\sum_{i=1}^{m} \frac{\partial}{\partial x_i}\, a_i(x, u, p_j) = a(x, u, p_j), \qquad (40.7)$$

which we shall call *equations of divergence form*. These equations constitute a particular class of quasi-linear equations, in which are included the EULER equations of integrals from the calculus of variations. For these it is possible to present the notion of *weak solution*, which considerably facilitates the study of the regularity properties of their solutions. If, for example, putting $|p| = \left(\sum_{j=1}^{m} p^2\right)^{1/2}$, we assume that:

$$\left.\begin{array}{l} |a_i| \le A_0 |p|^{k-1} + d |u|^{k-1} + \varphi \\[2mm] |a| \le e |p|^{k-1} + c |u|^{k-1} + f \end{array}\right\} \qquad (40.8)$$

with $A_0 \ge 0$ constant, $m > k > 1 > \varepsilon > 0$ and:

$$d, \varphi \in L^{\frac{m}{k-1}}(\Omega), \qquad e \in L^{\frac{m}{1-\varepsilon}}(\Omega), \qquad c, f \in L^{\frac{m}{k-\varepsilon}}(\Omega), \qquad (40.9)$$

we can define as a *weak solution* in the region Ω of (40.7) a function u which, for every bounded and measurable domain $D \subset \Omega$, is of class $H^{1,\,k}(D)$ for which, $a_i(x, u, p_j)$ and $a(x, u, p_j)$ being measurable, we have:

$$\int_\Omega \left[\sum_{i=1}^m a_i(x, u, p_j) \frac{\partial v}{\partial x_i} + a(x, u, p_j) v \right] dx = 0 , \; \forall \, v \in H_0^{1,\,k}(D) . \qquad (40.10)$$

Indeed, under the hypotheses posed, the known inclusion theorems of S. L. SOBOLEV [5] ensure that the integral on the left side of (40.10) is meaningful. The definition posed is correct even for $k = m$ if the first of (40.9) is replaced by the hypothesis:

$$d, \varphi \in L^{\frac{m}{m-2-\varepsilon}} (\Omega) . \qquad (40.9')$$

The systematic study of equations of type (40.7) was begun by O.A. LADYŽENSKAJA and N.N. URAL'CEVA [1, 2] by posing more restrictive hypotheses than those now indicated, and by assuming among other things that (40.7) was elliptic. This research is interesting above all for the study of boundary value problems, and hence we shall occupy ourselves with it in subsequent sections, which are dedicated precisely to this argument. Here we wish instead to occupy ourselves with some of J. SERRIN's research [2, 6, 7, 8] in which the solutions of (40.7) were deeply studied under the hypotheses indicated above and, moreover, assuming that:

$$\sum_{i=1}^m p_i \, a_i \geq A_0^{-1} \, | \, p \, |^k - d \, | \, u \, |^k - g , \quad g \in L^{\frac{m}{k-2}} (\Omega) . \qquad (40.11)$$

We note that this condition does not imply that (40.7) is elliptic, nor is it implied by the possible ellipticity of (40.7). This hypothesis turns out always to be satisfied with $k = 2$ if (40.7) is reduced in particular to a linear elliptic equation of the type (30.9) and SERRIN succeeded in proving that a great portion of the results indicated in § 30 for weak solutions of this latter equation can be extended to weak solutions of (40.7), when (40.11) is satisfied. In particular, for every domain $D \subset \Omega$ we can give a bound for $\| u \|_{C^{1,\alpha}(D)}$, α being appropriately chosen, a result which, for a class of elliptic equations of the type (40.7), had been already obtained previously by O.A. LADYŽENSKAJA and N.N. URAL'CEVA [1, 2]. Moreover, we can give an extension of HARNACK's theorem, as well as theorems 30, V and VI. Here we shall limit ourselves to reporting the following theorem:

40, X. *Let hypotheses* (40.8), (40.9) [*and* (40.9') *if* $k = m$], (40.11) *be satisfied, and moreover let* $\varphi \in L^{\frac{m}{k-1-\varepsilon}} (\Omega)$. *If Q is a compact set in Ω*

with zero s-capacity for $2 \leq s \leq m$, then every function $u \in L^q(\Omega)$ of class $C^{(0)}$ in $\Omega - Q$ with $q > s\,(k-1)/(s-k)$, which is a weak solution in $\Omega - Q$ of (40.7), can be extended to a function of class $C^{(0)}$ in Ω which is a weak solution of (40.7) in Ω.

Then J. SERRIN [5, II] studied the case in which the hypotheses of theorem 40, X are satisfied with $k = 1$, $\varepsilon = 0$, and Q has zero $(m-1)$-dimensional HAUSDORFF measure. Under this hypothesis, the singularity on Q is then removable, in the sense that u is a weak solution of (40.7) in Ω, without, however, being possible to ensure the continuity on Q of the extension of u. Of the same type, then, is the following further theorem which refers to the equation:

$$\sum_{i=1}^{m} \frac{\partial}{\partial x_i}\, a_i(p_j) = a(u) \,. \tag{40.12}$$

40, XI. *Let the bounds:*

$$\left. \begin{array}{c} \sum_{i=1}^{m} (p_i - q_i)\,[a_i(p_j) - a_i(q_j)] < 0 \quad for \quad |p - q| > 0 \,, \\[2mm] (u - v)\,[a(u) - a(v)] \geq 0 \end{array} \right\} \tag{40.13}$$

be satisfied and let Q be a compact set in Ω with zero $(m-1)$-dimensional HAUSDORFF measure. Then every function u of class $C^{(0)}$ in $\Omega - Q$ which with $k = 1$ is a weak solution in $\Omega - Q$ of (40.12) and for which:

$$a_i(p_j) \in L^\infty(\Omega) \,, \quad a(u) \in L^m(\Omega) \,, \tag{40.14}$$

can be extended into a function of class $C^{(0)}$ in Ω which is a weak solution in Ω of (40.12). If, moreover, (40.12) is elliptic, and the a_i and a are in class $C^{(2)}$ and $a_u \geq 0$, then every solution of class $C^{(2)}$ in $\Omega - Q$ of (40.12) is also of class $C^{(2)}$ in Ω.

We note that this theorem can be applied in particular to the equation of minimal surfaces in m variables:

$$\sum_{i=1}^{m} \frac{\partial}{\partial x_i}\, \frac{p_i}{\sqrt{1 + |p_i|^2}} = 0 \,. \tag{40.15}$$

In this case, in fact, all the conditions are satisfied: those relating to a because $a = 0$, the first of (40.14) because $|a_i| < 1$, the first of (40.13) as a consequence of the ellipticity of the equation. We thus rediscover a theorem which was somewhat earlier proved by JOHANNES NITSCHE [9] for $m = 2$ and by E. DE GIORGI and G. STAMPACCHIA [1] in the general case. In the case in which $m = 2$ and Q is a single point, this result was proved many years earlier by L. BERS [5] for minimal surfaces and by R. FINN [1] in the case of an elliptic equation of type (40.12) with $a = 0$ and the a_i's bounded functions. Other proofs of the theorems of BERS and FINN were given by JOHANNES NITSCHE [6], JOACHIM and JOHANNES NITSCHE [4], and L. BERS [15] and R. FINN [4, 12, 15] themselves. These results spotlight a notable difference in behavior between the solutions of the equations considered by R. FINN in [1], which we shall call *equations of minimal surface type*, and the solutions of linear equations. For these latter, for example, an isolated singular point x is removable (theorem 19, XI) if a bound of the type $u(x) = o(|x - x_0|^{2-m})$ is valid for the solution $u(x)$ as $x \to x_0$, while in the former case every isolated singularity is removable without our being obliged to require a particular behavior of $u(x)$. A difference of behavior is verified even in regard to the maximum and minimum properties. In fact, it has been observed that for the solutions $u(x)$ of equations of minimal surface type, a bound for $|u(x)|$ on a portion of ∂T implies an analogous bound for $|u(x)|$ in every domain D interior to T, and sometimes even in all of T. Results in this sense were established in order of time by H. JENKINS and J. SERRIN [1], R. FINN [15], H. JENKINS [1] for $m = 2$ and then by H. JENKINS [2] for $m \geq 2$. In these works and in others by R. FINN [2, 4, 11, 13, 14] and N. MEYERS [4], bounds for the first derivatives of solutions in the interior of the domain were established. In the case of the equation of minimal surfaces for $m = 2$, particularly simple bounds were established by J. SERRIN [3] for the first derivatives and by E. HEINZ [1], E. HOPF [8, 9], R. OSSERMANN [3], R. FINN and R. OSSERMANN [1] for the second derivatives. For $m > 2$, integral bounds were obtained by M. MIRANDA [1] for the second derivatives. However, we refer to the report by JOHANNES NITSCHE [10] for everything which concerns these results in the case of minimal surfaces.

Note that in some of the cited works by R. FINN and by H. JENKINS and in the note [8] by JOHANNES NITSCHE, equations "of the minimal surface type" more general than those just considered are also studied along the same lines.

Finally, we recall that L. BERS [4, 7] and CHEN YU-WHY [1] have also studied the singularities of polydrome solutions of the minimal surface equation, and that JOHANNES NITSCHE [5] characterized the behavior at an isolated singular point of the solutions of the equation $\Delta u = e^u$.

41. Functional equations in abstract spaces[1]. In this section we shall summarize some notions relating to functional equations in an

---

[1] For better details on the contents of this section see also C. MIRANDA, Loc. Cit. in the first footnote of § 36, and M. A. KRASNOSEL'SKIĬ: Topologi-

abstract space which are fundamental in the study of boundary value problems for nonlinear elliptic equations. The origin of this research is found in a note by G. D. BIRKHOFF and O. D. KELLOGG[1] in 1922 and in another by P. LEVY in 1920[2]. The first of these works looks at the extension to abstract spaces of BROUWER's *fixed point theorem*, a question which was later studied deeply by J. SCHAUDER [1, 2, 4] and by other authors and to which we shall return further on. The results obtained, though useful in other fields of research, have found scant applications in the theory of elliptic equations, for which more refined results are necessary. Still, these were the point of departure for later research first by SCHAUDER himself [3, 7], then by J. LERAY and J. SCHAUDER [1], which led to the formulation of one of the methods upon which one can base the study of DIRICHLET's problem for nonlinear elliptic equations. Another method which, in some way, can be rejoined to P. LEVY's note, was then developed by R. CACCIOPPOLI [3, 4, 9]. All these works were published between 1927 and 1937. Of these the first is SCHAUDER's memoir [3] which is dated in 1927, followed then in 1932 by SCHAUDER's memoir [7] and CACCIOPPOLI's notes [3, 4], in 1934 by LERAY's and SCHAUDER's memoir, and in 1937 by CACCIOPPOLI's latest work [9]. For reasons of convenience we shall begin our exposition with CACCIOPPOLI's works [3, 4], in which this author proved how certain classical procedures adopted by various authors for the proof of existence and uniqueness theorems, the method of successive approximations and that of analytic continuation, could be led back to a few principles of general character, the abstract formulation of which renders their application to particular problems considerably more simple and systematic.

Let Σ and Σ' be two BANACH spaces and let

$$\sigma' = \mathfrak{T}(\sigma) \tag{41.1}$$

be a (continuous) transformation which causes a point σ' of Σ' to correspond to each point σ of Σ. We shall say that $\mathfrak{T}$ is invertible between two sets A and A' of Σ and Σ' if it places a one-to-one correspondence between the aforesaid sets; in particular $\mathfrak{T}$ will be called *locally invertible*

cal methods in the theory of nonlinear integral equations, Gos. Izdat Tehn.-Teor. Lit. Moscow (1956) 392 pp. (Russian); English transl. Pergamon Press (1964) 395 pp. For some new and very important results see also F. E. BROWDER: Local and global properties of nonlinear mappings in BANACH spaces, Ist Naz. Alta Mat. Roma, Symposia Math. II (1968).

[1] BIRKHOF, G.D., and O.D. KELLOGG: Invariant points in function space. Trans. Amer. Math. Soc. **23** (1922) 96—115.

[2] LEVY, P.: Sur les fonctions de lignes implicites. Bull. Soc. Math. de France **48** (1920) 13—27.

in a neighborhood of two corresponding points σ_0 and σ_0' if it is invertible between two neighborhoods of σ_0 and σ_0', and *completely invertible* if it places a one-to-one correspondence between Σ and Σ'. The following theorem holds:

41, I. *The transformation $\mathfrak{X}$ is completely invertible if the following two conditions are satisfied: A) $\mathfrak{X}$ is locally invertible in a neighborhood of every pair of corresponding points. B) Every set of Σ which is transformed by $\mathfrak{X}$ into a compact set of Σ' is itself compact.*

This theorem essentially affirms that for (41.1) considered as an equation in σ, an existence and uniqueness theorem for the solution holds under the hypotheses posed. We note that the existence of the solution is obvious since conditions A) and B) ensure that the image of Σ in Σ' is simultaneously open and closed in Σ' and therefore coincides with Σ'. More delicate is the proof of uniqueness, which we do not relate. The range of this theorem is then considerably extended by the following complements:

41, II. *If the properties A) and B) are satisfied only when σ' belongs to a closed and connected subset Σ_1' of Σ', and if: C) at least one point of Σ_1' is the correspondent of a unique point of Σ, then every point of Σ_1' is the correspondent of a unique point of Σ.*

41, III. *Let the transformation $\mathfrak{X}$ depend continuously on a parameter τ, which is also in a* BANACH *space S, and let the following properties be satisfied: A_1) for every fixed τ belonging to a closed, connected, and compact subset S_1 of S, $\mathfrak{X}$ is locally invertible in a neighborhood of every pair of corresponding points. B_1) Every set of Σ which, as τ also varies in S_1, is transformed by $\mathfrak{X}$ into a compact subset of Σ', is itself compact. C_1) For a particular value of $\tau \in S_1$, there exists at least one point of Σ' corresponding to a unique point of Σ. Then $\mathfrak{X}$ is completely invertible for every fixed $\tau \in S_1$. If the same properties are satisfied only for $\sigma' \in \Sigma_1'$, with Σ_1' closed and connected, then every point of Σ_1' is the correspondent of a unique point of Σ.*

By means of these theorems, the proof of the existence and uniqueness theorem for a given functional equations is brought back to the verification of the properties stated in the different theorems. Of these, B) or B_1) will require in general some a priori majorization of the possible solutions of the equation. The verification of A) instead requires a local study of the transformation and can often be reached by applying the following theorem of T. H. HILDEBRANDT and L. M. GRAVES[1]:

---

[1] HILDEBRANDT, T. H., and L. M. GRAVES, Implicit functions and their differentials in general analysis. Trans. Amer. Math. Soc. **29** (1927) 127—153.

41, IV. *Let the transformation $\mathfrak{T}$ be continuously differentiable in every bounded set, and denote $\mathfrak{D}(\sigma, \delta\sigma)$ its differential. If the linear transformation $\delta\sigma' = \mathfrak{D}(\sigma_0, \delta\sigma)$ is completely invertible, then $\mathfrak{T}$ is locally invertible in a neighborhood of the points σ_0 and $\sigma_0' = \mathfrak{T}(\sigma_0)$.*

It is now obvious that the existence theorems which we can attain in application of the general principles just stated are necessarily attached to a uniqueness theorem. Consequently it is interesting to extend the range of these procedures in a way which renders them applicable even in those cases in which the equation which we want to study admits a solution which is not necessarily unique. Now we can observe that for the transformations for which only property B) is valid, the set of points of Σ having the same image σ_0' is a compact set Σ_0. It is now clear that for the purpose of proving only the existence theorem, the property A) of theorem 41, I can be replaced by the following:

A') *For every point $\sigma_0' \in \Sigma'$, the image of a compact set Σ_0 in Σ, we can find a neighborhood, every point of which is the image of at least one point of a neighborhood of Σ_0.*

Therefore it is a question of establishing conditions of the type of those in theorem 41, IV under which A') turns out to be true. R. CACCIOPPOLI [9] succeeded in this under the following conditions. Suppose first of all that Σ and Σ' are embedded in two HILBERT spaces $\overline{\Sigma}$ and $\overline{\Sigma'}$ in such a way that it is valid to define for every linear manifold of Σ or Σ' an orthogonal manifold. Then, if σ_0 and $\sigma_0' = \mathfrak{T}(\sigma_0)$ are two corresponding points, we shall say that σ_0', as correspondent of σ_0, is *ordinary* or *singular* for the inverse transformation $\mathfrak{T}^{-1}$ according as $\mathfrak{D}(\sigma_0, \delta\sigma)$ is or is not completely invertible. A singular point is then called *regular* if $\delta\sigma' = \mathfrak{D}(\sigma_0, \delta\sigma)$ is invertible between two linear manifolds of Σ and Σ', whose orthogonal manifolds have an equal finite number of dimensions.

The following theorem holds:

41, V. *Let the transformation $\mathfrak{T}$ admit only regular singular points and satisfy property B). Moreover, suppose we have that: C') There exists at least one point σ_0' of Σ', the correspondent of a finite odd number of points of Σ with respect to each one of which σ_0' is ordinary for $\mathfrak{T}^{-1}$. Then every point of Σ' is the image of at least one point of Σ.*

The central point of the proof, somewhat complex, is essentially the following. If σ_0' is a regular singular point, the image of a compact set Σ_0 of Σ, then the inversion of $\mathfrak{T}$ between a neighborhood of σ_0' and a neighborhood of Σ_0 allows us to be led back to that of a transformation between two finite dimensional spaces R_n and R_n'. We can then define the topological order χ of σ_0', with respect to an appropriate manifold of R_n', in such a way that if this is different from zero, property A') turns out to be satisfied. The theorem then follows from the fact that, under hypothesis B'), we can prove that χ remains constant as σ_0' varies and is therefore always different from zero if it is in a particular case. And because, since Σ_0 is reduced to an isolated point and σ_0' is ordinary for $\mathfrak{T}^{-1}$, χ is ± 1, the condition C') ensures precisely that, at least in a particular case, χ is different from zero.

Naturally even the theorem now stated can be completed by observations of the type contained in theorems 41, II and 41, III, but we shall not insist on this. We wish rather to exhibit another procedure with which it is equally possible to obtain existence theorems in the case in which the solution is not necessarily unique. This procedure is contained in the following theorem due to J. LERAY and J. SCHAUDER [1], which was proved by these authors some years in advance of CACCIOPPOLI's note [9]:

41, VI. *Let $\mathfrak{S}(\sigma, \tau)$ be a transformation of the space Σ into itself, depending continuously on the parameter τ and completely continuous with respect to σ for every fixed τ. If the transformation $\sigma' = \mathfrak{T}(\sigma, \tau) = \sigma - \mathfrak{S}(\sigma, \tau)$ satisfies, at least for one value of τ, property C') of theorem 41, V, and if: B') the possible solutions of the equation:*

$$\sigma = \mathfrak{S}(\sigma, \tau) , \tag{41.2}$$

as τ varies also, describe a bounded set of Σ, then the aforesaid equation admits at least one solution for every value of τ.

The proof of this theorem also requires considerations of a topological character. It is in fact based on the extension to the transformations of the type $\sigma' = \sigma - \mathfrak{S}(\sigma, \tau)$ of the notion of topological degree. Indeed, if we consider the aforesaid transformation for a fixed τ, and as σ varies in a bounded domain Σ_1 of Σ, then we can associate to every point σ_0 of Σ not belonging to the image of $\partial \Sigma_1$ an integer $\Phi(\sigma_0, \Sigma_1)$, called precisely the *topological degree* of the transformation at the point σ_0, which enjoys the following properties:

1°) $\Phi(\sigma_0, \Sigma_1)$ is an additive function of Σ_1. 2°) If $\Phi(\sigma_0, \Sigma_1) \neq 0$, then the point σ_0 belongs to the image of $\Sigma_1 - \partial \Sigma_1$. 3°) $\Phi(\sigma_0, \Sigma_1)$ does not vary as σ_0, Σ_1, and τ vary continuously, as long as σ_0 does not reach the image of $\partial \Sigma_1$.

After this the proof of the theorem follows from the fact that under hypothesis B') there exists in Σ a hypersphere Σ_1 of radius so large that the origin ω of Σ is, for arbitrary τ, exterior to the image of $\partial \Sigma_1$. The topological degree of the transformation at the point ω is therefore preserved constant as τ varies and is certainly different from zero if it is for a particular value of τ. Because it is easy to prove that this last circumstance holds under hypothesis C'), ω will belong for all τ to the image of $\Sigma_1 - \partial \Sigma_1$ and therefore (41.1) will admit a solution for arbitrary τ.

From theorem 41, VI we could easily deduce the following theorem, which is a particular case[1] of SCHAUDER's fixed point theorem:

41, VII. *Every completely continuous transformation of a sphere K in a* BANACH *space into itself admits at least one fixed point.*

As another consequence, though not an immediate one, of theorem 41, VI, we have also that:

41, VIII. *If the transformation* $\sigma' = \sigma - \mathfrak{S}(\sigma)$, *with* $\mathfrak{S}$ *completely continuous, places a one-to-one correspondence between a bounded domain* Σ_1 *and its image* Σ_1', *then* Σ_1' *is also a domain, and the transformation maps interior points of* Σ_1 *into interior points of* Σ_1'. *In particular, if for the equation in* σ: $\sigma' = \sigma - \mathfrak{S}(\sigma)$ *a uniqueness theorem is valid, then the transformation being considered is locally invertible in a neighborhood of every pair of corresponding points.*

This theorem, which extends the so-called property of the *invariance of domains* to transformations in abstract spaces, was proved in the preceding form by J. LERAY [2]. Under more restrictive hypotheses, but ones sufficient for applications to elliptic equations, the same theorem was proved earlier with a direct procedure by J. SCHAUDER [3, 7].

For other results related to the contents of this section which are chiefly interesting for the study of questions of local invertibility, see L. V. KANTOROVIČ [3, 4, 5] and J. CRONIN [1, 2].

Recently other methods of a different nature have been added to the methods for study of nonlinear equations stated in this section. These new methods have great importance above all for the study of problems related to equations of higher order and to systems of equations; but they can find some application even in the theory of equations of second order.

A first method applied by M. I. VIŠIK [12, ..., 18] consists of approximating the functional equation which we want to solve by another posed in a finite dimensional space and then in passing to the limit, making

---

[1] According to the more general result obtained by J. SCHAUDER [4], theorem 41, VII holds under the sole hypothesis that K is bounded and convex. On this question see also R. CACCIOPPOLI [1, 2] and for a first generalization TICHONOFF: Ein Fixpunktsatz. Math. Ann. **111** (1935) 767—776. Other noteworthy extensions have now been obtained in recent years by F. E. BROWDER. For bibliographic indications relating to this we refer to BROWDER's more recent works on the argument: A further generalization of the SCHAUDER fixed point theorem, Duke Math. J. **32** (1965) 575—578. Fixed point theorems for non linear semicontractive mappings in BANACH spaces, Arch. Rat. Mech. Anal. **21** (1966) 259—269.

use of appropriate a priori majorizations in order to ensure the necessary compactness of the approximating solutions.

A second method, very much more unlike the traditional method than that of Višik, is applied to the study of equation (41.1) in the case where Σ is a reflexive BANACH space and Σ' is its dual, and is based on the idea that the compactness hypotheses which occur in the preceding theorems can be replaced by appropriate monotonicity hypotheses. Just to give an example of the results, we relate the following theorem which is among the simplest, even if not among the most significant, which can be obtained along these lines:

41, IX. *Let* (41.1) *be a transformation of a reflexive and separable* BANACH *space* Σ *into its dual* Σ', *which transforms bounded sets into bounded sets and is continuous in the weak topology of* Σ' *on every finite dimensional subset of* Σ.

If:

$$\lim_{\|\sigma\| \to \infty} \frac{<\mathfrak{T}\sigma, \sigma>}{\|\sigma\|} = +\infty, \quad \mathscr{R}e\left[<\mathfrak{T}(\sigma_1) - \mathfrak{T}(\sigma_2), \sigma_1 - \sigma_2 >\right] > 0,$$

then every point of Σ' *is the image of at least one point of* Σ.

F. E. BROWDER has dedicated a long series of works to this research, which draws its origin from some initial observations due to G. J. MINTY. Complete bibliographic indications in this regard can be extracted from MINTY's work [1] and from those by BROWDER [15, ..., 18, 20, 21, 22] cited in our Bibliography. We recall also an important contribution by J. LERAY and J. L. LIONS [1] which we can not relate and for which we refer also the volume [9] by C. B. MORREY. Theorem 41, IX is included among the results obtained by LERAY and LIONS in the work cited.

For other interesting results along the same lines, but regarding the existence of solutions no longer of equations but of functional inequalities, see P. HARTMAN and G. STAMPACCHIA [1] and F. E. BROWDER [23, 24]. Problems of this sort, for example, can present themselves when the unknown element which we seek is restricted not to leave a given convex set K of a BANACH space and to satisfy an equation or an inequality according as it is an interior point or a boundary point of K. This type of problem can be considered naturally also in the case of a linear equation, and to this particular case we intend to return in § 50, limiting ourselves for now to citing the works by G. STAMPACCHIA [18] and J. L. LIONS and G. STAMPACCHIA [1].

42. Dirichlet's problem for equations in m variables. DI-RICHLET's problem for the elliptic equation (40.1), in a given bounded domain T, consists of seeking a solution of the equation which is continuous in T and of class $C^{(2)}$ in $T - \partial T$ and which satisfies the boundary condition:

$$u(x) = \varphi(x) \quad \text{for} \quad x \in \partial T. \tag{42.1}$$

In the study of this problem, the function F will be assumed continuous with all its derivatives up to those of order $n \geq 2$, the domain T will be assumed to be of class $A^{(n, \lambda)}$ and the function φ, at least for the moment, in class $C^{(n, \lambda)}(\partial T)$.

Let us put $\Sigma = C^{(n, \lambda)}(T)$ and denote by Σ' the space of pairs $\sigma' \equiv \{f, \varphi\}$ with $f \in C^{(n-2, \lambda)}(T)$ and $\varphi \in C^{(n, \lambda)}(\partial T)$, the norm of σ' being defined by:

$$\| \sigma' \| = F_0(T) + F_{n-2, \lambda}(T) + \Phi_0(\partial T) + \Theta_{n, \lambda}[\varphi; \partial T].$$

Between Σ and Σ' we can set up a correspondence $\sigma' = \mathfrak{T}(\sigma)$ by causing to correspond to every point $\sigma \equiv u(x)$ of Σ the element σ' of Σ' defined by:

$$\varphi = u \quad \text{for} \quad x \in \partial T, \qquad f = F(p_{ik}, p_i, u, x) \quad \text{for} \quad x \in T.$$

The existence theorem for Dirichlet's problem will be proved if we can show that $\mathfrak{T}$ is invertible between an appropriate manifold of Σ and the manifold Σ_1' of Σ' whose equation is $f = 0$. Now $\mathfrak{T}$ is continuously differentiable, and even more:

$$\mathfrak{D}(\sigma, \delta\sigma) = \left\{ \sum_{i, k=1}^{m} F_{p_{ik}} \frac{\partial^2 \delta\sigma}{\partial x_i \partial x_k} + \sum_{i=1}^{m} F_{p_i} \frac{\partial \delta\sigma}{\partial x_i} + F_u \delta\sigma, \delta\sigma \right\},$$

and therefore (theorem 41, IV) $\mathfrak{T}$ is locally invertible if the Dirichlet problem in the unknown function $\delta\sigma$:

$$\sum_{i, k=1}^{m} F_{p_{ik}} \frac{\partial^2 \delta\sigma}{\partial x_i \partial x_k} + \sum_{i=1}^{m} F_{p_i} \frac{\partial \delta\sigma}{\partial x_i} + F_u \delta\sigma = \delta f, \tag{42.2}$$

$$\delta\sigma = \delta\varphi \quad \text{for} \quad x \in \partial T,$$

is solvable *unconditionally*, for arbitrary δf and $\delta\varphi$. And we note that the linear equation (42.2) is elliptic as a consequence of the ellipticity of (40.1); it is customary to call this the *variational equation* of (40.1). After this, from theorem 41, II we immediately draw:

42, I. Dirichlet's *problem for equation* (40.1) *for arbitrary* $\varphi \in C^{(n, \lambda)}(\partial T)$ *with* $n \geq 2$, *admits one and only one solution of class* $C^{(n, \lambda)}(T)$ *if the following conditions are satisfied:* α) *For every possible solution in class* $C^{(n, \lambda)}(T)$ *of* (40.1), *the* Dirichlet *problem for the corresponding variational problem is unconditionally solvable.* β) *The norms of the possible solutions of* (40.1) *admit a bound as a function of* $\Phi_0 + \Theta_{n,\lambda}[\varphi; \partial T]$. *$\gamma$) For a particular φ the problem admits one and only one solution.*

Indeed, if suffices to prove, under condition β), that every family of solutions of (40.1) whose boundary values constitute a convergent sequence in Σ' turns out to be compact in Σ. But this is obvious. In fact, by hypothesis β) the aforesaid family will contain at least one sequence $\{u_k\}$ which, together with the sequences of its derivatives of order not greater than n, converges uniformly to a function $u(x)$ in class $C^{(n,\lambda)}(T)$ which is also a solution of (40.1); putting $u - u_k = v$, v satisfies linear equations of type (40.3) whose coefficients are endowed with uniformly HÖLDER continuous $(n-2)^{nd}$ derivatives, whence by theorem 35, I a bound of the following type will hold:

$$V_{n,\lambda} \leq K(V_2 + V_1 + V_0 + \Theta_{n,\lambda}[v; \partial T]), \tag{42.3}$$

from which it follows, as desired, that $\lim_{k \to \infty} \| u - u_k \| = 0$.

We note now, under the hypotheses admitted, that the alternative theorem (36, II) turns out to be valid for (42.2) and for the related DIRICHLET problem, by which, if this problem is not unconditionally solvable, then the associated homogeneous problem admits a finite number of linearly independent solutions, and the problem itself is solvable under the condition that an equal number of compatibility conditions are satisfied. It follows easily from this that every possible singular point of $\mathfrak{T}^{-1}$ is regular; by then applying theorem 41, V we have immediately that:

42, II. DIRICHLET's *problem for equation* (40.1), *for arbitrary $\varphi \in C^{(n,\lambda)}(\partial T)$, admits at least one solution in class $C^{(n,\lambda)}(T)$ with $n \geq 2$ under the sole hypothesis that the following conditions are satisfied: β) of the preceding theorem and: γ') for a particular φ the problem admits a finite number of distinct solutions, corresponding to each one of which DIRICHLET's problem for variational equation is unconditionally solvable.*

From theorem 41, III and from the analogous extension of theorem 41, V, it follows finally that:

42, III. *If* (40.1) *depends continuously on a parameter τ, then theorems 42, I and II hold for arbitrary τ under the sole hypothesis that condition γ) or γ') is satisfied for at least one value of τ and the other conditions are satisfied also as τ varies.*

It is now useful to observe that the proof of theorems 42, II and III can be developed also if we apply theorem 41, VI instead of theorem 41, V. And further, the first proof of these theorems was given precisely by LERAY and SCHAUDER [1] with this procedure. Here we limit ourselves to displaying the details under the hypothesis that the equation which we consider is of quasi-linear type (40.2), in which case the preceding theorems can be perfected as follows:

42, IV. *For the quasi-linear equation* (40.2), *assuming that the coefficients* a_{ik}, b_i, f *are of class* $C^{(n-2,\lambda)}$, *theorems* 42, I, II, III *hold even if hypothesis* β) *is replaced by:* β') *there exists a* $\beta \in {]}0,1{[}$ *such that, as* $u(x)$ *varies in the family of all solutions of the equation, the quantities* $U_0 + U_{n-1,\beta}$ *permit a bound as a function of* $\| \varphi \|_{C^{n,\lambda}(\partial T)}$.

We begin with the observation that under the hypotheses posed, with a fixed upper bound for $\| \varphi \|_{C^{n,\lambda}(\partial T)}$, (40.2) can be considered as a linear equation in u whose coefficients vary in families of functions endowed with uniformly $\beta\lambda$-Hölder continuous $(n-2)^{nd}$ derivatives. It follows from this, by theorem 35, II, that the $U_{n,\beta\lambda}$ remain bounded; but then the coefficients of (40.2) are endowed with uniformly λ-Hölder continuous $(n-2)^{nd}$ derivatives, and theorem 35, II ensures a bound of $U_{n,\lambda}$. Consequently β') bears β) as a consequence. From this, if the functions a_{ik}, b_i, f were of class $C^{(n)}$, with which the variational equation would have coefficients in class $C^{(n-2,\lambda)}(T)$, we could then arrive at the assertion with the method followed for the preceding theorems. Under the considerably reduced hypotheses of theorem 42, IV, we can attain the same end by using the procedure of LERAY-SCHAUDER [1]. We denote this time by Σ the space $C^{(n)}$ and for every element $\sigma = u$ of Σ we now put:

$$\bar{a}_{hk}[\sigma] = a_{hk}(x, u, p_i), \qquad \bar{f}(\sigma) = f(x, u, p_i) - \sum_{k=1}^{m} b_k(x, u, p_i)\, p_k ,$$

and let $\sigma' = u'$ be the solution of the DIRICHLET problem:

$$\sum_{h,k=1}^{m} \bar{a}_{hk}[\sigma] \frac{\partial^2 u'}{\partial x_h\, \partial x_k} = \bar{f}(\sigma) , \qquad u' = \varphi \quad \text{for} \quad x \in \partial T .$$

We can evidently write $\sigma' = \mathfrak{S}(\sigma)$; the transformation $\mathfrak{S}$ turns out to be completely continuous, establishing that every bounded set of Σ is transformed by $\mathfrak{S}$ into a family of functions u' for which the quantities $U_0' + U_{n,\lambda}'$ remain bounded, and this bound obviously ensures the compactness of the family in Σ. From this, the application of theorem 41, VI leads immediately to the assertion, where we consider as parameter τ the function φ which occurs in the boundary condition.

We shall mention finally that, by making use of 41, VIII, J. SCHAUDER [7, 8] proved that:

42, V. *Theorem* 42, I *is valid even if hypothesis* α) *is replaced by one which says that a uniqueness theorem is valid for the* DIRICHLET *problem for the more general equation* $F = f(x)$.

This theorem is evidently not equivalent to theorem 42, I; still, it is useful to observe that there is a particular case in which equivalence holds. Indeed, such is the case when we assume $F_u \leq 0$, because this hypothesis ensures both the unconditional solvability of the variational problem (theorem 36, I) and, as an obvious consequence of theorem 40, I, the uniqueness of the solution of DIRICHLET's problem for the equation $F = f(x)$.

The results just stated constitute everything which can be obtained by using the principles of functional analysis summarized in the first eight theorems of § 41; however, further results can be obtained by determining conditions, which are always less restrictive, under which the hypotheses β) or β') of the preceding theorems turn out to be essentially true. Along these lines we begin to see what can be obtained from results related to linear equations established in § 35. The four theorems which follow are due to R. CACCIOPPOLI [6].

42, VI. *Theorems* 42, I, II, III, *and* V *hold even if hypothesis* β) *is replaced by the following:* δ) *the possible solutions of* (40.1) *and their first and second derivatives remain equicontinuous and uniformly bounded as* φ *varies in every bounded set of* $C^{(n, \lambda)}(\partial T)$.

Evidently, it suffices to prove, for every sequence of solutions of (40.1) whose boundary values constitute a convergent sequence in Σ', that hypothesis β) is a consequence of δ). Now if this were not so, we could find a sequence $\{u_k\}$ of these solutions of (40.1) which is uniformly convergent in T together with the sequences of its derivatives of order not greater than n and for which $\| u_k \| \geq 2 \| u_{k-1} \|$. On the other hand, the functions $v = u_k - u_{k-1}$ satisfy a linear homogeneous equation of type (40.3) for whose coefficients the quantities denoted in § 35 by $A_{n-2, \lambda}$, $B_{n-2, \lambda}$, $C_{n-2, \lambda}$, by dint of theorem 33, IV, turn out to be bounded by a linear function of $\| u_k \|$ while the A_0, B_0, C_0 remain uniformly bounded. Then, by applying theorem 35, I we would arrive at an inequality of the type:

$$\frac{1}{2} \| u_k \| \leq \| u_k - u_{k-1} \| \leq K [\| u_k \| + 1] [V_2 + V_1 + V_0 + \Theta_{n, \lambda}[v; \partial T]],$$

$$(42.4)$$

which is patently absurd, because the quantity in the last set of brackets is infinitesimal for k going to infinity.

Theorem 42, IV relating to quasi-linear equations can also be perfected as follows:

42, VII. *For the quasi-linear equation* (40.2), *theorems* 42, I, II, III, V *hold even if hypothesis* β) *is replaced by the following:* δ') *the possible solutions of* (40.2) *and their first derivatives turn out to be equicontinuous and uniformly bounded as* φ *varies in every bounded set of* $C^{(n, \lambda)}(\partial T)$.

The proof is analogous to that of the preceding theorem, the sequence $\{u_k\}$ turning out to be convergent together with those of its first derivatives, except that now, because the coefficients of $\partial^2 v/\partial x_i\, \partial x_j$ in (40.3) depend only on v and $\partial v/\partial x_h$, it is possible, also by dint of theorem 33, IV, to majorize the quantities $A_{n-1,\lambda}$, $B_{n-2,\lambda}$, $B_0^{n-1+\lambda}$, $C_0^{\frac{n+\lambda}{2}}$, $C_{n-2,\lambda}$, $(A_{n-2,\lambda}+1)\cdot(B_0+C_0+1)$ with a linear function of $\|\,u_k\,\|$, while the A_0 remain uniformly bounded. Then, applying theorem 35, III, we arrive at an inequality similar to (42.4) in which, however, the quantity V_2 no longer occurs in the last parentheses, which is sufficient to conclude the reasoning, as above.

We then have:

42, VIII. *If the coefficients a_{ij}, b_i, f of (40.2) depend only on x and u, and if, moreover, the a_{ij} are endowed with continuous derivatives of order $n \geq 3$, then hypothesis δ') of the preceding theorem can be restricted to say only that the solutions of the equation remain uniformly bounded and equicontinuous. The sole hypothesis of uniform boundedness for u will suffice when the a_{ij} depend only on x.*

The proof is analogous to that of the preceding theorem where we bear in mind, under the new hypotheses and always taking account of theorem 33, IV, that it is also possible to majorize the quantities $A_{n,\lambda}$, $B_{n-1,\lambda}$, $C_{n-2,\lambda}$, $(A_{n-2,\lambda}+1)\,C_0$, $C_0^{\frac{n+\lambda}{2}}$ with a linear function of $\|\,u_k\,\|$, while the A_0 and B_0 remain uniformly bounded, by which, having recourse to (35.9), we will be able also to eliminate the term in V_1 from (42.5), which permits the conclusion, under the hypothesis of uniform convergence of the single sequence of the u_k. Then, when the a_{ik} do not depend on u, the majorization of $\|\,u\,\|$ can be done by applying (35.9) directly to the equation (40.2) and bearing in mind that in this case $A_{0,\lambda}$ and $A_{n,\lambda}$ are bounded, C_0 is zero, and $B_{n-1,\lambda}$ and $F_{n-2,\lambda}$ can be bounded with a linear function of $\|\,u\,\|^{\frac{n-1+\lambda}{n+\lambda}}$. For every family of uniformly bounded solutions of (40.2) we thus arrive at an inequality of the type:

$$\|\,u\,\| \leq K\left[\|\,u\,\|^{\frac{n-1+\lambda}{n+\lambda}}+1\right]\cdot[U_0+\Theta_{n,\lambda}\,[\varphi;\,\partial T]]\,,$$

sufficient to ensure a bound for $\|\,u\,\|$.

By way of example, we observe that for the semilinear equation of the form:

$$\sum_{i,k=1}^{m} a_{ik}(x)\,\frac{\partial^2 u}{\partial x_i\,\partial x_k}+\alpha\sum_{i=1}^{m} b_i(x,u)\,\frac{\partial u}{\partial x_i}=\alpha\,f(x,u)\,,$$

where α is a non-negative parameter and $uf(x, u) > 0$ for $|u| > M$, we certainly have $U_0 \leq M + \Phi_0$. By theorems 42, III, IV, and VIII it follows from this that DIRICHLET's problem for this equation admits at least one solution for arbitrary φ and $\alpha > 0$. In fact, it suffices to observe that for $\alpha = 0$, hypothesis γ') is satisfied for arbitrary φ.

In the particular case that the b_i do not depend on u and f_u is non-negative, the aforesaid solution is unique.

The following complements to the preceding theorems can turn out to be quite useful when we wish to study the solvability of a DIRICHLET problem with boundary data which are only of class $C^{(2)}$ or $C^{(1)}$.

42, IX. *If $\{u_k\}$ is a sequence of solutions of class $C^{(2, \lambda)}$ of (40,1) uniformly convergent in T together with the sequences of the $\{\partial u_k / \partial x_i\}$ and $\{\partial^2 u / \partial x_i \, \partial x_j\}$, then the limit function of this sequence is of class $C^{(2, \lambda)}$ in $T - \partial T$. If the equation is of quasi-linear type (40,2), then the same conclusion holds under the sole hypothesis that the sequences $\{u_k\}$ and $\{\partial u_k / \partial x_i\}$ converge uniformly, and in this case also, the limit function is a solution of the equation.*

Evidently, it suffices to prove that the second derivatives of the u_k are uniformly λ-Hölder continuous in every domain interior to T, and for this it is sufficient to bound from above the quantity $\mathfrak{H}_\mu^{(3)}[u_k]$ defined in § 35. We arrive at this in the general case by taking up again the proof of theorem 42, VI with obvious modifications and making use no longer of theorem 35, I but rather of theorem 35, V. In this way, if we negate the assertion, we easily establish the inequality:

$$\mathfrak{H}_0^{(3)}[u_k] \leq K \left[\mathfrak{H}_0^{(3)}[u_k] + 1\right] \left[V_2 + V_1 + V_0\right],$$

from which absurdity the theorem follows. In the case of the quasi-linear equation, we proceed in the same way except that the preceding inequality is replaced by the equally absurd:

$$\mathfrak{H}_1^{(3)}[u_k] \leq K \left[\mathfrak{H}_1^{(3)}[u_k] + 1\right] \left[V_1 + V_0\right].$$

CACCIOPPOLI's note [6], in which the last four theorems were essentially established, dates from 1933, and for about twenty-five years it represented the most advanced point reached in the theory of nonlinear equations with $m > 2$ variables. In fact, it was only in 1958 that DE GIORGI's fundamental result [3] opened the way to a whole series of new research. The progress of the theory of nonlinear elliptic equations is in fact nearly simultaneous with that of the theory of linear equations of which we have spoken in § 30. This research had as its aim at the first moment the study of DIRICHLET's problem for equations of the type (40,7). A propos of this the first of all to be cited are the works of O. A. LADYŽENSKAJA and N. N. URAL'CEVA [1, ..., 7] and of N. N.

URAL'CEVA [2], the results of which are reported in definitive form in Chapter IV of the volume [8] by the first two of these authors, and partially in §§ 1.10 and 5.11 of the volume [9] by C. B. MORREY. In these works (40.7) is considered under the following hypothesis:

A) The functions $a_i(x, u, p_j)$ and $a(x, u, p_j)$ are of classes $C^{(1, \lambda)}$ and $C^{(0, \lambda)}$ respectively, and there exists a number $k > 1$ and two positive functions $\mu(t)$ and $\nu(t)$ defined and continuous for $t \geq 0$ and such that:

$$|a_i|(1 + |p|) + |a| \leq \mu(|u|)(1 + |p|)^k,$$

$$\left|\frac{\partial a_i}{\partial p_j}\right|(1 + |p|) + \left|\frac{\partial a_i}{\partial x_j}\right| + \left|\frac{\partial a_i}{\partial u}\right| \leq \mu(|u|)(1 + |p|)^{k-1},$$

$$\left|\frac{\partial a}{\partial p_j}\right|(1 + |p|) + \left|\frac{\partial a}{\partial x_j}\right| + \left|\frac{\partial a}{\partial u}\right| \leq \mu(|u|)(1 + |p|)^k,$$

$$\nu(|u|)(1 + |p|)^{k-2} \sum_{i=1}^m \xi_i^2 \leq \sum_{i,j=1}^m \frac{\partial a_i}{\partial p_j} \xi_i \xi_j \leq \mu(|u|)(1 + |p|)^{k-2} \sum_{i=1}^m \xi_i^2.$$

The most important results can be summarized in the following theorems, in which by a weak solution of equation (40.7) we mean a function $u \in H^{1, k}(T)$ which satisfies (40.10) for every $v \in H_0^{1, k}(T)$:

42, X. *Let* $T \in A^{(2, \lambda)}$ *and* $\varphi \in C^{(2, \lambda)}(\partial T)$. *Under hypothesis* A) *it is possible to associate to every number* $M > 0$ *two positive numbers* $\alpha < 1$ *and* K *such that every weak solution* $u(x)$ *of* (40.7), *satisfying* (42.1) *and the condition* $|u(x)| \leq M$ *for* $x \in T$, *is in class* $C^{(1, \alpha)}(T)$ *and*:

$$\|u\|_{C^{(1, \lambda)}(T)} \leq K.$$

42, XI. *Let the hypotheses of the preceding theorem be satisfied, and moreover let it be possible to bound with a function of an upper bound of* $\|\varphi\|_{C^{(2, \lambda)}(T)}$ *the maximum modulus of all the possible weak solutions of the equation:*

$$\tau\left[\sum_{i=1}^m \frac{\partial a_i}{\partial x_i} - a\right] + (1 - \tau)\left[k \sum_{i=1}^m \frac{\partial}{\partial x_i}\left(p_i(1 + |p|^2)^{\frac{k-2}{2}}\right) - 2u\right] = 0$$

$$(42.5)$$

satisfying (42.1) *with* $\tau \in [0,1]$. *Then for every* $\tau \in [0,1]$, *equation* (42.5) *admits at least one solution* $u \in C^{(2, \lambda)}(T)$ *satisfying* (42.1). *In particular there exists at least one solution of* (40.7) *satisfying* (42.1).

Of these two theorems, the fundamental one is the first, because from it the second follows immediately by reasoning of the type described at the start of this section. For example, if the a_i and a are of class $C^{(3)}$, then theorem 42, XI is an immediate corollary of theorem 42, VII, because theorem 42, X ensures that hypothesis δ') is satisfied, while it is easy to verify by taking account of theorem 40, I that for $\tau = 0$ and $\varphi = 0$ problem (42.5), (42.1) admits only the solution $u = 0$, in correspondence with which DIRICHLET's problem for the variational equation is unconditionally solvable. But even if a_i and a satisfy only hypothesis A), we arrive, in an equally smooth way, at the conclusion, by making use of the procedure of LERAY-SCHAUDER.

We note that hypothesis A) ensures that the functions a_i and a have a type of growth similar to that of polynomials with respect to the variables p_j. Consequently it is also interesting to consider the case in which we allow the a_i and a to have a stronger type of growth. Results of this sort were obtained by T.B. SOLOMJAK [4, 5]; others can be deduced from a work by M.I. VIŠIK [18] concerning, more generally, equations of higher order also. Some preceding notes by A.I. KOŠELEV [2] and T.B. SOLOMJAK [1, 2, 3] are along the same lines as the works of O.A. LADYŽENSKAJA and N.N. URAL'CEVA but refer to more particular equations.

Of considerable interest is a work by D. GILBARG [2] relating to the case in which $a = 0$ and the a_i depend only on the p_j. In this case and if, moreover, the a_i are of class $C^{(2)}$, it is possible to establish results analogous to theorems 42, X and XI, always assuming $T \in A^{(2, \lambda)}$ and replacing hypothesis A) by one of the following:

$$A')\quad \lambda_1(u) \sum_{i=1}^{m} \xi_i^2 \leq \sum_{i,j=1}^{m} \frac{\partial a_i}{\partial p_j}\, \xi_i\, \xi_j \leq \lambda_2(u) \sum_{i=1}^{m} \xi_i^2 ,$$

where $\lambda_1(u)$ and $\lambda_2(u)$ are functions positive and continuous in $(-\infty, +\infty)$ such that:

$$\lambda_2(u)/\lambda_1(u) \leq \gamma < \infty ,$$

γ being a constant.

A") T is uniformly convex[1] and:

$$\sum_{i,j=1}^{m} \frac{\partial a_i}{\partial p_j}\, \xi_i\, \xi_j > 0 \quad \text{for} \quad \sum_{i=1}^{m} \xi_i^2 > 0 .$$

We note that case A") includes the equation of minimal surfaces in $m \geq 2$ variables. More precisely, the following theorem holds:

42, XII. *Let* $T \in A^{(2, \lambda)}$, $\varphi \in C^{(2, \lambda)}(\partial T)$, $a_i \in C^{(1, \lambda)}$. *If hypothesis* A') *or* A'') *is also satisfied, then the equation:*

$$\sum_{i=1}^{m} \frac{\partial a_i(p_j)}{\partial x_i} = 0$$

admits one and only one solution in class $C^{(2, \lambda)}(T)$ *satisfying* (42.1).

---

[1] The definition of a uniformly convex domain is given in § 35 after theorem 35, VIII.

That the solution, if it exists, is unique, is established immediately. Indeed, if two solutions u and u' exist, then we easily have, by integrating by parts, that:

$$\int_T \sum_{i=1}^m \left[a_i(p_j) - a_i(p_j')\right] \left[p_i - p_i'\right] dx = 0$$

from which:

$$\int_0^1 \frac{dt}{t} \int_T \sum_{i,j=1}^m \frac{\partial a_i(p_j' + t(p_j - p_j'))}{\partial p_j} (p_i - p_i')(p_j - p_j') dx = 0,$$

which is absurd as much under hypothesis A') as under hypothesis A'') if u does not equal u'. Concerning the existence, in trying to apply theorem 42, IV, it will be necessary to establish a bound with an appropriate β for $U_{1,\beta}$ for the possible solution u of the equation, U_0 being obviously already bounded by $\sup |\varphi|$. For this purpose it is first of all essential to establish a bound for U_1. Because we recognize easily that each one of the p_j is a weak solution of the equation:

$$\sum_{i,k=1}^m \frac{\partial}{\partial x_i} \left(\frac{\partial a_i}{\partial p_k} \frac{\partial p_j}{\partial x_k}\right) = 0,$$

we shall obtain a bound on the p_j in T as soon as an analogous bound has been established on ∂T (see § 30). Therefore it is definitely a question of bounding du/dn on ∂T, and this can be done by referring to theorem 35, VI under hypothesis A') and to theorem 35, VII under hypothesis A''). Once U_1 is bounded, the bound for $U_{1,\beta}$ for an appropriate β can be obtained by taking account anew of the fact that each of the p_j is a weak solution of an elliptic equation, making use of techniques analogous to those which permit the proof of theorem 30, IV and also bearing in mind the hypotheses made on φ. Upon this, on the other hand, we can not enter into details.

We have already observed that among quasi-linear equations of the type (40.7) are included the EULER equations of multiple integrals; if the integral which we consider is regular then the equation is of elliptic type. Hence the great importance for the calculus of variations of everything which concerns equations of type (40.7). We note that in general the proof of the existence of the minimum for regular integrals does not present any grave difficulties. Thus, likewise, it is easy to prove that the minimizing function is a weak solution of the EULER equation. From the point of view of the calculus of variations, the results which are of interest are therefore the regularization theorems. Along these lines, O. A. LADY-

ŽENSKAJA and N.N. URAL'CEVA in the works already cited and C.B.
MORREY [8] have deeply studied the quasi-linear elliptic equations which
are presented in the calculus of variations. In order of precedence, less
advanced results were obtained by MORREY himself [4] and by G. STAM-
PACCHIA [3]. In the volume [8] by LADYŽENSKAJA and URAL'CEVA
(Chapter V) and in the volume [9] by MORREY (Chapters 1 and 5) the
results obtained are presented systematically.

We shall also mention an interesting work by G. STAMPACCHIA [11]
in which, by the methods of calculus of variations, he gives existence
theorems for solution of DIRICHLET's problem for the equation:

$$\sum_{i=1}^{m} \frac{\partial}{\partial x_i} f_{p_i} (p_j) = a (x, u) .$$

For going the detailed statements of these theorems, we shall mention
only that he assumes that the ellipticity condition:

$$\sum_{i,k=1}^{m} f_{p_i p_k} \xi_i \xi_k > 0 \quad \text{for} \quad \sum_{i=1}^{m} \xi_i^2 > 0$$

is satisfied and that the domain $T \in A^{(3)}$ is uniformly convex. These
theorems are thus to be considered similar to GILBARG's results. In a
later work, P. HARTMAN and G. STAMPACCHIA [1] considered more
generally the equation:

$$\sum_{i=1}^{m} \frac{\partial a_i (p_j)}{\partial x_i} = F [u] ,$$

where $F [u]$ is a mapping of $C^{(0,1)} (T)$ to $L^1 (T)$ satisfying appropriate
hypotheses[1]. It is interesting to note that in the case $F = 0$, for the
proof of the existence of a weak solution $u \in C^{(0,1)}$ of DIRICHLET's problem
in a uniformly convex domain, then for the usual ellipticity hypothesis,
for example type A''), we can substitute a weakened ellipticity hypothesis
of the type:

$$\sum_{i=1}^{m} \left(a_i(p_j) - a_i(q_j) \right) (p_i - q_i) \geq 0 .$$

If now, we assume that:

$$\sum_{i=1}^{m} \left(a_i(p_j) - a_i(q_j) \right) (p_i - q_i) > 0 \quad \text{for} \quad |p - q| > 0 ,$$

---

[1] For the more general equation $\sum_{i=1}^{m} \frac{\partial a_i (x, u, p_j)}{\partial x_i} = F [u]$ see P. HART-
MAN [5].

then the solution is unique. When, however, we wish to attain the regularization of u and therefore an existence theorem in the ordinary sense, we can not do without the usual ellipticity hypothesis. These results of HARTMAN and STAMPACCHIA are obtained with the methods which we mentioned at the end of § 41 and are included in other more general ones relating to the study of certain functional inequalities. The preceding results of STAMPACCHIA [11] and of GILBARG [2] can also be obtained with analogous procedures; C.B. MORREY [9, § 4.2], for example, has proved that this can be done by resorting to the theorem of LERAY and LIONS [1]. We still note that these methods, if they allow the proof of the existence of weak solutions under weakened hypotheses, do not give any new result in regard to the regularization.

Let us see now what course of action can be drawn from theorems 42, X and XI for the study of equations of type (40.2), which, without loss of generality, can be written assuming $b_i = 0$. We shall consider first the equation:

$$\sum_{i,k=1}^{m} a_{ik}(x, u)\, p_{ik} = f(x, u, p_j) \qquad (42.6)$$

under the following hypothesis:

B) The functions a_{ik} and f are of classes $C^{(1,\lambda)}$ and $C^{(0,\lambda)}$, respectively, and there exist two positive functions $\mu(t)$ and $\nu(t)$ defined and continous for $t \geq 0$ and such that:

$$\left| \frac{\partial a_{ik}}{\partial x_j} \right| + \left| \frac{\partial a_{ik}}{\partial u} \right| + |f|\, (1 + |p|)^{-2} \leq \mu\,(|u|)\,,$$

$$\nu\,(|u|) \sum_{i=1}^{m} \xi_i^2 \leq \sum_{i,k=1}^{m} a_{ik}\, \xi_i\, \xi_k \leq \mu\,(|u|) \sum_{i=1}^{m} \xi_i^2\,.$$

The following theorem holds:

42, XIII. *Let* $T \in A^{(2,\lambda)}$, $\varphi \in C^{(2,\lambda)}(\partial T)$ *and let hypothesis B) be satisfied. If, uniformly with respect to* $\tau \in [0,1]$, *it is possible to bound with a function of an upper bound of* $\|\varphi\|_{C^{2,\lambda}(\partial T)}$ *the maximum modulus of all possible solutions of the equation:*

$$\tau \left[\sum_{i,k=1}^{m} a_{ik}\, p_{ik} - f \right] + 2\,(1 - \tau)\,[\varDelta u - u] = 0 \qquad (42.7)$$

satisfying (42.1), *then* (42.7) *and in particular* (42.6), *admits at least one solution in class* $C^{(2,\lambda)}(T)$ *satisfying* (42.1).

This theorem, which is also due to O.A. LADYŽENSKAJA and N.N. URAL'CEVA [1] and which extends a previous result of M. NAGUMO [1] for the case in which the a_{ik} are functions only of x, is an immediate consequence of theorem 42, XI. In fact, this follows from the observation that (42.6) can be written in the form (40.7) with:

$$a_i = \sum_{k=1}^{m} a_{ik}(x, u)\, p_k, \quad a = f + \sum_{k=1}^{m} \left(\frac{\partial a_{ik}}{\partial x_i} + \frac{\partial a_{ik}}{\partial u}\, p_i \right) p_k,$$

hypothesis B) ensuring that A) is satisfied with $k = 2$. We note that under the hypotheses posed, uniqueness can also fail; for an example see N. MEYERS [2].

If now we seek to bring (40.2) back to (40.7) even in the case in which the a_{ik} depend also on the p_j, we see that this is not possible, and, in fact, even after the works by O. A. LADYŽENSKAJA and N. N. URAL'CEVA, (40.2) was studied only under particular conditions in which it is only slightly different from a linear equation or at least from an equation for which a solution is already known. Along these lines we shall recall some works by H. O. CORDES [2] and A. I. KOŠELEV [1, 3], connected with the works on linear equations by these authors, which are discussed in §§ 37 and 38, and a note by S. S. DYMKOV [1]. Also related to the works by CORDES are a note by P. SZEPTYCKI [1] and two works by K. AKÔ [4] and by I. HIRAI and K. AKÔ [1]; in these latter two a method of PERRON type is also used.

A substantial progress in the theory was obtained only in 1962 with a note by N. N. URAL'CEVA [1] in which some theorems of considerable interest were announced. The proofs were then given in Chapter VI of the volume [8] by O. A. LADYŽENSKAJA and N. N. URAL'CEVA. A first theorem which can be deduced from the work cited is the following:

42, XIV. *For the quasi-linear equation* (40.2), *assuming that the coefficients* a_{ik}, b_i, f *are of class* $C^{(n-2, \lambda)} \cap C^{(1)}$, *theorems* 42, I, II, *and* III *hold even if hypothesis* β) *is replaced by:* β'') *as* $u(x)$ *varies in the family of all solutions of the equation, the* $\| u \|_{C^1(T)}$ *admit a bound as a function of an upper bound* $\| \varphi \|_{C^{(n, \lambda)}(\partial T)}$.

In fact, this follows from theorem 42, IV and from a result by N. N. URAL'CEVA according to which hypothesis β'') implies that hypothesis β') of theorem 42, IV is satisfied. We could also obtain an existence theorem with analogous procedures under weakened hypotheses on the boundary data. Even hypothesis β'') can then be weakened if we assume that the equation satisfies a uniform ellipticity condition and some others which we shall now make precise.

We consider the equation:

$$\sum_{i,k=1}^{m} a_{ik}(x, u, p_j, \tau) = f(x, u, p_j, \tau), \tag{42.8}$$

where $\tau \in [0,1]$ is a paramer, under the hypothesis that:

C) There exist two positive functions $\nu(t)$, $\mu(t)$, defined and continuous for $t \geq 0$, such that:

$$\nu(|u|) \sum_{i=1}^{m} \xi_i^2 \leq \sum_{i,k=1}^{m} a_{ik}\, \xi_i\, \xi_k \leq \mu(|u|) \sum_{i=1}^{m} \xi_i^2, \qquad (42.9)$$

$$\left| \frac{\partial a_{ik}}{\partial p_j} \right| (1 + |p|^2) + \left| \frac{\partial f}{\partial p_j} \right| \leq \mu(|u|)(1 + |p|^2)^{1/2}, \qquad (42.10)$$

$$\left| \frac{\partial a_{ik}}{\partial u} \right| (1 + |p|^2) + |f| + \left| \frac{\partial f}{\partial u} \right| \leq \mu(|u|)(1 + |p|^2),$$

$$\left| \frac{\partial a_{ik}}{\partial x_j} \right| (1 + |p|^2) + \left| \frac{\partial f}{\partial x_j} \right| \leq \mu(|u|)(1 + |p|^2)^{3/2}.$$

The following theorem holds:

42, XV. *Let* $T \in A^{(2,\lambda)}$, a_{ik}, $f \in C^{(1,\lambda)}$ *and together with hypothesis* C) *let the following also be satisfied:* γ'') *For* $\tau = 0$ (42.8) *admits a finite odd number of distinct solutions vanishing on* ∂T, *in correspondence to each one of which* DIRICHLET's *problem for the variational equation is unconditionally solvable.* β''') *The possible solutions of* (42.8) *satisfying* (42.1) *turn out to be uniformly bounded and equicontinuous as* τ *varies in* $[0,1]$ *and as* φ *varies in every bounded set of* $C^{(2,\lambda)}(\partial T)$.

Then, for arbitrary $\tau \in [0,1]$ *and* $\varphi \in C^{(2,\lambda)}(\partial T)$, (42.8) *admits at least one solution in class* $C^{(2,\lambda)}(T)$ *satisfying* (42.1).

Let us now assume that the a_{ik} and f do not depend on u, that, namely, the equation is of the type:

$$\sum_{i,k=1}^{m} a_{ik}(x, p_j, \tau) = f(x, p_j, \tau) \qquad (42.11)$$

and that:

C′) There exist three positive functions $\nu(t)$, $\mu(t)$, $\mu_1(t)$ defined and continuous for $t \geq 0$ and $\mu_1(t)$ infinitesimal as $t \to \infty$, such that (42.9) and (42.10) and the following are satisfied:

$$\left| \frac{\partial a_{ik}}{\partial x_j} \right| (1 + |p|^2) + |f|(1 + |p|) + \left| \frac{\partial f}{\partial x_j} \right| \leq \mu_1(|p|)(1 + |p|^2)^{3/2}.$$

The following theorem holds:

42, XVI. *Let* $T \in A^{(2,\lambda)}$, a_{ik}, $f \in C^{(1,\lambda)}$, *and let hypotheses* C′) *and* γ'') *be satisfied. Then, for arbitrary* $\tau \in [0,1]$ *and* $\varphi \in C^{(2,\lambda)}(\partial T)$, (42.11) *admits one and only one solution satisfying* (42.1) *under the sole condition that:* β^{IV}) *the possible solutions of* (42.11) *satisfying* (42.1) *turn out to be uniformly bounded as* $\tau \in [0,1]$ *varies and as* φ *varies in every bounded set of* $C^{(2,\lambda)}(\partial T)$.

This theorem, as the rest of the procedure, holds even under the weakened hypothesis that a_{ik}, $f \in C^{(1)}$ on condition that hypothesis γ'') is formulated a bit differently. Theorem 42, XVI holds also in a more general formulation in which a_{ik} and f are allowed to depend on u but under the restriction that $\partial a_{ik}/\partial u$ and $\partial f/\partial u$ have quantitative bounds. We note also that if for $\tau = 0$ (42.8) or (42.11) reduces to a linear equation which is unconditionally solvable, then hypothesis γ'') is automatically satisfied.

For both theorems 42, XV and XVI, the proof is attained by establishing first a bound for $U_1(T)$ and then proceeding as for theorem 42, XIV.

Finally, if we consider the equation:

$$\sum_{i,k=1}^{m} a_{ik}(p_j)\, p_{ik} = 0 \qquad (42.12)$$

we have that:

42, XVII. *Let $T \in A^{(2,\lambda)}$, $a_{ik} \in C^{(1,\lambda)}$ and let:*

$$\sum_{i,k=1}^{m} a_{ik}\, \xi_i\, \xi_k > 0 \quad for \quad \sum_{i=1}^{m} \xi_i^2 > 0. \qquad (42.13)$$

If T is uniformly convex, then (42.12) *admits one and only one solution satisfying* (42.1) *for arbitrary $\varphi \in C^{(2,\lambda)}(\partial T)$.*

For the proof of the existence here we can go back to theorem 42, XIV after having established a bound for $\| u \|_{C^{(1)}(T)}$, which is done with the same procedure indicated a propos of theorem 42, XII. Uniqueness follows from theorem 40, I.

To these works, finally, are to be added a brief note by I. YA. BAKEL'-MAN [4] regarding equation (40.1) and a work by M. I. VIŠIK [8] relating to an equation of type (40.2) in which the coefficients depend functionally on u.

43. Dirichlet's problem for equations in two variables. As we have already said in the introduction to the chapter, the study of equations in two variables preceded by many years that of the general case, and results of great importance were established by S. BERNSTEIN [2, 3] at the end of 1906. Thus, for example, in BERNSTEIN's memoir [2] we already find theorem 42, XIV proved for equations in two variables with analytic coefficients independent of u. Without this last hypothesis, the same theorem for $n \geq 3$ was later proved by J. SCHAUDER [8] and R. CACCIOPPOLI [8][1]. We now state two other theorems which can also be rejoined with the preceding results:

---

[1] For an exposition of CACCIOPPOLI's work see also the first edition in Italian of this monograph [8]. I take this occasion to point out that on page 154 the text starting with line 25 is in error. The error has been corrected in the Russian edition (p. 180, from line 6 to line 29).

43, I. *Let* $T \in A^{(2,\lambda)}$, a_{ik}, $f \in C^{(0,\lambda)}$. *Moreover let* (42.9) *be satisfied and let:*

$$|f| \leq \mu_1 (|u|) [1 + |p|^{2-\varepsilon}], \tag{43.1}$$

with $\varepsilon > 0$ *and* $\mu_1(t)$ *a positive and continuous function for* $t \geq 0$. *If, as* τ *varies in* $[0,1]$ *and* φ *varies in every bounded set of* $C^{(2,\lambda)}(\partial T)$, *the possible solutions of the equation:*

$$\tau \left[\sum_{i,k=1}^{2} a_{ik}(x, u, p_j) p_{ik} - f(x, u, p_j) \right] + (1 - \tau) \Delta u = 0 \tag{43.2}$$

satisfying (42.1) *remain uniformly bounded, then* (43.2) *admits at least one solution* $u \in C^{(2,\lambda)}(T)$ *satisfying* (42.1) *for arbitrary* $\tau \in [0,1]$ *and* $\varphi \in C^{(2,\lambda)}$ (∂T).

43, II. *Let* $T \in A^{(2,\lambda)}$, $a_{ik} \in C^{(0,\lambda)}$. *If* T *is uniformly convex and if* (42.13) *is satisfied, then the equation:*

$$\sum_{i,k=1}^{2} a_{ik}(x, u, p_j) p_{ik} = 0 \tag{43.3}$$

admits at least one solution satisfying (42.1) *for arbitrary* $\varphi \in C^{(2,\lambda)}(\partial T)$.

Both of these theorems could be stated with some economy of hypotheses by placing under consideration solutions in class $H^{2,2}(T)$ instead of $C^{(2,\lambda)}(T)$. In the form in which we have stated it, theorem 43, I is included in slightly more general theorems proved by O.A. LADY-ŽENSKAJA and N.N. URAL'CEVA [8] and by A. ZITAROSA [2]. Previously, we ought to recall an existence theorem for generalized solutions given by J. SCHAUDER [11] under the hypothesis that f is bounded[1]. In the case in which the a_{ik} do not depend on u and in which the a_{ik} and f are continuous with their third derivatives and bounded with their first derivatives and satisfy (42.9), it was proved by R. CACCIOPPOLI [8][2] that the problem admits one and only one solution of class $C^{(2,\lambda)}$ in $T - \partial T$ even if φ is assumed only in class $C^{(1)}(\partial T)$. Finally, L. BERS and L. NIRENBERG [2] previously proved theorem 43, I under the hypothesis that:

$$f = b_1(x, u) p_1 + b_2(x, u) p_2 + f_1(x, u, p_1, p_2),$$

with b_1, b_2, and f λ-Hölder continuous bounded functions of their arguments. If the a_{ik}, b_i, and f_1 are assumed only continuous and bounded, we can then prove the existence of a solution in class $H^{2,2}(T)$.

Regarding theorem 43, II, its first formulation, in the analytic case, is due to S. BERNSTEIN [2]. Then J. LERAY and J. SCHAUDER [1] and

---

[1] See also C. MIRANDA [8] theorem 43, VI.
[2] See also C. MIRANDA [8] theorem 43, V.

R. CACCIOPPOLI [8] proved the existence of solutions in class $C^{(3,\lambda)}(T)$ under the hypothesis that $T \in A^{(3,\lambda)}(\partial T)$, $\varphi \in C^{(3,\lambda)}(\partial T)$ and that the a_{ik} have continuous third derivatives[1]. This result was then improved by C.B. MORREY [1]. Finally, in the form in which we have stated theorem 43, II, it was proved by L. NIRENBERG [1].

As for the proof of these theorems, we shall limit ourselves here to giving an idea. Now, regarding theorem 43, I, assuming for simplicity that $\varphi = 0$, we can observe that by dint of theorem 37, III, we have a bound of the type:

$$\int_T \sum_{i,k=1}^{2} p_{ik}^2 \, dx = O\left(\int_T \sum_{i=1}^{2} p_i^{2\,(2-\varepsilon)} \, dx + U_0^2 + 1 \right).$$

Because, on the other hand, it is true that $\varphi = 0$, we also have, by simple integration by parts and by applying SCHWARZ's inequality:

$$\int_T \sum_{i=1}^{2} p_i^4 \, dx = O\left(U_0^2 \int_T \sum_{i,k=1}^{2} p_{ik}^2 \, dx \right),$$

and we deduce easily a bound of the type

$$\int_T \sum_{i=1}^{2} p_i^4 \, dx = O\left(U_0^{\frac{4}{\varepsilon}} + U_0^2 \right).$$

From this formula we draw:

$$\int_T |f|^{\frac{4}{2-\varepsilon}} \, dx = O\left(1 + U_0^{\frac{4}{\varepsilon}} \right)$$

and therefore from theorem 38, II we draw a bound for $\| u \|_{C^{(1,\beta)}(T)}$ with an appropriate β. As for theorem 43, II, this one can be brought back to the preceding theorem after having established an a priori bound for U_1. This can be done based on geometrical considerations which we find already in S. BERNSTEIN's memoir and which are then stated precisely by other authors[2]. Roughly speaking, the solution surfaces

---

[1] See also C. MIRANDA [8] theorem 43, VII.

[2] See T. RADÒ: Geometrische Betrachtungen über zwei dimensionale reguläre Variationsprobleme. Acta Szeged (1924/26) 228—253. J. von NEUMANN: Über einen Hilfssatz der Variationsrechnung. Abh. Math. Sem. Hamburg 8 (1931) 28—31. For an extension to the case of m variables, which however does not seem to have been used in the theory of elliptic equations, see P. HARTMAN and L. NIRENBERG [1].

of (43.3) consist mainly of hyperbolic points since from (43.3) we draw:

$$a_{22}(p_{12}^2 - p_{11}p_{22}) = a_{11}p_1^2 + 2\,a_{12}\,p_{11}\,p_{12} + a_{22}\,p_{12}^2 \geq 0\,. \quad (43.4)$$

It follows from this that the tangent plane at a point of this surface cuts the boundary of it in at least three points. Now under the hypotheses posed for T and φ a result of SCHAUDER [1] ensures that φ satisfies the 3-point condition[1] and from this follows the desired bound for U_1.

In complement to theorem 43, I we recall also that if hypothesis (43.1) were not satisfied, then the theorem could hold only for sufficiently small domains T. On the bounds to impose on T in relation to the order of growth of f with respect to $|p|$ see I. YA. BAKEL'MAN [6, 9].

We observe finally that theorem 43, II is applicable to the study of the equation of minimal surfaces, which, however, is also included among equations of divergence form, to which theorem 42, XII is applicable. For this equation, nevertheless, a direct study allows the achievement of quite advanced results. A propos of this we can consult the works by S. BERNSTEIN [2, 3, 4, 5], CH. H. MÜNTZ [1], G. ZWIRNER [1], H. JENKINS [1], H. JENKINS and J. SERRIN [1], JOHANNES NITSCHE [9]. For other indications regarding the study of these equations we refer to T. RADÒ's monograph [3] and R. COURANT's volume [1]. For an up-to-date bibliography of the most recent works see also JOHANNES NITSCHE [10].

We pass on now to examine how we can make theorems relating to equation (40.1) precise in the case $m = 2$. We owe the following theorem to L. NIRENBERG [1]:

43, III. *Let F be continuous with its derivatives up to order $n - 2$ (≥ 1) and let its derivatives of order $n - 2$ be* LIPSCHITZ *continuous with respect to the variables p_{ik} and λ-Hölder continuous with respect to the variables p_j, u, x_j. If $m = 2$ and $T \in A^{(n)}$ then theorems 42, I, II, III, and V hold even if hypothesis β) is replaced by the following: θ) The possible solutions in class $C^{(n,\lambda)}(T)$ of (40.1) satisfying (42.1) and their first and second derivatives turn out to be bounded as φ varies in every bounded set of $C^{(n,\lambda)}(\partial T)$.*

NIRENBERG obtained this result by applying techniques similar to those which occur in the proof of theorem 38, II to the study of the elliptic equations satisfied by p_1 and p_2, which are deduced by differentiation from (40.1). In this way, he succeeds in bounding the norm of every solution $u \in C^{(3)}(T)$ in $C^{(2,\alpha)}(T)$ for an appropriate α. From this the equations satisfied by p_1 and p_2 have coefficients in class $C^{(0,\alpha)}(T)$ and, by making use of the results of § 35, this permits the bounding of the norms of p_1 and p_2 in $C^{(2,\alpha)}(T)$, and therefore of u in $C^{(3,\alpha)}(T)$. The

---

[1] See the observations which follows the proof of theorem 35, VII.

coefficients of the equations satisfied by p_1 and p_2 are then LIPSCHITZ continuous and consequently the norm of u can also be bounded in $C^{(3,\lambda)}(T)$. All this by assuming only that the hypotheses on F are satisfied for $n = 3$. For $n > 3$, the theorems of § 35 equally well allow the conclusion to be reached. Naturally the procedure which we have described in summary requires somewhat technically complex manipulations, whose description we will forego.

We shall mention also that, under hypotheses of analyticity for F, theorem 43, III was already proved by S. BERNSTEIN [2, 3] between 1906 and 1910. Under intermediate hypotheses, namely, assuming F continuous with its derivatives up to order n (≥ 4), the theorem was proved by J. SCHAUDER [8] in 1933, by making use of the results he obtained in the study of linear equations and of some of BERNSTEIN's procedures, appropriately modified. Another considerably simpler proof was given by R. CACCIOPPOLI [8] in 1935, with a method which opened the way to new developments. In fact, by suitably combining CACCIOPPOLI's method with BERNSTEIN's, J. LERAY [4] succeeded in 1938 in characterizing a large class of elliptic equations, more general than quasi-linear equations, for which the existence theorem is then valid under the sole hypothesis that it is possible to bound a priori the possible solutions and their first derivatives. In a second memoir, LERAY [5] then sought to characterize the class of elliptic equations for which the aforesaid a priori bound is effectively possible. Another class of equations of type (40.1) for which the existence theorem holds was indicated by I. YA. BAKEL'MAN [3].

We recall finally that the possibility of substituting hypothesis θ) for β) in the statement of theorem 42, I, when $m = 2$ and when in place of hypothesis α) we assume $F_u < 0$, was also proved by N. SIMONOV [1] with a procedure of the PERRON type and under slightly weakened hypotheses with respect to those of theorem 42, I.

44. Equations in the analytic field. In the introduction to this chapter we have already perceived the importance in the development of the modern theory of elliptic equations of the research directed at proving the analyticity of the solutions of analytic elliptic equations. We now wish to point out briefly the results of this research, specifying that a function of the point x in R_m is called analytic if it is defined in an open Ω and if it can be developed in a multiple power series in a neighborhood of every point of Ω.

The first studies of the question are due to É. PICARD[1] and concern linear equations, for which this author proved the following fundamental theorem, already mentioned in § 19:

---

[1] For the bibliography see L. LICHTENSTEIN [5].

44, I. *Every solution of class $C^{(2)}$ of a linear elliptic equation with analytic coefficients is analytic.*

This research by PICARD, which was in large part prior to HILBERT's communication to the congress of Paris, is based on the following procedure. Denoting by u a solution of class $C^{(2)}$ of the equation and by T a domain interior to the region of definition of u, we consider the DIRICHLET problem consisting of finding a solution v of the equation coinciding with u on ∂T. If T is sufficiently restricted then v can be constructed with a procedure of successive approximations, which permit recognition of its analytic character. A uniqueness theorem then permits the assertion that $u = v$ and therefore that u also is analytic.

Years later, E. E. LEVI [1] achieved the same result with an entirely different procedure, consisting essentially of proving the analyticity first of all of the fundamental solutions and then of every regular solution of the equation, making use for this last point of STOKES's formula. This formula, appropriately modified, can even furnish the analytic continuation of u into the complex field.

PICARD's research and that of LEVI regard the case $m = 2$; nevertheless at least LEVI's procedure seems to be able to apply in general. However, the theorem is true even in the case of arbitrary m, resulting as a particular case of the theorems relating to nonlinear equations. For these latter, the question appears much more difficult, and the treatments which have been given are all (more or less) extremely complex. The first research along these lines is that of S. BERNSTEIN [1] who in 1904 proved for the first time that *every elliptic solution of class $C^{(3)}$ of an analytic equations in two variables is analytic.* To this result, proved with a procedure which, at least in its point of departure, is inspired in some way by that followed by PICARD for linear equations, BERNSTEIN returned at different times [3, 6, 8] to make precise and to simplify the proof. A definitive systematization of BERNSTEIN's proof was then elaborated in 1926 by T. RADÒ [1].

Other proofs of BERNSTEIN's theorem were then given in 1918 by M. GEVREY [1] and in 1929 by H. LEWY [1, 2]. GEVREY's proof, later simplified by the same author [2] in 1926, is based on the majorization of the TAYLOR series development of the solution; LEWY's, on the continuation of the solution into the complex field, obtained through the study of a CAUCHY problem for a system of partial differential equations of first order[1].

LEWY's procedure requires, on the other hand, that the solution which we consider be of class $C^{(4)}$. A propos of this we observe, however, that

---

[1] A nice exposition of LEWY's result is also found in an appendix to J. HADAMARD's volume cited in the Bibliography [1].

VI. Nonlinear equations

once we have proved the analyticity of solutions of class $C^{(k)}$ with arbitrary $k \geq 3$, theorem 40, II permits us to conclude that all the solutions of class $C^{(2, \lambda)}$ are analytic.

Therefore, either from the theorem of BERNSTEIN and GEVREY or from the more restricted one of LEWY, it follows that at least in the case $m = 2$:

44, II. *If F is an analytic function of its arguments, then every elliptic solution of class* $C^{(2, \lambda)}$ *of* (40.1) *is analytic.*

And from theorem 40, III we have analogously:

44, III. *If the coefficients of the quasi-linear equation* (40.2) *are analytic, then every elliptic solution of class* $C^{(2)}$ *of this equation is analytic.*

It is useful to perceive that theorems 40, II and 40, III, on which we base the preceding extension of the results of BERNSTEIN, GEVREY, and LEWY, are later than the research of these authors[1], having been proved by E. HOPF only in 1930. HOPF's memoir [4], in which these results are obtained, contains also the proof of theorems 44, II and III for the case of arbitrary m. The method followed by HOPF, which can in some way be related to that followed by E.E. LEVI in the study of linear equations, consists of observing that every elliptic solution u of (40.1) together with certain of its derivatives assumed as auxiliary unknowns satisfies a system of nonlinear integral equations, the solutions of which can be extended even into the complex field, by means of a procedure of successive approximations; we then recognize, by means of an application of MORERA's theorem, that these solutions are holomorphic functions of the complex variables on which they happen to depend. From this, the analyticity of u in the real field.

Another proof of theorems 44, II and III was given, contemporaneously with HOPF, by G. GIRAUD [4] with a procedure like that of PICARD; some years earlier GIRAUD himself [1] had obtained, for the case of arbitrary m, a theorem of the same range as that established by BERNSTEIN and GEVREY in the particular case $m = 2$.

It is now naturally worthwhile to ask if, analogously to what happened for quasi-linear equations, it is not possible to prove even in the general case the analyticity of the solutions of class $C^{(2)}$. To this question, after the proof of theorems 40, IV and V, we can now respond in the affirmative. Namely, we have that:

44, IV. *If F is an analytic function of all its arguments, then every elliptic solution of class* $C^{(2)}$ *of* (40.1) *is analytic.*

---

[1] Indeed, some idea of the possibility of weakening the hypotheses in the sense indicated in theorems 44, II and III is already found in a note by M. GEVREY [4] regarding the first of these, and in the memoir [1] by G. GIRAUD regarding the second.

In fact, the theorems cited permit us to assert that every solution of class $C^{(2)}$ is also of class $C^{(n, \lambda)}$ for arbitrary n, and therefore its analyticity derives from the theorems of HOPF and of GIRAUD mentioned previously. This result, which R. CACCIOPPOLI [8] established in 1935 for $m = 2$, was obtained in the general case by C. B. MORREY and L. NIRENBERG [1] for linear equations and by C. B. MORREY [5] for nonlinear ones. On the other hand, these authors arrived at the result with a new procedure, which, without making any reference to earlier results, is based on a new proof of theorem 44, II, which is also reported in §§ 5.7 and 5.8 of the volume [9] by C. B. MORREY. The procedure followed consists of obtaining a majorization of the successive derivatives of a solution so as to ensure the convergence of its TAYLOR series development. This method is applicable even in the case of equations of higher order and of systems of equations (see §§ 52 J and 55 H).

Almost contemporaneously with MORREY, theorem 44, IV was also obtained by A. FRIEDMAN [1, 3] as a particular case of a more general theorem. FRIEDMAN considered a sequence of positive numbers M_0, M_1, ... such that, putting $M_{-i} = 1$ ($i = 1, 2, \ldots$), we have, with a positive constant A:

$$\binom{n}{i} M_i M_{n-i} \leq A M_n \ (i, n = 1, 2, \ldots) . \tag{44.1}$$

Denoting by $F(x, y)$ the left side of (40.1), having put $y = (u, p_i, p_{ik})$, we assume, for every compact set D contained in the region of definition of F, that there exist two numbers H_0 and H such that in D:

$$\left| D_x^\alpha D_y^\beta F \right| \leq H_0 H^{|\alpha| + |\beta|} M_{|\alpha|} M_{|\beta|} . \tag{44.2}$$

The theorem proved by FRIEDMAN asserts that:

44, V. *If the sequence of the M_i satisfies* (44.1) *and if F satisfies* (44.2), *then for every solution $u(x) \in C^{(2)}$ of* (40.1) *and for every compact D contained in its domain of definition, there exist two numbers K_0 and K such that in D we have:*

$$\left| D_x^\alpha u \right| \leq K_0 K^{|\alpha|} M_{|\alpha|} .$$

We obtain, in particular, theorem 44, IV by assuming $M_i = i!$, while for $M_i = (i!)^s$ with $s \geq 1$ we reobtain previous results of GEVREY's [1, 2]. Another interesting result was obtained by T. KOTAKE and M. S. NARASIMHAN [1], strengthening a preceding result of E. NELSON [1]:

44, VI. *Let $\mathfrak{M}$ be a linear elliptic operator with analytic coefficients in a region Ω of R_m, and let $\mathfrak{M}^k$ be its kth iterate. Then every function $u \in C^{(\infty)}(\Omega)$ which satisfies the inequality:*

$$\| \mathfrak{M}^k u \|_{L^2(\Omega)} \leq (2m)! \, H^{k+1} , \qquad k = 1, 2, \ldots$$

with an $H > 0$ independent of k, is analytic in Ω.

14*

Theorems 44, V and VI are also particular cases of theorems regarding equations of higher order.

We have just been occupied with analyticity in an open set. We wish now to point out the problem of analytic continuation of the solution of an elliptic equation beyond an analytic portion of hypersurface X on which it assumes analytic values. A propos of this the following theorem holds:

44, VII. *This continuation is certainly possible if the equation is linear even when, denoting by T a domain whose boundary contains X, the solution is assumed to be only regular in $T - \partial T$ and continuous in $(T - \partial T) \cup X$. In the case of a nonlinear equation the result holds under the final hypothesis that the solution is of class $C^{(2)}$ in $(T - \partial T) \cup X$.*

This theorem was proved by G. GIRAUD [1, 4] for solutions of class $C^{(4)}$ in $(T - \partial T) \cup X$ in the case of a general equation and of class $C^{(3)}$ in the case of a quasi-linear equation and by M. GEVREY [2] for solutions of class $C^{(3)}$ in the case of a general equation in two variables. From the proof of theorem 36, VI for linear equations and of theorems 40, V and VI for nonlinear ones, the theorem is therefore acquired in general. However, this is also contained as a particular case in analogous theorems by MORREY and NIRENBERG, MORREY, FRIEDMAN for equations of higher order and for which we refer to the bibliographic citations already given. A theorem of continuation along the lines of KOTAKE and NARISIMHAN was given by J. L. LIONS and E. MAGENES [3] to whom we also owe a later generalization of this result contained in a work [4] in which the point of view of KOTAKE and NARASIMHAN is combined with that of FRIEDMAN.

We observe finally that the problem of continuation can also be posed in another way. Namely, instead of assuming that the trace of u is analytic on X, we can make the hypothesis that on X u satisfies for example a NEUMANN condition with analytic coefficients and known terms. On this question and on other similar ones, regarding equations of higher order also, various results were obtained for linear equations by E. MAGENES and G. STAMPACCHIA [1] along the lines of MORREY and NIRENBERG, and by M. K. V. MURTHY [1] along the lines of FRIEDMAN. For nonlinear equations the study of the question was then resumed on a more general plane, namely for systems of equations, by C. B. MORREY [9, Chapter 6].

A final question connected with the problem of continuation is that of determining precisely the region into which it turns out to be possible to continue a solution of an elliptic equation. This question has been studied only in particular cases, namely for linear equations with the a_{ik} constant. A propos of this we refer to H. LEWY [9].

45. Equations in parametric form. Let us consider an equation of the form:

$$F(x, y, z, x_u, y_u, z_u, x_v, y_v, z_v, D, D', D'') = 0, \qquad (45.1)$$

in which the functions:

$$x = x(u, v), \qquad y = y(u, v), \qquad z = z(u, v) \qquad (45.2)$$

are assumed of class $C^{(2)}$ in a circle Γ in the uv plane and D, D', D'' denote the coefficients of the second differential quadratic form of the surface S with parametric equation (45.2).

Assuming that the left side of (45.1) is invariant with respect to arbitrary change of the variables u and v, (45.1) can well be called a *partial differential equation in parametric form*.

This is then called *of elliptic type* if $4 F_D F_{D''} - F_{D'}^2 > 0$. Obviously the simplest problem which can be posed for such an equation is that of seeking a solution surface S which is *regular* ($EG - F^2 > 0$) and which has as its boundary a given curve C.

A first question which is posed in this study is to see if, given a solution surface S_0 of (45.1) with boundary C_0 there exist solution surfaces of the same equation for a boundary curve C which is in a neighborhood, in some sense, of C_0. A propos of this C. MIRANDA [2] proved that the answer to this question is certainly affirmative on condition that DIRICHLET's problem for the variational equation of (45.1) is unconditionally solvable. This variational equation is an ordinary elliptic equation in a single unknown function η, which is deduced from (45.1) by considering an infinitesimal deformation of the surface S in the direction of the normal. The study of the problem of existence in the large was then attacked by R. CACCIOPPOLI [9] by means of a convenient extension to certain nonlinear abstract spaces of his theory of the inversion of functional correspondences (§ 41). The result which he obtained is that:

45 I. (45.1) *admits at least one solution surface bounded by an arbitrary curve in a continuous family Γ if the following hypotheses are satisfied: $1°$) For a particular curve in Γ the problem has a finite odd number of solutions and in correspondence the DIRICHLET problem for the variational equation is unconditionally solvable; $2°$) the solution surfaces having as boundaries the curves in a compact portion of Γ have uniformly bounded areas and have equicontinuous curvature or else, if (45.1) is linear in D, D', D'', equicontinuous normals.*

Naturally the continuity and the compactness in Γ is in reference to an appropriate metric, which essentially implies the λ-Hölder continuity of the curvature of the boundaries. Based on this theorem, CACCIOPPOLI also delineated a scheme for the treatment of PLATEAU's problem, with the purpose of proving that *however a closed curve C not interlacing with* *itself and having λ-Hölder continuous curvature is chosen, there exists at least one regular minimal surface, of the type of the circular disc, having C as its boundary.*

It will be useful to perceive that the novelty of this theorem, in comparison with all the preceding treatments of PLATEAU's problem, would be in the fact that the minimal surface of which it ensures the existence is *regular* in the sense of classical differential geometry. Still, according to some observations by CACCIOPPOLI himself, his proof is incomplete in one essential point. Neither, after CACCIOPPOLI, was this theorem proved by others; thus, establishing whether it is true or not remains one of the most interesting open questions of the theory of minimal surfaces.

Still, in other directions, the study of PLATEAU's problem has accomplished very considerable progress. This has been obtained generally by combining the direct methods of the calculus of variations with regularization procedures for weak solutions of elliptic equations of divergence form. In this way we could, among other things, also study PLATEAU's problem in the case in which we require the unknown surface to be of the topological type of a k-tuply connected plane domain and to have as its complete boundary a complex of k given curves. A picture of the results little by little obtained can be had by consulting the volumes by T. RADÒ [3] which dates from 1933, by R. COURANT [1] from 1950, by C.B. MORREY [9] from 1966, as well as the expository article by JOHANNES NITSCHE [10] from 1965. We also cite a work by M. SHIFFMAN [1] on the questions of regularization and a note by H. LEWY [8] on the problem of extending a minimal surface across an analytic boundary.

Important results have also been obtained regarding PLATEAU's problem for the EULER equation of the double integrals in parametric form and for the study of PLATEAU's problem relating to minimal surfaces in more dimensions. For these arguments and for the related bibliography we also refer to C.B. MORREY's volume [9]. Finally, we recall the notes by E. HEINZ [2] and H. WERNER [1] on a problem analogous to that of PLATEAU for surfaces with constant mean curvature.

We now observe, in relation to (45.1), that the problem of determining a solution surface which is closed and of a given topological type can sometimes be posed. Among problems of this type the simplest is probably the MINKOWSKI problem, which consists of seeking a closed surface of spherical type whose gaussian curvature is at every point a given positive function f of the direction cosines of its normal. A proof of the existence and uniqueness theorem for this problem was given by C. MIRANDA [2] with a method not dissimilar to that described above, under the hypothesis that f is of class $C^{(2,\lambda)}$. Previously H. LEWY [7] had proved the same theorem in the analytic case.

Later MIRANDA's result was proved again by L. NIRENBERG [2] with a method in which LEWY's procedures were combined with MIRANDA's. With this method, and also making use of theorem 40, V, NIRENBERG succeeded in completing the proof of the existence theorem under the less restrictive hypothesis that f is of class $C^{(2)}$. To MINKOWSKI's problem we also adjoin a note by J. J. STOKER [1] on the uniqueness theorem for this problem and a work by A. WINTNER [2] in which, among other things, he strengthens an inequality established by MIRANDA and proved again by NIRENBERG. For the study of a more general problem, in which the curvature is given as a function not only of the normal but also of the point on the surface, see A. V. POGORELOV [9, 11].

It would also be interesting to deepen, from the point of view of existence, the study of the problem analogous to MINKOWSKI's for an elliptic equation of the type $F(H, K, X, Y, Z) = 0$, in which H and K denote the mean and total curvature of the unknown surface and X, Y, and Z the direction cosines of the normal. For some problems of this type and for other analogous ones in more variables various uniqueness theorems are known. A propos of this see H. HOPF [1, 2], P. HARTMAN and A. WINTNER [1, 4], K. VOSS [1], A. D. ALEKSANDROV [3], P. HARTMAN [1, 3], A. AEPPLI [1].

The questions treated in this section will be rejoined with those which, in Chapter VII, will be the object of § 48.

46. The Neumann and oblique derivative problems. For a nonlinear elliptic equation, the NEUMANN problem was considered by G. GIRAUD [7, 8] but only in particular cases. In fact, GIRAUD considered a quasi-linear equation of the type (40.2) with the b_i vanishing and the a_{ik} functions only of u and x; to this equation he then associated a boundary condition of the type:

$$\sum_{i, k = 1}^{m} a_{ik} X_i \frac{\partial u}{\partial x_k} + \psi(u, x) = 0 .$$

Assuming that the a_{ik}, f, ψ depend also on a parameter τ, he proved that the problem admits a solution for every τ when 1°) *it is solvable for a particular value of τ;* 2°) *the variational problem is unconditionally solvable;* 3°) *u and its first derivatives possess an a priori bound.* Without having specified the qualitative hypotheses necessary for the validity of the theorem, we shall say only that the method followed by GIRAUD for the proof, without ever having recourse to the considerations of functional analysis, is analogous to that which we used to establish theorem 42, I.

The general study of the NEUMANN and regular oblique derivative problems with the methods of functional analysis and a priori majorization was begun only in 1954 by L. BERS and L. NIRENBERG [2] for an equation in two variables of type (40.1) or (40.2) to which is associated a boundary condition of the type:

$$\frac{du}{dn} = \varphi + k \quad \text{for} \quad x \in \partial T ,$$

where k is an unknown constant to be determined appropriately. In their treatment BERS and NIRENBERG assume that $F_u \leq 0$ for (40.1) and $f = a_0(x, u, p_j) u + a(x, u, p_j)$ with a_0, a, b_1, b_2 bounded and $a_0 \geq 0$ for (40.2) and they prove an existence theorem using SCHAUDER's fixed point theorem.

In 1959 other results were obtained by R. FIORENZA [1, 2]. In the work [1], FIORENZA considers the regular oblique derivative problem which is obtained by associating to the equation (40.1) a boundary condition of the type:

$$G(x, u, p_j) = 0 \quad \text{for} \quad x \in \partial T ,$$

under the hypothesis that $\sum_{i=1}^{m} X_i \, G_{p_i} \geq c > 0$ and he establishes results analogous to theorems 42, VI, VII, and VIII for these problems. In the work [2], taking into consideration the same problem for an equation in two variables, the analog to theorem 43, III is then proved for the case $m = 2$, under hypotheses of the type of those adopted by R. CAC-CIOPPOLI [8] in his proof of this theorem.

In 1962, N. N. URAL'CEVA [2] announced new results relating to the NEUMANN problem for equations of the type (40.7) and (42.6). These results analogous to theorems 42, XI and XIII concern the boundary value problem which are obtained by associating to the aforesaid equation boundary conditions respectively of the type:

$$\sum_{i=1}^{m} X_i \, a_i + \varphi(x, u) = 0 \quad \text{for} \quad x \in \partial T ,$$

$$\sum_{i,j=1}^{m} a_{ij} X_i \, p_i + \varphi(x, u) = 0 \quad \text{for} \quad x \in \partial T .$$

The proof, together with a reelaboration of FIORENZA's results [1] are contained in Chapter X of the volume [8] by O. A. LADYŽENSKAJA and N. N. URAL'CEVA. Along the same lines some partial results were previously obtained by T. B. SOLOMJAK [2, 3] who then [4, 5] also studied the case of (40.7) under the hypothesis that the a_i and a have with respect to the p_j a type of growth stronger than that of polynomials.

Finally R. FIORENZA [4] extended some of the results established for NEUMANN's problem by N. N. URAL'CEVA to the regular oblique deriva-tive problem for (40.7) and obtained in the case of two variables, new

results, one of which is to be considered as an extension of theorem 43, I.
For an oblique derivative problem relating to (42.6) see also D. Sadowska
[1]. For other works relating to semilinear equations see § 47.

47. Equations of particular type. In this section we wish to
occupy ourselves in the first place with semilinear equations[1] beginning
with the equation:

$$\Delta_2 u = f(x, u, p_i) . \tag{47.1}$$

This is the simplest of the nonlinear equations of elliptic type and as such
has been, since the first memoirs by Picard and Bernstein, the proving
ground for the various methods thought out for the study of more
general equations. Referring to L. Lichtenstein's monograph [5] for
the older literature, we shall cite those works by J. Schauder [1] which
studied Dirichlet's problem for (47.1) with the fixed point theorem,
by R. Caccioppoli [4] which indicated how the general theorems of § 42
are susceptible to various perfections for (47.1), by L. Lichtenstein [6]
which applied the theory of nonlinear integral equations to the study
of (47.1). Also, the following authors had recourse to the transformation
into integral equations: T. Satō [2, 4] for the study of Dirichlet's
problem, Satō himself [3] and W. Pogorzelski [1] for the study of
Neumann's problem, J. Wolska Bochenek [1, 2] for the study of an
oblique derivative problem. For the application of Perron's method we
instead adjoin two notes by S. Simoda [1, 2].

The case in which in (47.1) f does not depend on the p_i was studied
by A. Hammerstein [1], L. Lichtenstein [6], G.F.D. Duff [1, 2],
N. Levinson [2], while to the equation $\Delta_2 u = f(u)$ are dedicated some
notes by R. Iglisch [1], J.B. Keller [1], S.I. Hudjaev [1], Z. Nehari [3]. In particular, S.I. Pohozaev [1, 2] studied Dirichlet's problem
for the equation $\Delta_2 u = u^2$, while for the equation $\Delta_2 u = e^u$, Dirichlet's
problem was considered in the volume [6] by L. Lichtenstein and that
of Neumann in the note [2] by Joachim and Johannes Nitsche. These
latter authors have also studied [3] Neumann's problem for the equation $\Delta_2 u = u_x^2 + u_y^2$. Finally, for a linear equation of the type (47.1),
K. Nakamori [1] studied a boundary value problem with a nonlinear
condition of Neumann type.

Other works examine the general semilinear equation:

$$\mathfrak{M} u = f(x, u, p_j) \tag{47.2}$$

in which $\mathfrak{M}$ is a linear operator. A propos of this we shall first of all recall
a memoir by M. Nagumo [1] in which Dirichlet's problem for (47.2) is
studied under the hypothesis that $f = O(k + \varepsilon |p|^2)$ with ε sufficiently

---

[1] A first result on this equation is already contained in theorem 42, VIII.

small[1]. This type of behavior for f is met again in a work by C. MIRANDA [17], in which other than DIRICHLET's problem he also considers NEU-MANN's problem, and in various works by O. A. LADYŽENSKAJA, N. N. URAL'CEVA, and others, regarding quasi-linear equations (see §§ 42 and 43).

We recall also the works by Y. HIRASAWA [2, 4, 5] which can in some way be joined to those by T. SATŌ already cited a propos of (47.1), the research of W. POGORZELSKY [3, 5, 6] on certain oblique derivative problems[2], and a note by A. LASOTA [1] on the DIRICHLET and NEUMANN problems. While the works of these latter authors make use of the method of integral equations, those by S. SIMODA [4, 5], K. AKÔ [3], I. HIRAI and K. AKÔ [1] utilize procedures of the PERRON type. Instead, upon appropriate a priori bounds of integral type are based the works by V. G. MAZ'JA [1] on DIRICHLET's problem and by S. I. HUDJAEV [2] on NEUMANN's problem.

In the notes by M. A. DŽAFARDI [1] and D. P. ZERAGIJA [1], DIRICH-LET's problem for (47.2) is studied under the hypothesis that the right side depends functionally on u[3].

Finally, another equation of particular type in two variables which has been the object of considerable research is the so-called MONGE-AMPÈRE equation, which in its most general form can be written:

$$\sum_{i,k=1}^{2} a_{ik}(x, u, p_j)\, p_{ik} + p_{11}\, p_{22} - p_{12}^2 = f(x, u, p_j)\ . \qquad (47.3)$$

This equation presents itself in many questions from differential geometry, one of which, for example, is the MINKOWSKI problem already studied in § 45, and it also has interest as an example of an equation which is not itself elliptic, but which can admit both elliptic and non-elliptic solutions.

A propos of (47.3) we cite the works by F. RELLICH [1], H. LEWY [5], P. GILLIS [1], A. V. POGORELOV [9, ..., 12], P. HARTMAN and A. WINT-NER [2], K. JÖRGENS [1], I. YA. BAKEL'MAN [1, 2, 5, 7, 8, 10, 11], E. HEINZ [4, 6, 7, 9], I. JA. GUBERMAN [1], I. YA. BAKEL'MAN and I. JA. GUBERMAN [1, 2]. All of these works concern both the a priori major-ization and the existence of elliptic solutions of the DIRICHLET problem for equations of the type (47.3). Some extensions to the case of more variables have been considered by P. GILLIS [2, 3, 4], A. D. ALEKSAN-DROV [4], JU. A. VOLKOV [1], I. JA. GUBERMAN [2].

---

[1] For a preceding result of analogous type, see R. CACCIOPPOLI [7].

[2] These works of POGORZELSKY are at the origin of the works by WOLSKA BOCHENEK, LASOTA, SADOWSKA cited in this section and in the preceding one.

[3] For an analogous study in relation to more general equations see the works by P. HARTMAN and G. STAMPACCHIA [1] and by M. I. VIŠIK [8] al-ready refferred to in § 42.

CHAPTER VII

Other research on equations
of second order. Equations of higher order.
Systems of equations.

In this chapter we intend to point out various questions briefly, which have been postponed until now, regarding both elliptic equations of second order and those of higher order and systems of equations. Many of these questions have great importance. Still it is not possible here, given the character of this volume, to give them a detailed and systematic treatment. Consequently we shall limit ourselves here to reporting the arguments to which this research refers and to giving, when it is possible, some indication of the method followed. In all this, we shall scarcely ever enter into the details either of the statements or of the proofs. In other words, in this chapter we wish only to classify, in some way, the bibliographic material collected at the end of the volume and upon which we have not touched until now.

§§ 48, 49, 50, 51 refer, then, to equations of second order, §§ 52, 53, 55 to equations of higher order and to systems, §§ 54 and 56 to questions regarding both arguments.

48. Second order equations on a manifold. Let $\mathfrak{B}$ be an m dimensional riemannian manifold, closed and orientable, and let:

$$\sum_{i,k=1}^{m} g_{ik}\, dx^i\, dx^k$$

be its fundamental quadratic form. If a^{ik} and b^i are two contravariant tensors, of which the first is symmetric, and if c is a scalar, then for every function of class $C^{(2)}$ on $\mathfrak{B}$ the following second order differential operator is defined:

$$\mathfrak{M}u = \sum_{i,k=1}^{m} a^{ik} D_k \frac{\partial u}{\partial x^i} + \sum_{i=1}^{m} b_i \frac{\partial u}{\partial x^i} + cu\,, \tag{48.1}$$

D_k being the CHRISTOFFEL symbol for covariant differentiation. If f is a scalar, then the equation:

$$\mathfrak{M}u = f \tag{48.2}$$

is said to be of elliptic type if the quadratic form $\Sigma a^{ik} \lambda_i \lambda_k$ is positive definite. For such an equation it is possible to pose boundary value problems analogous to those considered in a euclidean space. Namely, given a domain T on $\mathfrak{B}$, it will be a question of determining a solution of (48.2) regular in $T - \partial T$ and satisfying given boundary conditions on ∂T, these latter being able to be either of the first, second, or third type.

Another problem, in a certain sense a simpler one, is the so-called PICARD's problem, consisting of seeking a solution of (48.2) which is regular on the entire manifold $\mathfrak{B}$. If we leave out of consideration some old works by É. PICARD[1] and D. HILBERT [1], relating to some particular cases, and the research of various authors concerning existence theorems for abelian integrals on RIEMANN surfaces, the first systematic treatment of these problems is due to G. GIRAUD, who extended to (48.2) a large part of the results he obtained (see Chapter III) for equations in euclidean space. The central part of this extension is contained in the memoir [14], other complementary results in the memoirs [17, 19, 21]. Other results of existence type were obtained by K. YOSIDA [2] for PICARD's problem and by S. ITO [1, 2, 3] for NEUMANN's problem.

Analogous problems can then be imposed even under the single hypothesis that $\mathfrak{B}$ is a closed and topologically defined manifold, namely by giving up the assumption that a metric is defined on $\mathfrak{B}$. Along these lines PICARD's problem, in the case $m = 2$, was treated by G. CIMMINO [3] with a method similar to that which is discussed in § 29 and whose application is here simplified by the lack of boundary conditions.

We now cite a memoir by O. NIKODYM [1] relating to the case of a two dimensional manifold endowed with a locally euclidean metric, two others by G. ASCOLI [2] and E. ROTHE [1] regarding the case of a manifold with constant curvature, and a note by L. MYRBERG [1] on a comparison theorem.

This research is now bound up with that on the so-called *invariant form* of elliptic equations; a propos of this we can usefully consult, other than the memoir by W. FELLER [1] which we have already pointed on in § 18, Appendix I to the volume [1] by J. HADAMARD and a memoir by T. Y. THOMAS and E. W. TITT [1].

Other questions for which the theory of equations on a manifold has much importance are those of differential geometry in the large, already considered from another point of view in § 45. For example, MINKOWSKI's problem was brought back by L. NIRENBERG [2] to the solution of an elliptic equation on a sphere. And as much can be said of the problem of

---

[1] PICARD, É.: Sur une équation aux dérivées partielles du second ordre relative à une surface fermée, correspondant à un équilibre calorifique. Ann. Éc. N. Sup. **26** (1909) 9–17.

determining a closed and convex surface with a given metric. This problem, posed and solved, but in an incomplete way, by H. WEYL[1] in 1916, has been the object of considerable and profound research. A first systematization in the analytic field of WEYL's procedure was described by H. LEWY [6] in 1938. For a given metric of class $C^{(4, \lambda)}$ a first complete proof of the existence theorem was given by R. CACCIOPPOLI [11] in 1940. Some years later, A.D. ALEKSANDROV [1] proved the existence of a generalized solution obtained as a limit of convex polyhedra. The regularization of ALEKSANDROV's solution was attained between 1949 and 1950 by A. V. POGORELOV [1, ..., 6]. Finally in 1953 L. NIRENBERG[2] gave a new direct proof of the existence theorem based on the results he obtained in a previous work [1] on nonlinear elliptic equations. Regarding the uniqueness theorem we mention the proofs given by E. COHN-VOS-SEN [1] and O.K. ŽITOMIRSKIĬ [1] in the analytic field, by G. HER-GLOTZ [2] in a larger class, by A.V. POGORELOV [7] for generalized solutions, and then by A.D. ALEKSANDROV [3]. For an exposition we refer to the volumes by A.D. ALEKSANDROV [1] and A.V. POGORELOV [8]. Finally, associated with the study of the problem are two works by A. WINTNER [2] and E. HEINZ [8] on a fundamental inequality due to WEYL and also a note by P. HARTMAN and A. WINTNER [1].

We mention finally that PLATEAU's problem for minimal surfaces with one or more boundaries can also be posed on a riemannian manifold instead of in a euclidean space. For the proof of the existence and regularity theorems see C.B. MORREY [9].

49. Second order equations in unbounded domains.
As has already been pointed out at the end of § 4, boundary value problems for elliptic equations can also be posed in unbounded domains, on condition that an appropriate behavior at infinity is assigned to the unknown function. The simplest case is that in which the unbounded domain T has a bounded boundary. In this case if the operator $\mathfrak{M}$, satisfying the hypotheses of theorem 20, I, is endowed with a principal fundamental solution, if the known term f of the equation is infinitesimal at infinity, and finally, if we require that u also be infinitesimal at infinity, the study of the various boundary value problems offers no difficulty. Indeed, by applying GIRAUD's methods described in §§ 21, 22, 23, the problem is transformed into a single integral equation on ∂T, the study of which easily leads to the existence theorem. On the other hand, in case all the hypotheses of theorem 20, I are satisfied except $c \leq 0$, GIRAUD's method is not immediately applicable because the problem is transformed into a system of integral equations in two unknown functions, one of them

---

[1] WEYL, H.: Über die Bestimmung einer geschlossenen Fläche durch ihr Linienelement. Vierteljahrsschrift natur. Ges. Zürich **61** (1916) 40–72.

defined in the unbounded domain T. Still, in the case when $T = R_m$ the problem is transformed into a single integral equation for which GIRAUD himself [17] proved that the FREDHOLM theorems are still valid; we thus arrive at the usual alternative theorem. In particular, we note that if $T = R_m$ and if $\mathfrak{M}$ admits a fundamental solution, then the problem is solved by (20.12). We can even affirm that if $f(x) = O(|x|^p)$ for a $p > 0$, then the problem still admits a solution given by (20.12), which is unique in the class of functions $u(x) = O(|x|^p)$.

Other than the procedure which we have just discussed, GIRAUD [17] also gave another method for the study of these problems. This method, which essentially consists of transforming the entire space into a spherical hypersurface by a change of variables, in a way which leads back to a problem of the type of those considered in § 48, brings a notable complement to the preceding results, because it is applicable under hypotheses profoundly different from those of theorems 20, I and II. Among the equations which can be studied thus is included, for example, $\Delta_2 u = f$, for which the preceding method fails since the hypothesis $c \leq -g^2$ (outside a bounded domain) is not satisfied. On the other hand, with this method and if $m = 2$ we could require u to be bounded at infinity and not, indeed, infinitesimal. It is then natural to ask for which equations DIRICHLET's problem in an unbounded domain with bounded boundary can be posed if we require u to be infinitesimal at infinity and for which equations, on the other hand, the problem turns out to be well posed by requiring u to be bounded. Agreeing to denote these two problems by problem I and II, we shall describe some of the results obtained on the argument by N. MEYERS and J. SERRIN [1], observing that for problem II another existence and uniqueness theorem was proved earlier by S. SIMODA and M. NAGUMO [1].

MEYERS and SERRIN consider an equation whose coefficients are of class $C^{(0,\lambda)}$ in every bounded portion of T and moreover assume that $c \leq 0$, $f(x) = O(|x|^{-2}\delta(|x|))$, $\delta(t) > 0$ being a DINI function, that is, a function such that $t^{-1}\delta(t)$ is integrable (not necessarily absolutely) on $[1, +\infty)$. Considering a hypersphere Γ of radius R containing ∂T, let u_R be the solution of DIRICHLET's problem in the domain $\Gamma - \complement T$, satisfying the boundary condition given on ∂T and vanishing on $\partial \Gamma$. If for example $T \in A^{(2,\lambda)}$ then theorem 35, IV ensures that the family of the u_R is compact in $C^{(2)}(D)$ for every bounded domain $D \subset T$. Because the functions in this family are also uniformly bounded, it follows that the family contains a sequence converging to a solution of problem II. In every case there therefore exists a solution of the problem bounded in T. The following theorems indicate in which cases this solution is also infinitesimal at infinity and give an answer to the question of uniqueness.

49, I. *Under the hypotheses indicated and if, moreover, there exists an* $\varepsilon > 0$ *such that:*

$$\mathfrak{A}^* = \frac{|x|^2 \sum\limits_{i=1}^{m} (a_{ii} + b_i x_i)}{\sum\limits_{i,k=1}^{m} a_{ik} x_i x_k} \geq 2 + \varepsilon, \tag{49.1}$$

then problem I admits one and only one solution. If instead there exists a DINI *function* $\varepsilon(t)$ *such that:*

$$\mathfrak{A}^* \leq 2 + \varepsilon(|x|),$$

then problem II admits one and only one solution.

We note that if $c(x) = O\left(x^{-2} \delta(|x|)\right)$, then we can also pose the problem of determining a solution of the equation for which $\lim\limits_{|x| \to \infty} u(x) = l$, with l arbitrary. In fact, $v(x) = u(x) - l$ should be the solution of a problem of the first type. We also have:

49, II. *If* $b_i = 0$, $\lim\limits_{|x| \to \infty} a_{ik}(x) = a_{ik}^0$ *and if rank* $\|a_{ik}^0\| \geq 3$, *then* (49.1) *is certainly satisfied.*

Moreover, MEYERS and SERRIN, generalizing previous results due to K. YOSIDA [1] and F. V. ATKINSON [1], also proved that:

49, III. *If* $0 \leq M \leq \mathfrak{A}^* \leq N, c \leq -g^2 < 0$, $f \in L^\infty(T)$, *then problem II admits one and only one solution, and this is unique not only in the class of bounded functions but also in the class of functions* $u(x)$ *such that:*

$$u = o\left(|x|^{\frac{1-N}{2}} e^{g|x|}\right) \quad \text{for} \quad |x| \to \infty.$$

Other results by J. SERRIN and H. F. WEINBERGER [1] concern the equation:

$$\sum_{i,k=1}^{m} \frac{\partial}{\partial x_i} \left(a_{ik} \frac{\partial u}{\partial x_k}\right) = 0. \tag{49.2}$$

Without pausing to specify the hypotheses on the a_{ik}, given that the problem could also be posed in the weak form, we can sum up in the following these authors' result:

49, IV. *If* (49.2) *is uniformly elliptic and if* $m \geq 3$ $(m = 2)$, *then problem I (problem II) admits one and only one solution.*

Obviously in the case $m \geq 3$, we can also prescribe that the solution converge at infinity to a given arbitrary value.

Another treatment of the exterior DIRICHLET problem in the case $m \geq 3$ for the general equation has been given by A. P. OSKOLKOV [1, 3, 4] based on a priori bounds of the SCHAUDER-CACCIOPPOLI type. This treatment, by requiring more restrictive hypotheses on the coefficients than that of MEYERS and SERRIN, has the advantage also of allowing the formulation of alternative theorems. Both the results of MEYERS and SERRIN and those of OSKOLKOV are extended to NEUMANN's problem.

Let us now point out some results regarding the behavior at infinity of a solution of the equation:

$$\sum_{i,k=1}^{m} a_{ik} p_{ik} + \sum_{i=1}^{m} b_i p_i = 0 , \qquad (49.3)$$

under the hypothesis that it is uniformly elliptic and that moreover $b_i(x) = O(|x|^{-1})$ for $|x| > R$. A propos of this D. GILBARG and J. SERRIN [1] proved the following theorem:

49, V. *Under the hypotheses indicated and moreover if in the case $m \geq 3$ the coefficients a_{ik} are convergent at infinity, then every bounded solution $u(x)$ of (49.3) is convergent at infinity. Denoting by u_∞ its limit and assuming $m \geq 3$, we have:*

$$u(x) - u_\infty = O(x^{2-m+\delta}) , \quad \forall \delta > 0 . \qquad (49.4)$$

Bearing in mind that every solution of (49.3) is devoid of positive maxima and of negative minima, the following extension of LIOUVILLE's theorem follows obviously from this:

49, VI. *Under the hypotheses indicated above, every bounded solution of (49.3) which is defined in all space is constant.*

J. SERRIN [1] for the case $m = 2$ and D. GILBARG and J. SERRIN [1] for the case $m \geq 3$ have proved that[1]:

49, VII. *Theorems 49, V and VI hold even if $u(x)$ is assumed only to be bounded on one side, provided that the a_{ik} are uniformly HÖLDER continuous Under this hypothesis, (49.4) holds even for $\delta = 0$.*

Now in the case that (49.3), being self-adjoint, assumes the form (49.2), the preceding theorems have been stated precisely and strengthened by R. FINN and D. GILBARG [2] and J. SERRIN and H.F. WEINBERGER [1]. We add also that for nonuniformly elliptic equations, theorem 49, VI is not true in general (see E. HOPF [2]). We have, however, that:

49, VIII. *If $m = 2$ and $b_i = 0$, then every bounded solution of (49.3) defined in the entire plane is constant, assuming (49.3) is only elliptic.*

---

[1] See also E. HOPF [5].

This can be established by using a geometric theorem due to S. BERN-STEIN [7] according to which[1]:

49, IX. *A function u* (x_1, x_2) *defined and of class* $C^{(2)}$ *in the entire plane and such that* $p_{11} p_{22} - p_{12}^2$ *is nonnegative and not identically zero can not be bounded.*

In fact, from (48.3) it follows that for a solution $u(x_1, x_2)$ of the equation considered in theorem 49, VIII, certainly $p_{11} p_{22} - p_{12}^2 \geq 0$. Therefore if $u(x_1, x_2)$ is bounded, it follows from the preceding theorem that $p_{11} p_{22} - p_{12}^2 = 0$, which always implies for (48.4) that $p_{11} = p_{12} = 0$. Because analogously it can be shown that $p_{22} = 0$ also, it follows from this that $u(x_1, x_2)$ is linear and therefore, being bounded, is constant.

A theorem of nonexistence of solutions in the entire plane which is not including in the preceding was given by Z. NEHARI [2] for the equation $\Delta_2 u + c u = 0$ with $c < 0$. Finally, still for equation (49.3) with $b_i = 0$ and with appropriate hypotheses for the a_{ik}, N. MEYERS [3] studied a type of development in series valid for every bounded solution of the equation.

Let us now pass on to occupy ourselves with the equation $\Delta_2 u + k^2 u = 0$, which, though being very particular, has been the object of a great deal of research in consideration of its importance for mathematical physics. Much of this research, assuming k real and positive, has been dedicated to complex solutions which satisfy SOMMERFELD's "radiation condition" at infinity, sometimes written in the pointwise form:

$$\lim_{\varrho \to \infty} \varrho^{\frac{m-1}{2}} \left[\frac{\partial u}{\partial \varrho} - i k u \right] = 0, \qquad \varrho = |x|, \tag{49.5}$$

and sometimes in the integral form:

$$\lim_{R \to \infty} \int_{|x| = R} \left| \frac{\partial u}{\partial \varrho} - i k u \right|^2 d\sigma = 0. \tag{49.6}$$

The results obtained are in regard to either the representation of solutions of the equation which satisfy (49.5) or (49.6) or the proof of existence or uniqueness theorems for solutions which other than satisfying these conditions satisfy a DIRICHLET or NEUMANN type condition on ∂T. We cite in this regard in order of time the works by W. MAGNUS [1, 2], F. RELLICH [2], F.V. ATKINSON [1], H. WEYL [2, 3], C. MÜLLER [1], W.L. MIRANKER [1, 2], C.H. WILCOX [1, 2], T. KATO [1], H. REICHARDT [1], G. HELLWIG [4], O.G. OWENS [1], R. LEIS [1], P. WERNER [2],

---

[1] On the proof of BERNSTEIN's theorem see the clarifications by E. HOPF [6] and E. J. MICKLE [1].

O. I. Panič [2]. For expository reports see the volumes by B. D. Kupradze [3] and C. Müller [5][1]. In Kupradze's volume the case of
complex k is also treated, in which case (49.5) and (49.6) must be appropriately modified. Along the same lines are the works by G. Zemach and
F. Odeh [1], L. G. Mihaĭlov [6], D. M. Ĕĭdus [2] which, however, look
at more general equations. Also related to the preceding works is some
research by P. Hartman [4] and by P. Hartman and C. Wilcox [1],
in which the condition at infinity is replaced by the following:

$$\overline{\lim_{\varrho \to \infty}} \; \varrho^{-1} \int_{a \le |x| \le \varrho} |u(x)|^2 \, dx < \infty .$$

Let us now pass on to occupy ourselves with the case in which ∂T is
not bounded. Existence and uniqueness theorems for solution of the boundary value problems have been established in this case by L. Amerio [1],
M. Krzyżański [1, 4], B. Pini [3]. The last of these works is related to
the procedures of § 29 and in it the condition at infinity is satisfied in the
sense of convergence in the mean. The others, on the other hand, draw
their origin from some uniqueness theorems due to M. Picone [7,
pp. 695–696 and 701–704]. In these works also, which are all previous
to the note by Meyers and Serrin, the solution is constructed by solving
the problem in a bounded domain $D \subset T$ and then passing to the limit
as $D \to T$.

For the equation $\Delta_2 u + k^2 u = 0$, we can also consider in this case a
problem with a condition at infinity like the radiation condition. This
condition is, however, appropriately modified, and for this we refer to the
works by J. J. Stoker [2] and by A. S. Peters and J. J. Stoker [1]. For
the same equation A. Weinstein [2] has studied solutions defined in a
strip and almost periodic functions of the free variable there.

Finally, let us take up, both in the case under examination and in the
case in which ∂T is bounded, the problems that can be posed if we impose on u, as a condition at infinity, that u belongs to $L^p(T)$ or to $H^{1,p}(T)$
for some convenient p. This particular statement appears especially
convenient when the coefficients of the equation depend on a parameter
and we pose the problem of determining the values of the parameter
(eigenvalues) in correspondence with which the homogeneous equation
admits a non-zero solution satisfying a given boundary condition which
is likewise homogeneous. Unfortunately, as we have already said in the
preface, the study of problems of this type has not been allowed a place
in this second edition of our monograph.

Let us now pass on to occupy ourselves with research concerning
the behavior at infinity of the solutious of a homogeneous elliptic

---

[1] See also R. Courant and D. Hilbert [2, pp. 315–318].

equation defined in a domain T with unbounded boundary. This research is all directed to extending to solutions of elliptic equations
the known theorem due to PHRAGMÈN-LINDELÖF for holomorphic
functions. We recall that this theorem affirms that a function $f(z)$
holomorphic and bounded in a sector T and infinitesimal at infinity on
∂T is also infinitesimal at infinity in T. For solutions $u(x)$ of elliptic
equations, various theorems have been given, assuming as hypotheses
a) that the unbounded domain T has certain properties; b) that $u(x)$ has
on ∂T a certain behavior at infinity; c) that $u(x)$ belongs to a certain
functional class. In these theorems it is proved as a consequence that
$u(x)$ has a certain behavior at infinity in T. The bibliography on this
argument is quite extensive. Here we shall limit ourselves to mentioning
the works by A. FRIEDMAN [2], E. M. LANDIS [2, 3], A. A. NOVRUZOV [1, 2],
M. L. GERVER [1], K. HABETHA [1], L. LITHNER [1], P. FIFE [3], referring
to the last of these for more ample references to other important works.

We shall conclude this section by mentioning some results relating to
nonlinear equations. A first group of works looks at the problem of
existence of solutions of a nonlinear equation defined in all space. Now it
is evident if we consider the quasi-linear equation (40.2) with $f = 0$,
that theorems 49, VI and VIII allow us to state theorems of LIOUVILLE
type even for nonlinear equations. An analogous result for the case
$f \neq 0$ was established by E. BOHN and L. K. JACKSON [1]. The most
interesting results, however, are in regard to the equation in two variables
of the type:

$$\sum_{i,k=1}^{2} a_{ik}(p_j)\, p_{ik} = 0\,. \tag{49.7}$$

Indeed, at the end of 1927, S. BERNSTEIN [7], making use of the geo-
metrical theorem 49, IX, proved that for the equation of minimal surfaces every solution defined in the whole plane is necessarily a linear
function. Other proofs of this theorem were given in the same year by
T. RADÒ [2] and in more recent times by L. BERS [5] and E. HEINZ [1].
Larger classes of equations to which BERNSTEIN's theorem is extended
have been characterized by L. BERS [13] and R. S. FINN [3]. On the
other hand, JOACHIM and JOHANNES NITSCHE [5] have pointed out a class
of equations of the type (49.7) which admit nonlinear solutions defined
in the whole plane. Finally, we point out that BERNSTEIN's theorem for
minimal surfaces has been extended to the case $m = 3$ by E. DE GIORGI [4]
and to the case $m = 4$ by F. I. ALMGREN [1]. For the minimal surface
equation in parametric form, the following theorem proved by R. OSSER-
MAN [2] following a conjecture due to L. NIRENBERG holds: *Every
complete minimal surface whose spherical image excludes the neighborhood
of a point is a plane.*

The nonexistence of entire solutions for the equation $\Delta_2 u = f(u)$ has been proved, under appropriate hypotheses for $f(u)$, by H. WITTICH [1] for $m = 2$, by E. K. HAVILAND [1] for $m = 3$, and by W. WALTER [1] in the general case[1]. We note that the hypotheses which are made on $f(u)$ in this theorem are such that the theorems can be applied to the equation $\Delta_2 u = e^u$. A theorem of nonexistence of solutions defined in the whole plane of the inequality $\Delta_2 u \geq f(u)$ has been proved by R. OSSERMAN [1].

Regarding boundary value problems for nonlinear uniformly elliptic equations, we shall mention a note by P. BESALA [1] in which the results of M. KRZYŻAŃSKI [1, 4] are extended to a nonlinear equation, and two works, one by T. KUSANO [2] and the other by K. AKÔ and T. KUSANO [1], in which they give conditions for the existence of a bounded solution of a semilinear or quasi-linear equation defined in the whole plane.

Then also of great interest are some results regarding the existence and uniqueness of a subsonic flow past a given profile. From the analytic point of view, the problem is the following. In an unbounded domain $T \in A^{(1, \lambda)}$, such that ∂T is bounded, we consider an equation of the type:

$$\frac{\partial}{\partial x_1}(\varrho \, p_1) + \frac{\partial}{\partial x_2}(\varrho \, p_2) = 0 , \qquad (49.8)$$

ϱ being a positive function of $|p|$, and we seek a solution $u(x)$ satisfying the conditions:

$$\frac{du}{dn} = 0 \quad \text{for} \quad x \in \partial T, \quad \lim_{|x| \to \infty} p_1 - i \, p_2 = w_\infty, \quad \int_{\partial T} p_1 \, dx_1 + p_2 \, dx_2 = \Gamma ,$$
$$(49.9)$$

w_∞ and Γ being given constants. u can even be multiple-valued but must have single-valued derivatives. u must moreover be an elliptic solution of (49.8). Elliptic solutions of (49.8), which is not in general an elliptic equation, are called *subsonic*. It is interesting also to consider the case in which ∂T presents at a point z (the trailing edge) a protruding corner of aperture $\varepsilon \, \pi$, with $0 < \varepsilon \leq 1$. In this case, instead of (49.9) we impose the KUTTA-JUKOWSKY condition which as $x \to z$ is:

$$|p| = \begin{cases} o(1) & \text{for} \quad \varepsilon = 1 \\ O(1) & \text{for} \quad 0 < \varepsilon < 1 . \end{cases}$$

This last problem was studied for the first time by F. I. FRANKL' and M. KELDYŠ [1] in the case where w_∞ is sufficiently small. Later the study of this problem was carried on by L. BERS [1, 6] in the particular case in

---

[1] See also E. HOPF [9].

which (49.8) reduces to the equation of minimal surfaces [1] and then in the general case [12]. BERS's theorem does not require any boundedness hypothesis for w_∞ if (49.8) is uniformly elliptic, while in the general case a q is determined such that the existence theorem is valid for $|w_\infty| < q$. The existence theorem under a hypothesis of smoothness for ∂T and with the condition (49.9) was proved by M. SHIFFMAN [2]. For the uniqueness theorem see also R. FINN and D. GILBARG [1]. These authors have also studied [2] an analogous problem in the three dimensional case. A propos of this see also some observations by L. E. PAYNE and H. F. WEINBERGER [2].

50. Other problems for second order equations. In this section we shall occupy ourselves with various boundary value problems different from the DIRICHLET, NEUMANN. and oblique derivative problems. Ous treatment will above all be turned over to the study of mixed problemt and of problems with boundary conditions of abstract type; however, we shall not fail also to point out briefly problems with higher order conditions, transmission problems, and so-called ambiguous problems.

According to the definition posed in § 4, we call mixed problems those boundary value problems in which the values of the unknown function u are given on a portion S_1 of ∂T and those of an operator $\mathfrak{P}u$ on the remaining part S_2, the latter defined by assuming $l \equiv \nu$. When S_1 and S_2 do not have points in common, the study of these problems offers no particular difficulties. Indeed, by representing u with the sum of a domain potential, of a double-layer potential over S_1, and of a single-layer potential over S_2, G. GIRAUD [7, 8] succeeded in giving a treatment of this problem similar to that developed for the other boundary value problems in Chapter III. On the other hand, the same problem when S_1 and S_2 have points in common is considerably more difficult. A first treatment of the problem under these more general conditions is due to GIRAUD [14] in which he proposes to seek the solutions which are continuous in all of T. GIRAUD's procedure, which we can not describe, consists of transforming the mixed problem into a NEUMANN problem considered on a certain riemannian manifold $\mathfrak{V}$ and then applying the usual methods (see § 48). For the validity of this procedure he requires, however, that at the common points of S_1 and S_2 the normal to S_1 and the conormal to S_2 be orthogonal; moreover, at the same points the given values of u and $\mathfrak{P}u$ must also satisfy connecting conditions.

Another procedure which reveals itself to be quite useful in the study of mixed problems is PICONE's method of transforming boundary value problems into systems of FISCHER-RIESZ equations.

---

[1] For another exterior problem related to the equation of minimal surfaces see CHEN YU-WHY [2].

The details of this transformation were in fact stated precisely by L. Amerio [6] with a method analogous to that which he followed in the study of the Dirichlet and Neumann problems (§ 31). Later G. Fichera [7] proved, through the study of the aforesaid Fischer-Riesz equations, written by making use of an appropriate fundamental solution, that we can definitely succeed in proving the existence theorem, at least in the case in which the equation is self-adjoint and $c < 0$, $\beta = 0$. And this theorem holds without any restrictive hypotheses about the behavior of ∂T or the boundary data at the points of $S_1 \cap S_2$. However, the solution for which Fichera proved the existence is not continuous in T, but is in class $\Gamma^{(2)}$ (in the sense of § 31) and satisfies the boundary condition on S_2 only almost everywhere. In the same class this solution is now unique, as can for example be proved by means of theorem 31, IX. We note that in the larger class of functions continuous in $T - S_2$ and endowed with first derivatives which are square integrable in T, the uniqueness theorem is no longer valid; this was proved by E. de Giorgi [1] with an example. G. Fichera [13, 31] also proved that the solution can not turn out to be in class $C^{(1)}(T)$, unless the boundary data satisfy certain conditions of a quantitative character.

Another treatment of the mixed problem was given by G. Stampacchia [4, 5] making use of the direct methods of the calculus of variations. The interest of these works lies above all in the fact that the results obtained also concern nonlinear equations. In the case of a self-adjoint linear equation, he arrives at an existence theorem by proving, under appropriate hypotheses on the data, that the functional:

$$\mathscr{I}(u) = \frac{1}{2} \int_T \left(\sum_{i,k=1}^{m} a_{ik}\, p_i\, p_k - c u^2 + f u \right) dx + \int_{S_2} u\, \psi\, d\sigma$$

admits a minimum in the class of functions $u \in H^{1,2}(T)$ which satisfy the condition $\gamma u = \varphi$ on S_1. The minimizing function is unique and satisfies the equation $\mathfrak{M}u = f$ and the condition $du/dv = \psi$ on S_2. We note that the solution of the mixed problem whose existence is thus proved is unique under the condition that u renders $\mathscr{I}(u)$ a minimum. We thus have a uniqueness theorem different from Fichera's. The solution whose existence Stampacchia proved is thus u class $H^{1,2}(T)$, as is Fichera's. As much can be said of the solution found by M. I. Višik [7] in the study of a mixed problem, in which the condition on S_2 is of an oblique derivative type, with the methods of § 30. On the other hand, E. Magenes [1, 2, 4, 5], using methods analogous to those of § 29, studied the problem posed under conditions on the data for which the preceding methods can not be applied, since the solution is not in class $H^{1,2}(T)$. Also, the solution found by Magenes is, naturally, continuous in $T - \partial T$, but not in T.

But the existence of a solution continuous in T was proved by G. FI-
CHERA [16] by making use of a procedure based in an essential way on
theorem 30, I. Another proof of the existence of a continuous — even
HÖLDER continuous — solution in T was given by C. MIRANDA [9] based
on some a priori bounds of SCHAUDER-CACCIOPPOLI type. MIRANDA's
hypotheses are less restrictive than FICHERA's regarding the known term
and the coefficients b_i and c, which can even grow infinite on $S_1 \cap S_2$.
On the other hand, regarding the domain T, MIRANDA's hypotheses are
more restrictive because he requires that along $S_1 \cap S_2$ the tangent planes
to S_1 and S_2 form an angle different from 0 or π, a hypothesis which is
not required in FICHERA's treatment. The existence of an α such that the
solution is α-Hölder continuous, under more general hypotheses for T
than both MIRANDA's and FICHERA's, was then proved by G. STAM-
PACCHIA [9] with a procedure analogous to that which allows the proof
of theorem 30, IV. For a deeper comparative analysis of the results obtained, see the lectures by E. MAGENES [6] and C. MIRANDA [11]. For
some mixed problems for equations with constant coefficients see
A. GHIZZETTI [4, 5].

We pass on now to point out other works regarding problems with
boundary conditions of order greater than one. For equations in two
variables the first results were obtained with the method of transformation
into integral equations by I. N. VEKUA [3], and to this the later works by
Z. I. HALILOV [1] A. DŽUAREV [1, 2, 4], R. A. KORDZADZE [1] are more
or less related. For $m \geq 2$ analogous research and other research regarding problems with boundary conditions of integro-differential character
have been attacked, with the methods of functional analysis, within the
purview of the theory of equations of higher order and of systems of
equations. For this reason we shall speak of this in §§ 52 D, 53, and 55 C,
even if the results obtained can be applied in particular to second order
equations. However, we can for now mention in anticipation that in this
research, among other things, a condition of algebraic character is given
for the a_{ik}'s and for the coefficients of the boundary condition, under
which the problem can be considered well posed in a certain sense. This
result is valid even for problems with a first order boundary condition,
and in the non-regular oblique derivative problems makes it possible to
recognize that these problems are to be considered well posed only in the
case $m = 2$ (see also § 23). It is also an interesting fact that this research
even takes second order equations with complex coefficients into consideration, which hitherto have been systematically excluded from our
treatise.

Let us now pass on to boundary value problems posed in an abstract
way. For this purpose we shall first occupy ourselves with a particular
procedure which draws its origin from a work by J. L. LIONS [1] which

was then developed by E. MAGENES and G. STAMPACCHIA [1] and by LIONS himself [6, 7]. We recall that in § 30 we saw how DIRICHLET's problem for equation (30.9) with the condition (30.2) could be formulated, in the weak form, as seeking a function $u \in H_0^{1,2}(T)$ which satisfies (30.8) for every $v \in H_0^{1,2}(T)$. Analogously, NEUMANN's problem in the weak form becomes that of seeking a function $u \in H^{1,2}(T)$ which satisfies (30.8) for every $v \in H^{1,2}(T)$. More generally, if V is a HILBERT space such that algebraically and topologically we have $H_0^{1,2}(T) \subseteq V \subseteq H^{1,2}(T)$, then seeking a function:

$$u \in V \tag{50.1}$$

satisfying (30.8) for arbitrary $v \in V$ can be interpreted as a weak formulation of a boundary value problem for (30.9). For example, if V is the subspace of $H^{1,2}(T)$ consisting of the functions u which have a vanishing trace on a portion S_1 of ∂T, then the boundary value problem for (50.1) is a weak form of the mixed problem which is obtained by associating to (30.9) the boundary conditions:

$$u(x) = 0 \quad \text{for} \quad x \in S_1 ,$$

$$\sum_{i=1}^{m} X_i \left[\sum_{k=1}^{m} a_{ik} \frac{\partial u}{\partial x_k} + d_i u + f_i \right] = 0 \quad \text{for} \quad x \in S_2 = \partial T - S_1 .$$

If V is such that a constant K exists for which we have:

$$| a(u, u) | \geq K \| u \|_V^2 \quad \forall u \in V ,$$

then we say that the problem being considered is V-elliptic, and it is easy to prove that every V-elliptic problem admits one and only one solution. Now if the problem is not V-elliptic then we can always prove by departing from (30.8) written for $v = u$ that the bilinear form $a(u, v)$ is *coercive* on V in the sense of ARONSZAJN [3], i.e., that there exist K and λ_0 such that we have for $\lambda > \lambda_0$:

$$| a(u, u) | + \lambda(u, u) \geq K \| u \|_V^2, \quad \forall u \in V ,$$

where (u, v) is the scalar product in L^2. From this it is easy to prove that FREDHOLM's alternative holds for the problem being considered.

Once the existence of a weak solution of the problem has been proved, the question of its regularization is posed; for this we refer to E. MAGENES and G. STAMPACCHIA [1] and to G. STAMPACCHIA [7, 9, 10, 14], limiting ourselves to mentioning that in the works cited, properties analogous to those expressed by theorem 30. IV are proved. Naturally this requires some hypotheses on V.

Another way of defining abstract boundary value problems is due to
M. I. Višik [5, 6][1]. He considered an elliptic operator $\mathfrak{M}$ and assumed
that $a_{ik} \in C^{(3,\lambda)}$, $b_i \in C^{(2,\lambda)}$, $c \in C^{(1,\lambda)}$ and $c(x) \leq 0$. We have already
said (§ 30) that under this hypothesis the strong and weak extensions of
$\mathfrak{M}$ coincide; their set of definition $\mathscr{H}$, considered as a subset of L^2, is called
the *maximal domain of definition* of the operator $\mathfrak{M}$ in L^2. Now if V is a
linear manifold of $C^{(2)}(T)$ and we denote by $\mathscr{H}_V$ the closure of V with
respect to the norm (30.20), then the mapping $f = \mathfrak{M}u$ of V into $C^{(0)}(T)$
is extended into a mapping of $\mathscr{H}_V$ into $L^2(T)$. In doing this we always
assume that $C_0^{(2)}(T) \subseteq V$, and the set $\mathscr{H}_0 = \mathscr{H}_{C_0^2(T)}$ is called *the minimal domain of definition* of the operator $\mathfrak{M}$ in L^2. Obviously we have
$\mathscr{H}_0 \subseteq \mathscr{H}_V \subseteq \mathscr{H} \subseteq L^2(T)$. The operator resulting from the extension of $\mathfrak{M}$
by V to $\mathscr{H}_V$ will be denoted by $\mathfrak{M}_V$, while $\mathfrak{M}_0$ will denote the extension
of $\mathfrak{M}$ by $C_0^{(2)}(T)$ to $\mathscr{H}_0$. We shall also say that $\mathfrak{M}_V$ is the *realization* of $\mathfrak{M}$
under the boundary condition (50.1).

Premising this and considering the abstract boundary value problem
consisting of seeking a solution of the equation $\mathfrak{M}u = f$ satisfying (50.1),
we can define every function $u \in \mathscr{H}_V$, such that $\mathfrak{M}_V u = f$, to be a generalized solution of this problem. In order to be able to illustrate at least
the design of Višik's research, let us recall some definitions from the
theory of operators in a Hilbert space[2].

Let H be a Hilbert space and A an operator defined in a set Ω_A of
H having as its range another set R_A of H. Let us denote by A^* the
adjoint operator of A, that is, an operator defined in a set Ω_{A^*} of H
such that we have $u \in \Omega_{A^*}$, $A^*u = f$ if and only if:

$$(A\,\varphi,\,u) = (\varphi,\,f) \qquad \forall\,\varphi \in \Omega_A .$$

The operator A is then called *closed* if for every sequence of elements
$u_n \in \Omega_A$ the relations $\lim_{n \to \infty} u_n = u$, $\lim_{n \to \infty} A\,u_n = f$ imply that $u \in \Omega_A$,
$A\,u = f$.

Assuming that A is closed, and given an arbitrary $f \in H$, we pose the
problem of seeing if f belongs to R_A, that is, whether there exists a solution of the equation:

$$A\,u = f, \qquad u \in \Omega_A . \tag{50.2}$$

A propos of this we pose the following definitions. The closed operator A
is called *normally solvable* if the necessary and sufficient condition for the
solvability of (50.2) is that f is orthogonal to every solution v of the
problem:

$$A^*v = 0, \qquad v \in \Omega_{A^*} . \tag{50.3}$$

---

[1] For some complementary results see also M. Š. Birman [1, 2].

[2] With some variations, analogous definitions can also be adopted in
the theory of operators in a Banach space.

Considering, besides (50.3), the equation:

$$A u = 0 , \quad u \in \Omega_A, \tag{50.4}$$

we say that A is an *index operator* if (50.3) and (50.4) each admit a finite number of linearly independent solutions, say ν for (50.3) and μ for (50.4). The difference $\varkappa = \mu - \nu$ is called the *index* of the operator. An operator A is then called *regularly solvable* if it is an index operator and moreover if this index is zero. Finally, A is called *solvable* if (50.2) admits one and only one solution $u = A^{-1}f$, *completely solvable* if, moreover, A^{-1} is completely continuous. These definitions, except that of an index operator, were adopted by M. I. Višik. Other authors call index operators normally solvable and those called by Višik normally solvable, operators with closed range. This because a necessary and sufficient condition that an operator A be normally solvable in the sense of Višik is that R_A be closed.

These definitions can be applied in the case that $H = L^2(T)$, $A = \mathfrak{M}_V$, and $\Omega_A = \mathscr{H}_V$, because $\mathfrak{M}_V$ considered as an operator defined on the subset $\mathscr{H}_V$ of $L^2(T)$ is obviously closed. In Višik's memoir [6], based on some general theorems from the theory of operators, he poses and solves the problem of characterizing which subspaces of L^2 are domains of definition of an extension $\mathfrak{M}_V$ of $\mathfrak{M}$ which is normally solvable or regularly solvable, or else solvable or completely solvable.

For this purpose he first of all proves that for every function $u \in \mathscr{H}$ it is possible to define a trace $\gamma(u)$ and a conormal derivative $du/d\nu$ on ∂T in a generalized sense. Since the hypothesis $c \leq 0$ then ensures that $\mathfrak{M}_0$ is completely solvable, we can then associate to each function $u \in \mathscr{H}$ a function $u_0 \in \mathscr{H}_0$ which is a solution of the equation $\mathfrak{M}_0 u_0 = \mathfrak{M}u$. Putting $\gamma'(u) = du_0/d\nu$ for every $u \in \mathscr{H}$, we can see that the inclusion of u in a subset $\mathscr{H}'$ of $\mathscr{H}$ is transformed into a linear relation between $\gamma(u)$ and $\gamma'(u)$, and the required characterization of $\mathscr{H}'$ can be expressed in terms of properties of this relation. When this linear relation has been determined for an abstract boundary condition $u \in V$, we say that the condition have been put *into canonical form*. For example, a necessary and sufficient condition that $\mathscr{H}'$ be the domain of definition of an extension of $\mathfrak{M}$ which is completely solvable is that the elements of $\mathscr{H}'$ are all the elements and only the elements of $\mathscr{H}$ which satisfy a relation of the type $\gamma(u) = \mathfrak{C}[\gamma'(u)]$ where $\mathfrak{C}$ is a completely continuous transformation of H' into H, H and H' being the appropriately normalized spaces described by $\gamma(u)$ and $\gamma'(u)$ as u varies in $\mathscr{H}$. In this way every abstractly defined boundary value problem can be interpreted in a concrete way, and vice versa: the study of a boundary value problem posed in an ordinary sense is simplified when we succeed in determining the canonical form of the boundary condition.

As an application of this method, Višik considered the boundary value problem which is obtained by associating to the equation $\mathfrak{M}u = f$ a boundary condition of the type:

$$\frac{du}{dn} = \mathfrak{R}u \quad \text{for} \quad x \in \partial T ,$$

where $\mathfrak{R}$ is a second order elliptic operator on ∂T, and he proved that this problem is regularly solvable. Then if $\mathfrak{M}$ is self-adjoint and $\mathfrak{R}$ is such that

$$\int_{\partial T} u \, \mathfrak{R}u \, d\sigma \leq -a^2 \int_{\partial T} u^2 d\sigma$$

then the problem is completely solvable. Another interesting question is the following: let $\mathfrak{R}_W$ be an extension of the operator $\mathfrak{R}$ adjoint to $\mathfrak{M}$ and let $\mathscr{K}_W$ be its domain of definition. If:

$$(u, \mathfrak{R}v) = (\mathfrak{M}u, v), \quad \forall u \in \mathscr{K}_V \text{ and } \forall v \in \mathscr{K}_W ,$$

we say that the two boundary value problems (50.2) and:

$$\mathfrak{R}_W v = g , \quad v \in \mathscr{K}_W \tag{50.5}$$

are formally adjoint, and we pose the question of seeing if the operator $\mathfrak{R}_W$ is also the adjoint of $\mathfrak{M}_V$ in the sense of the theory of operators. Along these lines Višik characterized the operators $\mathfrak{M}_V$ which are self-adjoint in the sense of the theory of operators.

Another way of defining abstract boundary conditions was recently proposed by F. E. Browder [19][1]. Let $\mathfrak{B}$ be a linear bounded mapping of $H^{2,2}(T)$ into a Hilbert space Γ such that the set $B = \{u \mid \mathfrak{B}u = 0\}$ is dense in $L^2(T)$. Moreover, let a constant K exist such that:

$$\| u \|_{H^{2,2}(T)} \leq K[\| u \|_{L^2(T)} + \| \mathfrak{M}u \|_{L^2(T)}], \quad \forall u \in B .$$

Under these hypotheses we say that $\mathfrak{B}$ is a *general elliptic boundary operator* and we can consider the abstract boundary value problem:

$$\mathfrak{M}u = f , \quad \mathfrak{B}u = \varphi ,$$

with $f \in L^2(T)$ and $\varphi \in \Gamma$. In his work Browder considers a problem of this type depending on a parameter τ, and he proves that certain properties, as for instance that of normal solvability in the sense of Višik, are preserved as τ varies when the problem depends continuously on τ. We shall conclude these references to problems with abstract boundary

---

[1] See also F. E. Browder [12].

conditions by mentioning that the first research along these lines is due to J. W. CALKIN [1].

Let us now pass on to say something about *transmission problems*, also called *diffraction problems* by some. An example typical of this genus of problems is the following. Let T be a bounded domain divided into two domains by a sufficiently regular hypersurface $S = T_1 \cap T_2$ and let $\mathfrak{M}$ be an elliptic operator whose coefficients are continuous in T_1 and T_2 separately but not in T. We then pose the problem of determining a solution of the equation $\mathfrak{M}u = f$ which satisfies, for example, a condition of DIRICHLET or NEUMANN type on ∂T, which is continuous in T and regular in $T - (\partial T \cup S)$ and which on S satisfies the further condition:

$$\sum_{i,k=1}^{m} \Delta(a_{ik}) X_i p_k = 0, \tag{50.6}$$

where by $\Delta(a_{ik})$ is denoted the jump of the a_{ik} in the passage across S, and X_i are as usual the direction cosines of the normal to S. Problems of this sort are of interest in questions of mathematical physics and in fact the first notification of a problem of this type was given by M. PICONE [13] in relation to a question from elasticity theory. Immediately after PICONE's notice, the problem we specified above was solved in a weak form by J. L. LIONS [3] and, with the methods of the calculus of variations and later regularization, by G. STAMPACCHIA [6]. Related to the preceding works, S. CAMPANATO [6] then treated a problem of the same type in which, however, (50.6) is replaced by a more general condition.

Independently of these works, transmission problems have also been the object of considerable research on the part of Russian mathematicians. We cite in order of time the works by O. A. OLEJNIK [5, 6] and by I. V. GIRSANOV [1], in which the solution of the problem is obtained as the limit of the corresponding solution for an equation with continuous coefficients, by V. A. IL'IN and I. A. SISMAREV [1] and by V. A. IL'IN [1], in which they use the method of integral equations, and by I. A. SISMAREV [1], in which he has recourse to a priori bounds of SCHAUDER-CACCIOPPOLI type. Finally, in the works by JA. A. ROĬTBERG and Z. G. SEFTEL' [1] and by O. A. LADYŽENSKAJA, V. JA. RIVKLIND and N. N. URAL'CEVA [1] the problem is studied with procedures analogous to those of LIONS and STAMPACCHIA, namely by first proving the existence of weak solutions and then passing on to a deepened regularization. An ample treatment of the problem along these lines is also contained in § 16 of Chapter III of the volume [8] by O. A. LADYŽENSKAJA and N. N. URAL'CEVA. We cite also a note by G. M. GASUMOV [1] relating to a particular equation and three other notes, one by GASUMOV himself [2] and two by M. N. BORSUK [1, 2] relating to transmission problems for nonlinear equations.

We shall terminate this section by taking into consideration a class of problems which we shall call *ambiguous* and which have some affinity with mixed problems. These problems are posed in the following way: We now impose on u that it satisfy the following boundary condition, which we assume to be homogeneous, for simplicity:

$$u = 0 \quad \text{for} \quad x \in S_1, \qquad \frac{du}{dv} = 0 \quad \text{for} \quad x \in S_2, \qquad (50.7)$$

but without assigning a priori the partition $\partial T = S_1 \cup S_2$. Leaving this partition among the unknowns of the problem, we impose instead that on ∂T:

$$u \geq 0, \qquad \frac{du}{dv} \geq 0. \qquad (50.8)$$

A problem of this type, for a self-adjoint equation with $c = 0$, was considered for the first time by G. FICHERA [30], in connection with an analogous problem from elasticity theory posed by SIGNORINI (see § 55G). FICHERA's treatment, developed with the calculus of variations, leads to an existence theorem for weak solutions if we assume that:

$$\int_T f dx > 0.$$

This condition becomes superfluous if we require that S_1 contain at least a given portion S_0 of ∂T, to which it is possible to associate a constant K such that for every function $u \in C^\infty (T)$ vanishing on S_0 we have:

$$\| u \|_{L^2(T)} \leq K_0 \Big[\sum_{|\alpha|=1} \| D^\alpha u \|_{L^2(T)} \Big]^{1/2}.$$

Stated in this way, the problem was studied and solved by J.L. LIONS and G. STAMPACCHIA [1, 2] even for non self-adjoint equations[1] and with procedures of functional analysis which can also be applied to other problems of the same type. It is interesting to observe that, if K denotes the closed and convex subset of $H^{1,2}(T)$ consisting of the functions which are zero on S_0 and non negative on $\partial T - S_0$, we define as a weak solution of the problem each function $u \in K$ which satisfies the equation:

$$\int_T \sum_{i,j=1}^m a_{ij} \frac{\partial u}{\partial x_i} \frac{\partial (v-u)}{\partial x_j} dx + \int_T f(v-u) dx \geq 0, \qquad \forall v \in K. \qquad (50.9)$$

---

[1] In fact, LIONS and STAMPACCHIA drop the hypothesis that $a_{ji} = a_{ij}$.

This definition is justified as usual by the fact that each sufficiently regular solution u of (50.9) satisfies the equation $\mathfrak{M}u = f$ and the boundary conditions:

$$u(x) = 0 \quad \text{for} \quad x \in S_0 ,$$

$$\left. u \geq 0 , \qquad \frac{du}{dv} \geq 0 , \qquad u \frac{du}{dv} = 0 \quad \text{for} \quad x \in \partial T - S_0 . \right\} \quad (50.10)$$

And it is obvious that if we put $S_1 = \{x \in \partial T : u(x) = 0\}$, $S_2 = \partial T - S_1$, then (50.10) implies (50.7) and (50.8).

The consideration of ambiguous problems thus leads naturally to the study of functional inequalities, which latter problem we have already pointed out at the end of § 41 and whose importance now comes to be better clarified by this example.

51. Inverse problems and axiomatic theory for second order equations. Generically, we shall call all those problems in which, given a priori a family of functions, we seek an elliptic equation which admits all the functions of the family as solutions, *inverse problems* of the theory of elliptic equations. A first result is due to F. JOHN [1] who showed that if all the functions in the family are of class $C^{(2)}$ and devoid of proper maxima and minima, and moreover if the family is a linear space, then there exists an operator $\mathfrak{M}$ lacking the term in u such that all the functions in the family are all solutions of $\mathfrak{M}u = 0$. Almost contemporaneously, N. AISENSTAT [1], along slightly different lines, proved that if $f = \mathfrak{C}(u)$ is a linear functional transformation of $C^{(2)}(T)$ into $C^{(0)}(T)$, such that f is nonpositive at every point at which u has a nonnegative maximum, then the given transformation is necessarily of the type $f = \mathfrak{M}u$, $\mathfrak{M}$ being an elliptic operator whose coefficient c is nonpositive.

Later G. TAUTZ [4, 5] established a system of necessary and sufficient conditions, given a linear functional transformation $u = \mathfrak{C}(\varphi)$ which causes a function $u \in C^{(0)}(T)$ which equals φ on ∂T to correspond to every function $\varphi \in C^{(0)}(T)$, so that an elliptic operator $\mathfrak{M}$ exists such that for every φ we have $\mathfrak{M}(\mathfrak{C}(\varphi)) = 0$. A result of the same type was also obtained by K. YOSIDA [3]. In another work by G. TAUTZ [6], the same question is treated by also taking nonlinear elliptic operators into consideration.

Axiomatic theory, which has inspired some research in recent years, is put from a different point of view. This is concerned with studying families of functions satisfying a system of axioms which are satisfied when the family reduces to the family of solutions of an elliptic equation, and with deducing uniquely from these axioms the properties of the functions in the family. The functions which are considered are for the most part defined in sets of a locally compact topological space, and the questions which are most deeply studied are those relating to the generalization of PERRON's method, to the regularity of boundary points in the sense of WIENER, and to the MARTIN boundary.

A first attempt along these lines is due to L. JACKSON [1], but the first important results were established by J. L. DOOB [1, 2, 3] and a little later by M. BRELOT [15, 16, 18]. DOOB's research is inspired by the idea of making use of the methods of the calculus of probability in the study of these questions; the results obtained are also applied to the treatment of parabolic equations. On the other hand, BRELOT's research is more similar to the classical foundation of the theory of elliptical equations and is related to previous works by the same author in the compass of the theory of harmonic functions. For information on this previous research see BRELOT [14, 17]. BRELOT's research was carried on by R. M. HERVÉ [1] and by K. GOWRISANKARAN [1], DOOB's by H. BAUER [1, 2]. BAUER's research, which like DOOB's is also applicable to parabolic equations, can be considered an attempt to synthesize DOOB's and BRELOT's ideas into one single theory. A variant of BAUER's theory was proposed by M. BRELOT in a work [19] which also contains a comparative examination of all the preceding theories.

Other contributions to the study of the question are contained in volume XV of the Annales de L'Institut Fourier, in which are contained the lectures held at the Colloque International sur la Théorie du Potentiel (Paris-Orsay, 1964).

52. Equations of higher order. The theory of elliptic equations of higher order has made enormous progress in the last fifteen or twenty years, so much so that some parts have reached an almost definitive systematization, which has allowed it to be given an exposition in treatise form. And a propos of this it will be necessary to mention some chapters from the volumes by J. L. LIONS [7], L. HÖRMANDER [2], J. L. LIONS and E. MAGENES [5], Part II of the volume *Partial Differential Equations* by L. BERS and M. SCHECHTER [1], the books by L. SCHWARTZ [2], S. AGMON [7], G. FICHERA [33], and the monographs by G. GEYMONAT and P. GRISVARD [1] and J. PEETRE [5]. In order to give a general idea of the development of the theory, the proceedings from various conferences, reports, and lectures, of which from time to time the results obtained have been the subject, can also be useful. We cite in order of time L. NIRENBERG [5], M. I. VIŠIK and O. A. LADYŽENSKAJA [1], J. L. LIONS [6], L. GÅRDING [6], E. MAGENES and G. STAMPACCHIA [1], P. C. ROSENBLOOM [1], L. NIRENBERG [7], G. STAMPACCHIA [8], E. MAGENES [9]. It is also to be observed that the research on elliptic equations of higher order has been developed parallel to that relating to elliptic systems, on which we refer to the following sections, and to that concerning the general theory of partial differential equations, for which we refer to the volume [2] by L. Hörmander.

We shall now enter into summary, observing that for greater clarity we shall divide the exposition into various parts, each with its subtitle marked by a letter.

A. Elliptic operators of higher order

A linear operator with partial derivatives of order $2r$ and with real or complex coefficients:

$$\mathfrak{M}u = \sum_{|\alpha|=0}^{2r} a_\alpha(x)\, D^\alpha u \qquad (52.1)$$

is called *elliptic* in a domain T if, putting $\xi^\alpha = \xi_1^{\alpha_1} \ldots \xi_m^{\alpha_m}$, the condition:

$$M(x,\xi) = \sum_{|\alpha|=2r} a_\alpha(x)\, \xi^\alpha \neq 0 \qquad (52.2)$$

is satisfied for arbitrary real non-vanishing ξ and $x \in T$. The polynomial $M(x,\xi)$ is called the *characteristic form* of the operator $\mathfrak{M}$. (52.2) is in particular satisfied if there exists a function $\gamma(x)$ such that we have for real, nonzero ξ:

$$\mathscr{R}e\,[\gamma(x)\,M(x,\xi)] > 0 \qquad (52.3)$$

in which case the operator $\mathfrak{M}$ is called *strongly elliptic*. This distinction has an obvious reason to exist only for operators with complex coefficients. The operator $\mathfrak{M}$ is then called *uniformly elliptic* (*strongly elliptic*) in T if there exists a constant $a > 0$ such that we have for every $x \in T$:

$$|M(x,\xi)| \geq a\,|\xi|^{2r} \qquad (\mathscr{R}e\,[\gamma(x)\,M(x,\xi)] \geq a\,|\xi|^{2r}). \qquad (52.4)$$

We shall now consider the equation of degree $2r$ in the unknown z:

$$M(x,\xi + z\eta) = 0, \qquad (52.5)$$

in which ξ and η are two real orthogonal unit vectors. If for arbitrary x, ξ, η satisfying the indicated condition, (52.5) always has an equal number of roots with positive imaginary parts $\tau_k^+(x,\xi,\eta)$ and with negative imaginary parts $\tau_k^-(x,\xi,\eta)$, we say that $\mathfrak{M}$ is *properly elliptic* and by $M^+(x,\xi,\eta,z)$ and $M^-(x,\xi,\eta,z)$ we denote the polynomials in z having the τ_k^+ and the τ_k^- as roots, respectively. On the other hand, it is immediately seen that *if $m \geq 3$, then every elliptic operator is also properly elliptic*. In fact, if x and ξ are fixed, and we denote by $p(\eta)$ and $n(\eta)$ the number of roots of (52.5) which have positive real part and negative real part, respectively, then we have $p(-\eta) = n(\eta)$. In addition, because (52.5) does not admit real roots by hypothesis, the functions p and n

are constant as η varies in every connected set. Consequently, since the set $\{\eta \mid \xi \cdot \eta = 0, \mid \eta \mid = 1\}$ is connected for $m \geq 3$, we have in this case $p(\eta) = p(-\eta)$, and therefore $p(\eta) = n(\eta)$. In the case $m = 2$, on the other hand, an operator can be elliptic without being properly elliptic. However, we do have that *every strongly elliptic operator is properly elliptic*. It will suffice to prove this for $m = 2$. In this case, for every ξ there are only two vectors η and $\eta' = -\eta$ which satisfy the conditions $\xi \cdot \eta = 0$, $\mid \eta \mid = 1$. It will therefore suffice to show that the two equations:

$$M(x, \xi + z\,\eta) = 0, \quad M(x, \xi - z\,\eta) = 0$$

have an equal number of roots with positive imaginary part. And this will be proved if we can show that for $\tau \in [0, 1]$ the number of roots with positive imaginary part of the equation:

$$\gamma(x) \left[\tau M(x, \xi + z\,\eta) + (1 - \tau)\, M(x, \xi - z\,\eta)\right] = 0 \tag{52.6}$$

remains constant. But this is evident, otherwise at least for one value of τ (52.6) would have to have a real root, which is impossible for real z the left side has, by hypothesis, a positive real part. Analogously, as done for $r = 1$, if the coefficients of $\mathfrak{M}$ are sufficiently regular, then we can define an *adjoint operator*:

$$\mathfrak{N}v = \sum_{|\alpha|=0}^{2r} (-1)^{\alpha}\, D^{\alpha} \left(\bar{a}_{\alpha}\, v\right),$$

where $\bar{a}_{\alpha}$ is the function conjugate to a_{α}. We also note that, always under hypotheses of sufficient regularity on the coefficients, the operator $\mathfrak{M}$ can be put by various means in the form:

$$\mathfrak{M}u = \sum_{|\beta|,|\gamma|=0}^{r} D^{\beta} \left(a_{\beta\gamma}\, D^{\gamma}\, u\right), \tag{52.7}$$

in which case the adjoint operator is written:

$$\mathfrak{N}v = \sum_{|\beta|,|\gamma|=0}^{r} (-1)^{|\beta|+|\gamma|}\, D^{\gamma} \left(\bar{a}_{\beta\gamma}\, D^{\beta}\, v\right). \tag{52.8}$$

B. Green's formula. Fundamental solutions. Cauchy's problem and the unique continuation property

As in the case $r = 1$, a Green's formula:

$$\int_{T} \left(\bar{v}\, \mathfrak{M}u - u\, \overline{\mathfrak{N}v}\right) dx = \int_{\partial T} \mathfrak{L}(u, \bar{v})\, d\sigma \tag{52.9}$$

holds, wheie $\mathfrak{L}(u, \bar{v})$ is a bilinear form in u and $\bar{v}$ and their derivatives of order less than $2r$. In order to be able to make (52.9) precise in the sense of N. Aronszajn and A.N. Milgram [1] and M. Schechter [5], we introduce the notion of a *normal system of boundary operators*. Given k linear differential operators:

$$\mathfrak{B}_j(u) = \sum_{|\alpha|=0}^{r_j} b_{j\alpha}(x)\, D^\alpha u, \qquad (j = 1, 2, \ldots, k),$$

with real or complex coefficients defined on ∂T, we shall say that these constitute a normal system if:

1) $r_i \neq r_j$ for $i \neq j$.

2) For a real, nonzero vector η orthogonal to ∂T at the point x we have:

$$B_j(\eta) = \sum_{|\alpha|=r_j} b_{j\alpha}(x)\, \eta^\alpha \neq 0.$$

Having premised this, if we assign an arbitrary normal system of $2r$ boundary operators $\mathfrak{B}_j$, all of order less than $2r$, another normal system of operators $\mathfrak{B}'_j$ of order $2r - r_j - 1$ is determined, so that (52.9) can be written:

$$\int_T \left(\bar{v}\, \mathfrak{M}u - u\, \overline{\mathfrak{N}v} \right) dx = \int_{\partial T} \sum_{j=1}^{2r} \mathfrak{B}_j(u)\, \overline{\mathfrak{B}'_j(v)}\, d\sigma. \qquad (52.9')$$

It can also be shown that (52.9') holds if we arbitrarily fix the operators $\mathfrak{B}'_{j_1}, \ldots, \mathfrak{B}'_{j_k}$ on the condition that we modify appropriately all the $\mathfrak{B}_j$ of order $r_j > \inf r_{j_k}$. Thus in particular (52.9') can be written:

$$\int_T \left(\bar{v}\, \mathfrak{M}u - u\, \overline{\mathfrak{N}v} \right) d\sigma = \int_{\partial T} \sum_{j=0}^{r-1} \left[\frac{d^j u}{d n^j}\, \overline{\mathfrak{B}'_j(v)} - \frac{d^j \bar{v}}{d n^j}\, \mathfrak{B}_j(u) \right] d\sigma,$$

where the $\mathfrak{B}_j$ and $\mathfrak{B}'_j$ are operators of order $2r - j - 1$. It is then obvious that Green's formula can bear numerous consequences similar to those established in the case $r = 1$, when we know a *fundamental solution* of the equation $\mathfrak{M}u = 0$.

   If $\mathfrak{M}$ has constant coefficients and contains only derivatives of order $2r$, we mean by a fundamental solution in a region Ω a function $G(x, y)$ which, as a function of x, is in class $C^{(2r)}(\Omega - y)$, satisfies in $\Omega - y$ the equation $\mathfrak{M}_x G(x, y) = 0$, and, finally, is such that its derivatives of order $2r - 1$ are homogeneous functions of degree $1 - m$ of the point

$x - y$. If such a function exists, it depends on a multiplicative constant which we shall always assume is determined in such a way that:

$$\int\limits_{\partial \Gamma(y,\varrho)} \mathfrak{L}\left(G(x,y)\,,\,1\right) d_x\sigma = 1\,,$$

which is possible because the integral on the left side depends neither on y nor on ϱ. From (52.9) we then deduce that if $T \subset \Omega$, we have for every $v \in C^{(2r)}(T)$ and any $y \in T - \partial T$:

$$v(y) = -\int\limits_T G(x,y)\,\overline{\mathfrak{N}v(x)}\,dx - \int\limits_{\partial T} \mathfrak{L}\left(G(x,y)\,,\,\overline{v}(x)\right) d_x\sigma\,. \qquad (52.10)$$

In the general case, we shall say that $G(x,y)$ is a fundamental solution of the equation $\mathfrak{M}u = 0$ if it is in class $C^{(2r)}(\Omega - y)$ and if moreover, (52.10) holds for every $v \in C^{(2r)}(T)$. We note also that if $\mathfrak{M}$ has constant coefficients, and more generally, if $\mathfrak{M}$ is self-adjoint, then we can write $\mathfrak{M}$ in place of $\mathfrak{N}$ in (52.10).

We shall say immediately that in the first case considered, the construction of a fundamental solution already presents notable difficulty. Indeed, while in the case of two variables, we succeed there in expressing the fundamental solution in an elementary way, this is no longer possible for $m > 2$, having instead in this case to have resort to certain expressions formed with definite integrals. A propos of this, the first interesting results are due to C. SOMIGLIANA [1] for $m = 2$, to I. FREDHOLM[1] for $m = 3$, to G. HERGLOTZ [1] for $m = 3$ and $m = 4$, to F. BUREAU [1, 2, 3, 4] for $m < 2r$ and also BUREAU [5, 6] and to H. G. GARNIR [1] for certain classes of operators of fourth order. Finally the problem was solved in general by F. JOHN [2], who proved that $G(x,y)$ can be calculated by means of quadratures operating on the solution of an appropriate CAUCHY problem for the equation $\mathfrak{M}u = 0$ with initial data given on a variable hyperplane. The expression for $G(x,y)$ found by JOHN in the case of an equation with constant coefficients containing only derivatives of order $2r$ is of the type:

$$G(x,y) = |\,x - y\,|^{\,2r-m}\,\psi(x - y) + q(x - y)\log|\,x - y\,|\,,$$

where $\psi(z)$ is an analytic function of the point z homogeneous of degree zero and $q(z)$ is a polynomial of degree $2r - m$ for m even and $m \leq 2r$ and is zero in the other cases. JOHN's method, which was then repeated by F. BUREAU [7], is applicable to equations with constant coefficients even when $\mathfrak{M}$ contains derivatives of order less than $2r$. Then, if $\mathfrak{M}$ has

---

[1] FREDHOLM, I.: Sur l'intégrale fondamentale d'une équation differentielle elliptique à coefficients constants. Red. Circ. Mat. Palermo **25** (1908) 346–351.

analytic coefficients, the method is still applicable if the fundamental solution is sought in a sufficiently small domain T, this last limitation being removable only in particular cases. The results of the work [2] were described and developed systematically by JOHN in the volume [6]. For a certain relation between the fundamental solutions of two equations in m and $m - 1$ variables see A. FRIEDMAN [5]. For the study of domain and boundary potentials which can be constructed by departing from fundamental solutions of equations with constant coefficients see S. AGMON [1] and C. MIRANDA [19].

Passing on now to consider the case of an operator with sufficiently regular coefficients, but not analytic ones, we can think of applying the method of E. E. LEVI which was proposed by this author [1] exactly for solving the problem indicated now in two variables. This procedure, however, suggested by JOHN in [3], strikes upon the same difficulties pointed out in § 19 for the case $r = 1$. That is, it leads to the construction of the fundamental solution if the measure of T is sufficiently small, but in the general case it is applicable only if the uniqueness theorem holds for CAUCHY's problem[1]. Now the proof of this theorem has been given by A. P. CALDERON [2] and L. HÖRMANDER [1, 2] only for equations with simple characteristics and under appropriate hypotheses on the manifolds holding the data. L. NIRENBERG [6], R. N. PEDERSON [1], S. MIZOHATA [1, 2], L. HÖRMANDER [1], T. SHIROTA [1], I. S. BERNSTEIN [1], H. KUMANO-GO [1] have characterized particular classes of equations with multiple characteristic for which the uniqueness theorem for CAUCHY's problem holds, and in certain cases, even the unique continuation property. However, some examples of equation with double characteristic are known for which this validity does not hold; a propos of this see A. PLIŚ [1]. Moreover, PLIŚ himself [2] also constructed an example of an equation $\mathfrak{M}u = 0$ with coefficients of class $C^{(\infty)}$ which has a solution u vanishing outside a sphere Γ. It follows from this that the equation $\mathfrak{N}v = f$ has a solution in Γ only if f is orthogonal to u. From this, the nonexistence of a fundamental solution of the equation $\mathfrak{N}v = 0$ defined in Γ^2.

---

[1] For a sufficiently simple proof of this fact, in the case of a strongly elliptic equation, we refer to P. DE LUCIA [2], and for a critical examination of the question in the purview of the general theory of differential equations, to F. E. BROWDER [9, II].

[2] We mention also that the existence of a fundamental solution in T is equivalent to the existence in T of at least one solution of the equation $\mathfrak{M}u = f$ for arbitrary $f \in C^{(0,\lambda)}(T)$. If $\mathfrak{M}$ has simple characteristic, the existence of at least one weak solution of the equation $\mathfrak{M}u = f$ for each $f \in L^2(T)$ has been proved by I. N. KATZ [1]. This is related to what we have premised, and to some aspects of the general theory of partial differential equations related by B. MALGRANGE [1] and P. LAX [2].

We must therefore conclude, given an elliptic equation of higher order with coefficients as regular as we wish in a domain T, that it can not be said that it admits a fundamental solution in T unless appropriate hypotheses are satisfied relating to the characteristics of the operator. The fact is thus the more notable that in certain cases, independently of hypotheses of this sort, it is possible instead to prove the existence of a fundamental solution defined in all of R_m, or in addition, of a principal fundamental solution defined in a way more or less analogous to that indicated in § 20 for second order equations. A propos of this see S. BOCHNER [2], S. AGMON [7, § 14], V. V. GRUŠIN [1], A. AVANTAGGIATI [3], G. FICHERA [28, 29].

Finally, we mention that as in the case $r = 1$ (see § 19) the validity of the RUNGE approximation property is related to that of the unique continuation property. For this question and for other similar ones we refer to the works by B. MALGRANGE [1], P. LAX [2], F. E. BROWDER [9 II, 14].

C. LOCAL PROPERTIES OF THE SOLUTIONS

The results obtained on this topic are rather limited. We owe first of all to L. BERS [16] the extension of theorem 19, XI on removable singularities and of theorem 27, I on zeros of finite order to elliptic equations of higher order. For the order of the zeros of one-signed solutions see A. FRIEDMAN [4] and R. N. PEDERSON [1] and for the study of non-isolated zeros (nodal lines) J. PAWLOWSKA [1].

For equations with analytic coefficients, the study of isolated polar singular points was deepened by F. JOHN [2, 3] with results similar to those recorded in § 27 for the case $r = 1$, which have now been extended to the nonanalytic case by M. MARCUS [1]. For essential singular points see M. WACHMAN [1] and for singularities on manifolds S. P. GAVELJA [1].

D. DIRICHLET'S AND GENERAL BOUNDARY VALUE PROBLEMS

A boundary value problem for an elliptic equation $\mathfrak{M} u = f$ of order $2r$ considered in a bounded domain T is posed in general by associating to it r boundary conditions:

$$\mathfrak{B}_j(u) = \sum_{|\alpha|=0}^{r_j} b_{j\alpha}(x)\, D^\alpha u = \varphi_j(x) \quad \text{for} \quad x \in \partial T. \qquad (52.11)$$

The operator $\mathfrak{M}$ is assumed to be strongly or properly elliptic, and as for the operators $\mathfrak{B}_j$, we assume that they *cover* the operator $\mathfrak{M}$, meaning by this that the following „complementing condition" is satisfied: for every $x \in \partial T$ and for every pair of real unit vectors ξ and η, tangent and

normal to ∂T at the point x, respectively, the polynomials $\sum\limits_{|\alpha|=r_j} b_{j\alpha}(x)$
$(\xi + z\eta)^\alpha$ are linearly independent modulo $M^+(x, \xi, \eta, z)$. This definition
is meaningful for arbitrary r_j, but in the exposition which follows we
assume, without explicit notice to the contrary, that $r_j < 2r$. Under the
indicated hypotheses we also say that the boundary value problem being
considered is *elliptic*[1]. Then if the operators $\mathfrak{B}_j$ satisfy the complementing
condition for every arbitrary properly elliptic operator $\mathfrak{M}$ we say that
the boundary value problem being considered is *absolutely elliptic*[2].

Naturally for the study of elliptic problems it is necessary every time
to specify the functional classes to which the functions f and φ_j belong
and the class in which we seek u. Secondary to the hypotheses which are
made on the coefficients of $\mathfrak{M}$ and of the $\mathfrak{B}_j$, limiting ourselves for simplicity to the case $\varphi_j = 0$, we can, for example, assume that $f \in H^{s,p}$ and
seek u in the class $H^{2r+s,p}$ or else assume that $f \in C^{(s,\lambda)}$ and seek u in the
class $C^{(2r+s,\lambda)}$. If these spaces are chosen in a reasonable way, then the
result which we achieve is that the homogeneous problem admits a finite
number μ of linearly independent solutions and that the inhomogeneous
problem is solvable if and only if a finite number ν of compatibility conditions are satisfied by f and φ_j. This is expressed by saying that the
problem is an *index problem* and the different $\varkappa = \mu - \nu$ is called precisely
the *index* of the problem. This will be better specified in what follows,
in conformity with what has already been said in § 50 about second order
equations. It can also be shown that in varying the coefficients of the
operators $\mathfrak{M}$ and $\mathfrak{B}_j$ a little, the index does not vary. The index is also
independent of any arbitrary variation of the coefficients of the terms
of $\mathfrak{M}$ and $\mathfrak{B}_j$ or order less than $2r$ and r_j[3], respectively. This last property
is often exploited to establish that a given problem has FREDHOLM
character, i.e., that its index is zero. Indeed, if for a certain λ_0 the boundary value problem $\mathfrak{M}u + \lambda_0 u = f$, $\mathfrak{B}_j u = \varphi_j$ admits one and only one
solution for arbitrary f and φ_j, then the index of this problem is zero,

---

[1] The definition of elliptic boundary value problems has been posed by
different authors in various ways, all, however, equivalent; For formally
different definitions see, for example, F. E. BROWDER [6] and L. HÖR-
MANDER [2]. Problems of this type were considered for the first time by
YA. B. LOPATINSKIĬ [4] and Z. YA. ŠAPIRO [4].

[2] For a characterization of these problems, see L. HÖRMANDER: On the
regularity of the solutions of boundary value problems, Acta Math. **99**
(1958) 225–264.

[3] See for example L. HÖRMANDER [2, p. 261] and for the general theory
of the index of a linear transformation I. C. GOHBERG and M. G. KREIN: The
basic propositions on defect numbers, root numbers, and indices of linear
operators. Usp. Mat. Nauk **12** no. 1 (1957) 43–118; English transl. Amer.
Math. Soc. (2) **13** (1960) 185–264.

and then the index of the problem $\mathfrak{M}u + \lambda u = f$, $\mathfrak{B}_j u = \varphi_j$ is zero for arbitrary λ.

More significant results are obtained in the case in which the operators $\mathfrak{B}_j$ are assumed to be normal. In this case we can define a (*formal*) *adjoint problem* which is obtained by associating to the equation $\mathfrak{N}v = g$ the boundary conditions:

$$\mathfrak{B}_j'(v) = \psi_j$$

the operators $\mathfrak{B}_i'$ $(j = r + 1, \ldots, 2r)$ being those which occur in (52.9) written by choosing the operators $\mathfrak{B}_{r+1}, \ldots, \mathfrak{B}_{2r}$ arbitrarily but in such a way that together with the $\mathfrak{B}_1, \ldots, \mathfrak{B}_r$, they constitute a normal system. It is obvious that in this way the adjoint problem is not uniquely defined, but it can be shown that two homogeneous adjoint problems are equivalent, in the sense that any solution of one is also a solution of the other. The result which we achieve is that, denoting by μ the number of linearly independent solutions of the homogeneous problem and μ' the analogous number for the adjoint problem, we have that the indices $\varkappa$ and $\varkappa'$ of the two problems are $\varkappa = -\varkappa' = \mu - \mu'$. Moreover, a necessary and sufficient condition that the given problem be solvable is that for every solution v of the homogeneous adjoint problem:

$$\int_T f\bar{v}\, dx = \int_T \sum_{j=1}^{r} \varphi_j(x)\, \overline{\mathfrak{B}_j'(v)}\, dx\,,$$

the $\mathfrak{B}_j'$ $(j = l, \ldots, r)$ now being the operators which occur in (52.9'). We shall now illustrate briefly the research which has led to these results, adhering as a rule to its chronological order.

The simplest of the elliptic boundary value problems with conditions of normal type is the DIRICHLET problem which is obtained by putting:

$$\mathfrak{B}_j = \frac{d^{j-1}}{dn^{j-1}}\,.$$

For an equation with real coefficients $a_\alpha \in C^{(|\alpha|+1)}(T)$ and for a domain $T \in A^{(2r+1)}$ this problem was studied by E. E. LEVI [2] in 1910 in the case $m = 2$ and under the hypothesis that the roots of the equation $M(x, 1, z) = 0$ have constant multiplicity as x varies. The proof of the FREDHOLM alternative theorem, which is developed in its general lines according to the procedure of § 24, is otherwise laborious and quite difficult reading. After LEVI's memoir, forty years passed without new contributions of a general character being brought to the question. In fact, it was only between 1950 and 1952 that the study of DIRICHLET's problem for a strongly elliptic equation with complex coefficients was taken up again by L. Gårding [4, 5], M. I. Višik [3], F. E. Browder [1], with procedures which were inspired by Weyl's [1] (see also § 30) and by a fundamental memoir by S. L. Sobolev [3] relating to boundary value problems for polyharmonic functions[1].

According to these procedures, which are extensions of those described in § 30 for second order equations, writing the operator $\mathfrak{M}$ in the form (52.7), every function $u \in H_0^{r,\,2}(T)$ which for every $v \in H_0^{r,\,2}(T)$ satisfies the equation:

$$a(u, v) = \int_T \sum_{|\beta|,\,|\gamma|\,=\,0}^{r} (-1)^{|\beta|}\, a_{\beta\gamma}\, D^{\gamma} u\, \overline{D^{\beta} v}\, dx = \int_T f\bar{v}\, dx \quad (52.12)$$

is called a *weak solution* of the Dirichlet problem for the homogeneous boundary condition:

$$\frac{d^j u}{d n^j} = 0 \quad \text{for} \quad x \in \partial T \; (j = 0, 1, \ldots, r - 1)\,.$$

We moreover assume without loss of generality that (52. 4) holds with $\gamma(x) = (-1)^r$. The theorems of functional analysis from § 30 and others of the same sort easily allow the proof of the alternative theorem on condition that $a(u, v)$ is coercive on $H_0^{r,\,2}(T)$, that is, on condition that there exist k and λ_0 such that for $\lambda > \lambda_0$ we have for every $u \in H_0^{r,\,2}(T)$:

$$\| u \|_{H^r,\,2(T)}^2 \leq k \left[\mathfrak{Re}\, a(u, u) + \lambda(u, u)\right]\,. \quad (52.13)$$

We note that (52.13), which is called Gårding's inequality[2], is not an immediate consequence of the strong ellipticity, inasmuch as this latter hypothesis in the case $r > 1$ does not imply that the quadratic form $\sum_{|\beta|\,=\,|\gamma|\,=\,r} a_{\beta\gamma}\, \bar{\lambda}_\beta\, \lambda_\gamma$ has a definite real part. However, it has been proved by the authors cited that if the coefficients $a_{\beta\gamma}$ are continuous for $|\beta| = |\gamma| = r$ and bounded for $|\beta| < r$ or $|\gamma| < r$, then (52.13) holds. This is sufficient to establish that for $\lambda > \lambda_0$ Dirichlet's problem for the strongly elliptic equation:

$$\mathfrak{M}u + \lambda\,(-1)^r\, u = f \quad (52.14)$$

and its adjoint admit one and only one solution and that for general λ, Fredholm's alternative holds (problem with zero index)[3]. A propos of

---

[1] For equations with constant coefficients see also some slightly earlier works by L. Gårding [1, 3] and M. I. Višik [2].

[2] A propos of this see also L. van Howe: Sur l'éxtension de la condition de Legendre du calcul des variations aux intégrales multiples à plusieures fonctions inconnues. Indag. Math. **9** (1947) 3–8.

[3] For some results of the same type for the case of complex λ see M. Z. Solomjak [1, 3].

this, it is well to point out explicitly that the adjoint problem of DI-RICHLET's problem coincides with DIRICHLET's problem for the adjoint equation, and that the problem with homogeneous boundary conditions which we are considering is solvable when and only when we have $(f, v) = 0$ for every solution v of the homogeneous adjoint problem. We can now add that if the diameter of T is sufficiently small or else if $\mathfrak{M}$ has constant coefficients and contains only derivatives of order $2r$, then (52.13) holds even with $\lambda = 0$ and therefore, for the problem for the equation $\mathfrak{M}u = f$, the existence and uniqueness theorem also holds.

Once the existence of weak solutions is proved, we pose the problem of their regularization both in $T - \partial T$ and on ∂T. Concerning the first question (regularization in the interior), one discusses proving that under appropriate hypotheses we have $u \in C^{(2r)}(T - \partial T)$ or at least $u \in H^{2r, 2}(T)$ and that the equation $\mathfrak{M}u = f$ is satisfied everywhere or almost everywhere in $T - \partial T$. Results along these lines have been obtained under different hypotheses and with various methods both by GÅRDING and BROWDER and by J. SCHWARZ [1], K. O. FRIEDRICHS [6], F. JOHN [3, 4, 5, 6], P. LAX [1], L. NIRENBERG [4]. Of the various procedures adopted, we shall limit ourselves to illustrating briefly that of NIREN-BERG which, in contrast to the others, has nothing in common with that indicated in § 30 for second order equations and which can comfortably be adapted also to obtain regularization on ∂T. We note still that this is applicable only to strongly elliptic equations, while some of the works cited further above also look at more general equations.

As in the proof of theorem 40, II we consider the incremental expressions $w_j(x, h)$ of u and, denoting by $\omega(x)$ a function with compact support in $T - \partial T$ and equal to 1 in a domain $D \subset T - \partial T$, we put $z_j(x, h) = \omega(x) w_j(x, h)$. It is easily established that each one of these functions is a weak solution in T of an elliptic equation, by which the related inequality of GÅRDING is applicable to each of these. By various artifices we thus succeed in establishing a bound for $\| z_j(x, h) \|_{H^{r, 2}(T)}$ and therefore also for $\| w_j(x, h) \|_{H^{r, 2}(D)}$. Because we recognize that this bound is uniform with respect to h, we draw from this for $h \to 0$ a bound for $\| \partial u / \partial x_j \|_{H^{r, 2}(D)}$. Departing from the equations satisfied by the $\partial u / \partial x_j$ and iterating the procedure, we arrive at a bound in $H^{r, 2}(D)$ of the norms of all the $D^\alpha u$ with $|\alpha| \leq r$ and therefore at a bound of $\| u \|_{H^{2r, 2}(D)}$. We then pass to the regularization on ∂T with an artifice of the type which served us for an analogous purpose in the proof of theorem 40, II.

Another question of great interest is seeking regularization theorems for weak solutions with the minimum of hypotheses on the coefficients, even when these hypotheses are not sufficient to ensure that $u \in H^{2r, 2}$.

In other words, it would be of great interest to establish for higher order equations theorems analogous to those which for second order equations were obtained by DE GIORGI, NASH, STAMPACCHIA, LADYŽENSKAJA and URAL'CEVA, MORREY, etc. (see § 30). For regularization in the interior, partial results along these lines have been obtained by J. KADLEC and J.NEČAS [1] and S.CAMPANATO [8] with a procedure which is related to that of which CAMPANATO himself made use [7] in the study of second order equations.

What we have said until now concerns DIRICHLET's problem with homogeneous boundary conditions. In the general case, if there exists a function $u_0 \in H^{r,2}(T)$ satisfying the conditions $d^{j-1} u/dn^{j-1} = \varphi_j$, then we can reduce it to the previous case by putting $u = u_0 + v$ with $v \in H_0^{r,2}(T)$. However, the existence of u_0 is ensured only under hypotheses for the φ_j which are certainly superabundant if we require only that u be of class $H^{r,2}$ in every domain contained in $T - \partial T$. For example, if we assume only that $\varphi_j \in C^{(r-j)}(\partial T)$ we can not ensure the existence uf u_0. However, it has been proved by C. MIRANDA [12] in the case $m = 2$ and for an equation with real and sufficiently regular coefficients, (52.13) being assumed valid for $\lambda = 0$, that there exists one and only one function $u(x)$ which for every domain $D \subset T - \partial T$ is in class $C^{(r-1)}(T) \cap H^{r,2}(D)$, which satisfies (52.12) for every $v \in C_0^{(r)}(T)$ and satisfies the boundary condition. For this function $u(x)$, a bound of the following type holds:

$$\| u \|_{C^{(r-1)}(T)} \le K \left[\sup | f | + \sum_{j=1}^{r} \| \varphi_j \|_{C^{(r-j)}(\partial T)} \right]. \qquad (52.15)$$

This result, which is called the *maximum modulus theorem*, reduces essentially to (35.15) in the case $r = 1$. S. AGMON [4] then extended and finally generalized this result for the case $m \ge 2$.

Under other hypotheses for the φ_j the DIRICHLET problem with inhomogeneous boundary conditions was considered foɪ $m = 2$ by B. PINI [16] and E. MAGENES [8] along the same lines as CIMMINO (see § 29). J. NEČAS [2, 3, 6, 8] instead treated the problem for arbitrary m with a method similar to that illustrated further above but making use of spaces different from the $H^{r,2}$.

We pass on now to consider (always for strongly elliptic equations) boundary value problems more general than DIRICHLET's. Proceeding as in § 50 we consider, following an idea by J.L.LIONS [1, 2, 6, 7], a HILBERT space V such that we have, algebraically and topologically, $H_0^{r,2}(T) \subseteq V \subseteq H^{r,2}(T)$. We can then consider the abstractly defined boundary value problem consisting of seeking a function $u \in V$ satisfying (52.12) for $v \in V$. As for DIRICHLET's problem, if the weak problem admits solutions, we shall then pass on to study their regularization; a propos

of this see also F. E. BROWDER [5], S. AGMON [2], E. MAGENES and
G. STAMPACCHIA [1]. We also observe that, if the weak problem admits
a sufficiently regular solution, then this satisfies not only the equation
$\mathfrak{M}u = f$ but also *natural boundary conditions* which will be added to those
in any case expressed by the condition $u \in V$. Thus for example if V
consists of the functions which satisfy a normal system of $k \leq r$ homo-
geneous boundary conditions $\mathfrak{B}_j(u) = 0 \; (j = 1, \ldots, k)$ with $\mathfrak{B}_j$ of order
$r_j < r$, it is possible to determine the operators $\mathfrak{B}_{k+1}, \ldots, \mathfrak{B}_r, \mathfrak{F}_1, \ldots, \mathfrak{F}_r$
in such a way that the two systems of operators $\mathfrak{B}_j$ and $\mathfrak{F}_j \; (j = 1, \ldots, r)$
are together normal and we have for $u, v \in C^{(2r)}(T)$:

$$a(u, v) = \int_T \bar{v} \, \mathfrak{M}u \, dx + \sum_{j=1}^{r} \int_{\partial T} \mathfrak{F}_j(u) \, \overline{\mathfrak{B}_j(v)} \, d\sigma \, .$$

It is then obvious that every solution of our problem will satisfy, other
than the conditions $\mathfrak{B}_j(u) = 0 \; (j = 1, \ldots, k)$, also (in a generalized sense)
the others $\mathfrak{F}_j(u) = 0 \; (j = k+1, \ldots, r)$.

Naturally, to obtain the existence or alternative theorems, it will be
necessary to establish that $a(u, v)$ is coercive on V, that is, that for every
$u \in V$ an inequality of the type (52.13) still holds. For this, however, it
is not true, contrary to what happens for $r = 1$ (see § 50), that the hypothesis of strong ellipticity is sufficient. For example, if we assume
$V = H^{r,2}(T)$, with which we have an abstract statement of a problem
which can be called a NEUMANN problem, we shall have to assume that:

$$\mathcal{R}e \sum_{|\alpha| = |\beta| = r} a_{\alpha\beta} z_\alpha z_\beta \geq k \sum_{|\alpha| = r} |z_\alpha|^2$$

for every system of complex numbers z (J. L. LIONS [1]). Thus even in the
other example mentioned above, a necessary and sufficient condition
that (52.13) be valid is that the system of operators $\mathfrak{B}_1, \ldots, \mathfrak{B}_k, \mathfrak{F}_{k+1},$
$\ldots, \mathfrak{F}_r$ cover $\mathfrak{M}$ (S. AGMON [2, 5], M. SCHECHTER [16]).

In every case in which the validity of (52.13) is recognized, it also
stands proved, as for DIRICHLET's problem, that the problem has zero
index and that the analogous problem for (52.14) admits a unique solution
for $\lambda > \lambda_0$.

We pass on now to occupy ourselves with boundary value problems
for properly elliptic equations. The study of these problems has had such
rapid development in the circuit of so few years, namely between 1957
and 1961, that it seems somewhat difficult to attribute the priority of
certain results to this rather than that author. It still seems that the
origin of the notion of properly elliptic equations is implicitly contained
in a work by N. ARONSZAJN [3] and was put in evidence for the first
time by M. SCHECHTER [1] in 1957.

The problem which is reported in Aronszajn's work is that of determining under what conditions a system of p linear differential operators $A_1, A_2, \ldots, A_p$ of order q is coercive on a linear manifold V of $H^{q,2}(T)$ in the sense that a constant K exists such that we have for every $u \in V$:

$$\| u \|_{H^{q,2}(T)} \leq K \left[\sum_{k=1}^{p} \| A_k u \|_{L^2} + \| u \|_{L^2} \right].$$

Schechter, deepening Aronszajn's research, established that in the case $p = 1$ and if $A_1 = \mathfrak{M}$ is a properly elliptic operator, then we have for $u \in H_0^{r,2}(T)$:

$$\| u \|_{H^{2r,2}(T)} \leq K \left[\| \mathfrak{M}u \|_{L^2(T)} + \| u \|_{L^2(T)} \right]. \tag{52.16}$$

A formula of this type, which constitutes an extension to the case $r > 1$ of the results of Caccioppoli and Ladyženskaja (see § 37), had already been taken up by O.A. Guseva [1, 2], L. Nirenberg [4], F.E. Browder [5] in the case in which $\mathfrak{M}$ is strongly elliptic, but the interest of Schechter's work is precisely in having enlarged the field of validity of this formula. It is not possible to enter into details in the later developments of the general question of coerciveness posed by Aronszajn, which was deeply studied by M. Schechter [2, 3, 16], S. Agmon [2, 5, 7], K.T. Smith [1]. We shall only say that (52.16) has been variously generalized both by Schechter and by Agmon in the works cited as well as by F.E. Browder [8, 10, 11], S. Agmon, A. Douglis, L. Nirenberg [1], A.I. Košelev [5, 6], J. Peetre [1]. The various generalizations consist both of ensuring its validity when V is the subspace of $H^{2r,2}$ consisting of all the functions which satisfy a system of homogeneous boundary conditions which cover $\mathfrak{M}$, and of replacing $H^{2r,2}$ and L^2 by $H^{2r+k,p}$ and $H^{k,p}$, respectively, with $k \geq 0$. (52.16) and its generalizations are at the basis of the various existence results obtained by the aforesaid authors and briefly summarized at the beginning. In particular, M. Schechter [4, 5] occupied himself with Dirichlet's problem and with boundary value problems with conditions of normal type, S. Agmon [3] with the existence of solutions of class $H^{2r,p}$ of Dirichlet's problem, F.E. Browder [8, 10, 11] both with problems with conditions of normal type and with general elliptic problems, J. Peetre [2] with general elliptic problems[1]. A systematic treatment of the existence question which is related in some way to the works by Browder [8] and by Peetre [2] was given by L. Hörmander in Chapter X of his volume [2]. In its major lines the procedure followed by these authors is the following. They consider the strong extension of the

---

[1] For some complements to Peetre's work see M. Schechter [15].

operator $\mathfrak{M}$ in the class V of functions satisfying the conditions $\mathfrak{B}_{j}(u) = 0$
(see § 50). (52.16) then allows the proof that the realization $\mathfrak{M}_V$ of the
operator $\mathfrak{M}$ is not only a closed operator but also an operator with closed
range and that the homogeneous problem admits only a finite number of
linearly independent solutions (see, for example, BROWDER [8]). From
this, a necessary and sufficient condition that the problem be solvable
is that $(f, v) = 0$ for every solution v of the problem $\mathfrak{M}_V^* v = 0$, $\mathfrak{M}_V^*$ being
the adjoint operator, in the sense of the theory of operators, of $\mathfrak{M}_V$. In
other words, (52.16) permits proof that $\mathfrak{M}_V$ is normally solvable in the
sense of VISIK. It is then more difficult to prove that the compatibility
conditions are finite in number. This question can be attacked in various
ways. A first procedure, which is more or less that followed by BROW-
DER [8], consists of seeking to construct an operator $\mathfrak{S}$ (a parametrix in
HÖRMANDER's terminology) which is linear and continuous from L^2
into $\mathscr{H}_V$, $\mathscr{H}_V$ being the domain of definition of $\mathfrak{M}_V$, such that we have
$\mathfrak{M} \, \mathfrak{S} = \mathfrak{I} + \mathfrak{C}$, having denoted by $\mathfrak{I}$ the identity operator in L^2 and by $\mathfrak{C}$
a completely continuous operator from L^2 into itself. If this is possible
then the linearly independent solutions of the equation $\mathfrak{M}_V^* \, v = 0$ are
finite in number because they are also solutions of the equation $v +$
$\mathfrak{C}^* v = 0$, by which the assertion follows from RIESZ theory (see § 29).

Referring to G. FICHERA [21] and to S. G. MIHLIN [10] for other, though
more obscure, procedures of the same type, we observe that to attain the
same object another method proposed and applied by J. PEETRE [2]
can consist of establishing bounds analogous to (52.16) for the solutions
of the problem $\mathfrak{M}_V^* \, v = g$. For the consideration of this problem we can
then substitute that of the formal adjoint problem in the case of a system
of boundary conditions of normal type. But we can not digress on this.
We shall add that more or less all the cited works contain regularization
theorems. Even more, the works by KOŠELEV and by AGMON, DOUGLAS,
and NIRENBERG are almost exclusively dedicated to this question. The
works by KOŠELEV look at the regularization in $H^{2r,p}$; the memoir of
AGMON, DOUGLAS, and NIRENBERG, which is the most complete on the
argument, contains regularization theorems both with respect to integral
norms and in $C^{(k,\lambda)}$ spaces valid also for $r_j \geq 2r$. The theorems of the
second type constitute a complete extension of the results of SCHAUDER-
CACCIOPPOLI for second order equations. For regularization in the interior
alone, theorems of equal generality are also found in the memoir [9, II]
by F. E. BROWDER. For other regularization theorems see also A. P.
CALDERON and A. ZYGMUND [2, 3] and P. FIFE [2], and for the case of a
nonregular domain, V. A. KONDRAT'EV [1, 2, 3].

Many of the works which we have just discussed and, in a special
way, the memoir by AGMON, DOUGLAS, and NIRENBERG, are also occupied
with problems with inhomogeneous boundary conditions considered

under hypotheses for the boundary data which are very natural but not of great generality. A deep study of boundary value problems with inhomogeneous boundary conditions under hypotheses of great generality on the data, or even of problems in which distributions take the place of f and the φ_j, was developed in a series of memoirs by J.L. LIONS and E. MAGENES [1, 2] and by M. SCHECHTER [12, 17]. In these works, which look at questions both of regularization and of existence and which are related to a general statement of the boundary value problems of M.I. VIŠIK and S.L. SOBOLEV [1] and of J.L. LIONS [5, 7], considerable use is made of the theory of functional interpolations, thus developing an idea also formulated contemporaneously by N. ARONSZAJN [5]. Along the same lines also are the works by C. BAJOCCHI [1], G. GEYMONAT [1, 2] M. BEREZANSKIĬ, S.G. KREIN,   and   JA.A. ROĬTBERG   [1],   M. BEREZANSKIĬ and JA.A. ROĬTBERG [1], JA.A. ROĬTBERG [1, 2], J. BARROS NETO [1], G. GEYMONAT and P. GRISVARD [2]. For a systematic treatment of this aspect of the theory see Volume I of the book by J.L. LIONS and E. MAGENES [5], and for a summary exposition of the results known up to 1963, the lecture [9] by E. MAGENES.

For properly elliptic equations also it is of great interest to consider equations containing a parameter λ and of determining a set of values of λ for which a certain boundary value problem admits a unique solution. Important results on the argument have been obtained by S. AGMON [6] and by M.S. AGRANOVIČ and M.I. VIŠIK [1] by considering the case in which the operators $\mathfrak{M}$ and $\mathfrak{B}_j$ are polynomials in λ and determining the conditions under which, as λ varies in an angular region with $|\lambda|$ sufficiently large, the solution exists and is unique.

We shall now add some observations. First of all, it is to be perceived that for equations of higher order the method of continuation, which we used in Chapters V and VI for the study of boundary value problems of second order, has not been mentioned. The reason for this is in the fact perceived both by AGMON, DOUGLIS, and NIRENBERG [1] and by I.M. GEL'FAND [1] that in general it is not possible, given two elliptic problems, to pass from one to the other with a continuous deformation of the coefficients of $\mathfrak{M}$ and of the $\mathfrak{B}_j$ which preserves the elliptic character of the problem. GEL'FAND therefore called two problems *homotopic* for which this was true, and he posed the problem of determining all the homotopic invariants of elliptic problems, among which the index of the problem is naturally to be included. On the other hand examples of homotopic problems are not lacking. Such, for example, are two absolutely elliptic problems with the same boundary conditions for two operators $\mathfrak{M}_1$ and $\mathfrak{M}_2$ both strongly elliptic, because for $\tau \in [0,1]$, $\tau \, \mathfrak{M}_1 + (1 - \tau) \, \mathfrak{M}_2$ is also strongly elliptic. And it is precisely this observation which has permitted the use of the method of continuation in the case of second order equations with real coefficients. For the case $m = 2$ the question posed by GEL'FAND was at least in part solved by A. I. VOL'PERT [5, 10, 11] and was then resumed by M. Z. SOLOMJAK [2] and J. F. LABROUSSE [1]. For the general case some partial results, associated chiefly with the calculation of the index, were established by A. I. VOL'PERT [9], A. S. DYNIN [2], M. S. AGRANOVIČ [1]. A general theorem of strictly topological nature relating to the calculation of the index is due to M. F. ATIYAH and R. BOTT [1].

Another method which has been almost abandoned in this research is that of integral equations. Still, it is useful to mention that this method is utilized in the memoir [4] by YA. B. LOPATINSKIĬ in which the notion of an elliptic problem is introduced for the first time. DIRICHLET's problem for equations in two variables was then treated with the method of integral equations by D. GRECO [5] for $r = 2$ and by S. AGMON [1] for arbitrary r in the case of equations with constant coefficients and by G. FICHERA [28] in the general case. The treatments by these authors are different from E. E. LEVI's since they use fundamental functions and not quasi-GREEN's functions.

Finally, we wish to indicate the study of nonelliptic boundary value problems. A development of the theory in this direction appears very difficult because it has been proved by various authors[1] that the hypothesis of ellipticity for a problem is not only sufficient for the validity of formulas such as (52.16) but also necessary. Moreover, it suffices to notice that the nonregular oblique derivative problem for second order equations is precisely, in the case $m \geq 3$, an example of a nonelliptic problem. On the study of this question from a general point of view we shall shortly give some bibliographic indications concerned with equations on a manifold. For now, we shall limit ourselves to pointing out some notes by M. SCHECHTER [11, 13] in which new facts are illustrated which can be presented when we leave the foundation to which we have adhered until now. We mention also that in the last paragraph of the report by E. MAGENES and G. STAMPACCHIA [1] they indicate the way of treating some particular problems for elliptic equations which are not included among elliptic problems, using a procedure due to J. L. LIONS [1, 2]. For some applications of this method see also G. PULVIRENTI [1, 2].

E. EQUATIONS ON MANIFOLDS. PSEUDO-DIFFERENTIAL OPERATORS. NON LOCAL BOUNDARY PROBLEMS.

Where an elliptic equation of order $2r$ is considered on a closed and topologically defined manifold, we can pose, as in the case $r = 1$, PICARD's

---

[1] See for example S. AGMON, A. DOUGLIS, and L. NIRENBERG [1].

problem of determining a solution which is regular on the whole manifold. In the case $m = 2$ for an equation with real coefficients and therefore strongly elliptic, B. PINI [4, 9] proved the validity of the alternative
theorem, extending the procedure followed by G. CIMMINO in the case
$r = 1$ (see § 48). Previously G. ZWIRNER [3] had studied the same equation
in a particular case. In the case of a properly elliptic equation the problem
has an index which can be different from zero. A propos of this see
M. F. ATIYAH and I. M. SINGER [1]. For the questions on regularization
see also H. MARCINKOWSKA [2]. Naturally, as in the case $r = 1$, it is
possible to consider boundary value problems on a manifold endowed
with a boundary. Regarding the study of the indices of these problems
see M. F. ATIYAH and R. BOTT [1].

Also concerned in a major part of the cases with functions defined on
a manifold is the theory of singular integral differential operators or
pseudodifferential operators, which in recent years has progressed,
acquiring an ever larger importance in the field of partial differential
equations. Without any pretense of completeness, we shall cite the works
by A. P. CALDERON and A. ZYGMUND [1], A. S. DYNIN [1], L. HÖR-
MANDER [3], J. J. KOHN and L. NIRENBERG [1], R. J. SEELEY [1] for the
general part, and those by J. J. KOHN and L. NIRENBERG [2], L. HÖR-
MANDER [4] for applications to the study of nonelliptic problems. On the
other hand, M. S. AGRANOVIČ [2] and M. I. VIŠIK and G. I. ESKIN [1, 2]
have amply developed a theory, which includes as a particular case that
of elliptic problems, in which singular integro-differential operators
replace the operators $\mathfrak{M}$ and $\mathfrak{B}_j$. Previously A. S. DYNIN [2] had studied
a boundary value problem for an elliptic equation with integro-differen-
tial boundary conditions. DYNIN's work just cited gives an example of
problems with boundary conditions of nonlocal type. From other points
of view problems of this type have been studied also by R. BEALS [1],
F. E. BROWDER [19], M. SCHECHTER [18].

F. EQUATIONS IN UNBOUNDED DOMAINS

In the works just cited are contained some results which preserve
their validity when T is not bounded. Thus, for example, in some of the
works summarized in B), the existence of fundamental solutions defined
in all space is proved, and in the works by S. AGMON [2], S. AGMON,
A. DOUGLIS, and L. NIRENBERG [1], F. E. BROWDER [10], certain bounds
of the type (52.13) and (52.16) are proved, in a preliminary way, in a
half-space for equations with constant coefficients containing only the
leading terms. Moreover, under the same hypotheses for the equation
and the domain, general elliptic problems with nonhomogeneous boundary conditions are considered. For these problems AGMON, DOUGLIS,
and NIRENBERG [1] also gave an explicit solution formula, by construct-

ing by means of definite integrals the so-called POISSON kernels. The work [7] by BROWDER also contains some results for unbounded domains. To these indications we shall now add the citation of a work by V.G. MAZ'JA [3] and of another by JU.S. NIKOL'SKIĬ [1], which are expressly dedicated to the exterior DIRICHLET problem. Some works by H. MARCINKOWSKA [1, 3, 4, 5], on the other hand, consider strongly elliptic equations which admit a solution defined in all space.

Of a little different nature are the works by A. FRIEDMAN [7] and C. MIRANDA [14]. In the first one the nonexistence of solutions of equations with constant coefficients defined in all space and having a certain exponent of integrability is proved. In the second theorems of LIOUVILLE type are given for strongly elliptic equations.

G. EQUATIONS IN REGIONS WITH DEGENERATE BOUNDARY

For elliptic equations of order $2r > 2$ it is possible to pose a boundary value problem by giving conditions also on a portion Σ of the boundary of Ω of dimension $n - s$ with $s > 1$. The first to study the question was S.L. SOBOLEV, who in his research on the equation for polyharmonic functions [1, 2, 3] proved that the DIRICHLET problem can be stated by giving on Σ the values of u and all its derivatives up to order $r - \left[\dfrac{s}{2}\right] - 1$. For DIRICHLET's problem for a strongly elliptic equation, the study of the problem was resumed by M.I. VIŠIK [3] from the weak point of view and by M. MARCUS [2] for solutions in the ordinary sense. More general boundary value problems with boundary conditions of integrodifferential type have been studied by B.JU. STERNIN [1, 2].

A work by C. MIRANDA [15] on DIRICHLET's problem for equations with constant coefficients is unfortunately invalidated by an error.

H. MIXED AND TRANSMISSION PROBLEMS

Mixed problems, namely with differential conditions of different types on different parts of the boundary, have been posed and studied for higher order equations by M. SCHECHTER [8]. The case of equations in two variables was later deepened by J. PEETRE [3], E. SHAMIR [1], M. TROISI [4]. For mixed problems with conditions also of integro-differential type see M.I. VIŠIK and G.I. ESKIN [1].

We also owe to M. SCHECHTER [10] a first study of transmission problems for higher order equations. Other results are due to JA.A. ROĬT-BERG and Z.G. SEFTEL' [2, 3], and to Z.G. SEFTEL' [1, 2, 3]. Transmission problems between two elliptic equations of different order have been studied by M. TROISI [1, 2, 3].

I. Polyharmonic Functions

We wish now to occupy ourselves with the equation:

$$\Delta_{2r} u = f, \tag{52.17}$$

where by Δ_{2r} we denote the operator Δ_2 repeated r times. A propos of this we shall begin by mentioning some works by M. Nicolescu [1], M. Picone [4, 5], J. Privaloff [1], B. Pini [10] relating to the study of mean-value properties of polyharmonic (or r-hyperharmonic) functions, i.e., solutions of the equation $\Delta_{2r} u = 0$, a study which leads to the establishment of important criteria for the convergence of series of these functions. Analogous, even if less precise, results also hold for the solutions of the more general equation $\Sigma a_k \Delta_{2k} u = 0$ with the a_k constant; a propos of this see the works by M. Nicolescu [1], J. P. Robert [1, 2], M. Ghermanesco [1, 2]. Other works in which older research by E. Almansi[1] and T. Boggio[2] is resumed, look instead at the problem of expressing polyharmonic functions by means of harmonic ones. The most general results a propos of this are due to C. Tolotti [2] to whom we refer for other bibliographic indications on the preceding results by Nicolescu, Picone, Colucci, Fichera. Finally, regarding fundamental solutions of the operator Δ_{2r}, we refer to F. John [6]. Many works then examine Dirichlet's problem for (52.17) considered first in a circular or spherical region and then in a general domain T. The case of a circular (or spherical) domain already considered in the past by G. Lauricella[3] and J. Hadamard[4] for $m = 2$ and $r = 2$, by V. Volterra[5] for $m = 3$ and $r = 2$, by T. Boggio[6] in the general case, has been the object of new studies on the part of M. Picone [4] for $m = 2, 3, r = 2$, of K. Schröder [1] for $m = 3, r = 2$, of H. Bremerkamp [6, 8] and H. Bremerkamp and O. Bottema [1] for $m = 2, 3$ and r arbitrary. The case of a plane

---

[1] Almansi, E.: Sull'integrazione dell' equazione $\Delta^{2n}u = 0$. Ann. Mat. pura appl. **2** (1899) 1–51.

[2] Boggio, T.: Sulla deformazione delle piastre elastiche soggette al calore. Atti Acc. Sci. Torino **40** (1904/5) 219–240. Integrazione dell' equazione $\Delta_2 \Delta_2 = 0$ in una corona circolare e in uno strato sferico. Atti Ist. Veneto **59** (1899/1900) 497–508.

[3] Lauricella, G.: Integrazione dell' equazione $\Delta^2(\Delta^2) = 0$. Atti Acc. Sci. Torino **31** (1895/1896) 1010–1018.

[4] Hadamard, J.: Sur l'équilibre des plaques élastiques circulaires libres ou appuyées et celui de la sphère isotrope. Ann. Éc. N. Sup. **18** (1901) 313–342.

[5] Volterra, V.: Osservazioni sulla nota precedente del Prof. Lauricella e sopra una nota di analogo argomento dell' Ing. Almansi. Atti Acc. Sci. Torino **31** (1895/1896) 1018–1021.

[6] Boggio, T.: Sulle funzioni di Green di ordine m. Rend. Circ. Mat. Palermo **20** (1905) 97–135.

elliptic region has been treated for $r = 2$ by H. BREMERKAMP [9] and by F. ROSATI [1]. For a plane domain of strip form the DIRICHLET problem for $r = 2$ was treated by A. GHIZZETTI [2], while general boundary value problems for arbitrary r were studied by J. LERAY [6]. The case of a domain of general form was then studied with many diverse methods.

A first method is that of transformation into integral equations of second kind, whose application has been limited at least until now to the case of two variables. To the classic memoirs by J. HADAMARD[1] G. LAURICELLA[2], A. KORN[3] relating to the equation $\Delta_4 u = 0$ we have added the works by N. MUSHELIŠVILI [1], L. AMERIO [2], H. BREMER-KAMP [6] concerning the same equation and others by H. BREMER-KAMP [4, 5], I.N. VEKUA [1], A.I. KALANDJA [1, 2, 3, 4] concerning (52.17) in the case of arbitrary r. H. BREMERKAMP [2, 3, 7], I.N. VEKUA [2], B.P. PANEYAL [1], O.I. PANIČ [1] have also considered the case of fourth order equations of more general type, in which, however, the terms of highest order are reduced to $\Delta_4 u$.

Regarding the application of the method of minima we shall mention the fundamental works by S.L. SOBOLEV [1, 2, 3] of which we have spoken several times (see D and G). Other interesting results were achieved by K.O. FRIEDRICHS [1] for $m = 2$ and $r = 2$ and by G. FICHERA [10] for $r = 2$ and arbitrary m. FICHERA's results, which give a constructive method for the solution, were extended to the general case by T. VIOLA [1]. A variational method along the lines of TREFFTZ [1] was applied by M.Š. BIRMAN [3, 4] for the study of a mixed problem in the case $r = 2$.

A third procedure, particularly adapted to furnish numerical evaluations of the solution, is that proposed by M. PICONE (see § 31) of the transformation into FISCHER-RIESZ integral equations. The study of this method, initiated by L. AMERIO [4] in the general case, was later completed and deepened by G. FICHERA [1, 5, 9] in the case $r = 2$. To Cimmino's method (see § 29), on the other hand, we can relate some works by B. PINI [11, 13, 14, 15].

A new proof of the existence theorem for the case $r = 2$, $m = 2$ was given by C. MIRANDA [5] by making use of procedures similar to those in § 29, but modified in a way to permit the consideration of ordinary and not generalized solutions of the problem. The interest of this research is in the fact that it is based on a previous proof of the maximum modulus theorem, i.e., on the formula (52.15), and this is the only case in which

---

[1] HADAMARD, J.: Mémoire sur le problème d'analyse relativ à l'équilibre des plaques élastiques encastrées. Mém. Acad. Sci. Paris **33** (1908) 9–22.

[2] LAURICELLA, G.: Sur l'intégration de l'équation relative à l'équilibre des plaques élastiques encastrées. Acta Math. **32** (1909) 201–256.

[3] KORN, A.: Sur l'équilibre des plaques élastiques encastrées. Ann. Ec. N. Sup. **25** (1908) 529–583.

success has been achieved in proving this theorem by elementary means for a higher order equation without making use of the existence theorems. The proof of the existence and uniqueness theorem and a theorem on approximation by means of polynomials follow easily with the minimum of hypotheses[1].

Also related to the maximum modulus theorem is some research by R. J. Duffin [1, 2], P. R. Garabedian [1], Z. Nehari [1], R. J. Duffin and Z. Nehari [1] meant to verify the validity of a conjecture of Hadamard according to which the Green's function for Dirichlet's problem for Δ_4 would have positive sign. The response to this question is negative, however. For other majorizations for biharmonic functions we shall mention a work by K. Schröder [2] relating to the behavior on the boundary of the derivatives of biharmonic function, which anticipates the results in the following established for a general elliptic equation, and various integral majorization formulas of M. Picone [3], J. B. Diaz and H. J. Greenberg [2], G. Fichera [15], L. E. Payne and H. F. Weinberger [1], J. I. Synge [1], J. H. Bramble and L. E. Payne [1] which are significant also from the numerical point of view, and from which pointwise bounds can be deduced in the interior of the domain.

J. Nonlinear equations. Questions of analyticity

A solution of a nonlinear partial differential equation of higher order is called *elliptic* if the related variational equation is such. An equation is called *elliptic* if all its solutions are such.

The difficulty of making use of methods of continuation for elliptic equations of higher order, already stated for linear problems, has very much limited the application of the techniques followed in the case of second order equations to general nonlinear problems of higher order. A propos of this, in fact, we can cite only a brief allusion contained in the memoir by Agmon, Douglis, and Nirenberg [1] and some works by F. E. Browder [25], Bui an Ton [1], J. Nečas [10, 11, 12].

Instead, important results have been obtained for equations in divergence form, namely for the equation $\mathfrak{M}u = f$ in which $\mathfrak{M}$ has the form (52.7) and the $a_{\beta\gamma}$ and f depend also on u and on certain of its derivatives. A propos of this we shall cite a first group of works by M. I. Višik [12, ..., 16, 18] in part summarized in [17], which were followed by a series of works by F. E. Browder [15, ..., 18, 20, 21] and a work by J. Leray and J. L. Lions [1]. The procedures followed in these works are those of which we have given an idea at the end of § 41.

Other works, almost all prior to those cited above, examine equations or problems of a particular type. In some works by A. Rosenblatt [1, 3]

---

[1] For analogous results for Δ_2 see C. Miranda [3, 4].

the method of successive approximations is applied to the study of the
equation $\Delta_{2r}u = f(x_1, x_2, u, \ldots, D^{2r-2}u)$. To the study of some equations
from mathematical physics of the type $\Delta_4 u = f(x_1, x_2, u, \ldots, D^3 u)$,
on the other hand, are dedicated the works by A. K. NIKITIN [1], N. F.
MOROSOV [1], P. FIFE [1], while H. PACHALE [1, 2] considered some
problems for biharmonic functions with nonlinear boundary conditions.

The methods of the calculus of variations have been applied by
G. STAMPACCHIA [2] to the study of boundary value problems for the
EULER equation of an m-tuple integral of the type $\int_T f(x, u, \Delta_2 u)\, dx$,
the operator Δ_2 being able to be replaced, at least for $m = 2$, by an
arbitrary elliptic operator.

Finally, W. WALTER [2] is occupied with the existence of entire
solutions of the equation $\Delta_{2r}u = f(u)$.

Regarding questions of analyticity, we shall mention that the works
cited in § 44 by C. B. MORREY and L. NIRENBERG [1], C. B. MORREY [5, 9],
A. FRIEDMAN [1, 3], E. NELSON [1], T. KOTAKE and M. S. NARASIM-
HAN [1], J. L. LIONS and E. MAGENES [3, 4], E. MAGENES and G. STAM-
PACCHIA [1], M. K. V. MURTHY [1] all examine equations of higher order.
With these works theorems 44, II and III and also, but limited to quasi-
linear equations, theorems 44, V and VII have been completely extended
to equations of order $2\,r$; naturally consideration of solutions of class $C^{(2r)}$
or $C^{(2r,\lambda)}$ replaces that of solutions of class $C^{(2)}$ or $C^{(2,\lambda)}$. The extension
of theorems of the type 44, VI and of the analogous continuation theo-
rems is also complete. At least for now, on the other hand, the extension
of theorem 44, IV is lacking, while the extension of theorems 44, V and
VII and of the other continuation theorems of MORREY in the case of
nonlinear equations of general type has been obtained only for solutions
of class $C^{(2r,\lambda)}$. To these citations we can add those of a note by S. BOCH-
NER [1] and of the works by F. JOHN [2, 6] on the analytic character of
the solutions of a linear equation with analytic coefficients and by
F. E. BROWDER [13] on the analytic dependence of solutions on a parameter.

53. Systems of equations of the first order. A system of two
linear equations of the first order:

$$\sum_{k=1}^{2}\left(a_{ik}\frac{\partial u}{\partial x_k} + b_{ik}\frac{\partial v}{\partial x_k}\right) = c_{i1}u + c_{i2}v + f_i, \qquad (i = 1, 2) \qquad (53.1)$$

with the a_{ik}, b_{ik}, c_{ik}, f_i real functions of x_1 and x_2, is called *elliptic* if the
quadratic form in ξ_1 and ξ_2:

$$\begin{vmatrix} a_{11}\xi_1 + a_{12}\xi_2 & b_{11}\xi_1 + b_{12}\xi_2 \\ a_{21}\xi_1 + a_{22}\xi_2 & b_{21}\xi_1 + b_{22}\xi_2 \end{vmatrix}$$

is definite, for example positive definite, *uniformly elliptic* if this form is minorized by $k \mid \xi \mid^2$ with k a positive constant. We easily see that if the system is elliptic we must have $b_{11} b_{22} - b_{12}^2 \neq 0$, by which this system can always be considered written in the form:

$$\left. \begin{array}{l} \dfrac{\partial v}{\partial x_2} = a_{11} \dfrac{\partial u}{\partial x_1} + a_{12} \dfrac{\partial u}{\partial x_2} + c_{11} u + c_{12} v + f_1 \\[3mm] -\dfrac{\partial v}{\partial x_1} = a_{21} \dfrac{\partial u}{\partial x_1} + a_{22} \dfrac{\partial u}{\partial x_2} + c_{21} u + c_{22} v + f_2 \end{array} \right\} \tag{53.2}$$

and will be uniformly elliptic if we have:

$$4 a_{11} a_{22} - (a_{12} + a_{21})^2 > a > 0, \tag{53.3}$$

with a constant. In particular, therefore, the system of CAUCHY-RIEMANN equations occurs among elliptic systems.

We observe also that by introducing the complex notation:

$$w = u + iv, \qquad z = x_1 + ix_2,$$

$$\frac{\partial}{\partial z} = \frac{1}{2}\left(\frac{\partial}{\partial x_1} - i \frac{\partial}{\partial x_2} \right), \qquad \frac{\partial}{\partial \bar{z}} = \frac{1}{2}\left(\frac{\partial}{\partial x_1} + i \frac{\partial}{\partial x_2} \right),$$

the system (53.2) can be condensed into a single equation of the form:

$$w_{\bar{z}} = \mu w_z + \nu \overline{w_{\bar{z}}} + \alpha w + \beta \bar{w} + \gamma, \tag{53.4}$$

while it is not difficult to see that (53.3) is changed into a condition of the type:

$$\mid \mu \mid + \mid \nu \mid < k < 1. \tag{53.5}$$

Let us begin with the observation that if in the system (53.2) we assume that the c_{ik} are zero and the other coefficients are sufficiently regular, then it is possible to eliminate v, obtaining an elliptic equation of the second order in the single unknown u, such that it and its adjoint are devoid of terms in u. A boundary condition of the type:

$$u(x) = \varphi \quad \text{for} \quad x \in \partial T \tag{53.6}$$

therefore determines u uniquely and in consequence v also stands determined, if T has a single boundary, at least up to an additive constant. It therefore turns out to be quite natural to ask if this result holds in general even when, dropping the hypothesis of regularity of the coefficients, it is impossible to effect the aforesaid elimination of v. Questions of this sort are interesting in view of various applications to the theory of conformal mapping and to the calculus of variations; very general results, always for the case $c_{ik} = 0$, are due to L. LICHTENSTEIN [2], E. HOPF [3], C.B. MORREY [1]. The latter in particular has succeeded in establishing the existence theorem for solutions $u, v \in H^{1,2}$ under the hypothesis that the a_{ik} and f_i are measurable and bounded. Later Z. YA. ŠAPIRO [1] proved an analogous theorem for a quasi-linear system under the hypotheses that the c_{ik} are zero and that the a_{ik} are continuous bounded functions of x_1, x_2, u, v. All these results, which go back to more than twenty-five years ago, could be obtained today more simply by making use of the theory of weak solutions of linear and nonlinear equations in divergence form. More recently, later extensions of these results even for the case in which the c_{ik} are not zero were obtained by F.G. DRESSEL and J.J. GERGEN [1, 3], L. BERS and L. NIRENBERG [1], B.V. BOYARSKII [4]. Notable in the last two of the works cited is the fact that the proof is based on a representation formula for the solutions of (53.4) which has a great interest also for other applications. This formula, if w is defined in a subset T of the circle $|z| \leq 1$, is of the type:

$$w = e^{s(z)} f(\chi(z)) + s_0(z) ,$$

where $\zeta = \chi(z)$ is a homeomorphism of $|z| \leq 1$ into itself, $f(\zeta)$ is holomorphic in $\chi(T)$ and $s(z)$, $s_0(z)$, $\chi(z)$, $\chi^{-1}(\zeta)$ are in class $C^{(0,\lambda)}$. We can moreover require that s and s_0 be zero for $z = 1$ and that $\chi(0) = 0$, $\chi(1) = 1$. Moreover, if $\gamma = 0$ we have $s_0 = 0$; if $\mu = \nu = 0$ we have $\chi(z) = z$; if $\alpha = \beta = 0$ we have $s = 0$.

Another problem which is posed for the system (53.2) is that of the extension of Riemann's mapping theorem. Assuming, that is, that the coefficients of the system are defined in a simply connected domain T and given another domain D in the uv plane, also simply connected, it is a question of determining a solution of the system $u = u(x_1, x_2)$, $v = v(x_1, x_2)$ which establishes a homeomorphism of T into D, in which to three given points of ∂T correspond three given points of ∂D. The existence theorem for this problem was first proved by L. LICHTEN-STEIN [2] and C.B. MORREY [1] under the hypotheses that $c_{ik} = f_i = 0$, $a_{12} = a_{21}$, $a_{11} a_{22} - a_{12}^2 = 1$. In this case (53.4), which reduces to the equation:

$$w_{\bar{z}} = \mu w_z, \qquad |\mu| < k < 1 , \tag{53.7}$$

is called the BELTRAMI equation.

The problem considered through (53.7) presents itself in various important questions such as the reduction of an elliptic equation of the second order to canonical form (see § 54) and the conformal mapping from a plane domain to a surface of the space. For the case of a quasi-linear equation in which we have $c_{ik} = f_i = 0$, but the a_{ik} are continuous functions also of u and v, the existence theorem has been proved by Z. Ya. Šapiro [1]. Other proofs, under hypotheses even more general than Šapiro's, are due to L. Bers and L. Nirenberg [1], B. V. Boyarskiǐ [4], I. I. Danyluk [3] and for the linear case to F. G. Dressel and J. J. Gergen [1, 3]. Some uniqueness theorems have been proved by Boyarskiǐ [4], Dressel and Gergen [2], R. M. McLeod and F. G. Dressel [1]. Finally we mention that M. A. Lavrent'ev [1, 2, 3] has extended Riemann's mapping theorem to the solutions of a nonlinear system:

$$F_k \left(x_1, x_2, u, v, u_{x_1}, u_{x_2}, v_{x_1}, v_{x_2} \right) = 0 \qquad (k = 1, 2) ,$$

under appropriate hypotheses for this system which he calls "strong ellipticity". A result intermediate between those of Šapiro and Lavrent'ev is due to B. V. Šabat [1]. For summary statements of the results with which we have just been occupied, see the works by F. G. Dressel and J. J. Gergen [4] and by B. V. Šabat [2] and the monograph by M. M. Lavrent'ev [5]. For some other results related with the preceding see also G. N. Polozii [4, 5], B. V. Boyarskiǐ [2, 3], R. Finn and J. Serrin [1], L. Ahlfors and L. Bers [1], Chen Kien-Kwong [1].

Another general form of the system (53.1) is the following[1]:

$$\left. \begin{aligned} \frac{\partial u}{\partial x_1} - \frac{\partial u}{\partial x_2} &= c_{11}\, u + c_{12}\, v + f_1 \\[2mm] \frac{\partial u}{\partial x_2} + \frac{\partial v}{\partial x_1} &= c_{21}\, u + c_{22}\, c + f_2 \end{aligned} \right\} \tag{53.8}$$

to which we can then associate the boundary condition (53.6) already considered by D. Hilbert [1] or else the other more general one:

$$a u + b v = \varphi \quad \text{for} \quad x \in \partial T , \tag{53.9}$$

where a and b are functions defined on ∂T. This latter problem, which besides could be considered even for the system (53.1), is called the *Riemann-Hilbert problem*. Some results in this regard can already be found in an old memoir by W. A. Hurwitz [1] and in another by F. Nöther [1]. It is a question of an index problem for which the method best adapted for its study is that of transformation into a singular integral equation and of the application of the general theory of these equations. On the argument we shall cite the works by N. K. Usmanov [1, 2, 3, 4], W. Haack and G. Hellwig [1], W. Haack [2], G. Hellwig [1, 2], I. N. Vekua [5], Joachim Nitsche [1], Johannes Nitsche [1, 4],

---

[1] In fact (see § 54) a system of the type (53.1) can always be reduced to the form (53.8).

JOACHIM and JOHANNES NITSCHE [1], J. JAENNICKE [1, 2, 3, 4], V.S. VINOGRADOV [1, 2, 3, 4]. For equations with singular coefficients see L.G. MIHAĬLOV [1, 3, 7]. Other problems with a differential or integro-differential boundary condition have been treated by B.V. BOYARSKIĬ [1] and I.I. DANYLUK [5, 6] and with conditions of other types by L.G. MIHAĬLOV [2] and HOU TSUNG-YI [1]. For the nonlinear case see W. HAACK and G. HELLWIG [1], H. BECKERT [1], G. HELLWIG [3], JOACHIM NITSCHE [2], JOHANNES NITSCHE [3, 4], V.S. VINOGRADOV [4], I.I. DANYLUK [8, 9], K. NAKAMORI [2] and for some problems concerning free boundaries M. SCHECHTER [7] and M.A. LAVRENT'EV [4]. We also mention the works by I.I. DANYLUK [1, 2, 4, 7], A.B. JUDANINA [1, 2, 3], JU. L. RODIN [1], W.L. KOPPELMAN [1] relating to equations on a manifold and those by C.B. MORREY [2], F.G. DRESSEL and J.J. GERGEN [3], H. BECKERT [2, 3, 4, 5], B.V. BOYARSKIĬ [5, 7], A. AVANTAGGIATI [2] relating to boundary value problems for systems of $2n$ linear and nonlinear equations in $2n$ unknowns. For these latter systems, A.I. VOL'PERT [4, 8] has studied problems with integro-differential boundary conditions and B.S. SIKORA [1], problems with differential boundary conditions but of higher order. Systematic expositions of the theory are found in the volumes by I.N. VEKUA [9], F.D. GAHOV [1], W. HAACK and W. WENDLAND [1], to which we refer also for all the indications for the particular case of boundary value problems for the CAUCHY-RIEMANN equations.

Another direction of research which has had great importance in the study of first order elliptic systems is the extension of the properties of analytic functions to the functions $w(z) = u(x_1, x_2) + iv(x_1, x_2)$ where (u, v) is a solution of the system (53.8) with $f_1 = f_2 = 0$ or else of the system (53.2) with $c_{ik} = f_i = 0$, $a_{11} = a_{22}$, $a_{12} = -a_{21}$. The first works in this direction, for particular cases, are by L. BERS and A. GELBART [1, 2, 3] and by G.N. POLOZIĬ [1, 2, 3] and the question is also discussed in a memoir by I.G. PETROWSKIĬ [3]. Later L. BERS [2, 3, 8, 10, 11, 15, 17, 18] and I.N. VEKUA [5, 7, 8], independently of one another, studied the general case deeply, bringing to light that the analogy between these functions and ordinary analytic functions goes very much beyond what one might expect.

BERS, who called these functions *pseudo analytic functions*, has also given in addition various systematic expositions of his theory [9, 14, 19] of which the most recent is contained in the volume by R. COURANT and D. HILBERT [2] under the form of a supplement to Chapter IV. To the works already cited by L. BERS is to be added a note by S. AGMON and L. BERS [1]. For a possible generalization to the case of a system with more than two unknown functions see also A. DOUGLIS [1, 3].

VEKUA, who called these functions *generalized analytic functions*, has written abundantly on his theory in the volume [9] which also contains an extensive treatment of the applications to the general theory of infinitesimal bendings of surfaces and to the membrane theory of shells. We mention also that pseudo-analytic functions enjoy the strong unique continuation property[1]. This is to be considered as an extension of a classical result due to T. CARLEMAN [1, 2] then generalized by A. DOUGLIS [2, 4].

Finally, regarding the connection between this theory and that of quasi-conformal mappings we refer to the monograph by H.P. KÜNZI [1].

Passing on to consider the case of more variables, the most natural extension of the system (53.2) is given by the system in $(m^2 - m + 2)/2$ unknowns:

$$\sum_{k=1}^{m} \frac{\partial v_{ik}}{\partial x_k} = \sum_{k=1}^{m} a_{ik} \frac{\partial v}{\partial x_i} + f_i \quad (i = 1, 2, \ldots, m), \qquad (53.10)$$

where the v_{ik} are components of a double antisymmetric tensor ($v_{ik} = -v_{ki}$). For this system also, the elimination of the v_{ik} leads to an equation of second order in v alone, which is elliptic if the quadratic form $\sum_{i,k=1}^{m} a_{ik} \xi_i \xi_k$ is definite. Therefore, if T has a single boundary, by associating the boundary condition (53.6) to the system (53.10) v stands uniquely determined and the v_{ik} at least up to the addition of the coefficients of a closed differential form of degree $m - 2$. A propos of this C. MIRANDA [6] proved that the existence theorem (for generalized solutions) holds even when the a_{ik} are assumed continuous, φ of class $C^{(0, \lambda)}$, and the f_i of class $L^{2, \mu}$ with $\mu - m + 2 \geq 2 \lambda > 1$. The solution is then ordinary, that is, of class $C^{(1, \lambda)}$ if $a_{ik}, f_i \in C^{(0, \lambda)}$. We arrive at this result by using in an essential way theorem 30, V and a series of majorization formulas connected to this. Some extension of these results and some interesting applications to questions from the calculus of variations are found in the memoir [3] by G. STAMPACCHIA. The case in which T has more boundaries and the a_{ik} are assumed only in class L^{∞} has been treated by G. ARUFFO [1, 2], that of a quasi-linear system in which the a_{ik} and f depend also on the v_{rs} and v by A. ZITAROSA [1].

Another class of elliptic systems of first order equations with constant coefficients which can also be considered as extensions of the Cauchy-Riemann system to the case of more variables has been deeply studied by A.A. DEZIN [1]. The memoir by N. ARONSZAJN, A. KRZYWICKI and J. SZARSKI [1], already mentioned in § 19, is dedicated to the proof of the

---

[1] See for example R. COURANT and D. HILBERT [2] p. 382.

unique continuation property for the exterior differential forms defined on a riemannian manifold which satisfy there certain systems of differential equations or inequalities of the first order.

We shall now consider the system with real or complex coefficients:

$$\mathfrak{M}\, u = \sum_{i=1}^{m} a_i(x)\, \frac{\partial u}{\partial x_i} + b(x)\, u = f(x)\,, \tag{53.11}$$

in which $u(x)$ is an unknown vector with $2n$ components, the a_i and b are $2n \times 2n$ matrices, and f is a known vector with $2n$ components. We shall always assume that the components of f are defined in a bounded domain T and that, on the other hand, the elements of a_i and b are defined in R_m. The system (53.11) is called *elliptic* if for every real nonzero ξ we have:

$$a(x, \xi) = \det\left(\sum_{i=1}^{m} a_i(x)\,\xi_i\right) \neq 0\,,$$

uniformly elliptic if there exists a real number M such that we have

$$M^{-1}\, |\, \xi\, |^{2n} \leq |\, a(x, \xi)\, | \leq M\, |\, \xi\, |^{2n},$$

properly elliptic if for every pair of real orthogonal vectors ξ and η the equation in z: $a(x, \xi + z\,\eta) = 0$ admits as many roots with positive imaginary part $\tau_j^{+}(x, \xi, \eta)$ as with negative imaginary part $\tau_j^{-}(x, \xi, \eta)$. As in the case of a single equation of order $2r$, if $m \geq 3$, then every elliptic system is also properly elliptic. In what follows we shall denote by $a^{+}(x, \xi, \eta, z)$ the monic polynomial which has as roots the $\tau_j^{+}(x, \xi, \eta)$ and by $\alpha(x, \xi)$ the matrix adjoint to $\sum_{i=1}^{m} a_i(x)\,\xi_i$.

Now, let a covering of ∂T by N open sets $A_1, A_2, \ldots, A_N$ be fixed, each A_k homeomorphic to a sphere in R_{m-1}, such that no A_k is contained in any of the $N-1$ others. If on every A_k a matrix c_k of type $n \times 2n$ is defined, then we can consider the problem of determining a solution of the system (53.11) which on each A_k satisfies the condition:

$$c_k u = \varphi_k\,. \tag{53.12}$$

Naturally it is necessary to assume that the matrices c_k and the vectors φ_k are bound by conditions of compatibility such that on every $A_h \cap A_k$ the conditions $(53.12)_h$ and $(53.12)_k$ are equivalent. Moreover, we must assume that the problem is *elliptic* in the sense that for every $x \in A_k$ and for every pair of unit vectors ξ and η tangent and normal to ∂T at the point x, respectively, the following complementing condition is satisfied:

Every vector λ with n components, for which the vector $\lambda c_k(x) \alpha(x, z\eta - \xi)$ with $2n$ components has all components zero mod $a^+(x, \xi, \eta, z)$, vanishes.

This problem, for a large class of systems of the type (53.11), was stated and deeply studied by A. AVANTAGGIATI with the method of transformation into a system of singular integral equations, which is achieved by representing the solution by means of potentials constructed by departing from a principal fundamental matrix of which, under aproppriate hypotheses, he had previously proved the existence in the memoir [3]. By principal fundamental matrix of the matrix operator $\mathfrak{M}$ we mean a $2n \times 2n$ matrix $G(x, y)$ infinitesimal of exponential type as $|x - y| \to \infty$ and such that for every vector v with $2n$ components of class $C^{(1,\lambda)}$ and of compact support in R_m we have:

$$\bar{v}(y) = -\int_T G(x, y)\, \overline{\mathfrak{N}v(x)}\, dx \,, \tag{53.13}$$

$\mathfrak{N}$ being the *adjoint operator* of $\mathfrak{M}$ (see § 55).

Avantaggiati's work [1] concerns the particular case of a system with constant coefficients with $m = 3$ and $n = 2$. The work [2] treats the case $m = 2$. The definitive results for the general case are found in the memoir [4], to which we also refer for a more complete bibliography of the works of this author. The fundamental result is that we are dealing with an index problem, this index being equal to that of a certain system of singular integral equations. Some cases in which the solution exists and is unique are also characterized. In the general case the problem is solvable when and only when f and the φ_k satisfy certain conditions of integral type, which are written by making the solutions of the adjoint homogeneous problem take part. A propos of this, the fact is interesting that, even if we consider a problem in which we impose on u that it satisfy a single condition $cu = \varphi$ on the whole boundary, the adjoint problem is in general a problem for the differential system adjoint to the system (53.11) with conditions of the type $(53.12)_k$. This way of posing a boundary value problem for a system of the type (53.11) appears therefore as the most natural. We note also that in every case a problem of the type considered can be elliptic only if the numbers m and n satisfy certain conditions determined by M. Z. SOLOMJAK [4].

We shall close this section with two other remarks. The first is in reference to a work by Z. YA. ŠAPIRO [2] on a system in three variables with conditions of particular type. The second refers to a group of works by C. MÜLLER [1, 3, 5], W. K. SAUNDERS [1], H. WEYL [4], A. P. CALDERON [1], P. WERNER [1, 3] concerning the interior and exterior boundary value problems for the systems satisfied by the solutions of MAXWELL's equations which depend on time with a sinusoidal law.

54. Canonical form of elliptic equations. The study of systems of the type (53.2) is closely related with the problem of reducing a second order elliptic equation to the *canonical form:*

$$\frac{\partial^2 u_2}{\partial y_1^2} + \frac{\partial^2 u_2}{\partial y_2^2} + \beta_1 \frac{\partial u}{\partial y_1} + \beta_2 \frac{\partial u}{\partial y_2} + \gamma u = \psi \,. \tag{54.1}$$

Indeed, if in an elliptic equation written with the usual symbols we execute the change of variables:

$$y_1 = y_1(x_1, x_2) \,, \qquad y_2 = y_2(x_1, x_2) \,, \tag{54.2}$$

then we immediately see that a sufficient condition that the transformed equation, at least up to multiplication by a factor, be of the form (54.1) is that y_1 and y_2 satisfy the system:

$$\left. \begin{array}{l} \dfrac{\partial y_1}{\partial x_2} = - \dfrac{1}{\sqrt{A}} \left(a_{11} \dfrac{\partial y_2}{\partial x_1} + a_{12} \dfrac{\partial y_2}{\partial x_2} \right) \\[3mm] - \dfrac{\partial y_1}{\partial x_1} = - \dfrac{1}{\sqrt{A}} \left(a_{12} \dfrac{\partial y_2}{\partial x_1} + a_{22} \dfrac{\partial y_2}{\partial x_2} \right) \end{array} \right\} . \tag{54.3}$$

Naturally it is necessary that the (54.2) be the equations of a homeomorphism between the domain T of the $x_1 x_2$ plane in which the coefficients of the equation are defined and a domain D of the $y_1 y_2$ plane. If T is simply connected, then by assigning a simply connected D arbitrarily, we can determine a solution of the system (54.3), which is of BELTRAMI type, which realizes the required homoemorphism (see § 53). If, on the contrary, T is not simply connected, we can not assign D arbitrarily, but it is necessary to be contented with determining a solution of the system (54.3) such that the function $w = y_1 + iy_2$ is univalent. This problem, simpler than the preceding but not free from difficulties, has occupied various authors who have sought to solve it with a minimum of hypotheses on the a_{ik}. A propos of this we mention the works by L. BERS [11], P. HARTMAN and A. WINTNER [7], JOHANNES NITSCHE [4], I.N. VEKUA [8, 9], B.V. BOYARSKII [2], L. AHLFORS and L. BERS [1]. In many of these works the other question, strictly related to the preceding, of reducing an elliptic system of general type (53.1) to the canonical form (53.8), is treated, which can be realized with a simultaneous change of variables and of unknown functions and a linear substitution of the equations. A propos of this, other than the works cited, see also G. HELLWIG [2] and W. HAACK and W. WENDLAND [1].

Turning our consideration to the case of a second order equation, let us observe that if the equation being considered is self-adjoint, then this also turns out to be (54.1) at least up to multiplication by a factor, by which this equation comes to assume the form:

$$\frac{\partial}{\partial y_1}\left(p\,\frac{\partial u}{\partial y_1}\right) + \frac{\partial}{\partial y_2}\left(p\,\frac{\partial u}{\partial y_2}\right) + qu = \psi\,.$$

For other simplifications of the equation which can be obtained by also executing a change of unknown functions see A. FRIEDMAN [6]. Finally, regarding equations in $m \geq 3$ variables, we mention that L. BIANCHI[1] observed that at least in a sufficiently restricted domain it is always possible by multiplying by a factor to put an equation into the form:

$$\sum_{i,k=1}^{m}\frac{\partial}{\partial x_i}\left(\alpha_{ik}\,\frac{\partial u}{\partial x_k}\right) + qu = \psi \tag{54.4}$$

without, however, $\alpha_{ik} = \alpha_{ki}$. We wish now to observe how this can be done without imposing bounds on T. In fact, by multiplying the elliptic equation (2.3) by a factor $v > 0$ and denoting by v_{ik} a double antisymmetric tensor, this equation assumes the form:

$$\sum_{i,k=1}^{m}(v\,a_{ik} + v_{ik})\,p_{ik} + \sum_{i=1}^{m}v\,b_i\,p_i + v\,cu = v\,f\,,$$

which reduces to (54.4), with $\alpha_{ik} = v\,a_{ik} + v_{ik}$, $q = vc$, $\psi = vf$, on condition that for $i = 1, 2, \ldots, m$:

$$\sum_{i=1}^{m}\frac{\partial v_{ki}}{\partial x_k} = -\sum_{k=1}^{m}\frac{\partial}{\partial x_k}(v\,a_{ik}) + v\,b_i\,. \tag{54.5}$$

Now if T has a single boundary, then the system (54.5) is solved by assuming for v a solution of the equation:

$$\mathfrak{N}v - cv = 0\,,$$

after which the determination of the v_{ik} is brought back to the calculation of the primitive of an integrable differential form of degree $m - 1$. Then in order to obtain a positive solution in T of (54.6) it suffices to take a solution positive on ∂T. In fact, if such a solution assumes negative values in T, then it must vanish on the boundary of a domain $D \subset T$ and this is absurd because then also the equation $\mathfrak{M}u - cu = 0$ would have to have a solution not identically zero in D and zero on ∂D. Naturally this procedure requires that the coefficients of $\mathfrak{N}v$ be at least of class $C^{(0,\lambda)}$; however, we could seek to weaken this hypothesis by using the results acquired for the system (53.10) in the preceding section.

---

[1] BIANCHI, L.: Sulle equazioni lineari a derivate parziali del 2⁰ ordine. Rend. Acc. Lindi **5** (1889) 35–44.

55. Systems of higher order equations. As in § 52, we shall divide the exposition into several parts, each one with its own subtitle distinguished by a letter.

A. ELLIPTIC SYSTEMS

A system of partial differential equations of the type:

$$\sum_{|\alpha|=0}^{2r} a_\alpha(x) \, D^\alpha u = f(x), \tag{55.1}$$

in which u and f are vectors with N real or complex components and the a_α are $N \times N$ matrices with real or complex elements, is called *elliptic in the sense of* PETROWSKIĬ [1, 2] if for ξ real and nonzero we have:

$$a(x, \xi) = \det \left(\sum_{|\alpha|=2r} a_\alpha(x) \, \xi^\alpha \right) \neq 0,$$

properly elliptic in the sense of PETROWSKIĬ if for every pair of real unit orthogonal vectors ξ and η the equation $a(x, \xi + z\eta) = 0$ has as many roots with positive imaginary part as with negative imaginary part. Finally, the system (55.1) is called *strongly elliptic in the sense of* PETROWSKIĬ if, denoting by $a_{hk}(x, \xi)$ the elements of the matrix $\sum_{|\alpha|=2r} a_\alpha(x) \, \xi^\alpha$, we have:

$$\mathscr{R}e \sum_{h,k=1}^{N} a_{hk}(x, \xi) \, \eta_k \, \bar{\eta}_k \neq 0$$

for real ξ and real or complex η, both not zero.

Elliptic systems in the sense of PETROWSKIĬ occur in a more general class of systems which are called elliptic according to DOUGLIS and NIRENBERG [1]. We shall consider a system of N equations in N unknowns of the form:

$$\sum_{j=1}^{N} \mathfrak{M}_{ij} \, u_j = f_i, \qquad (i = 1, 2, \ldots, N), \tag{55.2}$$

where the $\mathfrak{M}_{ij}$ are linear differential operators of order α_{ij}. Let us consider two systems of integers $s_1, \ldots, s_N$ and $t_1, \ldots, t_N$ such that $\alpha_{ij} \leq s_i + t_j$; then we must intend that $\mathfrak{M}_{ij} = 0$ if $s_i + t_j < 0$. Denoting by $M_{ij}(x, \xi)$ the characteristic form of the operator $\mathfrak{M}_{ij}$, we put:

$$M'_{ij} = \begin{cases} M_{ij} & \text{if } \alpha_{ij} = s_i + t_j \\ 0 & \text{if } \alpha_{ij} < s_i + t_j. \end{cases}$$

The system (55.2) is called *elliptic in the sense of* Douglis *and* Niren-berg[1] if it is possible to choose the numbers s_i and t_j in such a way that for ξ real and nonzero:

$$a'(x, \xi) = \det\left(M'_{ij}(x, \xi)\right) \neq 0 . \tag{55.3}$$

Making $a'(x, \xi)$ assume the same office that $a(x, \xi)$ had for ellipticity in the sense of Petrowskiĭ, we define *proper ellipticity in the sense of* Douglis *and* Nirenberg. Then, if $s_i = t_i$ and moreover for real ξ and real or complex η not both zero we have:

$$\mathscr{Re} \sum_{h,k=1}^{N} a'_{hk}(x, \xi)\, \eta_h\, \overline{\eta}_k \neq 0 ,$$

then the system is called *strongly elliptic in the sense of* Douglis *and* Nirenberg and then *uniformly strongly elliptic* if a constant a exists such that:

$$\mathscr{Re} \sum_{h,k=1}^{N} a'_{hk}(x, \xi)\, \eta_h\, \overline{\eta}_k \geq a \sum_{i=1}^{N} |\,\xi\,|^{2s_i}\,|\,\eta_i\,|^2 ,$$

where with $a'_{hk}(x, \xi)$ we have denoted the elements of the matrix $a'(x, \xi)$. In particular, if $s_i = 2r$, $t_j = 0$ we reobtain elliptic systems according to Petrowskiĭ. We observe also that, if we put $t_j = 0$, $s_i = \sup \alpha_{ij}$, then we obtain a system of integers for which $\alpha_{ij} \leq s_i + t_j$; if the system (55.2) satisfies (55.3) with this choice of the s_i and t_j we say that it is *elliptic in the sense of* Lopatinskiĭ [2].

Analogously to what happens for a single equation, the strong ellipticity implies proper ellipticity, and for $m \geq 3$ every elliptic system is also properly elliptic.

Finally, we shall call the system:

$$\sum_{i=1}^{N} \mathfrak{N}_{ij} v_i = g_j \quad (j = 1, \dots, N) \tag{55.4}$$

the *adjoint* of the system (55.2), where $\mathfrak{N}_{ij}$ is the operator adjoint to $\mathfrak{M}_{ij}$.

B. Green's formula. Fundamental matrix. Cauchy's problem and the unique continuation property

If we agree to denote by u and v the vectors with components u_j and v_i, respectively, and by $\mathfrak{M}u$ and $\mathfrak{N}v$ the vectors which have as com-

---

[1] On an equivalent definition see S. Agmon, A. Douglis, L. Niren-berg [1, II] and L. R. Volevič [3].

ponents the left sides of (55.2) and (55.4), then a Green's formula holds which appears formally like (52.9) except for the interpretation of the terms $\bar{v}\,\mathfrak{M}u$ and $u\,\overline{\mathfrak{N}v}$ as scalar products and $\mathfrak{L}(u, v)$ as a bilinear form in the components of u and v and their derivatives.

We shall then say that an $N \times N$ matrix $G(x, y)$ is a *fundamental matrix* in a domain T of the system (55.2) if for every vector $v \in C^{(2r)}$ of compact support in $T - \partial T$ we have:

$$v(y) = -\int_T G(x, y)\, \overline{\mathfrak{N}v(x)}\, dx \,.$$

The first research on fundamental matrices looks at particular systems and is due to C. SOMIGLIANA [1], E.E. LEVI [1], G. GIRAUD [5]. For elliptic systems in the sense of LOPATINSKIĬ this study was then taken up in general by this author [2], first for equations with constant coefficients and then, with LEVI's method in a sufficiently restricted domain, for equations with variable coefficients. For elliptic systems in the sense of PETROWSKIĬ the case of constant coefficients was treated by F. JOHN [6] as a particular case of systems with analytic coefficients and by C.B. MORREY [3, 4] in the case $r = 1$. YA.B. LOPATINSKIĬ [3] proved, with LEVI's method, the existence in the small of fundamental matrices for systems with variable coefficients, while the existence in certain cases of *principal fundamental matrices* was established by A. AVANTAGGIATI [3]. For elliptic systems in the sense of DOUGLIS and NIRENBERG the construction of the fundamental matrix, in the case of constant coefficients, was done by these authors [1] with a procedure due to F. BUREAU [7]. Also for these systems, V.V. GRUŠIN [1] and A. AVANTAGGIATI [3] proved the existence of principal fundamental matrices in certain cases. For the case of variable coefficients, some indication relating to the existence in the small is given in § 10.6 of the volume [2] by L. HÖRMANDER. Finally, for certain systems of particular type see D. GRECO [6] and L.S. PARAS-JUK [5].

As for the case of a single equation, the problem of the existence of a fundamental matrix can be related on one hand with that of the existence for arbitrary f of at least one solution of the equation $\mathfrak{M}u = f$ and on the other hand with that of the validity of the unique continuation property. Each one of these questions has separately been an object for study, but the relationship between the results obtained had to be deepened later.

Regarding the first question we note that from theorems of existence in the small of a fundamental matrix it follows that in every sufficiently restricted domain an elliptic system always admits at least one solution. Besides, this can also be established directly without making use of the fundamental matrix, and for this we refer to J. PEETRE [5] and G. FI-

CHERA [33]. The existence in the large of at least one weak solution of every elliptic system with simple characteristics was then proved by N. KATZ [1].

Regarding the second question, we shall limit ourselves here to pointing out that for systems with simple characteristics some uniqueness theorems for Cauchy's problem have been given by YA. B. LOPATINSKIĬ[7] and T. MURAMUTU [1] and a theorem on the unique continuation property by K. HAYASHIDA [1]. An example of nonuniqueness for solutions of Cauchy's problem has been given by A. PLIŚ [1].

Finally, we mention that the study of fundamental matrices is also related with that of the singularities of the solutions of systems. A propos of this see YA. B. LOPATINSKII [1] for pointwise singularities and S. P. GAVELJA [1] for singularities on manifolds.

C. BOUNDARY VALUE PROBLEMS

For a properly elliptic system in the sense of DOUGLIS and NIRENBERG, the degree of the polynomial $a'(x, \xi)$ is necessarily even, say $2q$. For such a system we now pose a boundary value problem by assigning on ∂T q conditions of the type:

$$\sum_{j=1}^{N} \mathfrak{B}_{hj}(u_j) = \varphi_j \qquad (h = 1, 2, \ldots, q),$$

where $\mathfrak{B}_{hj}$ is a linear differential operator of order not larger than $r_h + t_j$, the t_j being the integers which already entered into the definition of ellipticity of systems, while the r_h constitute a new system of integers. We interpret $r_h + t_j < 0$ to mean $\mathfrak{B}_{hj} = 0$. In the particular case of an elliptic system according to PETROWSKIĬ, since we can assume the t_j all zero, it turns out that the $\mathfrak{B}_{hj}$ can have arbitrary order not larger than r_h; we have, moreover, $q = rN$.

As in the case of a single equation and of elliptic systems of the first order, the boundary value problem being considered is called *elliptic* if the $\mathfrak{B}_{hj}$ satisfy a "complementing condition" for the exact formulation of which we refer to S. AGMON, A. DOUGLIS, and L. NIRENBERG [1, II]. This condition, in the case of elliptic problems according to PETROWSKIĬ, was formulated for the first time by YA. B. LOPATINSKIĬ [4] and Z.YA. ŠAPIRO [4], who studied these problems in some particular cases with the method of integral equations. However, their treatment is bound to the hypothesis of the existence of a fundamental matrix for the system, the validity of which hypothesis is ensured only under particular conditions for the system and the domain. And this is one of the motives because of which preference is given to the methods of functional analysis in later developments of the theory.

If we leave less recent works by M. Gevrey [9], N. Simonov [2, 3], and M. Picone [9] out of consideration, the first elliptic problem taken into consideration is Dirichlet's problem for a strongly elliptic system in the sense of Petrowskiĭ. In this case the boundary conditions are written:

$$\frac{d^h u_j}{d n^h} = \varphi_{hj} \quad (h = 0, 1, ..., r-1; \quad j = 1, 2, ..., N). \quad (55.5)$$

Let us assume that the φ_{hj} are zero and that (55.1) is presented in the form (52.7) where this time the $a_{\beta\gamma}$ are $N \times N$ matrices and u is a vector with N components. We shall agree, moreover, to write $u \in H^{k,2}[H_0^{k,2}]$ if all the components of u are of class $H^{k,2}[H_0^{k,2}]$, likewise putting:

$$\| u \| = \left[\sum_{j=1}^{N} \| u_j \|^2 \right]^{1/2}$$

in these spaces with vectorial elements. We shall then say that the vector $u \in H_0^{r,2}(T)$ is a weak solution of Dirichlet's problem if it satisfies (52.12) for every $v \in H_0^{r,2}(T)$. M.I. Višik [3] has proved that, as for the case of a single equation, there exist a k and a λ_0 such that for $\lambda > \lambda_0$, Gårding's inequality (52.13) is satisfied. It follows from this that for $\lambda > \lambda_0$, Dirichlet's problem for the system (52.14) admits one and only one weak solution and that for a general λ, Fredholm's alternative (problem with zero index) holds. An analogous result was also established by F.E. Browder [4], who also gave regularization theorems of weak solutions in classes of the type $C^{(2m+s,\alpha)}$. Other regularization theorems in $H^{2m,2}$ are due to O.V. Guseva [1, 2]. On questions for strongly elliptic systems in the sense of Petrowskiĭ, we refer also to the works by L.N. Slobodeckiĭ [1] and J. Gobert [1].

Regarding strongly elliptic systems in the sense of Douglis and Nirenberg, Dirichlet's problem is posed by requiring that (55.5) be satisfied for $h = 1, 2, ..., s_j$. This problem turns out also to be elliptic and along these lines results analogous to Browder's, mentioned above, were established by L. Nirenberg [4]; of these, a detailed exposition can be found in § 6.5 of the volume [9] by C.B. Morrey. A maximum modulus theorem, which extends the analogous theorems obtained by C. Miranda [12] and S. Agmon [4] for the case of a single equation to the case of systems, has been proved by A. Canfora [1]. For the study of general boundary value problems, always for strongly elliptic systems, see M.I. Višik [9] and D.G. de Figueiredo [1].

Let us now pass on to occupy ourselves with boundary value problems for properly elliptic systems. In this regard we mention first of all a series of works directed at obtaining theorems on regularization and a priori bounds. Regarding the proof of a priori bounds in classes $H^{k,2}$ and $H^{k,p}$ of the solutions of elliptic systems according to PETROWSKIĬ, we shall mention the works by L. N. SLOBODECKII [2, ..., 6] and A. I. KOŠELEV [6]. Concerning elliptic systems in the sense of DOUGLIS and NIRENBERG, first to be mentioned is the memoir [1] by these authors dedicated to regularization in the interior of the domain in classes of the type $C^{(k,\alpha)}$. Regularization and a priori majorization in classes $C^{(k,\alpha)}(T)$ and $H^{k,p}(T)$ of the solutions of elliptic problems was then obtained by V. SOLONNIKOV [5, 6] and, in a more complete way, by S. AGMON, A. DOUGLIS, and L. NIRENBERG [1, II]. In this regard see also H. O. CORDES [5].

Regarding existence theory for elliptic problems for properly elliptic systems, the fundamental result, as for the case of a single equation, is that these are index pioblems. Concerning properly elliptic systems in the sense of DOUGLIS and NIRENBERG, we shall cite the works by L. R. VO-LEVIČ [2, 4] and by G. GEYMONAT [3] as well as paragraph 10.6 of the volume [2] by L. HÖRMANDER. For the particular case of DIRICHLET's problem see also the note [4] by M. SCHECHTER which is the first on the argument. The case of elliptic systems according to PETROWSKIĬ was deepened by G. GEYMONAT [4], who characterized a class of problems for which it is possible to define a formal adjoint[1].

Also for systems it is of gieat interest to consider the case in which the various equations and boundary conditions contain a parameter λ with the goal of determining a set of values for λ for which the problem admits a unique solution. Important results have been obtained in this regard by M. S. AGRANOVIČ and M. I. VIŠIK [1] in the memoir already cited in § 52 D foi the case of a single equation.

In the case of two variables, important results relating to the explicit determination of the index are due to A. I. VOL'PERT, first for Dirichlet's problem [5, 6, 10], then for geneial problems [7, 11]. In two other works [4, 8] VOL'PERT answers the question by transforming the problem into another for a first order system of equations[2]. Foi boundary conditions of even highei order than that of the equations in the system, the pioblem of the determination of the index has been studied by B. S. SIKORA [2]. For some properties of the index for $m > 2$ see M. S. AGRANOVIČ [1] for problems with differential boundary conditions and M. S. AGRANOVIČ and A. S. DYNIN [1] for problems with integro-differential boundary conditions.

---

[1] For problems of other types for which this is then possible see B. R. LAVRUK [5].

[2] This transformation is possible even in the case of more variables, but this does not seem ever to have been utilized. In this regard see S. AGMON, A. DOUGLIS, L. NIRENBERG [1, II].

D. Systems of second order equations

The theory of these systems has considerable analogy with that developed in the preceding chapters for the case of a single second order equation and therefore various results have been obtained for this which we shall not find recounted in the theory of general systems. Thus, for example, we know for these systems some comparison and uniqueness theorems for Dirichlet's problem of the types of those established in §§ 3, 5, 7 for the case of a single equation. In this regard see G. Cimmino [2], L. M. Kuks [2, 3, 4, 5], V. Ya. Skorobogatko [4, 5], T. Styš [1].

In the study of boundary value problems, the simplest is that in which the ith equation of the system contains the second derivatives of the ith unknown alone. Under these conditions both Dirichlet's problem and other more general ones can be treated with the methods of Chapter III or with similar methods; in this regard see M. Gevrey [6, 8] and G. Giraud [5, 6]. It was then observed by J. Schauder [9] that one could apply the method of Chapter V to the study of such systems, by basing it on the same a priori majorizations in classes $C^{(n + 2, \lambda)}$ already obtained for the case of a single equation. More advanced results for equations of this type were obtained by O. A. Ladyženskaja and N. N. Ural'ceva [8, Chapt. VII]. The study of a priori majorizations for elliptic systems, not satisfying the supplementary condition indicated above, was amply developed by C. B. Morrey [3, 4]. In the particular case of strongly elliptic systems these majorizations allowed Morrey to establish existence theorems for Dirichlet's problem also, under considerably reduced hypotheses on the coefficients of the system. A work by J. Nečas [9] then concerns a similar argument.

Let us also point out a work by G. Fichera [7] on the mixed problem for self-adjoint systems in $m \geq 2$ variables, two notes by B. Pini [7, 8] on Dirichlet's problem, understood in the sense of § 29, for certain self-adjoint strongly elliptic systems in two variables, a note by P. W. Berg [1] on the univalent mappings generated by a solution $u = u(x_1, x_2)$, $v = v(x_1, x_2)$ of a certain second order system.

All other works relating to these systems look at the problem of giving sufficient conditions in order that Fredholm's alternative holds for a boundary value problem, or else that of calculating the index of the problem. In almost all these works the method adopted is that of transforming the problem into a system of integral equations. In the development of this research, an observation by A. V. Bicadze, in which he proves with an example that for a (non properly elliptic) system Dirichlet's problem may not have a finite index, had considerable importance. Bicadze himself then gave sufficient conditions in order that Fredholm's alternative hold for problems of this kind. For all these questions we can consult the volume [1] by BICADZE which also summarizes the contents of previous works by the same author. The following works, listed chronologically, are all dedicated to the study of the questions pointed out in relation to boundary value problems of various types for systems of second order in two variables: A. I. VOL'PERT [1, 2, 3], YA. B. LOPATINSKIĬ [5], D. GRECO [4, 7], B. R. LAVRUK [1, 2, 3, 4], B. V. BOYARSKIĬ [6], E. V. ZOLOTAREVA [1, 2, 3], DING SHIA-KUAI, WANG KAN-TING, MA JU-NIEN, SHUN CHIAO-LI, CHANG-TONG [1], N. E. TOVMASJAN [1, ..., 5], C. KALIK and P. SZILÁGYI [1, 2], LI MING-ZHONG [1], A. DŽUAREV [5]. To the study of questions of the same sort for systems in three or more variables, then, are dedicated the works by Z. YA. ŠAPIRO [3, 4], A. I. VOL'PERT [9], B. V. BOYARSKIĬ [8], B. R. LAVRUK [6].

E. SYSTEMS ON MANIFOLDS. PSEUDO-DIFFERENTIAL OPERATORS

Analogously to what was said for a single equation (see §§ 48 and 52 E), it is also possible to consider problems for systems in which the unknown functions are defined on a manifold which can be either compact or endowed with a boundary. Thus, for example, B. PINI [4, 9] proved the existence theorem for a system of second order equations in two variables on a compact manifold, and A. B. JUDANINA [1] studied the same problem for a system of higher order equations. For a general statement of this problem see also § 10.6 of the volume [2] by L. HÖRMANDER. The question of the index has been studied by M. F. ATIYAH and M. I. SINGER [1] and W. KOPPELMAN [1] for compact manifolds and by M. F. ATIYAH and R. BOTT [1] for manifolds with boundary.

However, the most interesting research is that relating to systems of second order equations in which an exterior differential form is considered as unknown. Along these lines a first group of works is connected with the theory of *harmonic forms*. For the study of this theory, already considered previously by VOLTERRA[1], we refer to the books by W. V. D. HODGE [1], G. DE RAHM and K. KODAIRA [1], G. DE RAHM [1] and Chapter 7 of the volume [9] by C. B. MORREY. In the last text cited, we can find an ample bibliography on the argument, to which it is possible only to add the citation of some works by C. MIRANDA [7] and G. FICHERA [18, 26, 27].

A second group of works examines the so-called ∂̄-Neumann problem for complex linear differential forms. For the bibliography related to this and for an extensive treatment we refer to Chapter 8 of the volume [9] by C. B. MORREY. The interest of this last research is in part in the fact that the problem being considered is not elliptic and hence requires the employment of new techniques, for instance the theory of pseudo-differential operators. Referring for this argument to the works already

---

[1] VOLTERRA: Theory of functionals. Blackie, London (1930) 226 pp.

cited in § 52 E and in particular to those by J.J. Kohn and L. Nirenberg [2] and by L. Hörmander [4], we mention that this is also related to the theory of singular integro-differential operators of M.S. Agranovič [2] and of M.I. Višik and G.I. Eskin [1, 2] also referred to in § 52 E a propos of equations of higher order and which finds analogous application in the study of systems also.

Another interesting motive in research on the $\bar{\partial}$-Neumann problem is in the application to questions of differential geometry. For this see the expository article [8] by L. Nirenberg.

F. Systems on unbounded domains. Transmission problems

The first argument has been little studied up to now. Some theorems of Liouville and Phragmèn-Lindelöff type have been proved by D.F. Mel'nik [1], Ya.B. Lopatinskiĭ [8], I.S. Aršon and M.A. Evgrafov [1], I.S. Arson [1], C. Miranda [14], M.A. Efgrafov [1].

For the second argument see N.V. Žitarašu [1].

G. Systems of equations from mathematical physics

A first system of mathematical-physical interest, which has been the object of considerable research, is that of the equations of elasticity. For this system three fundamental problems are posed which are differentiated by the type of boundary conditions. Indeed, in the first problem the displacements are given on the boundary, in the second the stresses, in the third (mixed problems) the stresses on one portion of the boundary and the displacements on the remaining part.

The first and second problems were treated by A. Korn[1] in two now classical memoirs which go back to 1907/08. The first problem was also solved by G. Lauricella[2] and by L. Lichtenstein [3] by means of transforming it into a system of integral equations; K.O. Friedrichs [4] and B. Pini [2] then returned to this. Both of these authors consider boundary conditions of generalized type, along the lines of §§ 30 and 29 respectively, and Friedrichs makes use of a method of minimum, Pini of a procedure similar to that of § 29. The mixed problem was finally treated for the first time and in a very exhaustive way by G. Fichera [8], who later [14] made the interesting observation that, if we want the displace-

---

[1] Korn, A.: Sur les équations de l'élasticité. Ann. Éc. N. Sup. **24** (1907) 9–75. Solution générale du problème d'équilibre dans le théorie de l'élasticité dans le cas où les efforts sont données à la surface. Ann. Fac. Sci. Toulouse **10** (1908) 165–269.

[2] Lauricella, G.: Alcune applicazioni della teoria delle equazioni funzionali alla Fisica-matematica. Nuovo Cimento **13** (1907) 104–118, 155–174, 237–262, 501–518.

ments and the components of tension to turn out to be continuous in the whole domain, then it is necessary that the boundary values for the mixed problem satisfy quantitative conditions of integral type. In his memoir, FICHERA gave still a new treatment of the first and second boundary value problems in a domain whose boundary can even have singular points, as well as a procedure for the calculation of the solutions; the method which he followed, in its general lines, is that of M. PICONE of transformation of the various problems into systems of FISCHER-RIESZ integral equations (see § 31).

Also dedicated to the study of a mixed problem are a note by D.M. ÈĬDUS [2] and one by S. CAMPANATO [1]. We also mention that the study of transmission problems (see § 50 and § 52 H) had its origin in a note by M. PICONE [13] in which he pointed out the interest of problems of this type for the system of equations of elasticity. This problem was then studied by J.L. LIONS [3] and S. CAMPANATO [3].

Of great interest also are the works by G. FICHERA [30, 32] in which he gives the existence theorem for the boundary value problem with ambiguous conditions (see § 50) posed by A. SIGNORINI for the study of elastostatic problems with unilateral bonds.

Other works on the equations of elasticity, regarding the approximation or the majorization of the solutions, are due to S. BERGMAN [1], J.L. SYNGE [1], J.B. DIAZ and A. GREENBERG [1], K. WASHIZU [1]. To G. FICHERA [23] we owe the proof of a maximum modulus theorem along these lines. We mention also a note by B. PETTINEO [1] on the problem of analytic continuation of the solutions of the system of elasticity equations across a surface.

Finally, regarding plane elasticity problems, we refer to the classical treatise by N.I. MUSHELIŠVILI [2], which also contains an extensive bibliography in this regard, to which it is still necessary to add the citations of some works by L. SOBRERO [1, 2], G. FICHERA [6], S. CAMPANATO [2].

Other works by S. CAMPANATO [4, 5] and P. VILLAGGIO [1] are concerned with the study of more general systems, but ones quite close to those of elasticity theory. In the works [1, 2, 4] and in the monograph [3] by V.D. KUPRADZE, boundary value problems from the theory of free oscillations of elastic solids are studied. In two works by M.L. PRINCIVALLI [1, 2] a boundary value problem for a system of equations relating to the equilibrium of cylindrical domes is studied.

Another system which has interest for mathematical physics is that of the linearized NAVIER-STOKES equations, which has been the object of studies on the part of V.A. SOLONNIKOV [2, 4], L. CATTABRIGA [1], A. KRZYWICKI [1]. V.A. SOLONNIKOV [3] himself also studied the stationary boundary value problems for the equations of magnetohydrodynamics. To CHANG I-DEE and R. FINN [1] we owe, finally, the study of a system which includes as a special case both that of the elasticity equations and that of the linearized Navier-Stokes equations.

H. NONLINEAR SYSTEMS. QUESTIONS OF ANALYTICITY

Regarding the general theory of these two arguments quite little is to be added to the indications given in § 52J for the case of a single equation. Indeed, in the works cited there by VIŠIK and by LERAY and LIONS, the study of questions of existence is stated directly for the case of a system. It is necessary only to add the indications of some works by T.B. SOLOMJAK [2], N.N. URAL'CEVA [1, 2], A.P. OSKOLKOV [2], relating mainly to regularization questions of solutions of systems of special types. Then, regarding analyticity questions, the results obtained are summarized by the extension of theorems 44, II and III and, with the conditions indicated in § 52J in the case of a single equation, of theorems 44, V and VII and of other theorems of continuation for solutions satisfying general boundary conditions. In this regard see the works by C.B. MORREY and L. NIRENBERG [1], C.B. MORREY [5], A. FRIEDMAN [1, 3], and §§ 6.6 and 6.7 of the volume [9] by C.B. MORREY.

Other works then look at particular systems. We mention among these the works by F. STOPPELLI [1, 2, 3] on the nonlinear elasticity equations, a note by C. PUCCI [4] on a system for gas dynamics, some works by E. HEINZ [4, 5, 10] on certain second order systems in two variables which occur in the theory of univalent plane transformations.

Finally, of great interest is a group of works on the study of stationary solutions of the system of Navier-Stokes equations. This research, begun in 1930 with K.F. ODQVIST [1] and continued in a fundamental memoir by J. LERAY [1] in 1933, was resumed in more recent times by O.A. LADYŽENSKAJA [6, 7] and R. FINN [5, ..., 10, 16, 17]. The most advanced result was then obtained by H. FUJITA [1].

56. Degenerate elliptic equations. Questions of a small parameter. A linear differential equation with real coefficients defined in a domain T is called *elliptic-parabolic* in T if its characteristic form is (positive) semi-definite at every point of T, without being definite at every point of T. An equation which is elliptic in a region Ω and elliptic-parabolic in $\overline{\Omega}$ is called *degenerate* at the points of $\partial\Omega$ at which its characteristic form is not definite. Thus, for example, the equation:

$$\mathfrak{M}u = \frac{\partial^2 u}{\partial x^2} + y^\alpha \frac{\partial^2 u}{\partial y^2} + b_1 \frac{\partial u}{\partial x} + b_2 \frac{\partial u}{\partial y} + cu = f \quad (\alpha > 0)$$

is elliptic in the open half-plane $y > 0$, elliptic-parabolic in the closed half-plane $y \geq 0$, degenerate on the straight line $y = 0$. For this equation it is interesting to consider DIRICHLET's problem in a domain T whose boundary consists of a segment Σ of the axis $y = 0$ and of a curve Γ in the half-plane $y > 0$. In this regard M.V. KELDYŠ [1] proved that no new fact presents itself in the study of this problem if $\alpha < 1$; on the other hand, if $\alpha \geq 1$ and if b_2 satisfies in a neighborhood of Σ a further hypothesis, which we shall not make precise, then the problem admits one and only one solution by giving the boundary condition exclusively on Γ[1]. Analogous results for the second or third boundary value problem have been proved by O.A. OLEJNIK [4]. Later, various results for the same equation or slightly dissimilar ones were proved by S.G. MIHLIN [6], N.D. VVEDENSKAJA [1], A.M. IL'IN [1], S.A. TERSENOV [1], HOU CUN'-I [1], F.I. FRANKL' [1], M.M. SMIRNOV [1], DONG GUANG-CHANG [1, 2], YANG GUANG-JUN [1], CHEN LIANG-CHING [1].

The case of equations in more variables which degenerate on one of the coordinate hyperplanes has been treated quite generally by M.I. VIŠIK [10] and by S.G. MIHLIN [7, 8] and for particular equations by E.T. POULSEN [1] and L.S. PARASJUK [1, 2, 3, 4]. Some equations which degenerate at isolated points have been studied by H. HORNICH [1], A.M. IL'IN [1, 2], A.I. AČIL'DIEV [1]. The case of an equation which degenerates on a portion of the boundary of a region Ω has been studied by M.S. BAOUENDI [1].

Various types of degenerate equations of higher order have been studied by V.K. ZAHAROV [1], V.P. GLUŠKO [1], L.N. SLOBODECKIĬ and I.A. SOLOMEŠČ [1], I.A. KIPRJANOV [1], A.S. FOHT [1], A. NARČAEV [1], S.M. NIKOLSKIĬ and P.I. LIZORKIN [1], and some nonlinear degenerate equations have been considered by JU.A. DUBINSKIĬ [1] and K. GRÖGER [1, 2]. Some degenerate systems have been studied by V.P. DIDENKO [1] and CHEN LIANG-CHING [2].

Let us now pass on to examine general second order elliptic-parabolic equations. In this research, in contrast to what was done in the works cited above, we assume only that the characteristic form is, rather than definite, positive semidefinite, without making any hypotheses either on the location of the points at which the equation is degenerate or on the behavior of the a_{ik} at these points. A first group of results regarding the extension of the results in § 3 is due to M. PICONE [1, 6, 7], C. PUCCI [1, 5], P. HARTMAN and R. SACKSTEDER [1], A.D. ALEKSANDROV [5], R.M. REDHEFFER and E.G. STRAUSS [1]. Concerning boundary value problems, a first treatment is due to G. FICHERA [20, 22, 25]. If, for example, with the notation of §§ 4 and 6, we denote by Σ^0 the

---

[1] This phenomenon is analogous to that already related a propos of the equation (26.3). This is natural because if we write (26.3) multiplying by y^2, we obtain an equation which is also degenerate for $y = 0$.

set of ∂T on which $\cos(n\,v) = 0$ and by $\varSigma^1$ the subset of $\varSigma^0$ on which $b > 0$, we can pose a problem of DIRICHLET type for the elliptic-parabolic equation being considered by assigning the boundary condition:

$$u = \varphi \quad \text{for} \quad x \in (\partial T - \varSigma^0) \cup \varSigma^1.$$

If, then, $(\partial T - \varSigma^0)$ is closed, we can also consider a mixed problem by imposing on u the boundary conditions:

$$u = \varphi \quad \text{for} \quad x \in \varSigma^1, \quad \mathfrak{P}u = \psi \quad \text{for} \quad x \in \partial T - \varSigma^0,$$

and this problem is absolutely of NEUMANN type if $\varSigma^1$ is empty. In the works by G. FICHERA already cited, the existence of weak solutions of this problem is proved, in the case $c(x) < 0$. Later O.A. OLEJNIK [7, 8, 9] found these results again and gave uniqueness and regularization theorem. The case in which $\varSigma^0$ is empty was studied by M.I. FREĬDLIN [3, 4] with probabilistic procedures. Finally, S.N. KRUŽKOV [1, 3] and M.K.V. MURTHY and G. STAMPACCHIA [1] studied the weak solutions of an elliptic-parabolic equation under the hypothesis that:

$$\sum_{i,\,k=1}^{m} a_{ik}(x)\,\xi_i\,\xi_k \geq a_0(x)\sum_{i=1}^{m}\xi_i^2$$

with:

$$[a_0(x)]^{-1} \in L^t(T), \quad t \geq \frac{n}{2}.$$

The results obtained are in regard to the extension of HARNACK's theorem, theorems on removable singularities, and regularization theorems (see § 30).

In order to close this section, we wish to point out the so-called *questions of a small parameter* which are often connected with the study of degenerate equations. In order to give an idea of these questions we shall consider Dirichlet's problem in a domain for the equation:

$$\varepsilon\,\varDelta_2 u + b_1\,\frac{\partial u}{\partial x} + b_2\,\frac{\partial u}{\partial y} + cu = f \tag{56.1}$$

and let $u(x,\varepsilon)$ be the solution, which we shall assume exists and is unique. A small parameter problem consists, for example, of studying the behavior of $u(x,\varepsilon)$ as $\varepsilon \to 0$ and in particular of conditions under which $u(x,\varepsilon)$ is convergent for $\varepsilon \to 0$, and, putting $u_0 = \lim_{\varepsilon \to 0} u(x,\varepsilon)$, we have that u_0 is a solution of the equation:

$$b_1\,\frac{\partial u_0}{\partial x} + b_2\,\frac{\partial u_0}{\partial y} + cu_0 = f\,.$$

Naturally, if this happens, then the convergence of $u(x, \varepsilon)$ to u can take place in $T - \partial T$ or at most also on a portion of ∂T, but not in all of T, otherwise we would obtain an existence theorem for the solution of a Dirichlet problem for a first order equation, which is absurd. The fact that $u(x, \varepsilon) - u_0$ is infinitesimal only in $T - \partial T$ is expressed by saying that is "a function of boundary layer type" and the problem of estimating this function asymptotically as $\varepsilon \to 0$ is of great interest. For (56.1) this problem was studied by W. WASOW [1] in a particular case and then by. N. LEVINSON [1] and S. L. KAMENOMOSTSKAJA [1, 2]. Analogous results for the second or third boundary value problems have been established by O. A. OLEJNIK [3]. Other results along the same lines are due to V. J. MIZEL [1], R. B. DAVIS [2], L. R. VOLEVIČ [1], while M. ZLAMAL [1] and P. DE LUCIA [1] have studied the case of an elliptic equation which as $\varepsilon \to 0$ reduces to a parabolic equation. Analogous questions have been studied for second order equations in $m > 2$ variables by O. A. LADY-ŽENSKAJA [4], I. KOPAČEK [1], V. J. MIZEL [2], M. I. FREĬDLIN [1, 2], and for some particular equations of fourth or sixth order by D. MORGEN-STERN [1], R. B. DAVIS [1], M. Š. BIRMAN [5].

Finally, the case of a general equation of higher order was deeply studied by M. I. VIŠIK and L. A. LYUSTERNIK [1] and by D. HUET [1, 2, 3, 4] and V. SOLONNIKOV [1]. VIŠIK himself and LYUSTERNIK then extensively studied other small parameter questions also. For example, problems in which as $\varepsilon \to 0$ the equation degenerates only in a portion of the domain, small parameter questions for transmission problems, problems in which the parameter ε occurs in the boundary conditions. Thus, for example, for a second order elliptic equation in two variables not containing the parameter we can pose a Dirichlet condition of the type:

$$u = \varphi(s) \exp\left(\frac{is}{\varepsilon}\right),$$

s being the arc length along ∂T, and study the behavior of u as $\varepsilon \to 0$. Various results for all these questions and for other analogous ones are discussed in the memoir [3] by the aforesaid authors, while the memoir [2] is dedicated to the study of the general fundamentals of the method adopted.

Bibliography[1]

Ačil'diev, A. I.: [1] The first and second boundary-value problem for elliptic equations which are degenerate at a finite number of interior points. Dokl. Akad. Nauk SSSR **152** (1963) 13—16 (Russian).

Aeppli, A.: [1] On the uniqueness of compact solutions for certain elliptic differential equations. Proc. Amer. Math. Soc. **11** (1960) 826—832.

Agmon, S.: [1] Multiple layer potentials and the Dirichlet problem for higher order elliptic equations in the plane. Commun. pure appl. Math. **10** (1957) 179—239. — [2] The coerciveness problem for integro-differential forms. J. Anal. Math. **6** (1958) 182—223. — [3] The L^p approach to the Dirichlet problem. Ann. Sc. N. Sup. Pisa **13** (1959) 405—448. — [4] Maximum theorems for solutions of higher order elliptic equations. Bull. Amer. Math. Soc. **66** (1960) 77—80. — [5] Remarks on self adjoint and semibounded elliptic boundary value problems. Proc. Int. Symp. Linear Spaces Jerusalem (1961) 1—13. — [6] On the eigenfunctions and on the eigenvalues of general elliptic boundary value problems. Commun. pure appl. Math. **15** (1962) 119—147. — [7] Lectures on elliptic boundary value problems. Van Nostrand Math. Studies **2** (1965) 291 pp.

Agmon, S., and L. Bers: [1] The expansion theorem for pseudo-analytic functions. Proc. Amer. Math. Soc. **3** (1952) 757—764.

Agmon, S., A. Douglis and L. Nirenberg: [1] Estimates near the boundary for solutions of elliptic partial differential equations satisfying general boundary conditions, I, II. Commun. pure appl. Math. **12** (1959) 623—727; **17** (1964) 35—92; Russian transl. of I, Izdat Inost. Lit. Moscow (1962).

Agranovič, M. S.: [1] On questions concerning the index of elliptic operators. Dokl. Akad. Nauk SSSR **142** (1962) 983—985 (Russian). — [2] Elliptic singular integro-differential operators. Usp. Mat. Nauk **20**, n. 5 (1965) 1—122 (Russian); transl. as Russian Math. Surveys **20**, n. 5/6 (1965) 1—122.

Agranovič, M. S., and A. S. Dynin: [1] General boundary problems for elliptic systems in multidimensional domains. Dokl. Akad. Nauk SSSR **146** (1962) 511—514 (Russian).

---

[1] In this bibliography translations of Dokl. Akad. Nauk SSSR as Soviet Math. are not mentioned.

AGRANOVIč, M. S., and M. I. VIšIK: [1] Elliptic problems with a parameter and parabolic problems of general type. Usp. Mat. Nauk **19**, n. 3 (1964) 53—158 (Russian); transl. as Russian Math. Surveys **19**, n. 3 (1964) 53—158.

AHLFORS, L., and L. BERS: [1] Riemann's mapping theorem for variable metrics. Ann. of Math. **72** (1960) 385—404.

AISENSTAT, N.: [1] Un type d'opérateurs homogènes. Učen. Zap. Moscow Mat. **15** (1939) 95—112.

AKô, K.: [1] On the maximum principles of second order elliptic differential equations. Proc. Japan. Acad. **36** (1960) 492—494. — [2] On the Dirichlet problem concerning linear elliptic differential equations of second order. J. Fac. Sci. Univ. Tokyo **9** (1961) 29—58. — [3] On Perron's process associated with second order elliptic differential equations, I. II. J. Fac. Sci. Univ. Tokyo **9** (1962) 165—202; **10** (1964) 81—87. — [4] Subfunctions for quasilinear elliptic differential equations. J. Fac. Sci. Univ. Tokyo **9** (1962) 403—416.

AKô, K., and T. KUSANO: [1] On bounded solutions of second order elliptic differential equations. J. Fac. Sci. Univ. Tokyo **11** (1964) 29—37.

ALBERTONI, S.: [1] Sulla soluzione del problema di Neumann per l'equazione $\Delta_2 u - k u = f$. Rend. Ist. Lombardo **87** (1954) 400—432. — [2] Sulla risoluzione del problema di Neumann per l'equazione $\Delta_2 u + k u = f$ in un dominio con punti angolosi. Rend. Ist. Lombardo **90** (1956) 221—243.

ALEKSANDROV, A. D.: [1] Intrinsic geometry of convex surfaces. Gos. Izdat Tehn.-Teor. Lit. Moscow Leningrad (1948) 387 pp. (Russian); German transl. Berlin: Akademie-Verlag (1955) 522 pp. — [2] Some theorems on partial differential equations of the second order. Vestnik Leningrad Univ. **9** (1954) 3—17 (Russian). — [3] Uniqueness theorems for surfaces "in the large", I, II, III, IV, V. Vestnik Leningrad Univ. **11**, n. 19 (1956) 5—17; **12**, n. 7 (1957) 15—44; **13** (1958) n. 7, 14—26; n. 13, 27—34; n. 19, 5—8 (Russian). — [4] Dirichlet's problem for the equation Det $|| z_{ij} || = \varphi(z_1, \ldots, z_n, x_1, \ldots, x_n)$. Vestnik Leningrad Univ. **13** (1958) 5—24 (Russian). — [5] Investigations on the maximum principle, I, II, III, IV, V, VI. Izv. Vysš. Učebn. Zaved Mat. (1958) n. 5, 126—157; (1959) n. 3, 3—12 and n. 5, 16—32; (1960) n. 3, 3—15 and n. 5, 16—26; (1961) n. 1, 3—20 (Russian). — [6] Certain estimates for the Dirichlet's problem. Dokl. Akad. Nauk SSSR **134** (1960) 1001—1004 (Russian). — [7] Uniqueness conditions and bounds for the solution of the Dirichlet problem. Vestnik Leningrad Univ. **18** (1963) n. 13, 5—29 (Russian). — [8] A general method of bounding the solution of a differential equation. Sibirsk Mat. Ž. **7** (1966) 486—498 (Russian). — [9] Bounds for the solution of a linear second order equation. Vestnik Leningrad Univ. **21** (1966) n. 1, 5—25 (Russian). — [10] Bounds and uniqueness conditions for the solution of an elliptic equation. Vestnik Leningrad Univ. **21** (1966) n. 7, 5—20 (Russian).

ALMGREN, F. I.: [1] Some interior regularity theorems for minimal surfaces and an extension of Bernstein's theorem. Ann. of Math. **84** (1966) 277—292.

AMERIO, L.: [1] Teoremi di esistenza per le equazioni lineari del secondo ordine di tipo ellittico nei domini illimitati. Rend. Acc. Italia **4** (1943) 287–298. – [2] Sull'integrazione dell'equazione $\Delta_4 u = 0$ in due variabili. Rend. Ist. Lombardo **77** (1943–44) 377–419. – [3] Sull'integrazione dell'equazione $\Delta_2 u - \lambda^2 u = f$ in un dominio di connessione qualsiasi. Rend. Ist. Lombardo **78** (1944–45) 1–24. – [4] Sull'integrazione dell'equazione $\Delta_{2k} u = f$. Ann. Mat. pura appl. **24** (1945) 119–138. – [5] Sull'integrazione delle equazioni lineari a derivate parziali del secondo ordine di tipo ellittico. Acta Pont. Acad. Sci. **9** (1945) 213–228. – [6] Sul calcolo delle soluzioni dei problemi al contorno per le equazioni lineari del secondo ordine di tipo ellittico. Amer. J. Math. **69** (1947) 447–489. – [7] Sul calcolo delle autosoluzioni dei problemi al contorno per le equazioni lineari del secondo ordine di tipo ellittico, I, II. Rend. Acc. Lincei **1** (1946) 352–359 and 505–509. – [8] Teoremi di esistenza per i problemi di Dirichlet e di Neumann per l'equazione $\Delta_2 u + k u = 0$. Ricerche di Mat. Napoli **5** (1956) 58–96.

ARONSZAJN, N.: [1] Theory of reproducing kernels. Trans. Amer. Math. Soc. **68** (1950) 337–404. – [2] Green's functions and reproducing kernels. Proc. Symp. Spectral Th. Diff. Pr. Oklahoma College (1951) 355–411. – [3] On coercive integro-differential quadratic forms. Proc. Conf. Part. Diff. Eq. Univ. Kansas (1954) 94–106. – [4] A unique continuation theorem for solutions of elliptic partial differential equations or inequalities of second order. J. Math. pures appl. **36** (1957) 235–249. – [5] Associated spaces, interpolation theorems and the regularity of solutions of differential problems. Proc. Symp. Pure Math. Amer. Math. Soc. **4** (1961) 23–32.

ARONSZAJN, N., A. KRZYWICKI and J. SZARSKI: [1] A unique continuation theorem for exterior differential forms on Riemannian manifolds. Ark. Mat. **4** (1962) 417–453.

ARONSZAJN, N., and A. N. MILGRAM: [1] Differential operators on Riemannian manifolds. Rend. Circ. Mat. Palermo **2** (1953) 1–61.

ARŠON, I. S.: [1] A Phragmèn-Lindelöf theorem for a linear elliptic system. Dokl. Akad. Nauk SSSR **139** (1961) 271–274 (Russian). – [2] A Pragmèn-Lindelöf theorem for a linear elliptic system whose coefficients depend on a variable. Mat. Sbornik **61** (1963) 362–376 (Russian).

ARŠON, I. S., and M. A. EVGRAFOV: [1] Estimate of the growth of a solution of a system with inhomogeneous boundary conditions and Phragmèn-Lindelöf theorems. Dokl. Akad. Nauk SSSR **134** (1960) 507–510 (Russian).

ARUFFO, G.: [1] Sistemi ellittici di equazioni lineari del primo ordine in domini a connessione multipla. Ricerche di Mat. Napoli **13** (1964) 80–91. – [2] Sull'integrazione delle forme differenziali di grado $n - 1$ a coefficienti in $L^\alpha (\alpha > 1)$ in domini di R^n molteplicemente connessi. Ann. Mat. pura appl. **69** (1965) 89–106.

ASCOLI, G.: [1] Sull'unicità della soluzione nel problema di Dirichlet. Rend. Acc. Lincei **8** (1928) 348–351. – [2] Sull'equazione di Laplace nello spazio iperbolico. Math. Z. **31** (1929) 45–96.

Ascoli, G., P. Burgatti e G. Giraud: [1] Equazioni alle derivate parziali dei tipi ellittico e parabolico. Pubbl. Sc. N. Sup. Pisa (1936) 186 pp.

Atiyah, M. F., and R. Bott: [1] The index problem for manifolds with
boundary. Proc. Bombay Coll. Diff. Anal. (1964) 175–186.

Atiyah, M. F., and I. M. Singer: [1] The index of elliptic operators on
compact manifolds. Bull. Amer. Math. Soc. 69 (1963) 422–433.

Atkinson, F. V.: [1] On Sommerfeld's "radiation condition". Phil. Mag. 40
(1949) 645–651. – [2] On a theorem of K. Yosida. Proc. Japan. Acad. 28
(1952) 327–329.

Avantaggiati, A.: [1] Nuovi contributi allo studio di un problema al
contorno per un sistema ellittico di equazioni lineari alle derivate parziali del primo ordine in tre variabili. Ricerche di Mat. Napoli 10 (1961)
3–30. – [2] Sui sistemi ellittici del primo ordine a coefficienti costanti
in due variabili indipendenti. Rend. Acc. Lincei 34 (1963) 611–619. –
[3] Sulle matrici fondamentali principali per una classe di sistemi
ellittici e ipoellittici. Ann. Mat. pura appl. 65 (1964) 191–238. –
[4] Nuovi contributi allo studio dei problemi al contorno per i sistemi
ellittici del primo ordine. Ann. Mat. pura appl. 69 (1965) 107–170.

Bajocchi, C.: [1] Su alcuni spazi di distribuzioni nel problema di Dirichlet
per le equazioni lineari ellittiche. Ricerche di Mat. Napoli 13 (1964)
3–29.

Bakel'man, I. Ya.: [1] Regularity of solutions of Monge-Ampère equations.
Leningrad Gos. Ped. Inst. Uč. Zap. 166 (1958) 143–184 (Russian). –
[2] On the theory of Monge-Ampère equations. Vestnik Leningrad
Univ. 13 n. 1 (1958), 25–38 (Russian). – [3] The first boundary-value
problem for some non linear elliptic equations. Dokl. Akad. Nauk SSSR
124 (1959) 249–252 (Russian). – [4] A class of non linear differential
equations. Dokl. Akad. Nauk SSSR 126 (1959) 244–247 (Russian). –
[5] The Dirichlet problem for equations of Monge-Ampère type and
their n-dimensional analogues. Dokl. Akad. Nauk SSSR 126 (1959)
923–926 (Russian). – [6] The first boundary-value probleme for quasi-
linear elliptic equations. Dokl. Akad. Nauk SSSR 134 (1960) 1005–1008
(Russian). – [7] On the stability of Monge-Ampère equations of elliptic
type. Usp. Mat. Nauk 15, n. 1 (1960) 163–170 (Russian). – [8] A
variational problem related to the Monge-Ampère equation. Dokl. Akad.
Nauk SSSR 141 (1961) 1011–1014 (Russian). – [9] On the theory of
quasi linear elliptic equations. Sibirsk Mat. Ž. 2 (1961) 179–186 (Russian).
– [10] A variational problem associated with the Monge Ampère
equation. Leningrad Gos. Ped. Inst. Učen. Zap. 238 (1962) 119–131
(Russian). – [11] Regular solutions of the Monge-Ampère equation.
Dokl. Akad. Nauk SSSR 157 (1964) 247–249 (Russian).

Bakel'man, I. Ya., and I. Ja. Guberman: [1] The Dirichlet problem for
an equation with a Monge-Ampère operator. Dokl. Akad. Nauk SSSR
148 (1963) 247–250 (Russian). – [2] The Dirichlet problem for an
equation with a Monge-Ampère operator. Sibirsk Mat. Ž. 4 (1963)
1208–1220 (Russian).

BAOUENDI, M. S.: [1] Sur une classe d'opérateurs elliptiques dégénérés. Bull. Soc. Math. France **95** (1967) 45–87.

BARRAR, R. B.: [1] The Schauder's paper on linear elliptic differential equations. J. Math. Anal. Appl. **3** (1961) 171–195.

BARROS NETO, J.: [1] Inhomogeneous boundary value problems in a half space. Ann. Sc. N. Sup. Pisa **19** (1965) 331–365.

BATTEZZATI, M.: [1] Limitazioni del gradiente di soluzioni di equazioni ellittiche in due variabili. Rivista Mat. Parma **5** (1964) 41–48.

BAUER, H.: [1] Axiomatische Behandlung der Dirichletschen Problems für elliptische und parabolische Differentialgleichungen. Math. Ann. **146** (1962) 1–59. – [2] Weiterführung einer axiomatischen Potentialtheorie ohne Kern (Existenz von Potentialen). Z. Wahrscheinlichkeitstheorie und verw. Geb. **1** (1962–63) 197–229.

BEALS, R.: [1] Non-local boundary value problems for elliptic operators. Amer. J. Math. **87** (1965) 315–362.

BECKERT, H.: [1] Die Abhängigkeit der Lösungen quasi linearer elliptischer Systeme partieller Differentialgleichungen erster Ordnung mit zwei unabhängigen Variablen von einem Parameter. Math. Nach. **5** (1951) 111–121. – [2] Über lineare elliptische Systeme partieller Differentialgleichungen erster Ordnung mit zwei unabhängigen Variablen. Math. Nach. **5** (1951) 173–208. – [3] Systeme partieller linearer elliptischer Differentialgleichungen erster und höherer Ordnung mit zwei unabhängigen Variablen. Math. Nach. **12** (1954) 257–272. – [4] Abschätzungen bei linearen elliptischen Systemen I. Ordnung mit zwei unabhängigen Variablen. Math. Nach. **13** (1955) 327–342. – [5] Das Dirichletsche Problem des Systems der Jacobischen Gleichung eines zweidimensionalen Variationsproblems für n gesuchte Funktionen im linearen und quasilinearen Fall. Math. Nach. **15** (1956) 7–29. – [6] Eine bemerkenswerte Eigenschaft der Lösungen des Dirichletschen Problems bei linearen elliptischen Differentialgleichungen. Math. Ann. **139** (1960) 255–264.

BEREZANSKIĬ, M., S. G. KREIN and JA. A. ROĬTBERG: [1] A theorem on homeomorphism and local increase of smoothness up to the boundary for solutions of elliptic equations. Dokl. Akad. Nauk SSSR **148** (1963) 745–748 (Russian).

BEREZANSKIĬ, M., and JA. A. ROĬTBERG: [1] On the smoothness up to the boundary of the region of the kernel of the resolvent of an elliptic operator. Ukrain Mat. Ž. **15** (1963) 185–189 (Russian).

BERG, P. W.: [1] On univalent mappings by solutions of linear elliptic partial differential equations. Trans. Amer. Math. Soc. **84** (1957) 310–318.

BERGMAN, S.: [1] Über die Bestimmung der elastischen Spannungen und Verschiebungen in einem konvexen Körper. Math. Ann. **98** (1927)

248–263. – [2] The kernel function and conformal mapping. Math. Survey Amer. Math. Soc. **5** (1950) 161 pp. – [3] Integral operators in the theory of linear partial differential equations. Ergeb. Math., Vol. **23** (1961) 145 pp. Berlin-Heidelberg-New York: Springer-Verlag; Russian transl. Izdat "Mir" Moscow (1964) 305 pp.

BERGMAN, S., and M. SCHIFFER: [1] A majorant method for non linear partial differential equations. Proc. Nat. Acad. Sci. USA **37** (1951) 744–749. – [2] Kernel functions and elliptic differential equations in mathematical physics. New York: Academic Press Inc. (1953) 432 pp.

BERNSTEIN, I.S.: [1] On the unique continuation problem for elliptic partial differential equations. J. Math. Mech. **10** (1961) 579–606.

BERNSTEIN, S.: [1] Sur la nature analytique des solutions de certaines équationes aux dérivées partielles du second ordre. Math. Ann. **59** (1904) 20–76. – [2] Sur la généralisation du problème de Dirichlet. Math. Ann. **62** (1906) 253–271 and **69** (1910) 82–136. – [3] Investigation and integration of second order partial differential equations of elliptic type. Soobšč. Har'kov Mat. Obšč. **11** (1908) 1–164 (Russian). – [4] Sur les équations du calcul des variations. Ann. Éc. N. Sup. **29** (1912) 431–485. – [5] Sur l'intégration des équations aux dérivées partielles du type elliptique. Math. Ann. **95** (1926) 585–594 and **96** (1927) 633–647. – [6] Sur la nature analytique des solutions des équations aux dérivées partielles du type elliptique. Extrait d'une correspondance avec M. Radò. Math. Z. **25** (1926) 505–513. – [7] Über ein geometrisches Theorem und seine Anwendung auf die partiellen Differentialgleichungen vom elliptischen Typus. Math. Z. **26** (1927) 551–558. – [8] Démonstration du théorème de M. Hilbert sur la nature analytique des solutions des équations du type elliptique sans l'emploi des séries normales. Math. Z. **28** (1928) 330–348. – [9] On some a priori estimates in the generalized Dirichlet problem. Dokl. Adak. Nauk SSSR **124** (1959) 735–738 (Russian).

BERS, L.: [1] An existence theorem in two-dimensional gas dynamics. Proc. Symp. Appl. Math. Amer. Math. Soc. **1** (1949) 41–46. – [2] Partial differential equations and generalized analytic functions. Proc. Nat. Acad. Sci. USA **36** (1950) 130–136 and **37** (1951) 42–47. – [3] The expansion theorem for sigma monogenic functions. Amer. J. Math. **72** (1950) 705–712. – [4] Singularities for minimal surfaces. Proc. Inter. Congr. Math. Cambridge **2** (1950) 157–163. – [5] Isolated singularities for minimal surfaces. Ann. of Math. **53** (1951) 364–386. – [6] Boundary value problems for minimal surfaces with singularities at infinity. Trans. Amer. Math. Soc. **70** (1951) 465–491. – [7] Abelian minimal surfaces. J. Anal. Math. **1** (1951) 43–58. – [8] The expansion theorem for pseudo-analytic functions. Proc. Amer. Math. Soc. **3** (1952) 757–764. – [9] Theory of pseudo-analytic functions. Lecture notes (mimeographed) New York Univ. (1953) 187 pp. – [10] Partial differential equations and pseudo-analytic functions on Riemannian surfaces. Ann. of Math. Studies **30** (1953) 157–165. – [11] Univalent solutions of linear elliptic systems. Commun. pure appl. Math. **6** (1953) 513–526. – [12] Existence and uniqueness of a subsonic compressible flow past a given profile. Commun. pure appl. Math. **7** (1954) 441–504. – [13] Non linear elliptic equations without non linear entire solutions. J. Rat. Mech. **3** (1954) 767–787. – [14] Local theory of pseudo-analytic functions. Lectures on functions of a complex variable, Univ. of Michigan (1955) 213–244. – [15] Function-theoretical properties of solutions of partial differential equations. Ann. of Math. Studies **33** (1954) 69–94. – [16] Local behavior of solutions of general linear elliptic equations. Commun. pure appl. Math. **8** (1955) 473–496. – [17] Formal powers and power series. Commun. pure appl. Math. **9** (1956) 693–711. – [18] Remark on an application of pseudo-analytic functions. Amer. J. Math. **78** (1956) 486–496. – [19] An outline of pseudo-analytic functions. Bull. Amer. Math. Soc. **62** (1956) 291–331.

BERS, L., and A. GELBART: [1] On a class of differential equations in mechanics of continua. Quart. appl. Math. **1** (1943) 168–188. – [2] On a class of functions defined by partial differential equations. Trans. Amer. Math. Soc. **56** (1944) 67–93. – [3] On generalized Laplace transformations. Ann. of Math. **48** (1947) 342–357.

BERS, L., and L. NIRENBERG: [1] On a representation theorem for linear elliptic equations with discontinuous coefficients and its applications. Convegno Int. Eq. Lin. Der. Parz. Trieste (1954) 111–140. – [2] On linear and non linear elliptic boundary value problems in the plane. Convegno Int. Eq. Lin. Der. Parz. Trieste (1954) 141–167.

BERS, L., and M. SCHECHTER: [1] Elliptic equations. Partial differential equations. New York: Interscience Publ. (1964) 131–299.

BESALA, P.: [1] On solutions of non linear second order elliptic equations defined in unbounded domains. Atti Sem. Mat. Fis. Modena **13** (1964) 74–86.

BICADZE, A. V.: [1] Equations of the mixed type. Izdat Akad. Nauk SSSR (1959) (Russian); English transl. Pergamon Press (1964) 160 pp. – [2] On the homogeneous oblique derivative problem for harmonic functions in a three dimensional region. Dokl. Akad. Nauk SSSR **148** (1963) 749–752 (Russian).

BIRMAN, M.Š.: [1] On the theory of general boundary problems for elliptic equations. Dokl. Akad. Nauk SSSR **92** (1953) 205–208 (Russian). – [2] On minimal functionals for elliptic differential equations of second order. Dokl. Akad. Nauk SSSR **93** (1953) 953–956 (Russian). – [3] On Trefftz's variational method for the equation $\Delta^2 u = f$. Dokl. Akad. Nauk SSSR **101** (1955) 201–204 (Russian). – [4] Variational methods of solution of boundary problems analogous to the method of Trefftz. Vestnik Leningrad Univ. **11** n. 13 (1956), 68–89 (Russian). – [5] Method of quadratic forms in problems with small parameter in the highest derivatives. Vestnik Leningrad Univ. **12** n. 13 (1957), 9–12 (Russian).

BIRMAN, M.Š., and G. E. SKVORKOV: [1] On square summability of highest derivatives of the solution of the Cauchy problem in a domain with pieceweise smooth boundary. Izv. Vysš. Učebn. Zaved Mat. (1962) 11–21 (Russian).

Boboc, N., et P. Mustaţă: [1] Sur un problème concernant les domaines d'unicité pour le problème associé à un opérateur elliptique. Rend. Acc. Lincei **42** (1967) 181–186.

Bochenek, J.: [1] Some properties of solutions of elliptic partial differential equations of the second order. Ann. Polon. Math. **16** (1965) 149–152.

Bochner, S.: [1] Partial differential equations and analytic continuations. Proc. Nat. Acad. USA **38** (1952) 227–230. — [2] Zeta functions and Green's functions for linear partial differential operators of elliptic type with constant coefficients. Ann. of Math. **57** (1953) 32–56.

Bohn, E., and L. K. Jackson: [1] The Liouville theorem for a quasi linear elliptic partial differential equation. Trans. Amer. Math. Soc. **104** (1962) 392–397.

Bonsall, E. F.: [1] On generalized subharmonic functions. Proc. Cambridge Philos. Soc. **46** (1950) 387–395.

Borrelli, R. L.: [1] A priori estimates for a class of second order non-regular elliptic boundary problems. Thesis Univ. of California (1963).

Borsuk, M. V.: [1] On the solvability of Dirichlet's problem for quasi-linear elliptic operators of the second order with discontinuous coefficients. Dopovidi Akad. Nauk Ukrain. RSR (1965) 1259–1261 (Ukrainian). — [2] On the solvability of Dirichlet and Neumann problems for the weakly non linear operator div $[\varepsilon(u) \operatorname{grad} u]$ with discontinuous coefficients. Dopovidi Akad. Nauk Ukrain RSR (1965) 1570–1574 (Ukrainian).

Bouligand, G.: [1] Sur les ensembles impropres dans le problème de Dirichlet pour une équation elliptique à coéfficients singuliers. Bull. Acad. R. Belgique **17** (1931) 40–42. — [2] Sur certaines équations du type elliptique à coéfficients singuliers. Bull. Adac. R. Belgique **18** (1932) 840–857. — [3] Sur quelques cas singuliers du problème de Dirichlet. Bull. Acad. R. Belgique **19** (1933) 301–317.

Boyarskiĭ, B. V.: [1] On a boundary problem for a system of elliptic first order partial differential equations. Dokl. Akad. Nauk SSSR **102** (1955) 201–204 (Russian). — [2] Homeomorphic solutions of Beltrami's systems. Dokl. Akad. Nauk SSSR **102** (1955) 661–664 (Russian). — [3] On solutions of linear elliptic systems of differential equations in the plane. Dokl. Akad. Nauk SSSR **102** (1955) 871–874 (Russian). — [4] Generalized solutions of a system of differential equations of first order and of elliptic type with discontinuous coefficients. Mat. Sbornik **43** (1957) 451–503 (Russian). — [5] A general representation of the solutions of an elliptic system of $2n$ equations on a plane. Dokl. Akad. Nauk SSSR **122** (1958) 543–546 (Russian). — [6] On the first boundary value problem for a system of elliptic equations of second order in the plane. Bull. Acad. Sci. Polon. **7** (1959) 565–570 (Russian). — [7] Some boundary value problems for a system of 2 m equations of elliptic type on the plane. Dokl. Akad. Nauk SSSR **124** (1959) 15—18 (Russian). — [8] On Dirichlet's problem for elliptic systems in space. Bull. Acad. Sci. Polon. **8** (1960) 19—23 (Russian).

BRAMBLE, J. H., and L. E. PAYNE: [1] Pointwise bounds in the first biharmonic boundary value problem. J. Math. Phys. **42** (1963) 278—286. — [2] Bounds for solutions of second order elliptic partial differential equations. Contrib. Diff. Eq. **1** (1963) 95—127. — [3] Some integral inequalities for uniformly elliptic operators. Contrib. Diff. Eq. **1** (1963) 129—135. — [4] Upper and lower bounds in equations of forced vibration type. Arch. Rat. Mech. Anal. **14** (1963) 153—170. — [5] Bounds for derivatives in elliptic boundary value problems. Pacific J. Math. **14** (1964) 777—782.

BRELOT, M.: [1] Sopra il problema di Dirichlet generalizzato relativo al dominio limitato più generale e all'equazione: $\varDelta u = c u + f (c \geqq 0)$. Rend. Ist. Lombardo **63** (1930) 917—941. — [2] Sur l'équation $\varDelta u = c u$, $c \geqq 0$, quand c admet des points singuliers; et un'équation de Fredholm correspondante à noyau singulier. Rend. Circ. Mat. Palermo **55** (1931) 21—49. — [3] Étude de l'équation de la chaleur $\varDelta u = c u$, $c \geqq 0$, au voisinage d'un point singulier du coéfficient. Ann. Éc. N. Sup. **48** (1931) 153—146. — [4] Étude des intégrales bornées de l'équation $\varDelta u = c u$ $(c \geqq 0)$ au voisinage de singularités de c formant un ensemble de capacité nulle. Bull. Sci. Math. **55** (1931) 281—296. — [5] Über die Singularitäten der Potentialfunktionen und der Integrale der Differentialgleichungen vom elliptischen Typus. Sitzber. Berlin. Math. Ges. **31** (1932) 46—54. — [6] Quelques propriétés générales des intégrales bornées de $\varDelta u = c u$, sur un domaine borné ouvert où c est continu. Bull. Sci. Math. **56** (1932) 105—117. — [7] Sur un théorème de non existence relatif à l'équation $\varDelta u = c u$. Bull. Sci. Math. **56** (1932) 389—395. — [8] Étude à la frontière de la solution du problème de Dirichlet généralisé relatif à l'équation $\varDelta u = c u + f$, $c \geqq 0$, $|f|$ borné. Rend. Ist. Lombardo **65** (1932) 119—128. — [9] Sur l'allure à la frontière des intégrales bornées de $\varDelta u = c u$ $(c \geqq 0)$. Rend. Ist. Lombardo **65** (1932) 433—448. — [10] Einige neuere Untersuchungen über das Dirichletsche Problem. Jber. Deut. Math. Vereinig. **42** (1932) 111—119. — [11] Sur l'allure des intégrales bornées de $\varDelta u = c (M) u [c(M) \geqq 0]$ au voisinage d'un point singulier de $c(M)$. Bull. Sci. Math. **60** (1936) 112—128. — [12] Sur le principes des singularités positives et la notion de source pour l'équation $\varDelta u = c u$. Ann. Univ. Lyon Sect. A **11** (1948) 9—19. — [13] Topologies of R. S. Martin and Green's lines. Lectures on functions of a complex variable. Univ. of Michigan (1955) 105—121. — [14] Le problème de Dirichlet. Axiomatique et frontière de Martin. J. Math. pures appl. **35** (1956) 297—235. — [15] Une axiomatique générale du problème de Dirichlet dans les espaces localement compactes. Sém. de Théorie du potentiel Paris **1** (1957) 16 pp. — [16] Axiomatique des functions harmoniques et surharmoniques dans un espace localement compact. Sém. de théorie du potentiel Paris **2** (1958) 40 pp. — [17] Quelques développements récents sur le problème de Dirichlet. Abhand. Sem. Math. Hamburg **23** (1959) 48—59. — [18] Lectures on potential theory. Tata Inst. of Fundam. Research Bombay (1960) 155 pp. —

[19] Étude comparée de quelques axiomatiques des fonctions harmoniques et surharmoniques. Sém. de théorie du potential Paris **6** (1962) 13 pp.

BREMERKAMP, H.: [1] Sur les équations aux dérivées partielles du second ordre du type elliptique. Proc. K. Akad. Amsterdam **34** (1931) 390–398. — [2] Sur l'unicité des solutions de certaines équations aux dérivées partielles du quatrième ordre. Proc. Nederl. Akad. Wetensch. **45** (1942) 546–552. — [3] Sur l'existence et la construction des solutions de certaines équations aux dérivées partielles du quatrième ordre. Proc. Nederl. Akad. Wetensch. **45** (1942) 675–680. — [4] On the uniqueness of the solutions of $\Delta^k u = 0$. Indag. Math. **7** (1945) 27–33 (Dutch). — [5] On the existence of a solution of $\Delta^k u = 0$ which together with its $k - 1$ first normal derivatives takes given values at the points of a given closed curve. Indag. Math. **8** (1946) 82–90 and 171–187 (Dutch). — [6] On the solutions of the equation $\Delta\Delta u = 0$ which satisfy certain boundary conditions. Indag. Math. **8** (1956) 188–199 (Dutch). — [7] On the partial differential equations occuring in the elastic plate. Nieuw. Arch. Wiskunde **22** (1946) 189–199. — [8] Construction of the solution of $\Delta^k u = 0$ satisfying given boundary conditions on a circle or a sphere. Nieuw Arch. Wiskunde **22** (1948) 293–299. — [9] Construction of the solution of $\Delta\Delta u = 0$ in the case the boundary is an ellipse. Nieuw Arch. Wiskunde **22** (1948) 300–305.

BREMERKAMP, H., and O. BOTTEMA: [1] On the solution of the equation $\Delta^{\nu} u = 0$ which satisfy certain boundary conditions. Indag. Math. **8** (1946) 279–298, (Dutch).

BROUSSE, P.: [1] Sur un problème de Dirichlet singulier. C. R. Acad. Sci. Paris **236** (1953) 1731–1732. — [2] Quelques propriétés des intégrales d'une classe d'équations singulières dans certaines domaines. C. R. Acad. Sci. Paris **237** (1953) 1381–1383. — [3] Quelques propriétés de la solution d'un problème singulier à un paramètre. C. R. Acad. Sci. Paris **242** (1956) 2093–2094. — [4] Résolution de divers problèmes du type Stokes-Beltrami posés par la technique aéronautique. Publ. Sci. Tech. Ministère de l'Air Paris n. **323** (1956) 67 pp.

BROUSSE, P., et H. PONCIN: [1] Quelques résultats généraux concernant la détermination de solutions d'équations elliptiques par les conditions aux frontières. Memoirs offerts à D. P. Riabouchinsky, Paris (1954) 17–24.

BROWDER, F.E.: [1] The Dirichlet problem for linear elliptic equations of arbitrary even order with variable coefficients. Proc. Nat. Acad. USA **38** (1952) 230–235. — [2] Assumption of boundary values and the Green's function in the Dirichlet problem for the general elliptic equation. Proc. Nat. Acad. USA **39** (1953) 179–184. — [3] Linear parabolic differential equations of arbitrary order; general boundary-value problems for elliptic equations. Proc. Nat. Acad. USA **39** (1953) 185–190 and 1298. — [4] Strongly elliptic systems of differential equations. Ann. of Math. Studies **33** (1954) 15–51. — [5] On the regularity properties of solutions of elliptic equations. Commun. pures appl.

Math. **9** (1956) 351—361. — [6] Regularity theorems for solutions of partial differential operators. Proc. Nat. Acad. USA **43** (1957) 234—236. — [7] Les opérateurs elliptiques et les problèmes mixtes C. R. Acad. Sci. Paris **246** (1958) 1363—1365. — [8] Estimates and existence theorems for elliptic boundary-value problems. Proc. Nat. Acad. USA **45** (1959) 365—372. — [9] Functional analysis and partial differential equations, I, II. Math. Ann. **138** (1959) 55—79; **145** (1961/62) 81—226. — [10] A priori estimates and existence theorems for elliptic boundary-value problems, I, II, III. Indag. Math. **22** (1960) 145—159 and 160—169; **23** (1961) 404—410. — [11] A priori estimates for elliptic and parabolic equations. Proc. Symp. Pure Math. Amer. Math. Soc. **4** (1961) 73—81. — [12] A continuity property for adjoints of closed operators in Banach spaces and its application to elliptic boundary value problems. Duke Math. J. **28** (1961) 157—182. — [13] Analyticity and partial differential equations, I. Amer. J. Math. **84** (1962) 666—710. — [14] Approximation by solutions of partial differential equations. Amer. J. Math. **84** (1962) 134—160. — [15] Variational boundary-value problems for quasi-linear elliptic equations, I, II, III. Proc. Nat. Acad. USA **50** (1963) 31—37; 592—598; 794—798. — [16] Non linear elliptic boundary-value problems. Bull. Amer. Math. Soc. **69** (1963) 862—874. — [17] Non linear elliptic problems, II. Bull. Amer. Math. Soc. **70** (1964) 299—302. — [18] Non linear elliptic boundary-value problems, II. Trans. Amer. Math. Soc. **117** (1965) 530—550. — [19] Non local elliptic boundary-value problems. Amer. J. Math. **86** (1964) 735—750. — [20] Existence and uniqueness theorems for solutions of non linear boundary-value problems. Proc. Symp. Appl. Math. Amer. Math. Soc. **17** (1965) 24—49. — [21] Les problèmes non linéaires. Lecture Notes, Univ. of Montreal (1965) 153 pp. — [22] Non linear operators in Banach spaces. Math. Ann. **162** (1966) 280—283. — [23] On the unification of the calculus of variations and the theory of monotone non linear operators in Banach spaces. Proc. Nat. Acad. USA **56** (1966) 419—425. — [24] Existence and approximation of solutions of non linear variational inequalities. Proc. Nat. Acad. USA **56** (1966) 1080—1086. — [25] Topological methods of nonlinear elliptic equations of arbitrary order. Pacific J. Math. **17** (1966) 17—32.

BUI AN TON: [1] On non linear elliptic boundary-value problems. Bull. Amer. Math. Soc. **72** (1966) 307—313.

BUREAU, F.: [1] Sur les solutions élémentaires des équations linéaires aux dérivées partielles totalement elliptiques. C. R. Acad. Sci. Paris **202** (1936) 454—456. — [2] Essai sur l'intégration des équations linéaires aux dérivées partielles. Mém. Acad. R. Belgique Coll. in 8°, **15** (1936) 111 pp. — [3] Les solutions élémentaires des équations linéaires aux dérivées partielles. Mém. Acad. R. Belgique Coll. in 8°, **15** (1936) 37 pp. — [4] Sur l'intégration des équations linéaires aux dérivées partielles. Bull. Acad. R. Belgique **22** (1936) 156—174. — [5] Sur la solution élémentaire d'une équation linéaire aux dérivées partielles d'ordre quatre et à trois variables indépendentes. Bull. Acad. R. Belgique **33** (1947) 473—484. — [6] Sur l'intégration d'une équation linéaire aux dérivées partielles totalement hyperbolique d'ordre quatre et à trois variables indépendentes. Mém. Acad. R. Belgique **21** (1948) 64 pp. — [7] Divergent integrals and partial differential equations. Commun. pure appl. Math. **8** (1955) 143—202.

CACCIOPPOLI, R.: [1] Un teorema generale sull'esistenza di elementi uniti in una trasformazione funzionale. Rend. Acc. Lincei **11** (1930) 794—799. — [2] Sugli elementi uniti delle trasformazioni funzionali: un'osservazione sui problemi di valori ai limiti. Rend. Acc. Lincei **13** (1931) 498—502. — [3] Sugli elementi uniti delle trasformazioni funzionali: un teorema di esistenza e di unicità e alcune sue applicazioni. Rend. Sem. Mat. Padova **3** (1932) 1—15. — [4] Un principio d'inversione per le corrispondenze funzionali e sue applicazioni alle equazioni a derivate parziali. Rend. Acc. Lincei **16** (1932) 390—395 and 484—489. — [5] Problemi non lineari in analisi funzionale. Rend. Sem. Mat. Roma **1** (1931—32) 13—22. — [6] Sulle equazioni ellittiche non lineari a derivate parziali. Rend. Acc. Lincei **18** (1933) 103—106. — [7] Sulle equazioni ellittiche a derivate parziali con n variabili indipendenti. Rend. Acc. Lincei **19** (1934) 83—89. — [8] Sulle equazioni ellittiche a derivate parziali con due variabili indipendenti e sui problemi regolari del calcolo delle variazioni. Rend. Acc. Lincei **22** (1935) 305—310 and 376—379. — [9] Sulle corrispondenze funzionali inverse diramate: teoria generale e applicazioni ad alcune equazioni non lineari e al problema di Plateau. Rend. Acc. Lincei **24** (1936) 258—263 and 416—421. — [10] Sui teoremi di esistenza di Riemann. Ann. Sc. N. Sup. Pisa **6** (1937) 177—187. — [11] Ovaloidi di metrica assegnata. Comm. Pont. Acad. Sci. **4** (1940) 1—20. — [12] Limitazioni integrali per le soluzioni di un'equazione lineare ellittica a derivate parziali. Giorn. Mat. Battaglini **80** (1950—51) 186—212.

CALDERON, A. P.: [1] The multipole expansion of radiation fields. J. Rat. Mech. Anal. **3** (1954) 523—537. — [2] Uniqueness in the Cauchy problem for partial differential equations. Amer. J. Math. **80** (1958) 16—36.

CALDERON, A. P., and A. ZYGMUND: [1] Singular integral operators and differential equations. Amer. J. Math. **79** (1957) 901—921. — [2] A note on local properties of solutions of elliptic differential equations. Proc. Nat. Acad. USA **46** (1960) 1385—1389. — [3] Local properties of solutions of elliptic partial differential equations. Studia Math. **20** (1961) 171—225.

CALKIN, J. W.: [1] Abstract symmetric boundary conditions. Trans. Amer. Math. Soc. **45** (1939) 360—442.

CAMPANATO, S.: [1] Teoremi di completezza relativi al sistema di equazioni dell'equilibrio elastico. Rend. Sem. Mat. Padova **25** (1956) 122—137. — [2] Sui problemi al contorno relativi al sistema di equazioni differenziali dell'elastostatica piana. Rend. Sem. Mat. Padova **25** (1956) 307—342. — [3] Sul problema di Picone relativo all'equilibrio di un corpo elastico incastrato. Ricerche di Mat. Napoli **6** (1957) 125—149. — [4] Sui problemi al contorno per sistemi di equazioni differenziali lineari del tipo dell'elasticità, I, II. Ann. Sc. N. Sup. Pisa **13** (1959) 223—258 and 275—302. — [5] Proprietà di alcuni spazi di distribuzioni a valori vettoriali e loro applicazioni. Ann. Sc. N. Sup. Pisa **14** (1960) 363—376. — [6] Osservazioni sul problema di transmissione per equazioni differenziali lineari del secondo ordine. Ricerche di Mat. Napoli **9** (1960) 43—57. — [7] Equazioni ellittiche del secondo ordine e spazi $\mathscr{L}^{(2/\lambda)}$.

Ann. Mat. pura appl. **69** (1965) 321—382. — [8] Alcune osservazioni relative alle soluzioni di equazioni ellittiche di ordine 2 *m*. Convegno Eq. Der. Parz. Bologna [1967] 17—25.

CAMPANATO, S., e G. STAMPACCHIA: [1] Sulle maggiorazioni in L^p nella teoria delle equazioni ellittiche. Boll. Un. Mat. Ital. **20** (1965) 393—399.

CANFORA, A.: [1] Teorema del massimo modulo e teorema di esistenza per il problema di Dirichlet relativo ai sistemi fortemente ellittici. Ricerche di Mat. Napoli **15** (1966) 249—294.

CAPOULADE, J.: [1] Des ensembles impropres. C. R. Acad. Sci. Paris **194** (1932) 426—428. — [2] Sur les arcs frontières rendus impropres par les singularités des coefficients dans le problème de Dirichlet pour les équations du second ordre et du type elliptique à deux variables. Rend. Acc. Lincei **15** (1932) 844—849. — [3] Sur certaines équations aux dérivées partielles du second ordre à coefficients singuliers. Mathematica Cluj **8** (1934) 139—184.

CARLEMAN, T.: [1] Sur les systèmes linéaires aux dérivées partielles du premier ordre à deux variables. C. R. Acad. Sci. Paris **197** (1933) 471—474. — [2] Sur un problème d'unicité pour les systèmes d'équations aux dérivées partielle à deux variables indépendantes. Ark. Mat. Astr. Fys. **26 B**, n. 17 (1939) 1—9.

CATTABRIGA, L.: [1] Su un problema al contorno relativo al sistema di equazioni di Stokes. Rend. Sem. Mat. Padova **31** (1961) 308—340.

CHANG I-DEE, and R. FINN: [1] On the solution of a class of equations occurring in continuum mechanics with application to the Stokes paradox. Arch. Rat. Mech. Anal. **7** (1961) 388—401.

CHEN KIEN-KWONG: [1] The Hölder continuity of the general solutions of the linear elliptic system of partial differential equations. Sci. Sinica **10** (1961) 153—159.

CHEN LIANG-CHING: [1] A boundary value problem for a degenerate elliptic equation. Acta Math. Sinica **13** (1963) 332—342 (Chinese); transl. as Chinese Math. **4** (1964) 360—371. — [2] The Dirichlet problem for a class of systems of degenerate elliptic equations. Acta Math. Sinica **14** (1964) 379—386 (Chinese); transl. as Chinese Math. **5** (1964) 409—417.

CHEN YU-WHY: [1] Branch points, poles and planar points of minimal surfaces in R^3. Ann. of Math. **49** (1948) 790—806. — [2] Existence of minimal surfaces with a simple pole at infinity and conditions of transversality on the surface of a cylinder. Trans. Amer. Math. Soc. **65** (1949) 331—347.

CIMMINO, G.: [1] Sulle equazioni lineari ellittiche autoaggiunte alle derivate parziali di ordine superiore al secondo. Mem. Acc. d'Italia **1** (1930) 59—69. — [2] Teoremi di confronto fra equazioni o sistemi di equazioni differenziali lineari del secondo ordine. Rend. Sem. Mat. Roma **1** (1936) 31—52. — [3] Sulle equazioni lineari alle derivate parziali del secondo ordine di tipo ellittico sopra una superficie chiusa. Ann. Sc. N. Sup. Pisa **7** (1938) 73—96. — [4] Nuovo tipo di condizioni al contorno e nuovo metodo di trattazione per il problema generalizzato di Dirichlet. Rend. Circ. Mat. Palermo **61** (1938) 177—221. — [5] Sul problema generalizzato di Dirichlet per l'equazione di Poisson. Rend. Sem. Mat. Padova **11** (1940) 28—89. — [6] Nuove proprietà caratteristiche per le soluzioni delle equazioni lineari alle derivate parziali di tipo ellittico del secondo ordine. Atti 2° Congr. Un. Mat. Ital. (1940) 198—204.

CLARK, C., and C. A. SWANSON: [1] Comparison theorems for elliptic differential equations. Proc. Amer. Math. Soc. **16** (1965) 886—890.

COHN VOSSEN, E.: [1] Zwei Sätze über die Starrheit der Einflächen. Göttinger Nach. (1927) 125—134.

COLAUTTI, M. P.: [1] Sul problema di Neumann per l'equazione $\Delta_2 u - \lambda c u = f$ in un dominio piano a contorno angoloso. Mem. Acc. Sci. Torino **4** (1959—60) 1—83.

COOPERMAN, PH.: [1] On extension of a method of Trefftz for finding local bounds on the solutions of boundary value problems and on their derivatives. Quart. appl. Math. **10** (1953) 359—373.

CORDES, H. O.: [1] Über die eindeutige Bestimmtheit der Lösungen elliptischer Differentialgleichungen durch Anfangsvorgaben. Göttinger Nach. (1956) 239—258. — [2] Über die erste Randwertaufgabe bei quasilinearen Differentialgleichungen zweiter Ordnung in mehr als zwei Variabeln. Math. Ann. **131** (1956) 278—312. — [3] Vereinfachter Beweis der Esixtenz einer Apriori-Hölderkonstante. Math. Ann. **138** (1959) 155—178. — [4] Zero order a priori estimates for solutions of elliptic differential equations. Proc. Symp. Pure Math. Amer. Math. Soc. **4** (1961) 157—161. — [5] The basic a priori estimates for elliptic systems of arbitrary order. Lecture Notes, California Univ. Berkeley (1962).

COURANT, R.: [1] Dirichlet's principle, conformal mapping and minimal surfaces. New York: Interscience Publ. (1950) 330 pp.

COURANT, R., u. D. HILBERT: [1] Methoden der mathematischen Physik, I, II. Berlin: Springer-Verlag (1924 and 1937) 469 + 549 pp. — [2] Methods of Mathematical Physics, II. New York-London: Interscience Publ. (1962) 830 pp.; Russian transl. Izdat "Mir" (1964) 830 pp.

CRONIN, J.: [1] The existence of multiple solutions of elliptic differential equations. Trans. Amer. Math. Soc. **68** (1950) 105—131. — [2] The Dirichlet problem for non linear elliptic equations. Pacific J. Math. **5** (1955) 335—346.

DANYLUK, I. I.: [1] On some problems of elliptic systems of differential equations of the first order on surfaces. Dokl. Akad. Nauk SSSR **105** (1955) 11—14 (Russian). — [2] On integral representations of solutions of some elliptic systems of the first order on surfaces and applications to the theory of shells. Dokl. Akad. Nauk SSSR **109** (1956) 17—20

(Russian). — [3] Some problems of the theory of elliptic differential systems and quasi-conformal transformations. Nauch. Zap. Un-ta L'vov (1956) 75—89. — [4] On the mappings corresponding to solutions of elliptic equations. Dokl. Akad. Nauk SSSR **120** (1958) 17—20 (Russian). — [5] On a problem involving a skew derivative for first order elliptic systems. Dokl. Akad. Nauk SSSR **122** (1958) 9—12 (Russian). — [6] The use of Fredholm's system equations in investigating a problem involving a skew derivative. Dokl. Akad. Nauk SSSR **122** (1958) 175—178 (Russian). — [7] Über das System von Differentialgleichungen vom elliptischen Typus auf Riemannschen Flächen. Ann. Acad. Sci. Fenn. n. **251** (1958) 7 pp. — [8] On the oblique derivative problem for the general quasi-linear system of the first order. Dokl. Akad. Nauk SSSR **127** (1959) 953—956 (Russian). — [9] A problem with a directional derivative. Sibirsk Mat. Ž. **3** (1962) 17—55 (Russian).

DAVIS, R.B.: [1] Asymptotic solutions of the first boundary value problem for a fourth order elliptic partial differential equation. J. Rat. Mech. Anal. **5** (1956) 605—620. — [2] A reduction-of-order theorem. J. Math. Phys. **36** (1957) 164—166.

DE FIGUEIREDO, D.G.: [1] The coercivness problem for forms over vector-valued functions. Commun. pure appl. Math. **16** (1963) 63—94.

DE GIORGI, E.: [1] Osservazioni relative ai teoremi di unicità per le equazioni a derivate parziali di tipo ellittico con condizioni al contorno di tipo misto. Ricerche di Mat. Napoli **2** (1953) 183—191. — [2] Sulla differenziabilità e l'analiticità delle estremali degli integrali multipli regolari. Mem. Acc. Sci. Torino **3** (1957) 1—19. — [3] Un'estensione del teorema di Bernstein. Ann. Sc. Norm. Sup. Pisa **19** (1965) 79—85.

DE GIORGI, E., e G. STAMPACCHIA: [1] Sulle singolarità eliminabili delle ipersuperficie minimali. Rend. Acc. Lincei **38** (1965) 352—357.

DE LUCIA, P.: [1] Una questione di piccolo parametro per un'equazione ellittico-parabolica. Ricerche di Mat. Napoli **12** (1963) 121—139. — [2] Sull'esistenza in grande di una soluzione fondamentale per un'equazione fortemente ellittica. Rend. Acc. Sci. Fis. Mat. Napoli **34** (1967) 133—144.

DE RHAM, G.: [1] Variétés differentiables: formes, courants, formes harmoniques. Paris: Hermann (1955) 196 pp.; Russian transl. Izdat. Inost. Lit. Moscow (1956) 250 pp.

DE RHAM, G., and K. KODAIRA: [1] Harmonic integrals. Mimeographed Notes Inst. Adv. Stud. Princeton (1950) 114 pp.

DEZIN, A.A.: [1] Invariant differential operators and boundary-value problems. Trudy Mat. Inst. Steklov **68** (1962) 88 pp. (Russian); English transl. Amer. Mat. Soc. (2) **41** (1964) 215—317.

DIAZ, J.B., and H.J. GREENBERG: [1] Upper and lower bounds for the solution of the first boundary value problem of elasticity. Quart. appl.

Math. **6** (1948) 326—331. — [2] Upper and lower bounds for the solution of the first biharmonic boundary value problem. J. Math. Phys. **27** (1948) 153—201.

DIAZ, J.B., and A.WEINSTEIN: [1] On the fundamental solutions of singular Beltrami operator. Studies presented to R. von Mises. New York: Acad. Press Inc. (1954) 97—102.

DIDENKO, V.P.: [1] On certain elliptic systems which are degenerate on the boundary. Dokl. Akad. Nauk SSSR **143** (1962) 1250—1253 (Russian).

DING SHIA-KUAI, WANG KAN-TING, MA JU-NIEN, SHUN CHIAO-LI and CHANG-TONG: [1] Definition of ellipticity of a system of second order differential equations with constant coefficients. Acta Math. Sinica **10** (1960) 276—287 (Chinese); transl. as Chinese Math. **1** (1962) 288—299.

DONG GUANG-CHANG: [1] A boundary value problem for a degenerate elliptic equation. Acta Math. Sinica **11** (1961) 371—375 and **13** (1963) 620—630 (Chinese); transl. as Chinese Math. **2** (1962) 425—430 and **4** (1964) 675—685. — [2] Boundary value problems for degenerate elliptic partial differential equations. Acta Math. Sinica **13** (1963) 94—115 (Chinese); transl. as Chinese Math. **4** (1963) 105- 126.

DOOB, J.L.: [1] Probability methods applied to the first boundary value problem. Proc. Third Berkeley Symp. Math. Stat. Probab. Univ. California Press **2** (1956) 49—80. — [2] Probability theory and the first boundary problem. Illinois J. Math. **2** (1958) 19—36. — [3] The first boundary value problem. Coll. Lectures Summer Meeting Amer. Math. Soc. (1960) 49—80.

DOUGLIS, A.: [1] A function-theoretic approach to elliptic systems of equations in two variables. Commun. pure appl. Math. **6** (1953) 259—289. — [2] Uniqueness in Cauchy problems for elliptic systems of equations. Commun. pure appl. Math. **6** (1953) 291—298. — [3] Function-theoretic properties of certain elliptic systems of first order linear equations. Lectures on functions of a complex variable. Univ. of Michigan (1955) 335—340. — [4] On uniqueness in Cauchy problems for elliptic systems of equations. Commun. pure appl. Math. **13** (1960) 593—607.

DOUGLIS, A., and L.NIRENBERG: [1] Interior estimates for elliptic systems of partial differential equations. Commun. pure appl. Math. **8** (1955) 503—538.

DRESSEL, F.G., and J.J. GERGEN: [1] Mapping by p-regular functions. Duke Math. J. **18** (1951) 185—210. — [2] Uniqueness for p-regular mappings. Duke Math. J. **19** (1952) 435—444. — [3] Mapping for elliptic equations. Trans. Math. Soc. **77** (1954) 151—178. — [4] The extension of the Riemann mapping theorem to elliptic equations. Proc. Conf. Diff. Eq. Univ. of Maryland (1956) 183—195.

DUBINSKIĬ, JU.A.: [1] Some integral inequalities and the solvability of degenerate elliptic systems of differential equations. Mat. Sbornik **64** (1964) 458—480 (Russian).

DUFF, G.F.D.: [1] A quasi-linear boundary value problem. Trans. Roy. Soc. Canada **49** (1955) 7—17. — [2] Modified boundary value problem for a quasilinear elliptic equation. Canad. J. Math. **8** (1956) 203—219.

DUFFIN, R. J.: [1] On a question of Hadamard concerning super-biharmonic functions. J. Math. Phys. **27** (1949) 253—258. — [2] The maximum principle and biharmonic functions. J. Math. Anal. Appl. **3** (1961) 399—405.

DUFFIN, R. J., and Z. NEHARI: [1] A note on polyharmonic functions. Proc. Amer. Math. Soc. **12** (1961) 110—115.

DYMKOV, S.S.: [1] The first boundary value problem for quasi linear elliptic equations. Dokl. Akad. Nauk SSSR **115** (1957) 220—222 (Russian).

DYNIN, A.S.: [1] Singular operators of arbitrary orders on a manifold. Dokl. Akad. Nauk SSSR **141** (1961) 21—23. — [2] n-dimensional elliptic boundary value problems with a single unknown function. Dokl. Akad. Nauk SSSR **141** (1961) 285—287 (Russian).

DYNKIN, E.B.: [1] Martin boundaries and non negative solutions of boundary value problems with a directional derivative. Usp. Mat. Nauk **19,** n. 5 (1964) 3—50 (Russian); transl. as Russian Math. Surveys **19,** n. 5 (1964) 1—48.

DŽAFARDI, M.A.: [1] Solution of the first boundary problem for quasilinear elliptic equations. Izv. Akad. Nauk Azerb. SSR (1959) 13—28 (Azerbaijani).

DŽUAREV, A.: [1] A general boundary value problem for a second-order elliptic equation with non analytic coefficients. Dokl. Akad. Nauk Todzik SSR (1962) n. 1, 3—9 (Russian). — [2] A general linear boundary-value problem for the equation $\Delta u + \lambda c\,(x, y)\, u = 0$. Dokl. Akad. Nauk SSSR **142** (1962) 994—997 (Russian). — [3] The Poincaré problem for a second-order elliptic equation with singular coefficients. Dokl. Akad. Nauk SSSR **146** (1962) 995—998 (Russian). — [4] General boundary-value problems for elliptic equations with non analytic coefficients. Sibirsk Mat. Ž. **4** (1963) 539—561 (Russian). — [5] Investigation of a boundary problem in the plane for an elliptic equation. Sibirsk Mat. Ž. **6** (1965) 494—498 (Russian).

ĖĬDUS, D.M.: [1] On the solution of a mixed problem of the theory of elasticity. Dokl. Akad. Nauk SSSR **76** (1951) 181—184 (Russian). — [2] Some boundary-value problems in infinite regions. Izv. Akad. Nauk SSSR **27** (1963) 1055—1080 (Russian).

EVGRAFOV, M.A.: [1] Generalization of the Phragmèn-Lindelöf theorems for analytic functions to solutions of different elliptic systems. Izv. Akad. Nauk SSSR **27** (1963) 843—854 (Russian).

FAEDO, S.: [1] Sul metodo di Ritz e su quelli fondati sul principio dei minimi quadrati per la risoluzione approssimata dei problemi della fisica matematica. Rend. Sem. Mat. Roma **6** (1947) 73—94. — [2] Applicazione ai problemi di derivata obliqua di un principio esistenziale e di una legge di dualità fra le formule di maggiorazione. Rend. Mat. Appl. Roma **16** (1957) 515—532.

FELLER, W.: [1] Über die Lösungen der linearen partiellen Differential-gleichungen zweiter Ordnung vom elliptischen Typus. Math. Ann. **102** (1930) 633—649.

FICHERA, G.: [1] Teoremi di completezza connessi all'integrazione del-l'equazione $\Delta_4 u = f$. Giorn. Mat. Battaglini **77** (1947—48) 184—199. — [2] Teoremi di completezza sulla frontiera di un dominio per taluni sistemi di funzioni. Ann. Mat. pura appl. **27** (1948) 1—28. — [3] Sul-l'equilibrio di un corpo elastico isotropo e omogeneo. Rend. Sem. Mat. Padova **17** (1948) 9—28. — [4] Applicazione della teoria del potenziale di superficie ad alcuni problemi di analisi funzionale lineare. Giorn. Mat. Battaglini **78** (1948—49) 71—80. — [5] Teorema di esistenza per il problema bi-iperarmonico. Rend. Acc. Lincei **5** (1948) 319—324. — [6] Sui problemi analitici dell'elasticità piana. Rend. Sem. Mat. Cagliari **18** (1948) 1—22. — [7] Analisi esistenziale per le soluzioni dei problemi al contorno misti relativi alle equazioni e ai sistemi di equazioni del secondo ordine di tipo ellittico autoaggiunti. Ann. Sc. N. Sup. Pisa **15** (1946) 75—100. — [8] Sull'esistenza e sul calcolo delle soluzioni dei problemi al contorno relativi all'equilibrio di un corpo elastico. Ann. Sc. N. Sup. Pisa **4** (1950) 35—99. — [9] On some general integration methods employed in connection with linear differential equations. J. Math. Phys. **29** (1950) 59—68. — [10] Esistenza del minimo in un classico problema di calcolo delle variazioni. Rend. Acc. Lincei **11** (1951) 34—39. — [11] Interpretazione ed estensione funzionale di recenti metodi di integrazione delle equazioni differenziali lineari. Atti IV Congr. Un. Mat. Ital. Taormina **1** (1951) 45—67. — [12] Sulla "Kernel function". Boll. Un. Mat. Ital. **7** (1952) 4—15. — [13] Sul problema della derivata obliqua e sul problema misto per l'equazione di Laplace. Boll. Un. Mat. Ital. **7** (1952) 367—377. — [14] Condizioni perchè sia compati-bile il problema principale della statica elastica. Rend. Acc. Lincei **14** (1953) 397—400. — [15] Formule di maggiorazione connesse a una classe di trasformazioni lineari. Ann. Mat. pura appl. **36** (1954) 273—296. — [16] Alcuni recenti sviluppi della teoria dei problemi al contorno per le equazioni alle derivate parziali lineari. Convegno Inter. Eq. Lin. Der. Parz. Trieste (1954) 174—227. — [17] Su un principio di dualità per talune formule di maggiorazione relative alle equazioni differenziali. Rend. Acc. Lincei **19** (1955) 411—418. — [18] Sull'esistenza delle forme differenziali armoniche. Rend. Sem. Mat. Padova **24** (1955) 523—545. — [19] Sulla teoria generale dei problemi al contorno per le equazioni differenziali lineari, I, II. Rend. Acc. Lincei **21** (1956) 46—55 and 166—172. — [20] Sulle equazioni differenziali lineari ellittico-paraboliche. Mem. Acc. Lincei **5** (1956) 1—30. — [21] Un'introduzione alla teoria delle equazioni integrali singolari. Rend. Mat. Appl. Roma **17** (1958) 89—191. — [22] Premesse a una teoria generale dei problemi al contorno per le equazioni differenziali. Ist. Naz. Alta Mat. Roma (1958) 292 pp. — [23] Teorema di massimo modulo e unicità delle soluzioni generalizzate dei problemi al contorno. Rend. Acc. Lincei **29** (1960) 503—508. — [24] Sul concetto di problema ben posto per un'equazione differenziale. Rend. Mat. Appl. Roma **19** (1960) 95—124. — [25] On a unified theory of boundary value problems for elliptic-parabolic equations of second order. Boundary problems in differential euqations, Wisconsin Univ. Press (1960) 97—120. — [26] Spazi lineari di *k*-misure e di forme differenziali. Proc. Inter. Symp. Lin. Spaces Jerusalem (1961) 175—226. — [27] Teoria assiomatica delle forme armoniche. Rend. Mat. Appl. Roma **20** (1961) 147—171. — [28] Linear elliptic equations of higher order in two independent variables with applications to anisotropic inhomogeneous elasticity. Partial differential equations and continuum mechanics, Univ. Wisconsin Press (1961) 55—80. — [29] La soluzione fondamentale principale per un'equazione differenziale ellittica di ordine superiore. Bull. Math. Soc. Sci. Math. Phys. R. P. Roumaine **6** (1962) 139—149. — [30] Problemi elastostatici con vincoli unilaterali: problema di Signorini con ambigue condizioni al contorno. Mem. Acc. Lincei **7** (1964) 91—140. — [31] Sul problema misto per le equazioni lineari alle derivate parziali del secondo ordine di tipo ellittico. Rev. Roumaine Math. pures appl. **9** (1964) 3—9. — [32] Elastostatics problems with unilateral constraints: The Signorini problem with ambiguous boundary conditions. Sem. 1962/63. Ist. Naz. Alta Mat. Roma **2** (1965) 613—679. — [33] Linear elliptic differential systems and eigenvalue problems. Lecture Notes n. 8, Berlin-Heidelberg-New York: Springer-Verlag (1965) 176 pp.

FIFE, P.: [1] Non linear deflection of thin elastic plate under tension. Commun. pure appl. Math. **14** (1961) 81—112. — [2] Schauder estimates under incomplete Hölder continuity assumptions. Pacific J. Math **13** (1963) 511—550. — [3] Growth and decay properties of solutions of second order elliptic equations. Ann. Sc. N. Sup. Pisa **20** (1966) 675—701.

FINN, R.S.: [1] Isolated singularities of solutions of non linear partial differential equations. Trans. Amer. Math. Soc. **75** (1953) 385—404. — [2] On equations of minimal surface type. Ann. of Math. **60** (1954) 397—416. — [3] On a problem of type with applications to elliptic partial differential equations. J. Rat. Mech. Anal. **3** (1954) 789—799. — [4]Growth properties of solutions of non linear elliptic equations. Commun. pure appl. Math. **9** (1956) 415—423. — [5] On the steady state solutions of the Navier-Stokes partial differential equations. Arch. Rat. Mech. Anal. **3** (1959) 381—396. — [6] Estimates at infinity for stationary solutions of the Navier-Stokes equations. Bull. Math. Soc. Sci. Math. Phys. R. P. Roumaine **3** (1959) 387—418. — [7] Estimates at infinity for steady state solutions of the Navier-Stokes equations. Proc. Symp. Part. Diff. Eq. Amer. Math. Soc. **4** (1960) 143—148. — [8] An energy theorem for viscous fluid motions. Arch. Rat. Mech. Anal. **6** (1960) 371—381. — [9] On the steady-state solutions of the Navier-Stokes equations, III. Acta Math. **105** (1961) 197—244. — [10] On the Stokes paradox and related questions. Non linear problems, Univ. of Wisconsin (1963) 99—115. — [11] New estimates for equations of minimal surface type. Arch. Rat. Mech. Anal. **14** (1963) 337—375. — [12] On partial differential equations (whose solutions admit no isolated singularities). Scripta Math. **26** (1963) 107—115. — [13] On the inclination of a minimal surface $\varphi(x, y)$. Colloq. Math. **11** (1963/64) 195—201. — [14] Remarks on my paper "On equations of minimal surface type". Ann. of Math. **80** (1964) 158—159. — [15] Remarks relevant to minimal surfaces and to surfaces of prescribed mean curvature. J. Anal. Math. **14** (1965) 135—160. —

[16] On the exterior stationary problem for the Navier-Stokes equations and associated perturbation problems. Arch. Rat. Mech. Anal. **19** (1965) 393–406. — [17] Stationary solutions of Navier-Stokes equations. Proc. Symp. Appl. Math. Amer. Math. Soc. **17** (1965) 121–153.

FINN, R., and D. GILBARG: [1] Asymptotic behavior and uniqueness of plane subsonic flows. Commun. pure appl. Math. **10** (1957) 23–63. — [2] Three dimensional subsonic flows and asymptotic estimates for elliptic partial differential equations. Acta Math. **98** (1957) 265–296.

FINN, R., and R. OSSERMANN: [1] On the Gauss curvature of the non parametric minimal surfaces. J. Anal. Math. **12** (1964) 351–364.

FINN, R., and J. SERRIN: [1] On the Hölder continuity of quasi conformal and elliptic mappings. Trans. Amer. Math. Soc. **89** (1958) 1–15.

FIORENZA, R.: [1] Sui problemi di derivata obliqua per le equazioni ellittiche. Ricerche di Mat. Napoli **8** (1959) 83–110. — [2] Sui problemi di derivata obliqua per le equazioni ellittiche non lineari in due variabili. Ricerche di Mat. Napoli **8** (1959) 222–239. — [3] Sulla hölderianità della soluzioni dei problemi di derivata obliqua regolare del secondo ordine. Ricerche di Mat. Napoli **14** (1965) 102–123. — [4] Sui problemi di derivata obliqua per le equazioni ellittiche quasi lineari. Ricerche di Mat. Napoli **15** (1966) 74–186.

FOHT, A.S.: [1] A boundary estimate for the solution of an equation of elliptic type of arbitrary order with variable coefficients where a number of the coefficients are degenerate on the boundary. Dokl. Akad. Nauk SSSR **154** (1964) 1287–1290 (Russian).

FRANKL, F.I.: [1] Uniqueness theorem for the boundary-value problem for the equation $v_{xx} + (v_y/y)_y = 0$. Izv. Vysš. Učebn. Zaved. Mat. **1** (1959) n. 1, 212–217 (Russian).

FRANKL, F.I., and M. KELDYŠ: [1] The exterior Neumann problem for non linear elliptic differential equations with application to wing theory in compressible gas. Dokl. Akad. Nauk SSSR **4** (1934) 561–601 (Russian).

FRASCA, M.: [1] Un problema variazionale per operatori ellittici. Le Matematiche Catania **18** (1963) 1–11.

FREĬDLIN, M.I.: [1] A mixed boundary value problem for second order elliptic differential equations with a small parameter. Dokl. Akad. Nauk SSSR **143** (1962) 1300–1303 (Russian). — [2] A Dirichlet problem for an equation with a small parameter and discontinuous coefficients. Dokl. Akad. Nauk SSSR **144** (1962) 501–504 (Russian). — [3] Ito's stochastic equations and degenerate elliptic equations. Izv. Akad. Nauk SSSR **26** (1962) 653–676 (Russian). — [4] A priori bounds for solutions of degenerate elliptic equations. Dokl. Akad. Nauk SSSR **158** (1964) 281–283 (Russian).

FRIEDMAN, A.: [1] On classes of elliptic partial differential equations. Proc. Amer. Math. Soc. **8** (1957) 418–427. — [2] On two theorems of Phragmèn-Lindelöf for linear elliptic and parabolic differential equations of second order. Pacific J. Math. **7** (1957) 1563—1575. — [3] On the regularity of the solutions of non linear elliptic and parabolic systems of partial differential equations. J. Math. Mech. **7** (1958) 43—59. — [4] Uniqueness properties in the theory of differential operators of elliptic type. J. Math. Mech. **7** (1958) 61—67. — [5] On fundamental solutions of elliptic equations. Proc. Amer. Math. Soc. **12** (1961) 533—537. — [6] Simplifying the structure of second order partial differential equations. Trans. Amer. Math. Soc. **99** (1961) 303—307. — [7] Entire solutions of partial differential equations with constant coefficients. Duke Math. J. **31** (1964) 235—240.

FRIEDRICHS, K.O.: [1] Die Randwert- und Eigenwertprobleme aus der Theorie der elastischen Platten (Anwendung der direkten Methoden der Variationsrechnung). Math. Ann. **98** (1928) 205—247. — [2] On differential operators in Hilbert spaces. Amer. J. Math. **61** (1939) 523—544. — [3] The identity of weak and strong extensions of differential operators. Trans. Amer. Math. Soc. **55** (1944) 132—151. — [4] On the boundary value problems of the theory of elasticity and Korn's inequality. Ann. of Math. **48** (1947) 441—471. — [5] A theorem of Lichtenstein. Duke Math. J. **14** (1947) 67—82. — [6] On the differentiability of the solutions of linear elliptic differential equations. Commun. pure appl. Math. **6** (1953) 299—326.

FUJITA, H.: [1] The existence and regularity of the steady-state solutions of the Navier-Stokes equations. J. Fac. Sci. Univ. Tokyo **9** (1961) 59—102.

GAGLIARDO, E.: [1] Un'osservazione sul problema di Dirichlet per un'equazione lineare ellittica del secondo ordine. Rend. Acc. Sci. Fis. Mat. Napoli **25** (1958) 66—80.

GAHOV, F.D.: [1] Boundary problems. Gos. Izdat. Fiz.-Mat. Lit. Moscow (1958) 534 pp. (Russian).

GARABEDIAN, P.R.: [1] A partial differential equation arising in conformal mappings. Pacific J. Math. **1** (1951) 485—524.

GÅRDING, L.: [1] Le problème de Dirichlet pour les équations aux dérivées partielles elliptiques homogènes à coefficients constants. C. R. Acad. Sci. Paris **230** (1950) 1030—1032. — [2] On a lemma of Weyl. Kungl. Fysiog. Sällsk. Lund Förth **20** (1950) 250—253. — [3] Dirichlet's problem and vibration problem for linear elliptic partial differential equations with constant coefficients. Proc. Symp. Spectral Th. Diff. Probl. Oklahoma College (1951) 291—299. — [4] Le problème de Dirichlet pour les équations aux dérivées partielles elliptiques linéaires dans les domaines bornées. C. R. Acad. Sci. Paris **233** (1951) 1554—1556. — [5] Dirichlet's problem for linear elliptic partial differential equations. Math. Scand. **1** (1953) 55—72. — [6] Some trends and problems in linear partial differential equations. Proc. Int. Math. Congr. Edinbourgh (1958) 87—102; Russian transl. Usp. Mat. Nauk **15** (1960) 137—152.

GARNIR, H.G.: [1] Sur la solution élémentaire pour l'éspace indefini d'un opérateur elliptique décomposable du quatrième ordre. Bull. Acad. R.

Belgique **38** (1952) 1129—1141. — [2] Les problèmes aux limites de la physique mathématique. Basel-Stuttgart: Birkhäuser-Verlag (1958) 234 pp.

GASUMOV, G. M.: [1] Some linear boundary-value problems for elliptic equations. Izv. Akad. Nauk Azerb. SSR (1963) 33—46 (Russian). — [2] A non linear problem for elliptic equations. Izv. Akad. Nauk Azerb SSR (1964) 49—59 (Russian).

GAVELJA, S. P.: [1] On solutions of elliptic systems of differential equations with higher-dimensional sets of singularities. Ukrain. Mat. Ž. **14** (1962) 191—197 (Russian).

GEL'FAND, I. M.: [1] On elliptic equations. Usp. Mat. Nauk **15,** n. 3 (1960) 113—123 (Russian); transl. as Russian Math. Surveys **15,** n. 3 (1960) 113—123.

GERVER, M. L.: [1] On the possible rate of decrease of a solution of an elliptic equation. Dokl. Akad. Nauk SSSR **156** (1964) 13—16 (Russian).

GEVREY, M.: [1] Sur la nature analytique des solutions des équations aux dérivées partielles. Ann. Éc. N. Sup. **35** (1918) 127—190. — [2] Démonstration du théorème de Picard-Bernstein par la méthode des contours successifs; prolongement analytique. Bull. Sci. Math. **50** (1926) 113—128. — [3] Sur certaines propriétés des fonctions harmoniques et leur extension aux équations aux dérivées partielles. C. R. Acad. Sci. Paris **183** (1926) 544—546. — [4] Nature analytique et prolongement des solutions des équations non linéaires du type elliptique et parabolique. C. R. Acad. Sci. Paris **182** (1926) 754—756. — [5] Détermination et emploi des fonctions de Green dans les problèmes aux limites relatifs aux équations linéaires du type elliptique. J. Math. pures appl. **9** (1930) 1—80. — [6] Détermination des intégrales des systèmes d'équations linéaires aux dérivées partielles du type elliptique. C. R. Acad. Sci. Paris **193** (1931) 693—695. — [7] De quelques questions concernant les types elliptiques et paraboliques: usage de la médiation périphérique ou spatiale; unicité des solutions des systèmes d'équations. C. R. Acad. Sci. Paris **197** (1933) 296—298. — [8] Les quasi fonctions de Green et les systèmes d'équations aux dérivées partielles du type elliptique. Ann. Éc. N. Sup. **52** (1935) 39—108. — [9] Sur certains systèmes d'équations aux dérivées partielles à caractéristiques imaginaires multiples. C. R. Acad. Sci. Paris **203** (1936) 604—606. — [10] Sur une généralisation du principe des singularités positives de M. Picard. C. R. Acad. Sci. Paris **211** (1940) 581—584. — [11] Sur le problème de la dérivée oblique relatif aux équations aux dérivées partielles ou intégrodifferentielles du type elliptique canonique à deux variables. C. R. Acad. Sci. Paris **213** (1941) 635—637. — [12] Sur un procédé de résolution dans le plan du problème aux limites linéaire le plus général relatif aux équations intégrodifférentielles du type elliptique. C. R. Acad. Sc. Paris **214** (1942) 206—208. — [13] Sur les problèmes aux limites comportant une dérivée oblique et concernant le type elliptique à *m* variables. C. R. Acad. Sci. Paris **214** (1942) 854—855. — [14] Sur le cas irrégulier du problème de la dérivée oblique lorsque le nombre des variables est supérieur à deux. C. R. Acad. Sci. Paris **225** (1947) 1251—1253.

GEYMONAT, G.: [1] Sul problema di Dirichlet per le equazioni lineari ellittiche. Ann. Sc. N. Sup. Pisa 16 (1962) 225—284. — [2] Su un problema relativo alle soluzioni delle equazioni lineari ellittiche. Ann. Sc. N. Sup. Pisa 18 (1964) 87—110. — [3] Sui problemi ai limiti per i sistemi lineari ellittici. Ann. Mat. pura appl. 69 (1965) 207—284. — [4] Su alcuni problemi ai limiti per i sistemi lineari ellittici secondo Petrowsky. Le Matematiche Catania 20 (1965) 211—253.

GEYMONAT, G., et P. GRISVARD: [1] Problèmes aux limites elliptiques dans L^p. Exposés faits au Dép. Math. Fac. Sci. Univ. Paris d'Orsay (1964) 167 pp. — [2] Problemi ai limiti lineari ellittici negli spazi di Sobolev con peso. Le Matematiche, Catania 22 (1967) 212—249.

GHERMANESCO, M.: [1] Sur les fonctions n-métaharmoniques de p variables. Rend. Acc. Lincei 14 (1931) 252—259. — [2] Sur les fonctions n-méta-harmoniques. Rend. Acc. Lincei 14 (1931) 415—421.

GHIZZETTI, A.: [1] Sul metodo della trasformata parziale di Laplace a intervallo d'integrazione finito. Rend. Mat. Appl. Roma 6 (1947) 1—47. — [2] Ricerche analitiche sul problema dell'equilibrio di una piastra indefinita a forma di striscia, incastrata lungo i due lati. Rend. Mat. Appl. Roma 6 (1947) 145—187. — [3] Applicazione del metodo della trasformata parziale di Laplace per l'equazione $\Delta_2 u - \lambda^2 u = F$ in n variabili. Rend. Sem. Mat. Padova 17 (1948) 39—74. — [4] Su un partico-lare problema misto per un'equazione di tipo ellittico a coefficienti costanti. Rend. Acc. Lincei 5 (1948) 344—348. — [5] Flow in a not homogeneous and anisotropic medium. Ann. Soc. Polon. Math. 22 (1949) 195—200. — [6] Sulle equazioni alle derivate parziali del secondo ordine, in due variabili indipendenti, le cui soluzioni godono di proprietà integrali rispetto a una delle variabili. Ann. Mat. pura appl. 40 (1955) 41—59.

GILBARG, D.: [1] Some local properties of elliptic equations. Proc. Symp. Pure Math. Amer. Math. Soc. 4 (1961) 127—141. — [2] Boundary value problems for non linear elliptic equations in n variables. Symp. Non Linear Problems, Univ. Wisconsin Press (1962) 151—159.

GILBARG, D., and J. SERRIN: [1] On isolated singularities of solutions of second order elliptic differential equations. J. Anal Math. 4 (1955) 309—340.

GILLIS, P. P.: [1] Intégrales doubles du calcul des variations. Bull. Acad. R. Belgique 36 (1950) 403—412. — [2] Équations de Monge-Ampère à quattre variables indépendantes. Bull. Acad. R. Belgique 36 (1950) 474—484. — [3] Équations de Monge-Ampère à six variables indépen-dantes. Bull. Acad. R. Belgique 37 (1951) 229—240. — [4] Sur certaines équations de Monge-Ampère du calcul des variations. Bull. Soc. Math. Belgique (1952) 38—50.

GIRAUD, G.: [1] Sur le problème de Dirichlet généralisé; équations non linéaires à m variables. Ann. Éc. N. Sup. 43 (1926) 1—128. — [2] Sur le problème de Dirichlet généralisé (deuxième mémoire). Ann. Éc. N. Sup.

46 (1929) 131—245. — [3] Sur les équations aux dérivées partielles du type elliptique. Bull. Sci. Math. **53** (1929) 367—395. — [4] Sur différentes questions relatives aux équations du type elliptique. Ann. Éc. N. Sup. **47** (1930) 197—266. — [5] Sur certains problèmes concernant des systèmes d'équations du type elliptique. C. R. Acad. Sci. Paris **192** (1931) 471—473. — [6] Détermination des tenseurs par des équations aux dérivées partielles jointes à des conditions à la frontière. C. R. Acad. Sci. Paris **192** (1931) 1338—1340. — [7] Sur certains problèmes non linéaires de Neumann et sur certains problèmes non linéaires mixtes. Bull. Acad. Polon. Sci. (1931) 421—437. — [8] Sur certains problèmes non linéaires de Neumann et sur certains problèmes non linéaires mixtes. Ann. Éc. N. Sup. **49** (1932) 1—104 and 245—308. — [9] Généralisation des problèmes sur les opérations du type elliptique. Bull. Sci. Math. **56** (1932) 248—272, 281—312 and 316—352. — [10] Sur certains cas de données discontinues relatifs aux problèmes de valeurs à la frontière. C. R. Acad. Sci. Paris **194** (1932) 1142—1145. — [11] Sur quelques problèmes de Dirichlet et de Neumann. J. Math. pures appl. **11** (1932) 389—416. — [12] Problèmes de valeurs à la frontière relatifs à certaines données discontinues. Bull. Soc. Math. France **61** (1933) 1—54. — [13] Sur certaines équations de Fredholm à noyau non borné. Bull. Sci. Math. **57** (1933) 390—401. — [14] Problèmes mixtes et problèmes sur de variétés closes, relativement aux équations linéaires du type elliptique. Ann. Soc. Polon. Math. **12** (1933) 35—54. — [15] Équations à integrales principales. Étude suivie d'une application. Ann. Éc. N. Sup. **51** (1934) 251—372. — [16] Problèmes des types de Dirichlet et de Neumann dans certains cas où les données sont discontinues. C. R. Acad. Sci. Paris **201** (1935) 925—928. — [17] Nouvelle généralisation des problèmes relatifs aux opérations du type elliptique. Ann. Soc. Polon. Math. **14** (1935) 74—115. — [18] Existence de certaines dérivées des fonctions de Green; consequence pour les problèmes du type de Dirichlet. C. R. Acad. Sci. Paris **202** (1936) 380—382. — [19] Équations à integrales principales d'ordre quelconque. Ann. Éc. N. Sup. **53** (1936) 1—40. — [20] Définitions élargies des noyaux résolvants de Freholm et des fonctions de Green. Bull. Sci. Math. **61** (1937) 172—192.— [21] Généralisation d'un type de problèmes relatifs aux équations aux dérivées partielles du type elliptique. Ann. Sc. N. Sup. Pisa **7** (1938) 25—71. — [22] Sur les dérivées des fonctions qui répondent à un problème du type de Dirichlet. C. R. Acad. Sci. Paris **207** (1938) 956—958. — [23] Nouvelle méthode por traiter certains problèmes relatifs aux équations du type elliptique. J. Math. pures appl. **18** (1939) 111—143. — [24] Sur un type de problèmes relatifs aux équations du type elliptique à deux variables indépendantes. C. R. Acad. Sci. Paris **208** (1939) 1462—1465. — [25] Sur une classe d'équations linéaires où figurent des valeurs principales d'intégrales simples. Ann. Éc. N. Sup. **56** (1939) 119—172.

GIRSANOV, I. V.: [1] The solution of certain boundary problems for parabolic and elliptic equations with discontinuous coefficients. Dokl. Akad. Nauk SSSR **135** (1960) 1311—1313 (Russian).

GLUŠKO, V. P.: [1] The first boundary value problem for elliptic equations degenerating on manifolds. Dokl. Akad. Nauk SSSR **129** (1951) 492—495 (Russian).

GOBERT, J.: [1] Opérateurs matriciels de dérivation elliptique et problèmes aux limites. Mém. Soc. Roy. Sci. Liège 1, n. 2 (1961) 144 pp.

GOWRISANKARAN, K.: [1] Limites fines „à la frontière" dans la théorie axiomatique du potentiel de M. Brelot. C. R. Acad. Sci. Paris 255 (1962) 450—451.

GRAVES, L.M.: [1] The estimates of Schauder and their application to existence for elliptic differential equations. Univ. Chicago, Invest. Theory Part. Diff. Eq. Tech. Rep. n. 1 (1956) 67 pp.

GRECO, D.: [1] Nuove formule integrali di maggiorazione per le soluzioni di un'equazione lineare di tipo ellittico ed applicazioni alla teoria del potenziale. Ricerche di Mat. Napoli 5 (1956) 126—149. — [2] Un teorema di esistenza per il problema di Dirichlet relativo ad un'equazione lineare ellittica in m variabili. Ricerche di Mat. Napoli 5 (1956) 150—158. — [3] Un'osservazione sul problema die Dirichlet. Rend. Acc. Sci. Fis. Mat. Napoli 23 (1956) 73—80. — [4] Il problema di derivata obliqua per certi sistemi di equazioni a derivate parziali di tipo ellittico in due variabili. Ann. Mat. pura appl. 42 (1956) 1—24. — [5] Sul problema di Lauricella per una particolare equazione di quarto ordine. Boll. Un. Mat. Ital. 11 (1956) 394—401. — [6] Le matrici fondamentali di alcuni sistemi di equazioni lineari a derivate parziali di tipo ellittico. Ricerche di Mat. Napoli 8 (1959) 197—221. — [7] Su un problema al contorno per certi sistemi di equazioni ellittiche. Ricerche di Mat. Napoli 8 (1959) 271—299.

GRÖGER, K.: [1] Nichtlineare ausgeartete elliptische Differentialgleichungen. Math. Nach. 28 (1964) 182—205. — [2] Anwendung der Variationsverfahren von A. Langenbach auf die Lösung ausgearteter elliptischer Differentialgleichungen. Math. Nach. 29 (1965) 9—16.

GRUŠIN, V.V.: [1] Fundamental solutions of hypoelliptic equations. Usp. Mat. Nauk 16, n. 4 (1961) 147—153 (Russian).

GUBERMAN, I. JA.: [1] On the existence of several solutions of the Dirichlet problem for an equation with a Monge-Ampère operator. Leningrad Gos. Ped. Inst. Učen. Zap. 238 (1962) 132—140 (Russian). — [2] On the existence of several solutions of the Dirichlet problem for multi-dimensional equations of the Monge-Ampère type. Izv. Vysš. Učebn. Zaved Mat. (1965) n. 4, 54—63 (Russian).

GUSEVA, O.V.: [1] On the first boundary-value problem for strongly elliptic systems of differential equations. Dissertation, Trudy Mat. Inst. Steklov (1953) (Russian). — [2] On boundary problems for strongly elliptic systems. Dokl. Akad. Nauk SSSR 102 (1955) 1069—1072 (Russian).

HAACK, W.: [1] Allgemeine Randwertprobleme für Differentialgleichungen vom elliptischen Typus. Math. Nach. 7 (1952) 1—30. — [2] Randwertprobleme höherer Charakteristik für ein System zwei elliptischen Differentialgleichungen. Math. Nach. 8 (1952) 123—132.

HAACK, W., u. G. HELLWIG: [1] Die Überführung des Randwertproblems für Systeme elliptischer Differentialgleichungen auf Fredholmsche Integralgleichungen. Math. Nach. **4** (1950/51) 404—418.

HAACK, W., u. W. WENDLAND: [1] Systeme linearer partieller Differentialgleichungen. Hahn-Meitner Inst. Kernforschung Berlin, BM **29** (1966) 171 pp.

HABETHA, K.: [1] Zum Phragmèn-Lindelöf Prinzip bei partiellen Differentialgleichungen. Arch. Mat. **15** (1964) 324—331.

HADAMARD, J.: [1] Le problème de Cauchy et les équations aux dérivées partielles linéaires hiperboliques. Ed. Hermann Paris (1932) 542 pp.

HALILOV, Z.I.: [1] Boundary-value problems for the elliptic equation. Izv. Akad. Nauk SSSR **11** (1947) 361—362 (Russian).

HAMMERSTEIN, A.: [1] Nichtlineare Integralgleichungen mit Anwendungen. Acta Math. **54** (1930) 117—176.

HARTMAN, P.: [1] Remarks on a uniqueness theorem for closed surfaces. Math. Ann. **133** (1957) 426—430. — [2] Hölder continuity and non linear elliptic partial differential equations. Duke Math. J. **25** (1958) 59—65. — [3] On elliptic partial differential equations and uniqueness theorem for closed surfaces. J. Math. Mech. **7** (1958) 377—392. — [4] On solutions of $\Delta v + v = 0$ in an exterior region. Math. Z. **71** (1959) 251—257. — [5] On quasi linear differential equations. Proc. Inter. Symp. Diff. Eq. Dyn. Syst. Puerto Rico (1965) 393—407. — [6] On the bounded slope condition. Pacific J. Math. **18** (1966) 495—511.

HARTMAN, P., and L. NIRENBERG: [1] On spherical image maps whose jacobians do not change sign. Amer. J. Math. **81** (1959) 901—920.

HARTMAN, P., and R. SACKSTEDER: [1] On maximum priciple for non-hyperbolic differential operators. Rend. Circ. Mat. Palermo **6** (1957) 218—232.

HARTMAN, P., and G. STAMPACCHIA: [1] On some non linear elliptic differential-functional equations. Acta Math. **115** (1966) 271—310.

HARTMAN, P., and C. WILCOX: [1] On solutions of the Helmholtz equation in exterior domains. Math. Z. **75** (1961) 228—255.

HARTMAN, P., and A. WINTNER: [1] On the third fundamental form of a surface. Amer. J. Math. **75** (1953) 298—331. — [2] On elliptic Monge-Ampère equations. Amer. J. Math. **75** (1953) 611—620. — [3] On the local behavior of solutions of non parabolic partial differential equations, I, II, III. Amer. J. Math. **75** (1953) 449—476; **76** (1954) 351—361; **77** (1955) 329—354. — [4] Umbilical points and W-surfaces. Amer. J. Math. **76** (1954) 502—508. — [5] Partial differential equations and a theorem of A. Kneser. Rend. Circ. Mat. Palermo **4** (1955) 237—255. — [6] On a comparison theorem for self-adjoint partial differential equations of elliptic type. Proc. Amer. Math. Soc. 6 (1955) 862—865. — [7] Binary linear elliptic partial differential equations. Duke Math. J. 22 (1955) 515—524.

HAVILAND, E. K.: [1] A note on unrestricted solutions of the differential equation $\Delta u = f(u)$. J. London Math. Soc. 26 (1951) 210—214.

HAYASHIDA, K.: [1] Unique continuation theorem of elliptic systems of partial differential equations. Proc. Japan. Acad. 38 (1962) 630—635. — [2] A note on a weak subsolution. Proc. Japan. Acad. 39 (1963) 203—207.

HEINZ, E.: [1] Über die Lösungen der Minimalflächengleichung. Göttingen Nach. (1952) 151—156. — [2] Über die Existenz einer Fläche constanter mittlerer Krümmung bei vorgegebener Berandung. Math. Ann. 127 (1954) 258—287. — [3] Über die Eindeutigkeit beim Cauchyschen Anfangswertproblem einer elliptischen Differentialgleichung zweiter Ordnung. Göttingen Nach. (1955) 1—12. — [4] Über gewisse elliptische Systeme von Differentialgleichungen zweiter Ordnung mit Anwendungen auf die Monge-Ampèresche Gleichung. Math. Ann. 131 (1956) 411—428. — [5] On certain non linear elliptic differential equations and univalent mappings. J. Anal. Math. 5 (1956/57) 197—272. — [6] On elliptic Monge-Ampère equations and Weyl's embedding problem. J. Anal. Math. 7 (1959) 1—52. — [7] Über die Differentialungleichung $0 < \alpha < r t - s^2 \leqq \beta < \infty$ Math. Z. 72 (1959) 107—126. — [8] Neue a priori Abschätzungen für den Ortsvektor einer Fläche positive Gauss'cher Krümmung durch ihr Linienelement. Math. Z. 74 (1960) 129—157. — [9] Interior estimates for solutions of elliptic Monge-Ampère operators. Proc. Symp. Pure Math. Amer. Math. Soc. 4 (1961) 149—155. — [10] Existence theorems for one to one mappings associated with elliptic systems of second order, I. J. Anal. Math. 15 (1965) 325—352.

HELLWIG, G.: [1] Bemerkungen zu der Satzgruppe von Hilbert über Systeme elliptischer Differentialgleichungen. Math. Z. 55 (1952) 276—283. — [2] Das Randwertproblem eines linearen elliptischen Systems. Math. Z. 56 (1952) 388—408. — [3] Randwertprobleme nichtlinearer elliptischer Differentialgleichungssysteme erster Ordnung mit Anwendungen auf die Verbiegung von elliptisch gekrümmten Flächenstücken. Math. Nach. 8 (1952) 13—30. — [4] Eine Bemerkung zum Ausstrahlungsproblem. Math. Z. 78 (1962) 446—448.

HENRICI, P.: [1] A survey of I. N. Vekua's theory of elliptic partial differential equations with analytic coefficients. J. Appl. Math. Phys. 8 (1957) 169—203.

HERGLOTZ, G.: [1] Über die Integration linearer partieller Differentialgleichungen mit konstanten Koeffizienten. Ber. Sächs. Akad. Leipzig 78 (1926) 93—126 and 287—318. — [2] Über die Starrheit der Eiflächen. Abh. Sem. Hansischen Univ. 15 (1943) 127—129.

HERVÉ, R. M.: [1] Recherches axiomatiques sur la théorie des fonctions surharmoniques et du potentiel. Ann. Inst. Fourier Grenoble 12 (1962)

415—471. — [2] Un principe du maximum pour les sous-solutions d'une équation uniformement elliptique de la forme

$$Lu = -\sum_i \frac{\partial}{\partial x_j}\left(\sum_j a_{ij}\frac{\partial u}{\partial x_j}\right) = 0.$$

Ann. Inst. Fourier Grenoble **14** (1964) 493—507.

HILBERT, D.: [1] Grundzüge einer allgemeinen Theorie der linearen Integralgleichungen. 2 Auflage Teubner Leipzig (1924) 282 pp.

HIRAI, I., and K. AKÔ: [1] On generalized Peano's theorem concerning the Dirichlet problem for semilinear elliptic differential equations. Proc. Japan. Acad. **36** (1960) 480—485.

HIRASAWA, Y.: [1] On the bounded solutions of the partial differential equation $\Delta u = f(x, y, u, p, q)$. Comment. Math. Univ. St. Paul **7** (1959) 13—19. — [2] Principally linear partial differential equations of elliptic type. Funkcial Ekvac. **2** (1959) 33—94. — [3] On a uniqueness condition for the Dirichlet problem concerning a quasi-linear equation of elliptic type. Kōdai Math. Sem. Rep. **14** (1962) 162—168. — [4] On an estimate for semi-linear elliptic differential equations of the second order. Kōdai Math. Sem. Rep. **16** (1964) 55—68. — [5] On an estimate for semi-linear elliptic differential equations of the second order with Dini-continuous coefficients. Kōdai Math. Sem. Rep. **17** (1965) 10—26. — [6] A remark on the generalization of Harnack's first theorem. Kōdai Math. Sem. Rep. **15** (1963) 121—126.

HODGE, W.V.D.: [1] The theory and applications of harmonic integrals. Cambridge Univ. Press 2th Ed. (1952) 282 pp.

HOPF, E.: [1] Elementare Betrachtungen über die Lösungen partieller Differentialgleichungen zweiter Ordnung vom elliptischen Typus. Sitzungb. Preuss. Akad. Wiss. **19** (1927) 147—152. — [2] Bemerkungen zu einem Satze von S. Bernstein aus der Theorie der elliptischen Differentialgleichungen. Math. Z. **29** (1928) 744—745. — [3] Zum analytischen Charakter der Lösungen regulärer zweidimensionaler Variationsprobleme. Math. Z. **30** (1929) 404—413. — [4] Über den funktionalen, insbesondere den analytischen Charakter, der Lösungen elliptischer Differentialgleichungen zweiter Ordnung. Math. Z. **34** (1931) 194—233. — [5] Kleine Bemerkung zur Theorie der elliptischen Differentialgleichungen. J. reine angew. Math. **165** (1931) 50—51. — [6] On S. Bernstein's theorem on surfaces $z(x, y)$ of non positive curvature. Proc. Amer. Math. Soc. **1** (1950) 80—85. — [7] A remark on linear elliptic differential equations of second order. Proc. Amer. Math. Soc. **3** (1952) 791—793. — [8] On an inequality for minimal surfaces. J. Rat. Mech. Anal. **2** (1953) 519—522 and 801—802. — [9] On non linear partial differential equations. Lecture Series Symp. Part. Diff. Eq. Univ. California Berkeley (1955) 1—32.

HOPF, H.: [1] Über Flächen mit einer Relation zwischen den Hauptkrümmungen. Math. Nach. **4** (1950/51) 239—249. — [2] Zur Differentialgeometrie geschlossener Flächen im Euklidischen Raum. Convegno Inter. Geom. Diff. 1953, Ed. Cremonese Roma (1954) 44—54.

HÖRMANDER, L.: [1] On the uniqueness of the Cauchy problem, I, II. Math. Scand. **6** (1958) 213–225 and **7** (1959) 177–190. — [2] Linear partial differential operators. Berlin-Göttingen-Heidelberg: Springer-Verlag (1963) 284 pp. — [3] Pseudo-differential operators. Commun. pure appl. Math. **18** (1965) 501–517. — [4] Pseudo-differential operators and non elliptic boundary problems. Ann. of Math. **83** (1966) 129–209.

HORNICH, H.: [1] Zur Lösbarkeit von gewissen elliptischen Differentialgleichungen. J. reine angew. Math. **189** (1952) 204–206.

HOU CUN'-I: [1] Dirichlet problem for a class of linear elliptic second order equations with parabolic degeneracy on the boundary of the domain. Sci. Record Peking **2** (1958) 244–249 (Russian).

HOU TSUNG-YI: [1] A Carleman boundary-value problem for elliptic systems of first order equations. Sci. Sinica **12** (1963) 12–37 (Russian).

HUBER, A.: [1] On uniqueness of generalized axially symmetric potentials. Ann. of Math. **60** (1954) 351–385. — [2] Some results on generalized axially symmetric potential. Proc. Conf. Part. Diff. Eq. Univ. of Maryland (1955) 147–155.

HUDJAEV, S.I.: [1] A criterion for the solvability of the Dirichlet problem for elliptic equations. Dokl. Akad. Nauk SSSR **148** (1963) 44–46 (Russian). — [2] Boundary-value problems for certain quasi-linear elliptic equations. Dokl. Akad. Nauk SSSR **154** (1964) 787–790 (Russian).

HUET, D.: [1] Phénomènes de perturbation singulière, I, II. C. R. Acad. Sci. Paris **244** (1957) 1438–1440 and **246** (1958) 2096–2098. — [2] Perturbations singulières d'opérateurs elliptiques. Sém. Lelong Paris exp. n. 13 (1958/59) 7 pp. — [3] Phénomènes de perturbation singulière dans les problèmes aux limites. Ann. Inst. Fourier Grenoble **10** (1960) 61–150. — [4] Perturbations singulières, I, II, III. C. R. Acad. Sci. Paris **258** (1964) 6320–6322; **259** (1964) 4213–4215; **260** (1965) 6800–6801.

HURWITZ, W.A.: [1] Randwertprobleme bei Systemen von linearen partiellen Differentialgleichungen erster Ordnung. Diss. Göttingen (1910) 97 pp.

HVEDELIDZE, B.V.: [1] The Poincaré's problem for a linear second-order differential equation of elliptic type. Trudy Inst. Mat. Tbilissk **12** (1943) 44–47 (Georgian).

IGLISCH, R.: [1] Relle Lösungsfelder der elliptischen Differentialgleichung $\Delta u = F(u)$ und nichtlinearer Integralgleichungen. Math. Ann. **101** (1929) 98–119.

IL'IN, A.M.: [1] On Dirichlet's problem for an equation of elliptic type degenerating on some set of interior points of a region. Dokl. Akad. Nauk SSSR **102** (1955) 9–12 (Russian). — [2] Degenerate elliptic and parabolic equations. Mat. Sbornik **50** (1960) 443–498 (Russian).

Il'in, V.A.: [1] Solvability of the Dirichlet and Neumann problem for a linear elliptic operator with discontinuous coefficients. Dokl. Akad. Nauk SSSR **137** (1961) 28—30 (Russian).

Il'in, V.A., and I.A. Sismarev: [1] Some problems for the $L(u)$ = div$(p(x)$ grad$u) - q(x) u$ operator with discontinuous coefficients. Dokl. Akad. Nauk SSSR **135** (1960) 775—778 (Russian).

Ito, S.: [1] On Neumann problem for Laplace-Beltrami operators. Proc. Japan. Acad. **37** (1961) 267—272. — [2] On Neumann problem for non-symmetric second order partial differential operators of elliptic type. J. Fac. Sci. Univ. Tokyo **10** (1963) 20—28. — [3] On existence of Green functions and positive superharmonic functions for linear elliptic operators of second order. J. Math. Soc. Japan **16** (1964) 299—306.

Jackson, L.: [1] The principle of the maximum for generalized subharmonic functions. Portugal. Math. **11** (1952) 69—74.

Jaenicke, J.: [1] Nullstellenvorgaben für Lösungen linearer Randwert-probleme elliptischer Differentialgleichungssysteme. Math. Nach. **18** (1958) 106—119. — [2] Lösungen gegebener Charakteristik linearer Randwertprobleme elliptischer Differentialgleichungssysteme. Math. Nach. **21** (1960) 223—232. — [3] Zum Maximum-Minimum-Prinzip bei elliptischen Differentialgleichungssystemen. Math. Nach. **22** (1960) 271—277. — [4] Eine Beziehung zwischen Lösungen adjungierter Randwertprobleme bei elliptischen Differentialgleichungssystemen. Arch. Math. **12** (1961) 118—121.

Jenkins, H.: [1] On quasi-linear elliptic equations which arise from variational problems. J. Math. Mech. **10** (1961) 705—728. — [2] Super-solutions for quasi-linear elliptic equations. Arch. Rat. Mech. Anal. **16** (1964) 402—410. — [3] On the behavior of solutions of quasi-linear elliptic equations. Trans. Amer. Math. Soc. **119** (1965) 407—416.

Jenkins, H., and J. Serrin: [1] Variational problems of minimal surfaces. Arch. Rat. Mech. Anal. **12** (1963) 185—212.

John, F.: [1] A note on the maximum principle for elliptic differential equations. Bull. Amer. Math. Soc. **44** (1938) 268—271. — [2] The funda-mental solution of linear elliptic differential equations with analytic coefficients. Commun. pure appl. Math. **3** (1950) 273—304. — [3] General properties of solutions of linear elliptic partial differential equations. Proc. Symp. Spectral Th. Diff. Probl. Oklahoma College (1951) 113 to 175. — [4] Derivatives of continuous weak solutions of linear elliptic equations. Commun. pure appl. Math. **6** (1953) 327—335. — [5] Deriva-tives of solutions of linear elliptic partial differential equations. Ann. of Math. Studies **33** (1954) 53—61. — [6] Plane waves and spherical means applied to partial differential equations. New York: Interscience Publ. (1955) 172 pp.

Jörgens, K.: [1] Harmonische Abbildungen und die Differentialgleichung $rt - s^2 = 1$. Math. Ann. **129** (1955) 330—344.

JUDANINA, A. B.: [1] An elliptic system of differential equations on a closed orientable surface. Dopovidi Akad. Nauk Ukrain RSR (1962) 859—861 (Ukrainian). — [2] Boundary value problems for first order elliptic systems on oriented surfaces with boundary. Dopovidi Akad. Nauk Ukrain RSR (1964) 1569—1572 (Ukrainian). — [3] Normal solvability and the index of a boundary-value problem for a first-order elliptic system on an orientable surface with boundary. Ukrain. Mat. Ž. 16 (1964) 704 to 709 (Russian).

JUDOVIČ, V. I.: [1] Some estimates connected with integral operators and with solutions of elliptic equations. Dokl. Akad. Nauk SSSR 138 (1961) 805—808 (Russian). — [2] Some bounds for solutions of elliptic equations. Mat. Sbornik 59 (1962) 229—244 (Russian).

KADLEC, J.: [1] On the regularity of the solution of the Poisson problem on a region whose boundary is similar to that of a cube. Czech. Math. J. 13 (1963) 599—611. — [2] On the maximum principle for second order elliptic equations and the method of Wiener. Czech. Math. J. 14 (1964) 154—155. — [3] The regularity of the Poisson problem in a domain whose boundary is similar to that of a convex domain. Czech. Math. J. 14 (1964) 386—393.

KADLEC, J., e J. NEČAS: [1] Sulla regolaritá delle soluzioni di equazioni ellittiche negli spazi $H^{k,\lambda}$. Ann. Sc. N. Sup. Pisa 21 (1967) 527—545.

KALANDIJA, A. L.: [1] The solution of a fundamental boundary problem for the equation $\Delta^n u = 0$ in a doubly connected region. Trudy Tbilissk Mat. Inst. 17 (1949) 131—168 (Georgian). — [2] The solution of a fundamental n-harmonic problem in the case of an infinite region. Trudy Tbilissk Mat. Inst. 17 (1949) 169—189 (Russian). — [3] Remark on the uniqueness of a fundamental boundary problem for a class of elliptic equations. Soobšč. Akad. Nauk Gruzin. SSR 12 (1951) 321—325 (Russian). — [4] The fundamental n-harmonic problem for multiply connected regions. Izv. Akad. Nauk SSSR 15 (1951) 185—198 (Russian).

KALIK, C., et P. SZILÁGYI: [1] Une condition nécessaire et suffisante pour la validité de l'alternative de Freholm dans le cas de problèmes de Dirichlet et de systèmes elliptiques. C. R. Acad. Sci. Paris 257 (1963) 2231—2233. — [2] Sur la solution du problème de Dirichlet pour une classe de systèmes elliptiques. Mathematica Cluj 5 (1963) 199—213.

KAMENOMOSTSKAJA, S. L.: [1] On equations of elliptic and parabolic type with a small parameter in the highest derivatives. Mat. Sbornik 31 (1952) 703—708 (Russian.) — [2] The first boundary problem for equations of elliptic type with a small parameter in the highest derivatives. Izv. Akad. Nauk SSSR 19 (1955) 345—360 (Russian).

KANTOROVIČ, L. V.: [1] On the convergence of variational processes. Dokl. Akad. Nauk SSSR 30 (1941) 107—111 (Russian). — [2] On the convergence of the method of reduction to ordinary differential equations. Dokl. Akad. Nauk SSSR 30 (1941) 585—588 (Russian). — [3] Functional analysis and applied mathematics. Usp. Mat. Nauk 3, n.6 (1948) 89—185

(Russian). — [4] Approximate solution of functional equations. Usp. Mat. Nauk 11, n.6 (1956) 99—116 (Russian). — [5] Some further applications of Newton's method for functional equations. Vestnik Leningrad Univ. 12, n.7 (1957) 68—103 (Russian).

KAPILEVIČ, M.B.: [1] Uniqueness theorems of singular Dirichlet-Neumann problems. Dokl. Akad. Nauk SSSR 125 (1959) 23—26 (Russian). — [2] The theorems of the mean for solutions of singular elliptic differential equations. Izv. Vysš. Učebn. Zaved. Mat. (1960) n. 6, 114—125 (Russian).

KARABEGOV, V.K.I.: [1] On stability in a closed region of Dirichlet's problem for linear equations of elliptic type. Dokl. Akad. Nauk Armyan SSR 16 (1953) 65—71 (Russian).

KATO, T.: [1] Growth properties of solutions of the reduced wave equation with a variable coefficient. Commun. pure appl. Math. 12 (1959) 403—425.

KATO, Y.: [1] The space of bounded solutions und removable singularities of the equation $\Delta u + a u_x + b u_y + c u = 0$, ($c \leqq 0$). Proc. Japan. Acad. 36 (1960) 644—649.

KATZ, I.N.: [1] On the existence of weak solutions to linear partial differential equations. J. Math. Anal. Appl. 2 (1961) 111—144.

KELDYŠ, M.V.: [1] On some cases of degeneracy of an elliptic equation on the boundary of a region. Dokl. Akad. Nauk SSSR 77 (1951) 181—183 (Russian).

KELLER, J.B.: [1] On solutions of $\Delta u = f(u)$. Commun. pure appl. Math. 10 (1957) 503—510.

KELLOG, O.D.: [1] On the derivatives of harmonic functions on the boundary. Trans. Amer. Math. Soc. 33 (1931) 486—510.

KIPRIAJANOV, I.A.: [1] A variational method of solving a class of degenerate elliptic equations. Dokl. Akad. Nauk SSSR 152 (1963) 35—38 (Russian).

KODAIRA, K.: [1] Über die Rand- und Eigenwertprobleme der linearen elliptischen Differentialgleichungen zweiter Ordnung. Proc. Imp. Acad. Tokyo 20 (1944) 262—268.

KOHN, J.J., and L.NIRENBERG: [1] An algebra of pseudo-differential operators. Commun. pure appl. Math. 18 (1965) 269—305. — [2] Non coercive boundary value problems. Commun. pure appl. Math. 18 (1965) 443—492.

KONDRAT'EV, V.A.: [1] On the solvability of the first boundary-value problem for elliptic equations. Dokl. Akad. Nauk SSSR 136 (1961) 771—774 (Russian). — [2] Bounds for the derivatives of solutions of elliptic equations near the boundary. Dokl. Akad. Nauk SSSR 146 (1962) 22—25 (Russian). — [3] Boundary-value problems for elliptic equations in conical regions. Dokl. Akad. Nauk SSSR **153** (1963) 27—29 (Russian).

KOPAČEK, I.: [1] On the Dirichlet problem for elliptic equations with a small parameter in the highest derivatives. Usp. Mat. Nauk **12**, n.5 (1957) 211—220 (Russian).

KOPPELMAN, W.: [1] On the index of elliptic operators on closed surfaces. Amer. J. Math. **85** (1963) 423—448.

KORDZADZE, R.A.: [1] A general boundary value problem with shift for a second-order equation of elliptic type. Dokl. Akad. Nauk SSSR **155** (1964) 739—742 (Russian).

KOŠELEV, A.I.: [1] Newton's method and the generalized solutions of non linear equations of the elliptic type. Dokl. Akad. Nauk SSSR **91** (1953) 1263—1266 (Russian). — [2] Existence of a generalized solution of the elastic-plastic problem of torsion. Dokl. Akad. Nauk SSSR **99** (1954) 357—360 (Russian). — [3] Differential corresponding spaces and existence theorems. Dokl. Akad. Nauk SSSR **105** (1955) 22—25 (Russian). — [4] On boundedness in L^p of the derivatives of solutions of elliptic differential equations. Mat. Sbornik **38** (1956) 359—372 (Russian). — [5] On differentiability of solutions of elliptic differential equations. Dokl. Akad. Nauk SSSR **112** (1957) 806—809 (Russian). — [6] A priori estimates in L^p and generalized solutions of elliptic equations and systems. Usp. Mat. Nauk **13**, n. 4 (1958) 29—88 and **14**, n. 3 (1959) 235 (Russian); English transl. Amer. Math. Soc. (2) **20** (1962) 105—171.

KOTAKE, T., and M.S. NARASIMHAN: [1] Fractional powers of a linear elliptic operator. Bull. Soc. Math. France. **90** (1962) 449—471.

KRUŽKOV, S.N.: [1] Some properties of solutions of elliptic equations. Dokl. Akad. Nauk SSSR **150** (1963) 470—473 (Russian). — [2] A priori bounds for generalized solutions of second order elliptic and parabolic equations. Dokl. Akad. Nauk SSSR **150** (1963) 748—751 (Russian). — [3] A priori bounds and some properties of solutions of elliptic and parabolic equations. Mat. Sbornik **65** (1964) 522—570 (Russian).

KRUŽKOV, S.N., and L.P. KOPKOV: [1] Harnack's inequality for solutions of elliptic differential equations. Vestnik Moscow Univ. (1964) n. 3, 3—14 (Russian).

KRZYWCKI, A.: [1] An application of Weyl's method to a boundary value problem of the linearized equations of hydrodynamics. Prace Mat. 5 (1961) 15—26 (Polish).

KRZYŻAŃSKI, M.: [1] Sur le problème de Dirichlet pour l'équation linéaire du type elliptique dans un domaine non borné. Rend. Acc. Lincei **4** (1948) 408—416. — [2] Un problème aux limites relatif aux équations du type elliptique. Colloq. Math. 2 (1949) 71—72. — [3] Sur les solutions de l'équation linéaire du type elliptique discontinues sur la frontière du domaine de leur existence. Studia Math. **11** (1950) 95—125. —

[4] Sur le second problème aux limites pour les équations linéaires aux dérivées partielles du type elliptique et parabolique dans un domaine non borné. Ann. Univ. Mariae Curie-Sklodowska **5** (1952) 1—21.

KUKS, L.M.: [1] Some properties of the solutions of non linear elliptic equations. Izv. Vysš. Učebn. Zaved (1958) n. 4, 131—149 (Russian). — [2] Some geometric criteria for the uniqueness of the solutions of the Dirichlet problem for strongly elliptic systems of second order partial differential equations. Izv. Vysš. Učebn. Zaved (1959) n. 3, 168—172 (Russian). — [3] On solvability regions of the first boundary value problem for strongly elliptic systems of differential equations. Izv. Vysš. Učebn. Zaved (1961) n. 2, 90—99 (Russian). — [4] Sturm's theorem and the oscillation of solutions of strongly elliptic systems. Dokl. Akad. Nauk SSSR **142** (1962) 32—35 (Russian). — [5] Theorems in the qualitative theory of strongly elliptic second order systems. Usp. Mat. Nauk **17,** n. 3 (1962) 181—184 (Russian).

KUMANO-GO, H.: [1] On the uniqueness of the solution of the Cauchy problem and the unique continuation theorem for elliptic equations. Osaka Math. J. **14** (1962) 181—212.

KÜNZI, H. P.: [1] Quasikonforme Abbildungen. Ergeb. Math. Berlin-Göttingen-Heidelberg: Springer-Verlag **26** (1960) 182 pp.

KUPRADZE, V. D.: [1] A space problem on the oscillation of an elastic body with given displacements on the boundary. Dokl. Akad. Nauk SSSR **67** (1949) 233—236 (Russian). — [2] On the boundary value problems of the steady vibrations of elastic bodies. Usp. Mat. Nauk **5,** n.3 (1950) 190—193 (Russian). — [3] Boundary problems of the theory of vibrations and integral equations. Gos. Izdat Tehn.-Teor. Lit. Moscow-Leningrad (1950) 280 pp. (Russian); German transl. Berlin: VEB Deutscher Verlag Wiss. (1956) 239 pp. — [4] Boundary problems of the theory of steady elastic vibrations. Usp. Mat. Nauk **8,** n. 3 (1953) 21—74 (Russian).

KUSANO, T.: [1] On a maximum principle for quasi linear elliptic equations. Proc. Japan. Acad. **38** (1962) 78—92. — [2] On bounded solutions of elliptic partial differential equations of the second order. Funkcial Ekvac **7** (1965) 1—13.

LABROUSSE, J. F.: [1] Homotopy invariants associated with elliptic partial differential operators. Mimeographed Rep. Univ. California (1963) 147 pp.

LADYŽENSKAJA, O.A.: [1] On the closure of an elliptic operator. Dokl. Akad. Nauk SSSR **79** (1951) 723—725 (Russian). — [2] The mixed problem for the hyperbolic equation. Gos. Izdat. Tehn.-Teor. Lit. Moscow (1953) 279 pp. (Russian). — [3] A simple proof of the solvability of the fundamental problem and of a problem of eigenvalues for linear elliptic equations. Vestnik Leningrad Univ. **10,** n. 11 (1955) 23—29 (Russian). — [4] On equations with a small parameter in the higher derivatives in linear partial differential equations. Vestnik Leningrad

Univ. **12,** n.7 (1957) 104—120 (Russian). — [5] On integral estimates, convergence, approximate methods and solution in functionals for elliptic operators. Vestnik Leningrad Univ. **13,** n.7 (1958) 60—69 (Russian). — [6] Study of the Navier-Stokes equation in the case of stationary motion of incompressible fluids. Usp. Mat. Nauk **14,** n.3 (1959) 75—97 (Russian). English transl. Amer. Math. Soc. **(2) 25** (1963) 173—197. — [7] Mathematical problems in the dynamics of viscous incompressible flow. Gos. Izdat. Fiz.-Mat. Moscow (1961) 203 pp. (Russian); English transl. New York-London: Gordon and Breach Sci. Publ. (1963) 184 pp; German. transl. Berlin: Akademie-Verlag (1965) 180 pp.

LADYŽENSKAJA, O. A., V. JA. RIVKLIND and N. N. URAL'CEVA: [1] Classical solvability of diffraction problems for equations of elliptic and parabolic type. Dokl. Akad. Nauk SSSR **158** (1964) 513—515 (Russian).

LADYŽENSKAJA, O. A., and N. N. URAL'CEVA: [1] Quasilinear elliptic equations and variational problems with many independent variables. Usp. Mat. Nauk **16,** n. 1 (1961) 19—92 (Russian); transl. as Russian Math. Surveys **16,** n. 1 (1961) 17—92. — [2] On the smoothness of weak solutions of quasilinear equations in several variables and of variational problems. Commun. pure appl. Math. **14** (1961) 481—495. — [3] Differential property of bounded generalized solutions of multidimensional quasi-linear elliptic equations and variational problems. Dokl. Akad. Nauk SSSR **138** (1961) 29—32 (Russian). — [4] Regularity of generalized solutions of quasi-linear elliptic equations. Dokl. Akad. Nauk SSSR **140** (1961) 45—47 (Russian). — [5] Admissible extensions of the concept of solutions for linear and quasi-linear elliptic equations of second order. Vestnik Leningrad Univ. **18,** n. 1 (1963) 10—25 (Russian). — [6] Hölder continuity of solutions and their derivatives of linear and quasi-linear equations of elliptic and parabolic type. Dokl. Akad. Nauk SSSR **155** (1964) 1258—1261 (Russian). — [7] On the Hölder continuity of the solutions and their derivatives of linear and quasi-linear equations of elliptic and parabolic type. Trudy Mat. Inst. Steklov **73** (1964) 173—190 (Russian). — [8] Linear and quasi-linear equations of elliptic type. Izdat. Nauka Moscow (1964) 538 pp. (Russian).

LANDIS, E. M.: [1] On some properties of solutions of elliptic equations Dokl. Akad. Nauk SSSR **107** (1956) 640—643 (Russian). — [2] Some questions in the qualitative theory of elliptic and parabolic equations. Usp. Mat. Nauk **14,** n. 1 (1959) 21—85 (Russian); English transl. Amer. Math. Soc. **(2) 20** (1962) 173—238. — [3] Some problems in the qualitative theory of second order elliptic equations (case of several variables). Usp. Mat. Nauk **18,** n. 1 (1963) 3—62 (Russian); transl. as Russian Math. Surveys **18,** n. 1 (1963) 1—62.

LASOTA, A.: [1] Sur l'existence de solutions des problèmes de Neumann et de Dirichlet pour l'équation différentielle elliptique de second ordre. Bull. Acad. Polon. Sci. **11** (1963) 441—446.

LAVRENT'EV, M. A.: [1] Quasi conformal mappings and their derivatives systems. Dolk. Akad. Nauk SSSR **52** (1946) 287—289 (Russian). — [2] General problem of quasi conformal representation of plane regions.

Mat. Sbornik **21** (1947) 285–320 (Russian); English Transl. Amer. Math. Soc. **(1) 2** (1962) 481–532. — [3] A fundamental theorem of the theory of quasi-conformal mapping of plane regions. Izv. Akad. Nauk SSSR **12** (1948) 513–554 (Russian); English transl. Amer. Math. Soc. **(1) 2** (1962) 425–480. — [4] Certain boundary value problems for systems of elliptic type. Sibirsk Mat. Ž. **3** (1962) 715–728 (Russian). — [5] Variational methods in boundary value problems for systems of elliptic equations. Izdat. Akad. Nauk SSSR (1962) 136 pp. (Russian); English transl. Groningen: Noordhoff Ltd. (1963) 150 pp.

LAVRENT'EV, M.M.: [1] On the problem of Cauchy for linear elliptic equations of the second order. Dokl. Akad. Nauk SSSR **112** (1957) 195–197 (Russian).

LAVRUK, B.R.: [1] Condition of solubility of a boundary value problem for a system of linear differential equations of second order of elliptic type. Dokl. Akad. Nauk SSSR **111** (1956) 23–25 (Russian). — [2] On certain boundary problems for a system of elliptic equations. Dopovidi Akad. Nauk Ukrain. RSR (1956) 214–219 (Ukrainian). — [3] On regular solutions of boundary problems for elliptical systems of linear differential equations of the second order for a half plane. Dopovidi Akad. Nauk Ukrain. RSR (1957) 107–111 (Ukrainian). — [4] On two elliptic systems of second order linear differential equations. Bull. Acad. Polon. Sci. **8** (1960) 209–216. — [5] Parametric boundary-value problems for elliptic systems of linear differential equations. I, Construction of conjugate problem. II, A boundary value problem for a half space. III, Conjugate boundary problems in a half space. Bull. Acad. Polon. Sci. **99** (1963) 257–267 and 269–278; **13** (1965) 105–110 (Russian). — [6] On the unique solvability of a general boundary-value problem for homogeneous linear systems of second order differential equations of elliptic type with constant coefficients in a half space. Ann. Polon. Math. **14** (1963) 85–95 (Russian).

LAX, P.D.: [1] On Cauchy's problem for hyperbolic equations and the differentiability of solutions of elliptic equations. Commun. pure appl. Math. **8** (1955) 615–633. — [2] A stability theorem for solutions of abstract differential equations and its application to the study of local behaviour of solutions of elliptic systems. Commun. pure appl. Math. **9** (1956) 747–766.

LEIS, R.: [1] Zur Eindeutigkeit der Randwertaufgaben der Helmholtzschen Schwingungsgleichung. Math. Z. **85** (1964) 141–153.

LEJA, E.: [1] Remarques sur le travail précédent de M. Mauro Picone. Ann. Soc. Polon. Math. **21** (1949) 170–172.

LERAY, J.: [1] Étude de diverses équations integrales non linéaires et de quelques problèmes que pose l'hydrodynamique. J. Math. pures appl. **12** (1933) 1–82. — [2] Topologie des espaces abstraits de M. Banach. C. R. Acad. Sci. Paris **200** (1935) 1082–1084. — [3] Les problèmes non linéaires. Enseign. Math. **35** (1936) 139–149. — [4] Majoration des dérivées secondes des solutions d'un problème de Dirichlet. J. Math. pures appl. **17** (1938) 89–104. — [5] Discussion d'un problème de Dirichlet. J. Math. pures appl. **18** (1939) 249- 284. — [6] Calcul, par réflexions, des fonctions M-harmoniques dans une bande plane vérifiant aux bords M conditions différentielles, à coefficients constants. Arch. Mech. Stos. **16** (1964) 1041- 1090.

LERAY, J., et J.L. LIONS: [1] Quelques résultats de Višik sur les problèmes, elliptiques non linéaires par les méthodes de Minty-Browder. Bull. Soc. Math. France. **93** (1965) 97- 107.

LERAY, J., et J. SCHAUDER: [1] Topologie et équations fonctionnelles. Ann. Éc. N. Sup. **51** (1934) 45–78; Russian transl. Usp. Mat. Nauk. **1** n. 3/4 (1946) 71–95.

LEVI, E.E.: [1] Sulle equazioni lineari totalmente ellittiche alle derivate parziali. Rend. Circ. Mat. Palermo **24** (1907) 275–317. — [2] I problemi dei valori al contorno per le equazioni lineari totalmente ellittiche alle derivate parziali. Mem. Soc. Ital. dei XL **16** (1910) 1–112.

LEVINSON, N.: [1] The first boundary value problem fur $\varepsilon \Delta u + A u_x + B u_y + C u = D$ for small ε. Ann. of Math. **51** (1950) 428–445. — [2] Dirichlet problem for $\Delta u = f(u, P)$. J. Rat. Mech. Anal. **12** (1963) 567- 575.

LEWY, H.: [1] Über den analytischen Charakter der Lösungen elliptischer Differentialgleichungen. Göttingen. Nach. (1927) 178–186. — [2] Neuer Beweis des analytischen Charakters der Lösungen elliptischer Differentialgleichungen. Math. Ann. **101** (1929) 609–619 and **107** (1934) 804. — [3] Eindeutigkeit der Lösung des Anfangsproblems einer Differentialgleichung zweiter Ordnung in zwei Veränderlichen. Math. Ann. **104** (1931) 325–339. — [4] Sur une nouvelle formule dans les équations linéaires elliptiques et une application au problème de Cauchy. C. R. Acad. Sci. Paris **197** (1933) 112–113. — [5] A priori limitations for solutions of Monge-Ampère equations. Trans. Amer. Math. Soc. **37** (1935) 417–434 and **41** (1937) 365–374. — [6] On the existence of a closed convex surface realizing a given Riemannian metric. Proc. Nat. Acad. Sci. USA **24** (1938) 104–106. — [7] On differential geometry in the large (Minkowski's problem). Trans. Amer. Math. Soc. **43** (1938) 258–270. - [8] On the boundary behavior of minimal surfaces. Proc. Nat. Acad. Sci. USA **37** (1951) 103–110. — [9] On the reflection laws of second order differential equations in two independent variables. Bull. Amer. Math. Soc. **65** (1959) 37–58.

LIANG SHI-TING: [1] The regularity of the solutions of elliptic equations. Bull. Math. Soc. Sci. Math. Phys. R. P. Roumaine **4** (1961) 57–68.

LICHTENSTEIN, L.: [1] Über das Poissonsche Integrale und über die partiellen Ableitungen zweiter Ordnung des logaritmischen Potentials. J. reine angew. Math. **141** (1912) 12–42. — [2] Zur Theorie der konformen Abbildung. Konforme Abbildung nichtanalytischer singularitätenfreier Flächenstücke auf ebene Gebiete. Bull. Sci. Gracovie (1916)

192–217. — [3] Über die erste Randwertaufgabe der Elastizitätstheorie. Math. Z. **20** (1924) 21–28. — [4] Neue Beiträge zur Theorie der linearen partiellen Differentialgleichungen zweiter Ordnung vom elliptischen Typus. Math. Z. **20** (1924) 194–212. — [5] Neuere Entwicklung der Theorie partieller Differentialgleichungen zweiter Ordnung vom elliptischen Typus. Encykl. Math. Wiss. Bd. II, 3. Heft 8 (1924) 1277–1334. — [6] Vorlesungen über einige Klassen nichtlinearer Integralgleichungen und Integrodifferentialgleichungen. Berlin: Springer-Verlag (1931) 164 pp. — [7] Neue Untersuchungen in der Theorie der linearen partiellen Differentialgleichungen zweiter Ordnung vom elliptischen Typus. Bull. Acad. Polon. Sci. (1931) 571–598.

LIÉNARD, A.: [1] Problème plan de la dérivée oblique dans la théorie du potentiel. J. Éc. Polytech. **144** (1938) 35–158 and 178–226. — [2] Nouveau mémoire sur le problème de la dérivée oblique dans la théorie du potentiel. J. Éc. Polytech. **145** (1939) 55–84 and 85–137.

LI MING-ZHONG: [1] An existence theorem and a representation formula for generalized solutions of second-order elliptic differential equations. Acta Math. Sinica **14** (1964) 7–22; transl. as Chinese Math. **5** (1964) 8–24.

LIONS, J.L.: [1] Problèmes aux limites en théorie des distributions. Acta Math. **94** (1955) 13–153. — [2] Sur quelques problèmes aux limites relatifs à des opérateurs différentielles elliptiques. Bull. Soc. Math. France **83** (1955) 225–250. — [3] Contribution à un problème de M. M. Picone. Ann. Mat. pura appl. **41** (1955) 201–219. — [4] Sur les problèmes aux limites du type de la dérivée oblique. Ann. of Math. **64** (1956) 207–239. — [5] Conditions aux limites de Višik-Sobolev et problèmes aux limites. C. R. Acad. Sci. Paris **244** (1957) 1126–1128. — [6] Some questions on elliptic differential equations. Tata Inst. Fund. Research, Bombay (1957). — [7] Équations différentielles opérationelles et problèmes aux limites. Berlin-Göttingen-Heidelberg: Springer-Verlag (1961) 292 pp.

LIONS, J.L., e E. MAGENES: [1] Problemi ai limiti non omogenei (Problèmes aux limites non homogènes) I, II, III, IV, V, VI, VII. Ann. Sc. N. Sup. Pisa **14** (1960) 269–308; Ann. Inst. Fourier Grenoble **11** (1961) 137–178; Ann. Sc. N. Sup. Pisa **15** (1961) 39–101; ib. 311–326; ib. **16** (1962) 1–44; J. Anal. Math. **11** (1963) 165–178; Ann. Mat. pura appl. **63** (1963) 201–224. — [2] Remarques sur les problèmes aux limites linéaires elliptiques. Rend. Acc. Lincei **32** (1962) 873–883. — [3] Espaces de fonctions et distributions du type de Gevrey et problèmes aux limites paraboliques. Ann. Mat. pura appl. **68** (1965) 341–417. — [4] Espaces du type de Gevrey et problèmes aux limites pour diverses classes d'équations. Ann. Mat. pura appl. **72** (1966) 343–394. — [5] Problèmes aux limites non homogènes et applications, I, II, III. Dunod Éd. Paris (1968).

LIONS, J.L., et G. STAMPACCHIA: [1] Inéquations variationelles non coercives. C. R. Acad. Sci. Paris **261** (1965) 25–27. — [2] Variational inequalities. Commun. pure appl. Math. **20** (1967) 453–519.

LITHNER, L.: [1] A theorem of Phragmèn-Lindelöf type for second order elliptic operators. Ark. Mat. **5** (1964) 281—285.

LITTMAN, W.: [1] A strong maximum principle for weakly L-subharmonic functions. J. Math. Mech. **8** (1959) 761—770. — [2] Generalized subharmonic functions: Monotonic approximations and an improved principle. Ann. Sc. N. Sup. Pisa **17** (1963) 207—222.

LITTMAN, W., G. STAMPACCHIA and H.F. WEINBERGER: [1] Regular points for elliptic equations with discontinuous coefficients. Ann. Sc. N. Sup. Pisa **17** (1963) 43—77.

LIZORKIN, P.A.: [1] The Dirichlet principle for the Beltrami equation in a semi-space. Dokl. Acad. Nauk SSSR **134** (1960) 761—764 (Russian). — [2] The Green's E-function of a Beltrami operator and some variational problems. Dokl. Akad. Nauk SSSR **139** (1961) 1052—1055 (Russian).

LJUBIČ, JU. I.: [1] On the fundamental solutions of linear elliptic partial differential equations of elliptic type. Mat. Sbornik **39** (1956) 23—26 (Russian). — [2] On the existence „in the large" of fundamental solutions of linear second-order elliptic equations. Mat. Sbornik **57** (1962) 45—58 (Russian).

LOPATINSKIĬ, YA.B.: [1] On the behavior of solutions of a linear elliptic system in the neighborhood of an isolated singular point. Dokl. Akad. Nauk SSSR **79** (1951) 727—730 (Russian). — [2] A fundamental system of solutions of an elliptic system of differential equations. Ukrain. Mat. Ž. **3** (1951) 3—88 (Russian). — [3] Fundamental solutions of a system of differential equations of elliptic type. Ukrain. Mat. Ž. **3** (1951) 290—316 (Russian). — [4] On a method of reducing boundary problems for a system of differential equations of elliptic type to regular integral equations. Ukrain. Mat. Ž. **5** (1953) 123—151 (Russian). — [5] Conditions for existence of the solution of the first boundary problem for a system of linear differential equations of second order and elliptic type. Dopovidi Akad. Nauk Ukrain. RSR (1956) 5—9 (Ukrainian). — [6] Uniqueness of the solution of Cauchy's problem for an equation of the Schrödinger type. Dopovidi Akad. Nauk Ukrain. RSR (1958) 119—122 (Ukrainian). — [7] Uniqueness of the solution of the Cauchy problem for a class of elliptic equations. Dopovidi Akad. Nauk Ukrain. RSR (1958) 689—693 (Ukrainian). — [8] The behavior at infinity of solutions of a system of differential equations of elliptic type. Dopovidi Akad. Nauk Ukrain. RSR (1959) 931—935 (Ukrainian).

MAGENES, E.: [1] Sui problemi al contorno misti per le equazioni lineari del secondo ordine di tipo ellittico. Ann. Sc. N. Sup. Pisa **8** (1954) 93—120. — [2] Osservazioni su alcuni teoremi di completezza connessi con i problemi misti per le equazioni lineari ellittiche. Boll. Un. Mat. Ital. **10** (1955) 452—459. — [3] Sui problemi di derivata obliqua regolare per le equazioni lineari del secondo ordine di tipo ellittico. Ann. Mat. pura appl. **40** (1955) 143—160. — [4] Sul teorema dell'alternativa nei problemi misti per le equazioni lineari ellittiche del secondo ordine. Ann. Sc. N. Sup. Pisa **9** (1955) 161—200. — [5] Su alcune recenti impostazioni dei problemi al contorno, in particolare misti, per le equazioni lineari ellittiche del secondo ordine. Ann. Sc. N. Sup. Pisa **10** (1956) 75—84. — [6] Recenti sviluppi nella teoria dei problemi misti per le equazioni lineari ellittiche. Rend. Sem. Mat. Fis. Milano **27** (1957) 75—95. — [7] Il problema della derivata obliqua regolare per le equazioni lineari ellittico-paraboliche del secondo ordine in m variabili. Rend. Mat. Appl. Roma **16** (1957) 363—414. — [8] Sul problema di Dirichlet per le equazioni lineari ellittiche in due variabili. Ann. Mat. pura appl. **48** (1959) 257—279. — [9] Spazi di interpolazione ed equazioni a derivate parziali. Atti VII Congr. Un. Mat. Ital. Genova (1964) 134—197.

MAGENES, E., e G. STAMPACCHIA: [1] I problemi al contorno per le equazioni differenziali di tipo ellittico. Ann. Sc. N. Sup. Pisa **12** (1958) 247—358.

MAGNUS, W.: [1] Über Eindeutigkeitsfragen bei einer Randwertaufgabe von $\Delta u + k^2 u = 0$. Jber. Deutsch. Math. Verein. **52** (1942) 177—188. — [2] Fragen der Eindeutigkeit und des Verhaltens im Unendlichen für Lösungen von $\Delta u + k^2 u = 0$. Abh. Mat. Sem. Hamburg **16** (1949) 77—94.

MALGRANGE, B.: [1] Existence et approximation des solutions des équations aux dérivées partielles et des équations de convolution. Ann. Inst. Fourier Grenoble **6** (1955/56) 271—355.

MARCINKOWSKA, H.: [1] Elliptic differential equation with the right member being a tempered distribution. Bull. Acad. Polon. Sci. **12** (1964) 299—304. — [2] Differentiability theorem for elliptic equations considered on compact Riemannian manifolds. Colloq. Math. **12** (1964) 91—105. — [3] An energetic inequality for strongly elliptic differential operators depending on a parameter. Bull. Acad. Polon. Sci. **13** (1965) 303—307. — [4] On the differentiability and growth properties of distributional solutions of elliptic equations having a tempered distribution as the right member. Bull. Acad. Polon. Sci. **13** (1965) 309—312. — [5] Some remarks concerning the fundamental solutions of elliptic differential equations with infinitely differentiable coefficients depending on a parameter. Bull. Acad. Polon. Sci. **13** (1965) 313—316.

MARCUS, M.: [1] Local behaviour of singular solutions of elliptic equations. Ann. Sc. N. Sup. Pisa **19** (1965) 519—561. — [2] The Dirichlet problem in a domain whose boundary is partly degenerated. Ann. Mat. pura appl. **73** (1965) 159—194.

MAZ'JA, V. G.: [1] Solution of Dirichlet's problem for an equation of elliptic type. Dokl. Akad. Nauk SSSR **129** (1959) 257—260 (Russian). — [2] Some estimates of solutions of second-order elliptic equations. Dokl. Akad. Nauk SSSR **137** (1961) 1057—1059 (Russian). — [3] The Dirichlet problem for elliptic equations of arbitrary order in unbounded domains. Dokl. Akad. Nauk SSSR **150** (1963) 1221—1224 (Russian). — [4] On the boundary regularity of solutions of elliptic equations and of conformal mappings. Dokl. Akad. Nauk SSSR **152** (1963) 1297—1300 (Russian). — [5] The solubility in $\mathring{W}_2^{(2)}$ of the Dirichlet problem for a region with a smooth irregular boundary. Vestnik Leningrad Univ. **19** (1964) n. 7, 163—165 (Russian).

McLEOD, R.M., and F.G. DRESSEL: [1] Uniqueness of mapping pairs for elliptic equations. Duke Math. J. **24** (1957) 173—181.

McNABB, A.: [1] Strong comparison theorems for elliptic equations of second order. J. Math. Mech. **10** (1961) 431—440.

MEL'NIK, D.F.: [1] Theorems of Liouville type for certain elliptic systems of differential equations. Izv. Vysš. Učebn. Zaved. Mat. (1958) n. 3, 163—171 (Russian).

MEYERS, N.: [1] An L^p-estimate for the gradient of solutions of second order elliptic divergence equations. Ann. Sc. N. Sup. Pisa **17** (1963) 189—206. — [2] An example of non uniqueness in the theory of quasi-linear elliptic equations of second order. Arch. Rat. Mech. Anal. **14** (1963) 177—179. — [3] An expansion about infinity for solutions of linear elliptic equations. J. Math. Mech. **12** (1963) 247—264. — [4] On a class of non uniformly elliptic quasi-linear equations in the plane. Arch. Rat. Mech. Anal. **12** (1963) 367—391.

MEYERS, N., and J. SERRIN: [1] The exterior Dirichlet problem for second order elliptic partial differential equations. J. Math. Mech. **9** (1960) 513—538.

MICKLE, E.J.: [1] A remark on a theorem by S. Bernstein. Proc. Amer. Math. Soc. **1** (1950) 86—89.

MIHAĬLOV, L.G.: [1] Boundary value problem of the Riemann type for elliptic system of differential equations of the first order and some integral equations. Učen. Zap Trudy Fiz. Mat. Fak. Tadžit. Univ. **10** (1957) 32—79 (Russian). — [2] A boundary problem of the type of Riemann for systems of first order differential equations of elliptic type. Dokl. Akad. Nauk SSSR **112** (1957) 13—15 (Russian). — [3] An investigation of the generalized Cauchy-Riemann system where the coefficients have first order singularities. Dokl. Akad. Nauk SSSR **129** (1959) 507—510 (Russian). — [4] Elliptic equations with singular coefficients. Dokl. Akad. Nauk SSSR **139** (1961) 552—555 (Russian). — [5] Elliptic equations with singular coefficients. Izv. Akad. Nauk SSSR **26** (1962) 293—312 (Russian). — [6] Elliptic equations with singular coefficients. Trudy Tbilissk Mat. Inst. **28** (1962) 123—142 (Russian). — [7] A new class of singular integral equations and its application to differential equations with singular coefficients. Akad. Nauk Tadžik SSR, Dushanbe (1963)183 pp. (Russian).

MIHLIN, S.G.: [1] Singular integral equations. Usp. Mat. Nauk **3**, n. 3 (1948) 29—112 (Russian); English transl. Amer. Math. Soc. **(1) 10** (1962) 84—198. — [2] Direct methods in mathematical physics. Gos. Izdat. Tehn.-Teor. Lit. Moscow-Leningrad (1950) 428 pp. (Russian); English revised transl. as Variational methods in mathematical physics. Per-

gamon Press (1964) 582 pp. — [3] On equations of elliptic type. Dokl.
Akad. Nauk S R 77 (1951) 377—380 (Russian). — [4] On some estimates
connected with Green's functions. Dokl. Akad. Nauk SSSR 78 (1951)
443—446 (Russian). — [5] The problem of the minimum of a quadratic
functional. Gos. Izdat. Tehn.-Teor. Lit. Moscow-Leningrad (1952)
216 pp. (Russian); English transl. Holden Day San Francisco (1965)
155 pp. — [6] On a variational method for the solution of extremal
problems. Leningrad Gos. Univ. Uč. Zap. 23 (1952) 151—164 (Russian). —
[7] On the applicability of a variational method to certain degenerate
elliptic equations. Dokl. Akad. Nauk SSSR 91 (1953) 723—726 (Russian).
— [8] On the theory of degenerate elliptic equations. Dokl. Akad. Nauk
SSSR 94 (1954) 183—185 (Russian). — [9] Degenerate elliptic equations.
Vestnik Leningrad Univ. 9, n. 8 (1954) 19—48 (Russian). — [10] Multi-
dimensional singular integrals and integral equations. Gos. Izdat. Fiz.
Mat. Lit. Moscow (1962) 254 pp. (Russian); English transl. Pergamon
Press (1965) 255 pp.

MILJUTIN, A. A.: [1] A priori estimates for solutions of second order linear
elliptic equations. Mat. Sbornik 51 (1960) 459—474 (Russian).

MILLER, K.: [1] Non existence of an a priori bound at the center in terms
of an L_1 bound on the boundary for solutions of uniformly elliptic
equations on a sphere. Ann. Mat. pura appl. 73 (1966) 11—16.

MINTY, G. J.: [1] On the solvability of non linear functional equations of
„monotonic" type. Pacific J. Math. 14 (1964) 249—255.

MIRANDA, C.: [1] Sull'esistenza e sull'unicità di una superficie di assegnato
bordo verificante un'equazione a derivate parziali in forma para-
metrica. Mem. Acc. d'Italia 6 (1935) 1023—1045. — [2] Su un problema
di Minkowski. Rend. Sem. Mat. Roma 3 (1939) 96—108. — [3] Sul principio di Dirichlet per le funzioni armoniche. Rend. Acc. Lincei 3 (1947)
55—59. — [4] Sull'approssimazione delle funzioni armoniche. Rend. Acc.
Lincei 5 (1948) 530—533. — [5] Formule di maggiorazione e teorema di
esistenza per le funzioni biarmoniche di due variabili. Giorn. Mat. Battaglini 78 (1948—49) 97—118. — [6] Sui sistemi di tipo ellittico lineari
alle derivate parziali del primo ordine, in n variabili indipendenti.
Mem. Acc. Lincei 3 (1952) 85—121. — [7] Sull'integrazione delle forme
differenziali esterne. Ricerche di Mat. Napoli 2 (1953) 151—182. —
[8] Equazioni alle derivate parziali di tipo ellittico. Ergeb. Math. Berlin-
Göttingen-Heidelberg: Springer Verlag 2 (1955) 222 pp.; Russian
transl. Izdat. Inostr. Lit. Moscow (1957) 256 pp. — [9] Sul problema
misto per le equazioni lineari ellittiche. Ann. Mat. pura appl. 39 (1955)
279—303. — [10] Le soluzioni fondamentali delle equazioni ellittiche.
Conf. Sem. Mat. Bari n. 30 (1957) 14 pp. — [11] Su alcuni aspetti della
teoria delle equazioni ellittiche. Bull. Soc. Math. France 96 (1958)
331—354. — [12] Teorema del massimo modulo e teorema di esistenza e
di unicità per il problema di Dirichlet relativo alle equazioni ellittiche
in due variabili. Ann. Mat. pura appl. 46 (1958) 265—312. — [13] Alcune
limitazioni integrali per le soluzioni delle equazioni lineari ellittiche del
secondo ordine. Ann. Mat. pura appl. 49 (1960) 375—384. — [14] Teoremi
di unicità in domini non limitati e teoremi di Liouville per le soluzioni

dei problemi al contorno relativi alle equazioni ellittiche. Ann. Mat. pura appl. **59** (1962) 189–212. — [15] Su un problema di Dirichlet per le equazioni fortemente ellittiche a coefficienti constanti in un dominio a frontiera degenere. Bull. Soc. Roy. Sci. Liège **31** (1962) 614–636. — [16] Alcune osservazioni sulla maggiorazione in L^p delle soluzioni deboli delle equazioni ellittiche del secondo ordine. Ann. Mat. pura appl. **61** (1963) 151–170. — [17] Sulle equazioni ellittiche del secondo ordine di tipo non variazionale a coefficienti discontinui. Ann. Mat. pura appl. **63** (1963) 353–386. — [18] Su di una particolare equazione ellittica del secondo ordine a coefficienti discontinui. Anal. Ştii. Univ. Iaşi XI$_B$ (1965) 209–215. — [19] Sulle proprietà di regolarità di certe trasformazioni integrali. Mem. Acc. Lincei **7** (1965) 303–336.

MIRANDA, M.: [1] Una maggiorazione integrale per le curvature delle ipersuperficie minimali. Rend. Sem. Mat. Padova **38** (1967) 94–107.

MIRANKER, W.L.: [1] The reduced wave equation in a medium with a variable index of refraction. Commun. pure appl. Math. **10** (1957) 491–502. — [2] Uniqueness and representation theorem for solutions of $\Delta u + k^2 u = 0$ in infinite domains. J. Math. Mech. **6** (1957) 847–858.

MIZEL, V.J.: [1] A boundary layer problem for an elliptic equation in the neighborhood of a singular point. Proc. Amer. Math. Soc. **8** (1957) 62–67. — [2] A boundary layer result for an n-dimensional linear elliptic equation. Proc. Amer. Math. Soc. **10** (1959) 775–783.

MIZOHATA, S.: [1] Unicité du prolongement des solutions des équations elliptiques du quatrième ordre. Proc. Japan. Acad. **34** (1958) 687–692. — [2] Unicité dans le problème de Cauchy pour quelques équations différentielles elliptiques. Mem. Coll. Sci. Univ. Kyoto **31** (1958) 121–128.

MOGILEVSKIĬ, Š.I.: [1] Stability of the Dirichlet problem for elliptic differential equations. Kalinin Gos. Ped. Ist. Uč. Zap. **26** (1958) 103–128 (Russian).

MORGENSTERN, D.: [1] Singuläre Störungstheorie partieller Differentialgleichungen. J. Rat. Mech. Anal. **5** (1956) 203–216.

MOROSOV, N.F.: [1] On non linear theory of thin plates. Dokl. Akad. Nauk SSSR **114** (1957) 968–971 (Russian).

MORREY, C.B. JR.: [1] On the solutions of quasi linear elliptic partial differential equations. Trans. Amer. Math. Soc. **43** (1938) 126–166. — [2] Multiple integral problems in the calculus of variations and related topics Univ. California Publ. **1** (1943) 1–130. — [3] Second order elliptic systems of differential equations. Proc. Nat. Acad. USA **39** (1953) 201–206. — [4] Second order elliptic systems of differential equations. Ann. Math. Studies **33** (1954) 101–159. — [5] On the analyticity of the solutions of analytic non-linear elliptic systems of partial differential equations, I, II. Amer. J. Math. **80** (1958) 198–218 and 219–237. — [6] Second order elliptic equations in several variables and Hölder continuity. Math. Z. **72** (1959) 146–164. — [7] Multiple integral problems

332   Bibliography

in the calculus of variations and related topics. Ann. Sc. N. Sup. Pisa **14**
(1960) 1—61. — [8] Existence and differentiability theorems for variational problems for multiple integrals. Part. Diff. Eq. and Cont. Mech.
Univ. Wisconsin Press (1960) 241—270. — [9] Multiple integrals in the
calculus of variations. Berlin-Heidelberg-New York: Springer Verlag
(1966) 506 pp.

MORREY, C. B. JR., and L. NIRENBERG: [1] On the analyticity of the solutions of linear elliptic systems of partial differential equations. Commun.
pure appl. Math. **10** (1957) 271—290.

MOSER, J.: [1] A new proof of De Giorgi's theorem concerning the regularity problem for elliptic differential equations. Commun. pure appl.
Math. **13** (1960) 457—468. — [2] On Harnack's theorem for elliptic
differential equations. Commun. pure appl. Math. **14** (1961) 577—591. —
[3] On the regularity problem for elliptic and parabolic differential
equations. Part. Diff. Eq. and Cont. Mech. Univ. Wisconsin Press (1961)
159—169.

MÜLLER, C.: [1] Über die Beugung elektromagnetischer Schwingungen an
endlichen homogenen Körpern. Math. Ann. **123** (1951) 345—378. —
[2] Zur Methode der Strahlungskapazität von H. Weyl. Math. Z. **56**
(1952) 80—83. — [3] Randwertprobleme der Theorie elektromagnetischer
Schwingungen. Math. Z. **56** (1952) 261—270. — [4] On the behaviour
of the solutions of the differential equations $\Delta U = F(x, U)$ in the
neighborhood of a point. Commun. pure appl. Math. **7** (1954) 505—515. —
[5] Grundprobleme der mathematischen Theorie elektromagnetischer
Schwingungen. Berlin-Göttingen-Heidelberg: Springer Verlag (1957)
344 pp.

MÜNTZ, CH. H.: [1] Die Lösung des Plateauschen Problems über konvexen
Bereichen. Math. Ann. **94** (1925) 53—96.

MURAMUTU, T.: [1] On the uniqueness of the Cauchy problem for elliptic
systems. Sci. Papers College Gen. Ed. Univ. Tokyo **11** (1961) 13—23.

MURTHY, M. K. V.: [1] A remark on the regularity at the boundary for solutions of elliptic equations. Ann. Sc. N. Sup. Pisa **15** (1961) 355—370.

MURTHY, M. K. V., e G. STAMPACCHIA: [1] Equazioni ellittiche che dege-
nerano. Convegno Eq. Der. Parz. Nervi, Ed. Cremonese (1965) 90—96.

MUSHELIŠVILI, N. I.: [1] Recherches sur les problèmes aux limites relatifs
à l'équation biharmonique et aux équations de l'élasticité à deux dimensions. Math. Ann. **107** (1933) 282—312. — [2] Singular integral
equations. OGIZ Moscow-Leningrad (1946) 448 pp. (Russian); English
transl. Groningen: P. Noordhoff Ltd. (1953) 447 pp.; German transl.
Berlin: Akademie Verlag (1965) 564 pp. — [3] Some basic problems of
mathematical theory of elasticity. 3rd ed. Izdat. Akad. Nauk SSSR
(1949) 636 pp. (Russian); English transl. Groningen: P. Noordhoff
Ltd. (1953) 704 pp.

MYRBERG, L.: [1] Über subelliptische Funktionen. Ann. Acad. Sci. Fennicae n. **290** (1960) 9 pp.

MYŠKIS, A. D.: [1] On the transition from the usual first boundary problem to the modified one. Mat. Sbornik **31** (1952) 128—135 (Russian).

NAGUMO, M.: [1] On principally linear elliptic differential equations of the second order. Osaka Math. J. **6** (1954) 207—229.

NAKAMORI, K.: [1] On a non linear boundary value problem for the equation $\Delta u + cu = f(x, y)$. Mem. Fac. Sci. Kyusyu Univ. **5** (1951) 1—7. — [2] Fundamental theorems for an elliptic system of partial differential equations of first order in two independent variables. Yokohama Math. J. **9** (1961) 1—27.

NARASIMHAN, M. S.: [1] The identity of the weak and strong extensions of a linear elliptic differential operator. Proc. Nat. Acad. Sci. USA **43** (1957) 513—514 and 620.

NARČAEV, A.: [1] The first boundary-value problem for elliptic equations which degenerate at the boundary. Dokl. Akad. Nauk SSSR **156** (1964) 28—31 (Russian).

NASH, J.: [1] Continuity of solutions of parabolic and elliptic equations. Amer. J. Math. **80** (1958) 931—954.

NAZIROV, G.: [1] The first and second boundary problems for elliptic equations with singular coefficients. Dokl. Akad. Nauk Tadzik SSR **8**, n. 2 (1965) 3—6 (Russian).

NEČAS, J.: [1] On the solutions of elliptic partial differential equations of the second order with unbounded Dirichlet's integral. Czech. Math. J. **10** (1960) 283—298 (Russian). — [2] Sur une méthode pour resoudre les équations aux dérivées partielles du type elliptique, voisine de la variationelle. Czech. Math. J. **11** (1961) 362—363. — [3] Sur le problème de Dirichlet pour l'équation aux dérivées partielles du quatrième ordre du type elliptique. Rend. Sem. Mat. Padova **31** (1961) 198—231 — [4] On the regularity of solutions of second order elliptic partial differential equations with an unbounded Dirichlet integral. Arch. Rat. Mech. Anal. **9** (1962) 134—144. — [5] On the solution of partial differential equations with an unbounded Dirichlet integral. Differential equations and their applications (Proc. Conf. Prague 1962) New York: Academic Press (1963) 93—104. — [6] Sur l'existence de la solution classique du problème de Poisson pour les domaines planes. Ann. Sc. N. Sup. Pisa **16** (1962) 285—296. — [7] Sur une méthode pour resoudre les équations aux dérivées partielles du type elliptique voisine de la variationelle. Ann. Sc. N. Sup. Pisa **16** (1962) 305—326. — [8] Sur la coercivité des formes sesquilinéaires elliptiques. Rev. Roumaine Math. pures appl. **8** (1964) 47—69. — [9] L'application de l'égalité de Rellich sur les systèmes elliptiques du deuxième ordre. J. Math. pures appl. **44** (1965) 133—147. — [10] Sur une méthode générale pour la solution des problèmes aux limites non linéaires. Ann. Sc. N. Sup. Pisa **20** (1967)

655—674. — [11] Sur l'existence de la solution régulière pour le problème de Dirichlet de l'équation elliptique non linéaire d'ordre $2k$. Rend. Acc. Lincei **42** (1967) 347—354. — [12] Sur la régularité des solutions variationnelles des équations elliptiques non-linéaires d'ordre $2k$ en deux variables. Ann. Sc. N. Sup. Pisa **21** (1967) 427—457.

NEHARI, Z.: [1] On the biharmonic Green's function. Studies in Math. and Mech. presented to R. von Mises (1954) 111—117. — [2] A differential inequality. J. Anal. Math. **14** (1965) 297—302. — [3] Bounds for the solutions of a class of non linear partial differential equations of the elliptic type. Proc. Amer. Math. Soc. **14** (1963) 829—836.

NELSON, E.: [1] Analytic vectors. Ann. of Math. **70** (1959) 572—615.

NICOLESCU, M.: [1] Les fonctions polyharmoniques. Paris: Ed. Hermann (1936) 54 pp.

NIKITIN, A. K.: [1] On the problem of steady motion of viscous incompressible fluid between pin and bearing. Dokl. Akad. Nauk SSSR **108** (1956) 405—408 (Russian).

NIKODYM, O.: [1] Sur l'existence du potentiel uniforme sur une surface de Riemann. Bull. Soc. Math. France **61** (1933) 220—245. — [2] Sur un théorème de M. S. Zaremba concernant les fonctions harmoniques. J. Math. pures appl. **12** (1933) 95—109.

NIKOL'SKIĬ, JU.S.: [1] On the theory of weight classes of differentiable functions of several variables and its applications to boundary problems for elliptic equations. Dokl. Akad. Nauk SSSR **162** (1965) 510—512 (Russian).

NIKOL'SKIĬ, S.M., and P.I. LIZORKIN: [1] Some inequalities for functions of weight classes and boundary-value problems with strong degeneracy on the boundary. Dokl. Akad. Nauk SSSR **159** (1964) 512—515 (Russian).

NIRENBERG, L.: [1] On non linear partial differential equations and Hölder continuity. Commun. pure appl. Math. **6** (1953) 103—156. — [2] The Weyl and Minkowski problems in differential geometry in the large. Commun. pure appl. Math. **6** (1953) 337—394. — [3] On a generalization of quasiconformal mapping and its application to elliptic partial differential equations. Ann. of Math. Studies **33** (1954) 95—100. — [4] Remarks on strongly elliptic differential equations. Commun. pure appl. Math. **8** (1955) 649—675. — [5] Estimates and existence of solutions of elliptic equations. Commun. pure appl. Math. **9** (1956) 509—529. — [6] Uniqueness in Cauchy problem for differential equations with constant leading coefficients. Commun. pure appl. Math. **10** (1957) 89—105. — [7] On elliptic partial differential equations. Ann. Sc. N. Sup. Pisa **13** (1959) 115—162. — [8] Partial differential equations with applications in geometry. Lectures in Modern Math. New York: Wiley II (1962) 1—41.

NITSCHE, JOACHIM: [1] Das erste Randwertproblem eines linearen elliptischen Differentialgleichungssystems. Math. Nach. **7** (1952) 31—33. —

[2] Beiträge zum Randwertproblem quasi linearer Differentialgleichungssysteme. Math. Nach. **7** (1952) 35—54.

NITSCHE, JOHANNES: [1] Beitrag zum Randwertproblem einer linearen elliptischen Differentialgleichungssystems im grossen. Rend. Circ. Mat. Palermo **3** (1954) 109—114. — [2] Eine charakteristische Eigenschaft der Lösungen von Randwertproblemen elliptischer Differentialgleichungssysteme. Arch. Math. **6** (1954) 18—24. — [3] Über die linearen Randwertprobleme einer quasilinearer elliptischen Differentialgleichungssysteme. Math. Z. **61** (1954) 336—347. — [4] Untersuchungen über die linearen Randwertprobleme linearer und quasilinearer Differentialgleichungssysteme, I, II. Math. Nach. **14** (1955) 75—127 and 157—182. — [5] Über die isolierten Singularitäten der Lösungen von $\Delta u = e^u$. Math. Z. **68** (1957) 316—324. — [6] Zu einem Satze von L. Bers über die Lösungen der Minimalflächengleichung. Arch. Math. **9** (1958) 427—429. — [7] On the non solvability of Dirichlet's problem for the minimal surface equation. J. Math. Mech. **14** (1965) 779—788. — [8] On differential equations of mode 2. Proc. Amer. Math. Soc. **16** (1965) 902—908. — [9] Über ein verallgemeinertes Dirichletsches Problem für die Minimalflächengleichungen und hebbare Unstetigkeiten ihrer Lösungen. Math. Ann. **158** (1965) 203—214. — [10] On new results in the theory of minimal surfaces. Bull. Amer. Math. Soc. **71** (1965) 195—270.

NITSCHE, JOACHIM, u. JOHANNES NITSCHE: [1] Allgemeine Randwertprobleme für Systeme elliptischer Differentialgleichungen; die Zurückführung auf eine von F. Noether untersuchte Klasse singulärer Integralgleichungen. Rend. Circ. Mat. Palermo **2** (1953) 40—45. — [2] Das zweite Randwertproblem der Differentialgleichung $\Delta u = e^u$. Arch. Math. **3** (1952) 460—464. — [3] Bemerkungen zum zweiten Randwertproblem der Differentialgleichung $\Delta \varphi = \varphi_x^2 + \varphi_y^2$. Math. Ann. **126** (1953) 69—74. — [4] Über reguläre Variationsprobleme. Rend. Circ. Mat. Palermo **8** (1959) 346—353. — [5] Ein Kriterium für die Existenz nichtlinearer ganzer Lösungen elliptischer Differentialgleichungen. Arch. Math. **10** (1959) 294—297.

NÖTHER, F.: [1] Über eine Klasse singulärer Integralgleichungen. Math. Ann. **82** (1920) 42—63.

NOVRUZOV, A. A.: [1] Properties of solutions of elliptic equations. Dokl. Akad. Nauk SSSR **139** (1961) 1304—1307 (Russian). — [2] Theorems of Hadamard and Phragmèn-Lindelöf type. Izv. Akad. Nauk Azerb. SSR (1963) n. 2, 37—48 (Russian).

ODQVIST, F. K.: [1] Über die Randwertaufgaben der Hydrodynamik zäher Flüssigkeiten. Math. Z. **32** (1930) 329—375.

OLEJNIK, O. A.: [1] On the Dirichlet problem for equations of elliptic type. Mat. Sbornik **24** (1949) 3—14 (Russian). — [2] On properties of solutions of certain boundary problems for equations of elliptic type. Mat. Sbornik **30** (1952) 695—702 (Russian). — [3] On equations of elliptic type with a small parameter in the highest derivatives. Mat. Sbornik **31**

(1952) 104—117 (Russian). — [4] On equations of elliptic type which
degenerate on the boundary of a region. Dokl. Akad. Nauk SSSR **87**
(1952) 885—888 (Russian). — [5] Solution of fundamental boundary
value problems for second order equations with discontinuous coefficients. Dokl. Akad. Nauk SSSR **124** (1959) 1219—1222 (Russian). —
[6] Boundary value problems for elliptic and parabolic equations
with discontinuous coefficients. Izv. Akad. Nauk SSSR **25** (1961)
3—20 (Russian). — [7] On a problem of Fichera. Dokl. Akad. Nauk SSSR
157 (1964) 1297—1300 (Russian). — [8] On linear second order equations
with non negative characteristic form. Mat. Sbornik **69** (1966) 111—140
(Russian). — [9] Alcuni risultati sulle equazioni lineari e quasi lineari
ellittico-paraboliche a derivate parziali del secondo ordine. Rend. Acc.
Lincei **40** (1966) 775—784.

OOLEVSKIĬ, M. N.: [1] Solution of the Dirichlet problem for the equation
$\Delta v + \dfrac{p}{x_n}\dfrac{\partial v}{\partial x_n} = \varrho$ in a half sphere. Dokl. Akad. Nauk SSSR **64** (1949)
767—770 (Russian).

OSEEN, C. W.: [1] Contributions à la théorie analytique des marées. Ark.
Mat. Astr. Fys. **25** A n. 24 (1937) 1—39.

OSKOLKOV, A. P.: [1] On the solutions of a boundary value problem for
linear elliptic equations in an unbounded domain. Vestnik Leningrad
Univ. **16,** n. 7 (1961) 38—50 (Russian). — [2] Differential properties
of bounded generalized solutions of quasi-linear systems of variational
type. Vestnik Leningrad Univ. **18,** n. 2 (1963) 9 — 29 (Russian). —
[3] A class of solutions of boundary-value problems for linear elliptic
equations with three independent variables in an unbounded region.
Vestnik Leningrad Univ. **19,** n. 1 (1964) 150—156 (Russian). — [4] On
the solution of boundary-value problems for linear elliptic equations
in an infinite region. Dokl. Akad. Nauk SSSR **153** (1963) 34—37 (Russian).

OSSERMAN, R.: [1] On the inequality $\Delta u \geqq f(u)$. Pacific J. Math. **7** (1957)
1641—1647. — [2] Proof of a conjecture of Nirenberg. Commun. pure
appl. Math. **12** (1959) 229—232. — [3] On the Gauss curvature of nimi-
mal surfaces. Trans. Amer. Math. Soc. **96** (1960) 115—128.

OSSICINI, A., e F. ROSATI: [1] Sulla regolarità alla frontiera di soluzioni
di equazioni ellittiche. Atti Acc. Sci. Torino **98** (1963/64) 893—907.

OWENS, O. G.: [1] A uniqueness theorem for the Helmholtz equation.
Duke Math. J. **31** (1964) 91—98.

PACHALE, H.: [1] Über ein ebenes nichtlineares biharmonisches Randwertproblem. Math. Nach. **7** (1952) 187—212. — [2] Über ein räumliches
nichtlineares Randwertproblem. Math. Nach. **8** (1952) 79—91.

PANEYAL, B. P.: [1] Existence and uniqueness of the solution of the n-
metaharmonic equation on an unbounded space. Vestnik Moscow
Univ. (1959) 123—135 (Russian).

PANIČ, O.I.: [1] Solution of the fundamental boundary-value problem for the fourth order polyharmonic equation on the plane by the method of potentials, I, II, III, IV. Izv. Vyss̆. Učebn. Zaved. Mat. (1961) n. 3, 80—90; n. 4, 66—77; n. 6, 89—96; (1962) n. 1, 118—129 (Russian). — [2] On the solubility of exterior boundary-value problems for the wave equation and for a system of Maxwell's equations. Usp. Mat. Nauk **20** n. 1 (1965) 221—226 (Russian).

PARASJUK, L.S.: [1] Boundary problems for elliptical differential equations degenerating along the boundary of the region. Dopovidi Akad. Nauk Ukrain. RSR (1960) 144—147 (Ukrainian). — [2] Boundary-value problems for some self-adjoint second order differential equations degenerate on the boundary of the region. Ukrain. Mat. Ž. (1961) 75—85 (Russian). — [3] Boundary value problems for two elliptic second-order differential equations which degenerate on the boundary. Ukrain. Mat. Ž. **14** (1962) 215—217 (Russian). — [4] Radiation conditions for certain elliptic differential equations which are degenerate on the boundary of the region. Dopovidi Akad. Nauk Ukrain. RSR (1963) 441—443 (Ukrainian). — [5] Fundamental solutions of elliptic systems of differential equations with discontinuous coefficients. Dopovidi Akad. Nauk Ukrain. RSR (1963) 986—989 (Ukrainian). — [6] Behaviour of solution of elliptic system of differential equations with discontinuous coefficients. Dopovidi Akad. Nauk Ukrain. RSR (1963) 1127—1129 (Ukrainian).

PAWLOWSKA, J.: [1] On the nodal lines of solutions of certain elliptic equations of order $2\,p$. Prace Mat. **9** (1965) 9—18 (Polish).

PAYNE, L.E.: [1] Some explicit inequalities for uniformely elliptic operators. Duke Math. J. **31** (1964) 485—489.

PAYNE, L.E., and H.F.WEINBERGER: [1] Upper and lower bounds for harmonic functions, Dirichlet integrals and biharmonic functions. Univ. Maryland Inst. Fluid Dyn. Appl. Math. Tech. Note BN **21** (1954) 29 pp. — [2] Note on a lemma of Finn and and Gilbarg. Acta Math. **98** (1957) 297—299. — [3] New bounds for solutions of second order elliptic partial differential equations. Pacific J. Math. **8** (1958) 551—573. — [4] Bounds for solutions of second order elliptic equations in terms of arbitrary vector fields. Arch. Rat. Mech. Anal. **20** (1965) 95—106.

PEDERSON, R.N.: [1] On the unique continuation theorem for certain second order and fourth order elliptic equations. Commun. pure appl. Math. **11** (1958) 67—80. — [2] On the order of zeros of one signed solutions of elliptic equations. J. Math. Mech. **8** (1959) 193—196.

PEETRE, J.: [1] Théorèmes de régularité pour quelques classes d'opérateurs différentielles. Medd. Lunds Univ. Mat. Sem. **16** (1959) 122 pp. — [2] Another approach to elliptic boundary problems. Commun. pure appl. Math. **14** (1961) 711—731; Russian transl. as Mathematika **7**, n. 1 (1963) 43—65. — [3] Mixed problems for higher order elliptic equations in two variables, I, II. Ann. Sc. N. Sup. Pisa **15** (1961) 337—353 and **17** (1963) 1—12. — [4] On the differentiability of the solutions of quasi-linear partial differential equations Trans. Amer. Math. Soc. **104** (1962) 476—482. — [5] Elliptic partial differential equations of higher order. Univ. Maryland. Inst. Fluid Dyn. Appl. Math. Lecture Series n. 40 (1962).

PETERS, A. S., and J. J. STOKER: [1] A uniqueness theorem and a new solution for Sommerfeld's and other diffraction problems. Commun. pure appl. Math. **7** (1954) 565—585.

PETROWSKIĬ, I. G.: [1] Sur les systèmes d'équations différentielles dont toutes les solutions sont analytiques. Dokl. Akad. Nauk SSSR **17** (1937) 343—346. — [2] On the analyticity of solutions of systems of differential equations. Mat. Sbornik **5** (1939) 3—70 (Russian). — [3] On some problems of the theory of partial differential equations. Usp. Mat. Nauk **1**, n. 3/4 (1946) 44—70 (Russian); English transl. Amer. Math. Soc. (1) 4 (1962) 373—414.

PETTINEO, B.: [1] Sul prolungamento analitico delle soluzioni di talune equazioni a derivate parziali della Fisica Matematica. Atti Acc. Sci. Palermo **16** (1955/56) 27—33. — [2] Nuova dimostrazione dei teoremi di esistenza per i problemi al contorno regolari relativi alle equazioni lineari a derivate parziali di tipo ellittico. Rend. Acc. Lincei **23** (1957) 32—38. — [3] Sul problema della derivata obliqua non regolare relativo alle equazioni differenziali lineari di tipo ellittico. Atti Acc. Sci. Palermo **21** (1960/61) 155—174. — [4] Sul problema di derivata obliqua non regolare relativo alle equazioni differenziali lineari di tipo ellittico. Le Matematiche Catania **16** (1961) 75—79. — [5] Sul teorema dell'alternativa per talune equazioni integrali singolari. Rend. Acc. Lincei **40** (1966) 366—372.

PICONE, M.: [1] Maggiorazione degli integrali delle equazioni alle derivate parziali del secondo ordine ellittico-paraboliche. Rend. Acc. Lincei **5** (1927) 138—143. — [2] Sul metodo delle minime potenze ponderate e sul metodo di Ritz per il calcolo approssimato nei problemi della fisica matematica. Rend. Circ. Mat. Palermo **52** (1928) 225—253. — [3] Particolare formula di maggiorazione per le soluzioni di una classica equazione alle derivate parziali del $4°$ ordine della Fisica Matematica. Rend. Acc. Lincei **10** (1929) 16—20. — [4] Nuovi indirizzi di ricerca nella teoria e nel calcolo delle soluzioni di talune equazioni lineari alle derivate parziali della Fisica Matematica. Ann. Sc. N. Sup. Pisa **5** (1936) 213—288. — [5] Sulla convergenza delle successioni di funzioni iperarmoniche. Bull. Math. Soc. Roumaine Sci. **38** (1936) 105—112. — [6] Nuove formule di maggiorazione per gli integrali delle equazioni lineari a derivate parziali del secondo ordine ellittico-paraboliche. Rend. Acc. Lincei **28** (1938) 331—338. — [7] Appunti di analisi superiore. Ed. Rondinella Napoli (1940) 847 pp. I Ed. — [8] Nuovi metodi risolutivi per i problemi d'integrazione delle equazioni lineari a derivate parziali e nuova applicazione della trasformata multipla di Laplace nel caso delle equazioni a coefficienti costanti. Atti Acc. Sci. Torino **75** (1939/40) 413—426. — [9] Sulla traduzione in equazione integrale lineare di prima specie dei problemi al contorno concernenti i sistemi di equazioni lineari a derivate parziali. Rend. Acc. Lincei **2** (1947) 365—371, 485—492,

Bibliography 339

717—725. — [10] Esistenza e calcolo della soluzione di un certo problema al contorno per il sistema di equazioni della elasticità. Rend.
Acc. Lincei 3 (1947) 427—435. — [11] Intorno alla teoria di una classica
equazione a derivate parziali della Fisica Matematica. Ann. Soc.
Polon. Math. 21 (1948) 161—169. — [12] Sur la théorie d'une équation
aux dérivcés partielles classique de la physique mathématique. C. R.
Acad. Sci. Paris 226 (1948) 1945—1947. — [13] Sur un problème nouveau pour l'équation linéaire aux dérivées partielles de la théorie
mathématique classique de l'élasticité. Second Coll. Éq. Dér. Part.
Centre Belge Math. Bruxelles (1954) 9—11.

PICONE, M., u. G. FICHERA: [1] Neue funktionalanalytische Grundlagen
für die Existenzprobleme und Lösungsmethoden von Systemen linearer
partieller Differentialgleichungen. Monatsh. für Math. 54 (1950) 188—209.

PICONE, M., e C. MIRANDA: [1] La formula di Green per i problemi con
arbitraria derivata obliqua. Rend. Acc. Lincei 29 (1939) 160—165.

PINI, B.: [1] Sul problema di Dirichlet per le equazioni a derivate parziali
del secondo ordine di tipo ellittico. Rend. Acc. Lincei 11 (1951) 325 to
333. — [2] Sul primo problema di valori al contorno della teoria dell'-
elasticità. Rend. Sem. Mat. Padova 21 (1952) 345—369. — [3] Sul problema di Dirichlet per le equazioni lineari del secondo ordine di tipo
ellittico nei domini non limitati. Rend. Acc. Sci. Fis. Mat. Napoli 19
(1952) 157—170. — [4] Sulle equazioni lineari a derivate parziali di
ordine 2 n di tipo ellittico e sui sistemi ellittici di equazioni lineari
del secondo ordine sopra una superficie chiusa. Rend. Mat. Appl.
Roma 11 (1952) 1—20. — [5] Osservazioni su un teorema di Picone
relativo all'equazione $\Delta u + cu = 0$. Boll. Un. Mat. Ital. 8 (1953) 19 to
25. — [6] Sulle singolarità delle soluzioni dell'equazione $\Delta u + cu = 0$.
Rend. Acc. Lincei 14 (1953) 21—26. — [7] Sui sistemi di equazioni
lineari a derivate parziali del secondo ordine dei tipi ellittico e parabolico. Rend. Sem. Mat. Padova 22 (1953) 265—280. — [8] Osservazioni
sulle soluzioni dei sistemi di equazioni a derivate parziali lineari di
tipo ellittico. Rend. Sem. Mat. Padova 22 (1953) 366—379. — [9] Preci-
sazioni a un ragionamento contenuto in una mia nota sulle equazioni
a derivate parziali di tipo ellittico. Ricerche di Mat. Napoli 3 (1954)
3—12. — [10] Sulle funzioni sub e super-biarmoniche. Rend. Acc.
Lincei 16 (1954) 702—707. — [11] Teoremi di unicità per i problemi
generalizzanti i problemi biarmonici fondamentali esterno e interno.
Boll. Un. Mat. Ital. 10 (1955) 465—473. — [12] Osservazioni sopra un
problema generalizzato di Dirichlet per le equazioni lineari del secondo
ordine ellittiche e paraboliche. Rend. Acc. Lincei 19 (1955) 237—246. —
[13] Una generalizzazione del problema biarmonico fondamentale.
Rend. Sem. Mat. Padova 25 (1956) 196—213. — [14] Osservazioni sulla
soluzione di un problema biarmonico generalizzato. Scritti Mat. in
onore di F. Sibirani Bologna (1957) 219—224. — [15] Sul comporta-
mento alla frontiera delle derivate delle soluzioni dei problemi armonico
e biarmonico in due variabili. Rend. Sem. Fac. Sci. Univ. Cagliari 26
(1956) 17—29. — [16] Sul problema di Dirichlet per le equazioni a
derivate parziali lineari ellittiche in due variabili. Rend. Sem. Mat.
Padova 26 (1956) 177—200. — [17] Sull'unicità della soluzione del

problema di Dirichlet per le equazioni lineari ellittiche in due variabili. Rend. Sem. Mat. Padova **26** (1956) 223—231.

PLIŚ, A.: [1] Non uniqueness in Cauchy problem for differential equations of elliptic type. J. Math. Mech. **9** (1960) 557—562. — [2] A smooth differential equation without any solution in a sphere. Commun. pure appl. Math. **14** (1961) 599—617. — [3] Unique continuation theorem for solutions of partial differential equations. Proc. Inter. Congr. Math. (Stockholm 1962) Inst. Mittag Leffler Djursholm (1962) 397—402. — [4] On non uniqueness in Cauchy problem for an elliptic second order differential equation. Bull. Acad. Polon. Sci. **11** (1963) 95—100.

POGORELOV, A. V.: [1] Intrinsic estimates for the derivatives of the radius vector of a point on a closed regular convex surface. Dokl. Akad. Nauk SSSR **66** (1949) 805—808 (Russian). — [2] A priori estimates for the derivatives of a regular solution of a partial differential equation of elliptic type. Usp. Mat. Nauk **4**, n. 4 (1949) 179—182 (Russian). — [3] On the proof of Weyl's theorem on existence of a closed analytic surface realizing an analytic metric with positive curvature given on a sphere. Usp. Mat. Nauk **4**, n. 4 (1949) 183—186 (Russian). — [4] On the regularity of convex surfaces with regular metric. Dokl. Akad. Nauk SSSR **66** (1949) 1051—1053 (Russian). — [5] On convex surfaces with regular metric. Dokl. Akad. Nauk SSSR **67** (1949) 791—794 (Russian). — [6] On regularity of convex surfaces. Usp. Mat. Nauk **5**, n. 3 (1950) 188—189 (Russian). — [7] The rigidity of general convex surfaces. Dokl. Akad. Nauk SSSR **79** (1951) 739—742. (Russian) — [8] The bending of convex surfaces. Gos. Izdat. Moscow-Leningrad (1951) 184 pp. (Russian); German transl. Berlin: Akad. Verlag (1957) 134 pp. — [9] Regularity of convex surfaces with given Gaussian curvature. Mat. Sbornik **31** (1952) 88—103 (Russian). — [10] On a boundary problem for the equation $rt - s^2 = \varphi (x, y)$ and its geometric applications. Dokl. Akad. Nauk SSSR **83** (1952) 361—363 (Russian). — [11] On Monge-Ampère equations. Izdat. Har'kov. Ord. Trud. Krasn. Znam. A. M. Gorkogo Har'kov (1960) 111 pp. (Russian); English transl. Groningen: Noordhoff Ltd. (1964) 116 pp. — [12] On strongly elliptic Monge-Ampère equations. Dokl. Akad. Nauk SSSR **132** (1960) 535—536 (Russian).

POGORZELSKI, W.: [1] Les propriétés d'une fonction de Green et ses applications aux équations elliptiques. Ann. Polon. Math. **3** (1956) 46—75. — [2] Étude de la solution fondamentale de l'équation elliptique et des problèmes aux limites. Ann. Polon. Math. **3** (1957) 247—284. — [3] Problème aux limites aux dérivées tangentielles pour l'équation elliptique. Bull. Acad. Polon. Sci. **7** (1959) 205—212. — [4] Propriétés des dérivées tangentielles d'une integrale de l'équation elliptique. Ann. Polon. Math. **7** (1960) 321—339. — [5] Sur quelques propriétés des potentiels généralisées et un problème aux limites pour l'équation elliptique. Ann. Polon. Math. **11** (1961) 177—197. — [6] Problème aux dérivées tangentielles discontinues pour une équation elliptique. Ann. Polon. Math. **13** (1963) 33—56.

POHOŽAEV, S. I.: [1] On the Dirichlet problem for the equation $\Delta u = u^2$. Dokl. Akad. Nauk SSSR **134** (1960) 769—772 (Russian). — [2] The boundary value problem for the equation $\Delta u = u^2$. Dokl. Akad. Nauk SSSR **138** (1961) 305—308 (Russian).

Položiǐ, G. N.: [1] On p-analytic functions of a complex variable. Dokl. Akad. Nauk SSSR **58** (1947) 1275—1278 (Russian). — [2] A generalization of Cauchy's integral formula. Mat. Sbornik **24** (1949) 375—384 (Russian). — [3] Singular points and residues of p-analytic functions of a complex variable. Dokl. Akad. Nauk SSSR **60** (1949) 769—772 (Russian). — [4] The theorem on preservation of domain for a certain elliptic system of differential equations and its applications. Mat. Sbornik **32** (1953) 485—492 (Russian). — [5] A theorem on the correspondence of boundaries and variational theorems for certain elliptic systems of differential equations. Dokl. Akad. Nauk SSSR **95** (1954) 927—930 (Russian).

Poritsky, H.: [1] Generalizations of the Gauss law of the spherical mean. Trans. Amer. Math. Soc. **43** (1938) 199—225.

Poulsen, E. T.: [1] Boundary value properties connected with some improper Dirichlet integral. Math. Scand. **8** (1960) 5—14.

Princivalli, M. L.: [1] Sul sistema di equazioni lineari alle derivate parziali, relativo all'equilibrio delle volte cilindriche. Ann. Sc. N. Sup. Pisa **8** (1954) 157—291. — [2] Su un teorema di unicità per un problema al contorno relativo all'equilibrio delle volte cilindriche. Ann. Sc. N. Sup. Pisa **9** (1955) 233—245.

Privaloff, J.: [1] Sur la théorie générale des fonctions polyharmoniques. C. R. Acad. Sci. Paris **204** (1937) 328—330.

Prodi, G.: [1] Sull'equivalenza fra la seconda formula di Green e la corrispondente equazione di Fredholm per l'equazione $\Delta_2 u + \lambda u = 0$. Rend. Sem. Mat. Padova **24** (1955) 103—122. — [2] Sul primo problema al contorno per equazioni ellittiche e paraboliche con un secondo membro illimitato sulla frontiera. Rend. Ist. Lombardo **90** (1956) 189—208.

Protter, M. H.: [1] A comparison theorem for elliptic equations. Proc. Amer. Math. Soc. **10** (1959) 296—299. — [2] Unique continuation for elliptic equations. Trans. Amer. Math. Soc. **95** (1960) 81—91.

Pucci, C.: [1] Alcune limitazioni per gli integrali delle equazioni differenziali a derivate parziali, lineari, del secondo ordine di tipo ellittico-parabolico. Rend. Acc. Lincei **11** (1951) 334—339. — [2] Maggiorazione della soluzione di un problema al contorno di tipo misto, relativo a una equazione a derivate parziali, lineare, del secondo ordine. Rend. Acc. Lincei **13** (1952) 360—366. — [3] Discussione sul problema di Cauchy per le equazioni di tipo ellittico. Ann. Mat. pura appl. **46** (1958) 131—153. — [4] Studio di un sistema di equazioni differenziali della dinamica dei gas. Rend. Acc. Lincei **24** (1958) 653—657. — [5] Proprietà di massimo e di minimo delle soluzioni di equazioni a derivate parziali del secondo ordine di tipo ellittico e parabolico, I, II. Rend. Acc.

Lincei **23** (1957) 370—375 and **24** (1958) 3—6. — [6] Limitazioni per il gradiente di una soluzione di un'equazione di tipo ellittico. Le Matematiche Catania **16** (1962) 51—54. — [7] Un problema variazionale per i coefficienti di equazioni differenziali di tipo ellittico. Ann. Sc. N. Sup. Pisa **16** (1962) 159—172. — [8] Sulle funzioni barriera. Le Matematiche Catania **18** (1963) 102—107. — [9] Sulla regolarità interna delle soluzioni di alcune equazioni ellittiche. Boll. Un. Mat. Ital. **19** (1964) 334—342. — [10] Regolarità alla frontiera di soluzioni di equazioni ellittiche. Ann. Mat. pura appl. **65** (1964) 311—328. — [11] Operatori ellittici estremanti. Ann. Mat. pura appl. **72** (1966) 141—170. — [12] Limitazioni per soluzioni di equazioni ellittiche. Ann. Mat. pura appl. **74** (1966) 15—30. — [13] Su una limitazione per soluzioni di equazioni ellittiche. Boll. Un. Mat. Ital. **21** (1966) 228—233. — [14] Equazioni ellittiche con soluzioni in $W^{2,p}$, $p < 2$. Convegno Eq. Der. Parz. Bologna (1967) 145—148.

PULVIRENTI, G.: [1] Ancora sui problemi ai limiti per l'operatore di Laplace iterato. Le Matematiche Catania **18** (1963) 108—115. — [2] Su una classe di problemi ai limiti. Le Matematiche Catania **20** (1965) 87—99.

PÜSCHEL, W.: [1] Die erste Randwertaufgabe der allgemeinen selbstadjungierten elliptischen Differentialgleichung zweiter Ordnung im Raum für beliebige Gebiete. Math. Z. **34** (1932) 535—553.

RADÓ, T.: [1] Das Hilbertsche Theorem über den analytischen Charakter der Lösungen der partiellen Differentialgleichungen zweiter Ordnung. Math. Z. **25** (1926) 514—589. — [2] Zu einem Satze von S. Bernstein über Minimalflächen im Grossen. Math. Z. **26** (1927) 559—565. — [3] On the problem of Plateau. Ergeb. Math. 2, Berlin: Springer-Verlag (1933) 109 pp.

REDHEFFER, R.M.: [1] A Sturmian theorem for partial differential equation. Proc. Amer. Math. Soc. **8** (1957) 458—462. — [2] Maximum principles and duality. Monatsh. Math. **62** (1958) 56—75. — [3] Eindeutigkeitsätze bei nichtlinearen Differentialgleichungen. J. reine angew. Math. **211** (1960) 70—77. — [4] An extension of certain maximum principles. Monatsh. Math. **66** (1962) 32—42.

REDHEFFER, R.M., and E.G. STRAUSS: [1] Degenerate elliptic equations. Pacific J. Math. **14** (1964) 265—268.

REICHARDT, H.: [1] Ausstrahlungsbedingungen für die Wellengleichung. Abh. Math. Sem. Univ. Hamburg **24** (1960) 41—53.

REĬTBLAT, Z.V.: [1] The mean value theorem for linear elliptic equations with Lipschitz class coefficients. Dokl. Akad. Nauk SSSR **133** (1960) 1300—1302 (Russian).

RELLICH, F.: [1] Zur ersten Randwertaufgabe bei Monge-Ampère Differentialgleichungen vom elliptischen Typus. Math. Ann. **107** (1932) 505—513 and 804. — [2] Über das asymptotische Verhalten der Lösungen von $\Delta u + \lambda u = 0$ in unendlichen Gebieten. Jber. Deutsch. Math. Ver. **53** (1943) 57—65.

ROBERT, J. P.: [1] Médiation et fonctions métaharmoniques. C. R. Acad. Sci. Paris **192** (1931) 326—328. — [2] Sur quelques propriétés des fonctions n-métaharmoniques. C. R. Acad. Sci. Paris **192** (1931) 1146—1148.

RODIN, JU. L.: [1] Elliptic systems on non orientable surfaces. Perm. Gos. Univ. Učen. Zap. (1963) 199—200.

ROĬTBERG, JA. A.: [1] Local increase of smoothness up to the boundary for solutions of elliptic equations. Ukrain. Mat. Ž. **15** (1963) 444—448 (Ukrainian). — [2] Elliptic problems with non homogeneous boundary conditions and local increase of smoothness of generalized solutions up to the boundary. Dokl. Akad. Nauk SSSR **157** (1964) 798—801 (Russian).

ROĬTBERG, JA. A., and Z. G. ŠEFTEL': [1] On equations of elliptic type with discontinuous coefficients. Dokl. Akad. Nauk SSSR **146** (1962) 1275—1278 (Russian). — [2] Energy inequalities for elliptic operators with discontinuous coefficients. Dokl. Akad. Nauk SSSR **148** (1963) 531—533 (Russian). — [3] General boundary value problems for elliptic equations with discontinuous coefficients. Dokl. Akad. Nauk SSSR **148** (1963) 1036—1037 (Russian).

ROSATI, P.: [1] Un problema al contorno connesso con quello di Lauricella per l'ellisse. Le Matematiche Catania **19** (1964) 31—49.

ROSENBLATT, A.: [1] Sopra le equazioni m-armoniche non lineari a due variabili indipendenti. Rend. Acc. Lincei **19** (1934) 212—219 and 306—310. — [2] Sur les équations aux dérivées partielles du second ordre du type elliptique non linéaires. Ann. Mat. pura appl. **13** (1935) 191—208. — [3] Sull'equazione biarmonica non lineare a due variabili indipendenti in un'area generale semplicemente connessa. Ann. Mat. pura appl. **14** (1935) 17—39. — [4] Sur les équations non linéaires du second ordre du type elliptique à trois variables indépendantes. Bull. Sci. Math. **59** (1935) 274—288.

ROSENBLOOM, P. C.: [1] Linear partial differential equations. Surveys in Appl. Math. V, London: Wiley and Sons (1958) 43—204.

ROTHE, E.: [1] Über lineare elliptische Differentialgleichungen, deren zugeordnete Maßbestimmung von konstanter Krümmung ist. Math. Ann. **105** (1931) 672—693.

ROYDEN, H. L.: [1] The growth of a fundamental solution of an elliptic divergence structure equation. Studies Math. Anal. Related Topics, Stanford Univ. Press (1962) 333—340.

ŠABAT, B. V.: [1] On the generalized solutions of linear elliptic systems. Mat. Sbornik **17** (1945) 193—210 (Russian).

SADOWSKA, D.: [1] Problème aux limites aux dérivées tangentielles pour l'équation elliptique dont les coefficients dépendent d'une fonction inconnue. Ann. Polon. Math. **10** (1961) 7—33; Errata **17** (1965) 117.

Šapiro, Z. Ya.: [1] Sur l'existence des représentations quasi-conformes. Dokl. Akad. Nauk SSSR **30** (1941) 690—692. — [2] On elliptical systems of partial differential equations. Dokl. Akad. Nauk SSSR **46** (1945) 133—135. — [3] The first boundary problem for an elliptical system of differential equations. Mat. Sbornik **28** (1951) 55—78 (Russian). — [4] On general boundary problems for equations of elliptic type. Izv. Akad. Nauk SSSR **17** (1953) 539—562 (Russian).

Satō, T.: [1] A new method for plane stress problems. Jap. J. Math. **19** (1948) 233—262. — [2] Sur l'équation aux dérivées partielles $z = f(x, y, z, p, q)$, I, II. Compos. Math. **12** (1954) 157—177; **14** (1959) 152—171. — [3] Sur le problème de Neumann pour l'équation $\Delta u(P) = F(P, u, \partial u)$. Proc. Japan. Acad. **34** (1958) 107—109. — [4] Sur le problème de Dirichlet généralisé pour l'équation $\Delta u = f(P, u, \partial u)$. Compos. Maht. **14** (1960) 237—259.

Sauer, L.: [1] Parametrixmethode zur Lösung von Randwertproblemen. Math. Ann. **118** (1942) 385—440 and **119** (1943) 67—130.

Saunders, W. K.: [1] On solutions of Maxwell's equations in an exterior region. Proc. Nat. Acad. USA **38** (1952) 342—348.

Schauder, J.: [1] Zur Theorie stetiger Abbildungen in Funktionalräumen. Math. Z. **26** (1927) 47—65. — [2] Bemerkungen zu meiner Arbeit ,,Zur Theorie stetiger Abbildungen in Funktionalräumen". Math. Z. **26** (1927) 417—431. — [3] Invarianz des Gebietes in Funktionalräumen. Studia Math. **1** (1929) 123—139. — [4] Der Fixpunktsatz in Funktionalräumen. Studia Math. **2** (1930) 170—179. — [5] Potentialtheoretische Untersuchungen. Math. Z. **33** (1931) 602—640. — [6] Bemerkung zu meiner Arbeit ,,Potentialtheoretische Untersuchungen". Math. Z. **35** (1932) 536—538. — [7] Über den Zusammenhang zwischen der Eindeutigkeit und Lösbarkeit partieller Differentialgleichungen zweiter Ordnung von elliptischen Typus. Math. Ann. **106** (1932) 661—721. — [8] Über das Dirichletsche Problem im Großen für nichtlineare elliptische Differentialgleichungen. Math. Z. **37** (1933) 623—634 and 768. — [9] Über lineare elliptische Differentialgleichungen zweiter Ordnung. Math. Z. **38** (1934) 257—282. — [10] Sur les équations linéaires du type elliptique à coefficients continus. C. R. Acad. Sci. Paris **199** (1934) 1366—1368. — [11] Sur les équations quasi linéaires du type elliptique à coefficients continus. C. R. Acad. Sci. Paris **199** (1934) 1566—1568. — [12] Numerische Abschätzungen in elliptischen linearen Differentialgleichungen. Studia Math. **5** (1934) 34—42. — [13] Équations du type elliptique, problèmes linéaires. Enseign. Math. **35** (1936) 126—139.

Schechter, M.: [1] On estimating elliptic partial differential operators in the L_2 norm. Amer. J. Math. **79** (1957) 431—443. — [2] Coerciveness of linear partial differential operators for functions satisfying zero Dirichlet type boundary data. Commun. pure appl. Math. **11** (1958) 153—174. — [3] Integral inequalities for partial differential operators and functions satisfying general boundary conditions. Commun. pure appl. Math. **12** (1959) 37—66. — [4] Solution of the Dirichlet problem for systems not necessarely strongly elliptic. Commun. pure appl. Math. **12**

(1959) 241—247. — [5] General boundary-value problems for elliptic partial differential equations. Commun. pure appl. Math. **12** (1959) 457—486; Russian transl. as Matematika **4,** n. 5 (1960) 93—122. — [6] Remarks on elliptic boundary value problems. Commun. pure appl. Math. **12** (1959) 561—578; Russian transl. as Matematika **4,** n. 6 (1960) 3—21. — [7] A free boundary problem for pseudo-analytic functions. Proc. Amer. Math. Soc. **10** (1960) 881—887. — [8] Mixed boundary problems for general elliptic equations. Commun. pure appl. Math. **13** (1960) 183—201. — [9] On the Dirichlet problem for second order elliptic equations with coefficients singular at the boundary. Commun. pure appl. Math. **13** (1960) 321—328. — [10] A generalization of the problem of transmission. Ann. Sc. N. Sup. Pisa **14** (1960) 207—236. — [11] Various types of boundary conditions for elliptic equations. Commun. pure appl. Math. **13** (1960) 407—425. — [12] Negative norms and boundary problems. Ann. of Math. **72** (1960) 581—593. — [13] Some unusual boundary value problems. Proc. Symp. Pure Math. Amer. Math. Soc. **4** (1961) 109—113. — [14] A local regularity theorem. J. Math. Mech. **10** (1961) 279—287. — [15] Observations concerning a paper of PEETRE. Commun. pure appl. Math. **14** (1961) 733—736. — [16] Coerciveness in L^p. Trans. Amer. Math. Soc. **107** (1963) 10—29. — [17] On L^p estimates and regularity, I, II, III. Amer. J. Math. **85** (1963) 1—13; Math. Scand. **13** (1963) 47—69; Ricerche di Mat. Napoli **13** (1964) 192—206. — [8] Non local elliptic boundary value problems. Ann. Sc. N. Sup. Pisa **20** (1966) 421—441.

SCHRÖDER, K.: [1] Zur Theorie der Randwertaufgaben der Differentialgleichungen $\Delta\Delta u = 0$. Math Z. **48** (1943) 553—675. — [2] Über die Ableitungen biharmonischer Funktionen am Rande. Math. Z. **49** (1943) 110—147.

SCHWARTZ, L.: [1] Théorie des distributions. Ed. Hermann Paris, t. I 2me éd. (1957) 162 pp., t. II (1951) 169 pp. — [2] Variedadas analiticas complejas. Ecuaciones diferenciales parciales elipticas. Dep. Mat. Univ. Bogotà (1956) 88 + 82 pp.; Russian transl. Izdat Mir. Moscow (1964).

SEELEY, R.T.: [1] Integro-differential operators on vector bundles. Trans. Amer. Math. Soc. **117** (1965) 167—204.

ŠEFTEL', Z.G.: [1] Estimates in L_p of solutions of elliptic equations with discontinuous coefficients and satisfying general boundary conditions and conjugacy conditions. Dokl. Akad. Nauk SSSR **149** (1963) 48—51 (Russian). — [2] Solvability in L_p and classical solvability of general boundary-value problems for elliptic equations with discontinuous coefficients. Usp. Mat. Nauk **18,** n. 3 (1964) 230—232 (Russian). — [3] Power inequalities and general boundary value problems for elliptic equations with discontinuous coefficients. Sibirsk Mat. Ž. **6** (1965) 636—668 (Russian).

SERRIN, J.: [1] On the Harnack inequality for linear elliptic equations. J. Anal. Math. **4** (1955/56) 297—308. — [2] A Harnack inequality for non linear equations. Bull. Amer. Math. Soc. **69** (1963) 481—486. — [3] A priori estimates for solutions of the minimal surface equations.

Arch. Rat. Mech. Anal. **14** (1963) 376—383. — [4] Pathological solutions of elliptic differential equations. Ann. Sc. N. Sup. Pisa **18** (1964) 385—387. — [5] Removable singularities of solutions of elliptic equations, I, II. Arch. Rat. Mech. Anal. **17** (1964) 67—78; **20** (1965) 163—169. — [6] Local behavior of solutions of quasi-linear elliptic equations. Acta Math. **111** (1964) 247—302. — [7] Singularities of solutions of quasi-linear equations. Proc. Symp. Appl. Math. Amer. Math. Soc. **17** (1965) 68—88. — [8] Isolated singularities of solutions of quasi-linear equations. Acta Math. **113** (1965) 219—240.

SERRIN, J., and H. F. WEINBERGER: [1] Isolated singularities of solutions of linear elliptic equations. Amer. J. Math. **88** (1966) 258—272.

SHAMIR, E.: [1] Mixed boundary value problems for elliptic equations in the plane. Ann. Sc. N. Sup. Pisa **17** (1963) 117—139.

SHIFFMAN, M.: [1] Differentiability and analiticity of solutions of double integral variational problems. Ann. of Math. **48** (1947) 274—284. — [2] On the existence of subsonic flows of a compressible fluid. J. Rat. Mech. Anal. **1** (1952) 605—652.

SHIROTA, T.: [1] A remark on the unique continuation theorem for certain fourth order elliptic equations. Proc. Japan. Acad. **36** (1960) 571—573.

SIKORA, B. S.: [1] A boundary value problem for an elliptic system of first-order equations with high-order derivatives on the boundary. Dopovidi Akad. Nauk Ukrain. RSR (1963) 1428—1431 (Ukrainian). — [2] The index and normal solubility of a boundary-value problem for an elliptic system of equations of higher order. Dopovidi Akad. Nauk Ukrain. RSR (1964) 26—30 (Ukrainian).

SIMODA, S.: [1] Sur le problème de Dirichlet discontinu dans l'équation $\Delta u = F(x, u, \operatorname{grad} u)$. Mem. Osaka Univ. **3** (1954) 29—36. — [2] Sur le théorème d'existence dans les problèmes aux limites pour l'équation $\Delta u = F(x, u, \operatorname{grad} u)$. Osaka Math. J. **6** (1954) 243—268. — [3] Notes pour la théorie des équations aux dérivées partielles du type elliptique. Mem. Osaka Univ. **5** (1956) 5—15. — [4] Sur la condition frontière dans le problème de Dirichlet pour les équations semi-linéaires du type elliptique et du second ordre. Proc. Japan. Acad. **35** (1959) 115—119. — [5] Traité sur la théorie des équations elliptiques, I, II, III, IV. Mem. Osaka Univ. **9** (1960) 119—136; **10** (1961) 5—29; **11** (1962) 1—23; **12** (1963) 1—9. — [6] Sur la régularité des points frontières relative à l'équation linéaire du type elliptique et du second ordre. Mem. Osaka Univ. **13** (1964) 33—53.

SIMODA, S., et M. NAGUMO: [1] Sur la solution bornée de l'équation aux dérivées partielles du type elliptique. Proc. Japan Acad. **27** (1951) 334—339.

SIMONOV, N.: [1] Über die erste Randwertaufgabe der nichtlinearen elliptischen Gleichung. Bull. Math. Univ. Moscou Série Inter. **2** (1939) 1—18. — [2] Solution of some boundary problems for elliptical systems

of linear equations. Dokl. Akad. Nauk SSSR **44** (1944) 259–261. —
[3] Solution of boundary-value problems for linear elliptic systems of
arbitrary order. Moscow Gos. Univ. Učen. Zap. Mat. **1** (1946) 53—84
(Russian).

SISMAREV, I. A.: [1] A priori estimation of solutions to the Dirichlet problem
for an elliptic operator with discontinuous coefficients. Dokl. Akad.
Nauk SSSR **131** (1960) 269–272 (Russian). — [2] Uniform estimates of
the derivatives of the solutions of the Dirichlet and eigenfunctions problems for the operator $L(u) = \mathrm{div}(p[x]\ \mathrm{grad}\ u) + q(x)u$ with discontinuous coefficients. Dokl. Akad. Nauk SSSR **137** (1961) 45—47.

SKOROBOGAT'KO, V. YA.: [1] On domains of solvability of Dirichlet's problem for self-adjoint elliptic equations. Ukrain. Mat. Ž. **7** (1955) 91—95
(Russian). — [2] Theorem on differential inequalities for an elliptic
equation. Ukrain. Mat. Ž. **8** (1956) 335–338 (Russian). — [3] Theorems
on the qualitative theory of partial second order differential equations.
Ukrain. Mat. Ž. **8** (1956) 435—440 (Russian). — [4] An extremal principle for a system of second order differential equations. Sibirsk. Mat. Ž.
2 (1961) 746—758 (Russian). — [5] A study of qualitative theory of
partial differential equations. Izdat. L'vovsk. Univ. (1961) 125 pp.
(Russian).

SLOBODECKIǏ, L. N.: [1] On strongly elliptic differential operators. Dokl.
Akad. Nauk SSSR **89** (1953) 13—15 (Russian). — [2] Generalized solutions of parabolic and elliptic systems. Izv. Akad. Nauk SSSR **21**
(1957) 809—834 (Russian). — [3] Estimates of solutions of elliptic and
parabolic equations. Dokl. Akad. Nauk SSSR **120** (1958) 468—471
(Russian). — [4] Estimates in L^p of solutions of elliptic systems. Dokl.
Akad. Nauk SSSR **123** (1958) 616—619 (Russian). — [5] Generalized
Sobolev spaces and their application to boundary-value problems for
partial differential equations. Učen. Zap. Leningrad Gos. Ped. Inst. **197**
(1958) 468—471 (Russian). — [6] Estimates in L^2 for solutions of linear
elliptic and parabolic systems. Vestnik Leningrad Univ. **15** (1960)
n. 7, 28—47 (Russian).

SLOBODECKIǏ, L. N., and I. A. SOLOMEŠČ: [1] On the first boundary value
problem for certain degenerate elliptic equations. Izv. Vysš. Učebn.
Zaved Mat. (1961) n. 3, 116—126 (Russian).

SMIRNOV, M. M.: [1] A boundary-value problem for an elliptic equation
degenerating in part on the frontier of the domain. Vestnik Leningrad
Univ. **16,** n. 13 (1961) 73—78 (Russian).

SMITH, K. T.: [1] Inequalities for formally positive integro-differential
forms. Bull. Amer. Math. Soc. **67** (1961) 368—370.

SOBOLEV, S. L.: [1] Problème limite fondamental pour les équations polyharmoniques dans un domaine au contour dégénéré. Dokl. Akad. Nauk
SSSR **3** (1936) 311—314. — [2] Sur une méthode directe pour résoudre
les équations polyharmoniques. Dokl. Akad. Nauk SSSR **4** (1936)

351—353. — [3] On a boundary value problem for polyharmonic equations. Mat. Sbornik **2** (1937) 467—500 (Russian); English transl. Amer. Math. Soc. (2) **33** (1963) 1—40. — [4] On a theorem of functional analysis. Mat. Sbornik **4** (1938) 471—497 (Russian); English transl. Amer. Math. Soc. (2) **34** (1963) 39—68. — [5] Applications of functional analysis in mathematical physics. Izdat. Leningrad Gos. Univ. (1950) 255 pp. (Russian); English transl. Amer. Math. Soc. (1963) 239 pp.; German transl. Berlin: Akad. Verlag (1964) 218 pp. — [6] Sur les équations aux dérivées partielles hyperboliques non linéaires. Monografie C.N.R. Ed. Cremonese Roma (1961) 144 pp.

SOBRERO, L.: [1] Theorie der ebenen Elastizität unter Benutzung eines Systems hyperkomplexer Zahlen. Hamburg Math. Einzelschriften **17**, B.G. Teubner (1934) 51 pp. — [2] Algebra delle funzioni ipercomplesse e sue applicazioni alla teoria matematica dell'elasticità. Mem. Acc. d'Italia **6** (1934) 1—64.

SOLOMJAK, M.Z.: [1] Analyticity of a semigroup generated by an elliptic operator in L^p spaces. Dokl. Akad. Nauk SSSR **127** (1959) 37—39 (Russian). — [2] Elliptic operators on two dimensional manifolds. Dokl. Akad. Nauk SSSR **139** (1961) 37—39 (Russian). — [3] Evaluation of norm of the resolvent of elliptic operators in L^p-spaces. Usp. Mat. Nauk **15, 6** n. 6 (1960) 141—148 (Russian). — [4] On linear elliptic systems of first order. Dokl. Akad. Nauk SSSR **150** (1963) 48—51 (Russian).

SOLOMJAK, T.B.: [1] Solution of the first boundary problem for quasilinear equations of elliptic type containing powertype non linearities. Dokl. Akad. Nauk SSSR **127** (1959) 274—277 (Russian). — [2] Boundary value problems for quasi-linear elliptic equations and systems of elliptic type. Izv. Vysš. Učebn. Zaved. Mat (1959) n. 5, 184—196 (Russian). — [3] Boundary value problems for quasi-linear elliptic equations with non linearities obeying a power law. Dokl. Akad. Nauk SSSR **139** (1961) 824—826 (Russian). — [4] The solvability of boundary value problems for a class of quasi-linear elliptic equations with strong non linearities. Dokl. Akad. Nauk SSSR **146** (1962) 1282—1285 (Russian). — [5] The first boundary problem for quasi-linear elliptic equations with non uniform growth of the coefficients. Vestnik Leningrad Univ. **20,** n. 1 (1965) 159—160 (Russian).

SOLONNIKOV, V.: [1] Linear differential equations with a small parameter in the terms of highest order. Dokl. Akad. Nauk SSSR **119** (1958) 454—457 (Russian). — [2] On estimates of Green's tensor for certain boundary problems. Dokl. Akad. Nauk SSSR **130** (1960) 988—991 (Russian). — [3] Some stationary boundary-value problems of magnetohydrodynamics. Trudy Mat. Inst. Steklov **59** (1960) 174—187 (Russian). — [4] A priori estimates for certain boundary value problems. Dokl. Akad. Nauk SSSR **138** (1961) 781—784 (Russian). — [5] Estimates for solutions of general boundary value problems for elliptic systems. Dokl. Akad. Nauk SSSR **151** (1963) 783—785 (Russian).

SOMIGLIANA, C.: [1] Sui sistemi simmetrici di equazioni a derivate parziali. Ann. Mat. pura appl. **22** (1894) 143—156.

STAMPACCHIA, G.: [1] Problema di Dirichlet e proprietà qualitative della soluzione. Giorn. Mat. Battaglini **80** (1950/51) 226—237. — [2] Problemi al contorno per equazioni di tipo ellittico a derivate parziali e questioni di calcolo delle variazioni connesse. Ann. Mat. pura appl. **33** (1952) 211—238. — [3] Sistemi di equazioni di tipo ellittico a derivate parziali del primo ordine e proprietà delle estremali degli integrali multipli. Ricerche di Mat. Napoli **1** (1952) 200—226. — [4] Problemi al contorno misti per equazioni del calcolo delle variazioni. Ann. Mat. pura appl. **40** (1955) 193—209. — [5] Osservazioni sull'esistenza e sull'unicità della soluzione dei problemi al contorno misti per equazioni a derivate parziali del secondo ordine di tipo ellittico. Rend. Acc. Sci. Fis. Mat. Napoli **22** (1955) 144—148. — [6] Su un problema relativo alle equazioni di tipo ellittico del secondo ordine. Ricerche di Mat. Napoli **5** (1956) 3—24. — [7] Contributi alla regolarizzazione delle soluzioni dei problemi al contorno per le equazioni del secondo ordine ellittiche. Ann. Sc. N. Sup. Pisa **18** (1958) 223—245. — [8] I problemi al contorno per le equazioni di tipo ellittico. Atti VI Congr. Un. Mat. Ital. Napoli (1959) 21—44. — [9] Problemi al contorno ellittici, con dati discontinui, dotati di soluzioni hölderiane. Ann. Mat. pura appl. **51** (1960) 1—38. — [10] Régularisation des solutions de problèmes aux limites elliptiques à données discontinues. Proc. Inter. Symp. Linear Spaces Jerusalem (1960) 399—408. — [11] On some regular multiple integral problems in the calcules of variations. Commun. pure appl. Math. **16** (1963) 383—421. — [12] Some limit cases of L^p estimates for solutions of second order elliptic equations. Commun. pure appl. Math. **16** (1963) 505—510. — [13] Formes bilinéaires coercitives sur les ensembles convexes. C. R. Acad. Sci. Paris **258** (1964) 4413—4416. — [14] Équations elliptiques du second ordre à coefficients discontinues. Sém. éq. dér. part. Collège de France (1964) 77 pp. — [15] Le problème de Dirichlet pour les équations elliptiques du second ordre à coefficients discontinues. Ann. Inst. Fourier Grenoble **15** (1965) 189—258. — [16] Équations elliptiques du second ordre à coefficients discontinues. Lecture Notes Univ. of Montréal (1965) 326 pp.

STERNBERG, W.: [1] Über die lineare elliptische Differentialgleichung zweiter Ordnung mit drei unabhängigen Veränderlichen. Math. Z. **21** (1924) 286—311.

STERNIN, B. JU.: [1] General boundary-value problems for elliptic equations in a region whose boundary consist of manifolds of different dimensions. Dokl. Akad. Nauk SSSR **159** (1964) 992—994 (Russian). — [2] General boundary problems for elliptic equations in a region whose boundary is formed of manifolds of various dimensions. Vestnik Univ. Moscow (1965) n. 2, 16—21 (Russian).

STOKER, J. J.: [1] On the uniqueness theorems for the embedding of convex surfaces in three-dimensional space. Commun. pure appl. Math. **3** (1950) 231—257. — [2] On radiation conditions. Commun. pure appl. Math. **9** (1956) 577—595.

STOPPELLI, F.: [1] Un teorema di esistenza ed unicità relativo alle equazioni dell'Elastostatica isoterma per deformazioni finite. Ricerche di

350 Bibliography

Mat. Napoli **3** (1954) 247—267. — [2] Sulla sviluppabilità in serie di potenze di un parametro delle soluzioni delle equazioni dell'elastostatica isoterma. Ricerche di Mat. Napoli **4** (1955) 58—73. — [3] Sull'esistenza di soluzioni delle equazioni dell'Elastostatica isoterma nel caso di sollecitazioni dotate di assi di equilibrio, I, II. Ricerche di Mat. Napoli **6** (1957) 241—287; **7** (1958) 71—101.

STYŠ, T.: [1] Hopf's theorem for a certain elliptic system of linear second order equations. Prace Mat. **8** (1963/64) 143—146 (Polish).

ŠUR, M.G.: [1] The Martin boundary for a linear elliptic second order operator. Izv. Akad. Nauk SSSR **27** (1963) 45—60 (Russian).

SYNGE, J.I.: [1] The hypercircle in mathematical physics: method for the approximate solution of boundary value problems. New York: Cambridge Univ. Press (1957) 424 pp.

SZEPTYCKI, P.: [1] Existence theorems for a quasi linear elliptic equation. Bull. Acad. Polon. Sci. **7** (1959) 409—424.

TALENTI, G.: [1] Sopra una classe di equazioni ellittiche a coefficienti misurabili. Ann. Mat. pura appl. **69** (1965) 285—304. — [2] Equazioni lineari ellittiche in due variabili. Le Matematiche Catania **21** (1966) 339—376. — [3] Soluzioni a simmetria assiale di equazioni ellittiche. Ann. Mat. pura appl. **73** (1966) 127—158.

TAUTZ, G.: [1] Reguläre Randpunkte beim verallgemeinerten Dirichletschen Probleme. Math. Z. **39** (1935) 532—559. — [2] Zur Theorie der elliptischen differentialgleichungen. Math. Ann. **117** (1941) 694—726 and **118** (1943) 733—770. — [3] Zur Theorie der ersten Randwertaufgaben. Math. Nach. **2** (1949) 279—303. — [4] Zum Umkehrungsproblem bei elliptischen Differentialgleichungen. Arch. Mat. **3** (1952) 232—250 and 361—365. — [5] Zur Theorie des Dirichletschen Problems. Convegno Inter. Eq. Lin. Der. Parz. Trieste (1954) 97—102. — [6] Zur Theorie des Dirichletschen Problems bei nichtlinearen Differentialgleichungen. Rend. Sem. Mat. Padova **24** (1955) 421—442.

TERSENOV, S.A.: [1] An elliptical type of equation degenerating at the domain boundary. Dokl. Akad. Nauk SSSR **115** (1957) 670—673 (Russian).

TETEREV, A.G.: [1] Čaplyagin's theorem for elliptic equations. Dokl. Akad. Nauk SSSR **134** (1960) 1024—1026 (Russian).

THOMAS, T.Y., and G.W.TITT: [1] On the elementary solution of the general linear differential equation of the second order with analytic coefficients. J. Math. pures appl. **18** (1939) 217—248.

TOLOTTI, C.: [1] La formula di Green per i problemi al contorno con derivata obliqua in spazi curvi quali si vogliano. Rend. Acc. Lincei **29** (1939) 285—293. — [2] Sulla struttura delle funzioni iperarmoniche in più variabili indipendenti. Giorn. Mat. Battaglini **77** (1947/48) 61—117.

TOVMASJAN, N.E.: [1] The Dirichlet problem for an elliptic system of two second order differential equations. Dokl. Akad. Nauk SSSR **153** (1963) 53—56 (Russian). — [2] The Dirichlet problem for an elliptic system of second order differential equations. Dokl. Akad. Nauk SSSR **159** (1964) 995—998 (Russian). — [3] Some boundary value problems for systems of second order equations of elliptic type not satisfying the condition of Ya. B. Lopatinskiĭ. Dokl. Akad. Nauk SSSR **160** (1965) 1028—1031 (Russian). — [4] Some boundary value problems for systems of second-order equations of elliptic type on the plane. Dokl. Akad. Nauk **160** (1965) 1275—1278 (Russian). — [5] On a boundary problem for an elliptic system of second order differential equations in the plane. Dokl. Akad. Nauk Armjan. SSR **40** (1965) 65—69 (Russian).

TREFFTZ, E.: [1] Konvergenz und Fehlerabschäztung beim Ritzschen Verfahren. Math. Ann. **100** (1928) 503—521.

TRICOMI, F.: [1] Un teorema di media per certe equazioni di tipo ellittico Rend. Sem. Mat. Padova **22** (1953) 350—353.

TROISI, M.: [1] Su un problema di trasmissione. Ricerche di Mat. Napoli **11** (1962) 24—50. — [2] Sui problemi di trasmissione per due equazioni ellittiche di ordine diverso. Ricerche di Mat. Napoli **12** (1963) 216—247. — [3] Sulla regolarizzazione delle soluzioni di taluni problemi di trasmissione. Ricerche di Mat. Napoli **13** (1964) 281—315. — [4] Su alcuni problemi misti per un'equazione ellittica di ordine superiore in due variabili. Ricerche di Mat. Napoli **14** (1966) 187—200.

URAL'CEVA, N.N.: [1] General second order quasi-linear equations and certain classes of equations of elliptic type. Dokl. Akad. Nauk SSSR **146** (1962) 778—781 (Russian). — [2] Boundary-value problems for quasi-linear elliptic equations and systems with principal part of divergence type. Dokl. Akad. Nauk SSSR **147** (1962) 313—316 (Russian).

USMANOV, N.K.: [1] On boundary problems of partial differential equations of the first order of elliptic type. Akad. Nauk Latv. SSR, Trudy Inst. Fiz. Mat. **1** (1950) 41—100 (Russian). — [2] On boundary problems of functions satisfying a system of partial differential equations of the first order of elliptic type. Akad. Nauk Latv. SSR, Trudy Inst. Fiz. Mat. **2** (1950) 59—100 (Russian). — [3] A new method of solution of a boundary problem for a system of differential equations of the first order of elliptic type. Izv. Akad. Nauk Latv. SSR **4** (1950) 129—142 (Russian). — [4] An approximate method of solving the generalized Hilbert problem. Izv. Akad. Nauk Latv. SSR **19** (1965) 69—90.

VASARIN, A.A., and P.I. LIZORKIN: [1] Certain boundary-value problems for elliptic equations with a strong degeneration at the boundary. Dokl. Akad. Nauk SSSR **137** (1961) 1015—1018 (Russian).

VEKUA, I.N.: [1] Solution of the fundamental boundary value problem for the equation $\Delta^{n+1}u = 0$. Soobšč. Akad. Nauk Gruzin. SSR **3** (1942) 213—220 (Russian). — [2] On metaharmonic functions. Trudy Inst. Mat. Tbilissk **12** (1943) 105—174 (Russian). — [3] New methods for solving elliptic equations. OGIZ Moscow-Leningrad (1948) 296 pp. — [4] On a representation of solutions of elliptic differential equations. Soobšč. Akad. Nauk Gruzin SSR **11** (1950) 137—141 (Russian). — [5] Systems of differential equations of the first order of elliptic type and boundary problems with an application to the theory of shells. Mat. Sbornik **31** (1952) 217—314 (Russian); German transl. Math. Forschungsberichte, Berlin (1956) 107 pp. — [6] A boundary problem with oblique derivative for an equation of elliptic type. Dokl. Akad. Nauk SSSR **92** (1953) 1113—1116 (Russian). — [7] On certain properties of a system of equations of elliptic type. Dokl. Akad. Nauk SSSR **98** (1954) 181—184 (Russian). — [8] The problem of reduction to canonical forms of elliptic type and the generalized Cauchy-Riemann system. Dokl. Akad. Nauk SSSR **100** (1955) 197—200 (Russian). — [9] Generalized analytic functions. Gos. Izdat. Fiz.-Mat. Lit. Moscow (1959) 628 pp. (Russian); English transl. Pergamon Press (1962) 668 pp.; German transl. Berlin: Akad. Verlag (1963) 538 pp.

VERŽBINSKIĬ, G.M.: [1] On the defect index of the second and third boundary value problems in a region with pieceweise smooth boundary. Vestnik Leningrad Univ. **19,** n. 7 (1964) 161—162 (Russian).

VILLAGGIO, P.: [1] Sul problema al contorno per sistemi di equazioni differenziali lineari del tipo di stabilità dell'equilibrio elastico. Ann. Sc. N. Sup. Pisa **15** (1961) 25—40.

VINOGRADOV, V.S.: [1] On the Neumann problem for equations of elliptic type. Dokl. Akad. Nauk SSSR **109** (1956) 13—16 (Russian). — [2] On a boundary value problem for linear elliptic systems of differential equations of the first order on the plane. Dokl. Akad. Nauk SSSR **118** (1958) 1059—1062 (Russian). — [3] On boundedness of solutions of boundary value problems for linear elliptic systems on the plane. Dokl. Akad. Nauk SSSR **121** (1958) 399—402 (Russian). — [4] On a problem for quasi-linear systems of equations on the plane. Dokl. Akad. Nauk SSSR **121** (1958) 579—582 (Russian).

VIOLA, T.: [1] Sull'esistenza del minimo assoluto di taluni integrali multipli connessi con i problemi al contorno per le funzioni iperarmoniche. Ann. Sc. N. Sup. Pisa **6** (1952) 109—145.

VIŠIK, M.I.: [1] The method of orthogonal projection for self-adjoint differential equations. Dokl. Akad. Nauk SSSR **56** (1947) 115—118 (Russian). — [2] The method of orthogonal projection and direct decomposition in the theory of elliptic differential equations. Mat. Sbornik **25** (1949) 189—234 (Russian). — [3] On strongly elliptic systems of differential equations. Mat. Sbornik **29** (1951) 615—676 (Russian). — [4] On the stability of solutions of boundary problems for elliptic differential equations (relative to variations of the coefficients and right-hand sides). Dokl. Akad. Nauk SSSR **81** (1951) 717—720 (Russian). — [5] On a general form of solvable boundary problems for homogeneous and non homogeneous elliptic differential equations. Dokl. Akad. Nauk SSSR **82** (1952) 181—184 (Russian). — [6] On general boundary problems for elliptic differential equations. Trudy Moskow. Mat. Obšč. **1**

(1952) 187—246 (Russian); English transl. Amer. Math. Soc. (2) **24** (1963) 107—172. — [7] On boundary problems for systems of elliptic differential equations and on the stability of their solutions. Dokl. Akad. Nauk SSSR **86** (1952) 645—648 (Russian) — [8] On the first boundary value problem for elliptic differential equations with operator coefficients. Soobšč. Akad. Nauk Gruzin. SSR **13** (1952) 129—136 (Russian). — [9] On systems of elliptic differential equations and their general boundary problems. Usp. Mat. Nauk **8**, n. 1 (1953) 181- 187 (Russian). - [10] Boundary problems for elliptic equations degenerating on the boundary of a region. Mat. Sbornik **35** (1954) 513—568 (Russian). English transl. Amer. Math. Soc. (2) **35** (1964) 15—78. — [11] On the first boundary problem for elliptic equations in a new functional aspect. Dokl. Akad. Nauk SSSR **107** (1956) 781—784 (Russian). — [12] On the solvability of the first boundary value problem for non linear elliptic systems of differential equations. Dokl. Akad. Nauk SSSR **134** (1960) 749—752 (Russian). — [13] Solution of a system of quasi-linear equations having divergence form under periodic boundary conditions. Dokl. Akad. Nauk SSSR **137** (1961) 502—5C5 (Russian). — [14] Boundary value problems for quasi-linear strongly elliptic systems of equations having divergence form. Dokl. Akad. Nauk SSSR **138** (1961) 518—521 (Russian). — [15] Quasi linear elliptic systems of equations containing subordinate terms. Dokl. Akad. Nauk SSSR **144** (1962) 13—16 (Russian). — [16] Quasi-linear strongly elliptic systems of differential equations of divergence form. Trudy Moscow Mat. Obšč. **12** (1963) 125—184 (Russian). — [17] Sur les problèmes aux limites pour des équations quasi-linéaires elliptiques et paraboliques d'ordre supérieur. Les équations aux dérivées partielles. Centre Nat. Rech. Sci. Paris (1963) 213—218. — [18] On the solvability of the first boundary-value problem for quasi-linear equations with coefficients of rapid growth in Orlicz classes. Dokl. Akad. Nauk SSSR **151** (1963) 758—761 (Russian).

VIŠIK, M.I., and G.I. ESKIN: [1] General boundary-value problems with discontinuous boundary conditions. Dokl. Akad. Nauk SSSR **158** (1964) 25—28 (Russian). — [2] Equations in convolutions in a bounded region. Usp. Mat. Nauk **20**, n. 3 (1965) 89—152 (Russian); transl. as Russian Math. Surveys **20**, n. 3 (1965) 85—152.

VIŠIK, M.I., and O.A.LADYŽENSKAJA: [1] Boundary value problems for a partial differential equation and certain classes of operator equations. Usp. Mat. Nauk **11**, n. 6 (1956) 41—97 (Russian); English transl. Amer. Math. Soc. (2) **10** (1958) 223—281.

VIŠIK, M.I., and L.A. LYUSTERNIK: [1] Regular degeneration and boundary layer for linear differential equations with small parameter. Usp. Mat. Nauk **12**, n. 5 (1957) 3—122 (Russian); English transl. Amer. Math. Soc. (2) **20** (1962) 239—364. — [2] Solution of some perturbation problems in the case of matrices and self-adjoint and non self-adjoint equations. Usp. Mat. Nauk **15**, n. 3 (1960) 3—80 (Russian); transl. as Russian Math. Surveys **15**, n. 3 (1960) 1—74. — [3] The asymptotic behaviour of solutions of linear differential equations with large or quickly changing coefficients and boundary conditions. Usp. Mat. Nauk **15**, n. 4 (1960) 27—95 (Russian); transl. as Russian Math. Surveys **15**, n. 4 (1960) 23—91.

VIŠIK, M.I., and S.L.SOBOLEV: [1] General formulation of certain boundary problems for elliptic partial differential equations. Dokl. Akad. Nauk SSSR 111 (1956) 521—523 (Russian).

VOLEVIČ, L.R.: [1] The Dirichlet problem for quasi-linear equations of elliptic type with small parameter with highest derivatives. Naucn. Dokl. Vyss. Skoly Fiz. Mat. Nauki (1958) 9—18 (Russian). — [2] On the theory of boundary-value problems for general elliptic systems. Dokl. Akad. Nauk SSSR 148 (1963) 489—492 (Russian). — [3] A problem of linear programming arising in differential equations. Usp. Mat. Nauk 18, n. 3 (1963) 155—162 (Russian). — [4] The solvability of boundary-value problems for general elliptic systems. Mat. Sbornik 68 (1965) 373—416 (Russian).

VOLKOV, JU.A.: [1] Bounds for the difference of solutions of the equation $f(z_1, z_2, \ldots, z_n)$ det $\| z_{ij} \| = h(x_1, x_2, \ldots, x_n)$ in terms of the right side of the equation. Vestnik Leningrad Univ. 15, n. 13 (1960) 5—14 (Russian).

VOL'PERT, A.I.: [1] The Dirichlet problem in the plane for an elliptic system of linear differential equations of the 2 nd order. Dokl. Akad. Nauk SSSR 79 (1951) 185—187 (Russian). — [2] Dirichlet's problem for an elliptic system of linear differential equations of 2 nd order in the plane. Ukrain Mat. Ž. 3 (1951) 449—464 (Russian). — [3] Boundary value problems for an elliptic system of second-order linear differential equations in the plane. L'vov. Lesotehn. Inst. Nauč. Trudy 1 (1954) 191—212 (Russian). — [4] Investigation of boundary problems for elliptical systems of differential equations on a plane. Dokl. Akad. Nauk SSSR 114 (1957) 462—467. (Russian) — [5] Calculation of the index of Dirichlet's problem. Dopovidi Akad. Nauk Ukrain. RSR (1958) 1042—1044 (Ukrainian). — [6] The first boundary problem for elliptic systems of differential equations. Dokl. Akad. Nauk SSSR 127 (1959) 487—489 (Russian). — [7] Boundary value problems for elliptic systems of differential equations of higher order on a plane. Dokl. Akad. Nauk SSSR 127 (1959) 739—741 (Russian). — [8] On the reduction of boundary value problems for elliptical systems of higher order to problems for systems of first order. Dopovidi Akad. Nauk Ukrain. RSR (1960) 1162—1166 (Ukrainian). — [9] On the index in boundary problems of harmonics functions in three independent variables. Dokl. Akad. Nauk SSSR 133 (1960) 13—15 (Russian). — [10] On the index of Dirichlet's problem. Izv. Vysš. Učebn. Zaved. Mat. (1960) n. 5, 40—42 (Russian). — [11] On the index and normal solvability of boundary-value problems for elliptic systems of differential equations on the plane. Trudy Moscow Mat. Obšč. 10 (1961) 41—87 (Russian).

VOSS, K.: [1] Einige differentialgeometrische Kongruenzsätze für geschlossene Flächen und Hyperflächen. Math. Ann. 131 (1956) 180—218.

VVEDENSKAJA, N.D.: [1] On a boundary problem for equations of elliptic type degenerating on the boundary of a region. Dokl. Akad. Nauk SSSR 91 (1953) 711—714 (Russian).

VÝBORNÝ, R.: [1] On a certain extension of the maximum principle. Differential equations and their applications (Proc. Conf. Prague 1962) Acad. Press New York (1963) 223—228. — [2] Über das erweiterte Maximumprinzip. Czech. Math. J. **14** (1964) 116—121.

WACHMAN, M.: [1] Generalized Laurent series for singular solutions of elliptic partial differential equations. Proc. Amer. Math. Soc. **15** (1964) 101—108.

WALTER, W.: [1] Über ganze Lösungen der Differentialgleichung $\Delta u = f(u)$. Jber. Deutsch. Math. Verein. **57** (1955) 94—102. — [2] Ganze Lösungen der Differentialgleichung $\Delta^p u = f(u)$. Math. Z. **67** (1957) 32—37. — [3] Zur existenz ganzer Lösungen der Differentialgleichung $\Delta u = e^u$. Arch. Math. **9** (1958) 308—312.

WASHIZU, K.: [1] Bounds for solutions of boundary value problems in elasticity. J. Math. Phys. **32** (1953) 117—128.

WASOW, W.: [1] Asymptotic solution of boundary value problems for the differential equation $\Delta u + \lambda \, \partial u / \partial x = \lambda f(x, y)$. Duke Math. J. **11** (1944) 405—415.

WEINBERGER, H.F.: [1] Symmetrization in uniformly elliptic problems. Studies in Math. Anal. and related topics. Stanford Univ. Press (1962) 424—428.

WEINSTEIN, A.: [1] Generalized axially symmetric potential theory. Bull. Amer. Math. Soc. **59** (1953) 20—38. — [2] Sulle soluzioni quasi periodiche di una classe di equazioni ellittiche. Rend. Acc. Lincei **32** (1962) 863—866. — [3] Singular partial differential equations. Symp. Inst. Fluid Dyn. Appl. Math. Univ. of Maryland, Gordon and Breach; New York (1963) 29—49.

WERNER, H.: [1] Das problem von Douglas für Flächen konstanter mitt. lere Krümmung. Math. Ann. **133** (1957) 303—319.

WERNER, P.: [1] On the exterior boundary value problem of perfect reflection for stationary magnetic wave fields. J. Math. Anal. Appl. **7** (1963) 348—396. — [2] Über die Randwertprobleme der Helmholtzschen Schwingungsgleichung. Math. Z. **85** (1964) 226—240. — [3] Randwertprobleme für die zeitunabhängigen Maxwellschen Gleichungen mit variablen Koeffizienten. Arch. Rat. Mech. Anal. **18** (1965) 167—195.

WEYL, H.: [1] The method of orthogonal projection in potetial theory. Duke Math. J. **7** (1940) 411—444. — [2] Radiation capacity. Proc. Nat. Acad. USA **37** (1951) 832—836. — [3] Kapazität von Strahlungsfeldern. Math. Z. **55** (1952) 187—198. — [4] Die natürlichen Randwertaufgaben im Außenraum für Strahlungsfelder beliebiger Dimension und beliebigen Ranges. Math. Z. **56** (1952) 105—119.

WILCOX, C.H.: [1] A generalization of theorems of Rellich and Atkinson. Proc. Amer. Math. Soc. **7** (1956) 271—276. — [2] Spherical means and radiation conditions. Arch. Rat. Mech. Anal. **3** (1959) 133—148.

WINTNER, A.: [1] On the Hölder restrictions in the theory of partial differential equations. Amer. J. Math. **72** (1950) 731–738. — [2] On Weyl's identity in the differential geometry of surfaces. Ann. Mat. pura appl. **41** (1956) 257–268.

WITTICH, H.: [1] Ganze Lösungen der Differentialgleichung $\Delta u = e^u$. Math. Z. **49** (1943) 579–582.

WOLSKA-BOCHENEK, J.: [1] Un problème aux limites à dérivée tangentielle pour l'équation du type elliptique. Ann. Polon. Math. **4** (1958) 275 to 285. — [2] Probléme non linéaire à dérivée oblique. Ann. Polon. Math. **9** (1960/61) 253–264.

YANG GUANG-JUN: [1] The Dirichlet problem for a class of equations of degenerating elliptic type. Acta Math. Sinica **12** (1962) 40–46 (Chinese); transl. as Chinese Math. **3** (1963) 42–48.

YOSIDA, K.: [1] A theorem of Liouville's type for meson equation. Proc. Japan. Acad. **27** (1951) 214–215. — [2] On the existence of the resolvent kernel for elliptic differential operator in a compact Riemann space. Nagoya Math. J. **4** (1952) 63–72. — [3] A characterization of the second order elliptic differential operator. Proc. Japan. Acad. **31** (1955) 406–409.

ZAHAROV, V.K.: [1] The first boundary problem for an elliptical type of equation of order four, degenerating at the domain boundary. Dokl. Akad. Nauk SSSR **114** (1957) 694–697 (Russian).

ZAREMBA, S.: [1] Sur un problème toujours possible comprenant, à titre de cas particuliers, le problème de Dirichlet et celui de Neumann. J. Math. pures appl. **6** (1927) 127–163.

ZEMACH, C., and F. ODEH: [1] Uniqueness of radiative solutions of the Schrödinger wave equation. Arch. Rat. Mech. Anal. **5** (1960) 226–237.

ZERAGIJA, D.P.: [1] On the solution of the Dirichlet problem for certain non-linear equations of elliptic type. Soobšč. Akad. Nauk Gruzin SSR **31** (1963) 9–14 (Russian).

ZITARASU, N.V.: [1] Apriori estimates and solvability of general boundary value problems for general elliptic systems with discontinuous coefficients. Dokl. Akad. Nauk SSSR **165** (1965) 24–27 (Russian).

ZITAROSA, A.: [1] Su di una classe di sistemi di tipo ellittico di equazioni quasi lineari del primo ordine. Ricerche di Mat. Napoli **12** (1963) 151–178. — [2] Sul problema di Dirichlet per le equazioni ellittiche non lineari in due variabili. Le Matematiche Catania **19** (1964) 144–156.

ŽITOMIRSKIĬ, O.K.: [1] Sur la non-flexibilité des ovaloides. Dokl. Akad. Nauk SSSR **25** (1939) 347–349.

ZLAMAL, M.: [1] The parabolic equation as a limiting case of a certain elliptic equation. Ann. Mat. pura appl. **57** (1962) 143–150.

ZOLOTAREVA, E. V.: [1] The Dirichlet problem for a class of elliptic systems. Dokl. Akad. Nauk SSSR **132** (1960) 751—753 (Russian). — [2] A necessary and sufficient condition that the Dirichlet problem for a class of elliptic systems be of Fredholm character. Dokl. Akad. Nauk SSSR **145** (1962) 724—726 (Russian). — [3] On the Dirichlet problem for a certain class of elliptic systems. Dokl. Akad. Nauk SSSR **145** (1962) 983—985 (Russian).

ZWIRNER, G.: [1] Sull'equazione a derivate parziali delle superficie minime. Rend. Sem. Mat. Padova **4** (1933) 140—154. — [2] Su una proprietà di media relativa alle equazioni alle derivate parziali di tipo ellittico del secondo ordine con un numero qualsiasi di variabili. Rend. Sem. Mat. Padova **12** (1941) 22—29. — [3] Su una particolare classe di equazioni alle derivate parziali del quarto ordine sopra una superficie chiusa. Rend. Sem. Mat. Padova **17** (1948) 139—159.

Author Index

Subject Index

Subject Index 369

Nonlinear systems 285
Normal system of boundary operators 256
Normal to the boundary of a domain 3, 4
Normally solvable operator 237

Oblique derivative problem 9
 adjoint 15
 alternative theorems 88, 92
 compatibility conditions 16
 existence theorems 90
 for nonlinear equations 219—221
 generalized 101
 in the small 66
 transformation into integral equations 86—92
 uniqueness 11, 66
Ordinary point of a transformation 191
Orthogonal projections, method of 122, 132

Parameter, questions of 287,—288
Parametric form, equations in 217—219
PERRON process 109—110
PHRAGMÈN-LINDELÖF theorem 231
Physics, equations from 283—285
PICARD problem 259—260
PLATEAU problem 217—218
POISSON formula 31
Polyharmonic functions 262—264
Positive regular in sense of calculus of variations 144
Potential
 axially symmetric 103
 domain 28—32
 double layer 39—43
 method of 54—59
 single layer 32—39
Principal fundamental matrix 272—277
Principal fundamental solutions 67—73
Properly elliptic operator 244
Properly elliptic system 271
 in the sense of DOUGLIS and NIRENBERG 276
 in the sense of PETROWSKIĬ 275

Property of the mean 113
Pseudo analytic function 269
Pseudo-differential operators 260, 282—283

Quasi-GREEN's function 67, 92—96
Quasilinear equations 181, 231—233
 DIRICHLET problem 197—208, 209
 NEUMANN problem 219—220
 oblique derivative problem 220—221
Questions of small parameter 287—288

Region 1
Regular boundary point 108, 109, 132
Regular integral (calculus of variations) 144
Regular solution 5
Regular solution surface 217, 218
Regularity of solutions of DIRICHLET problem 167—169, 182—184
Regularization of weak or generalized solutions 121—112, 128, 129, 138—139, 175—179, 253, 257
Regularly solvable operator 238
Resolvent of a kernel matrix 51, 54
RIEMANN-HILBERT problem 268
RUNGE approximation property 61

s-capacity 131
SCHAUDER's existence theory 164—169
SCHAUDER's fixed point theorem 193
Second boundary value problem: see NEUMANN problem
Second order systems 281—282
Self-adjoint operator 12
Semilinear equations 181, 221—222
Single layer potential 32—39
Singular coefficients, equations with 100—104
Singular point of a transformation 191

Sellier OHG, Freising

Ergebnisse der Mathematik und ihrer Grenzgebiete